# Fundamentals of
# Transportation Engineering

# Fundamentals of Transportation Engineering

*Second Edition*

Robert G. Hennes
*Professor and Chairman*
*Department of Civil Engineering*
*University of Washington*

Martin Ekse
*Professor of Civil Engineering*
*University of Washington*

McGraw-Hill Book Company

*New York*    *St. Louis*    *San Francisco*    *Toronto*    *London*    *Sydney*

**Fundamentals of
Transportation Engineering**

*Library of Congress Catalog Card Number* 68-27507

28171

234567890  H D B P  7543210

# Preface

The authors have endeavored to present in this book a combined treatment of several types of transportation engineering within one volume, thus catering to a growing trend toward a general coordinated course in transportation engineering. The material in each section is adaptable to design and layout projects for both the student and the professional engineer concerned with such problems as design of a road to meet standards for given traffic volume and heavy-wheel-load repetitions; planning, design, and layout for a civil airport to standards required for given air-transport traffic in a given community; layout of railroad terminal facilities and coordination with other forms of transportation; river training for flood control and navigation; development of port and harbor facilities and structures; design and layout of pipeline facilities and belt lines.

Publications of many organizations and agencies active in the field of transportation engineering have been consulted freely in the preparation of this book, thus providing authentic and up-to-date planning, design, and construction standards and procedures of value to both student and practicing engineer. Every effort has been made to give appropriate recognition for material taken from such sources as the U.S. Bureau of Public Roads, American Association of State Highway Officials, Federal Aviation Administration, American Railway Engineering Association, The Asphalt Institute, Portland Cement Association, Hydrographic Office of the U.S. Navy, U.S. Army Corps of Engineers, American Society of Civil Engineers, and others. To engineers of these organizations goes the credit for a large number of the illustrations, charts, and tables in the text.

The material in the book has been developed in classroom use over the past two decades and has been tried and tested in actual classroom work for three courses during the past twelve years in the Civil Engineering Department of the University of Washington.

Robert G. Hennes
Martin Ekse

# Contents

# Introduction

The pursuit of happiness in America grows more and more dependent upon transportation. On the one hand there is a recreational exodus by motorbus, by cruise ship, by ski-train, and by air coach; on the other hand, by rail and water and highway the products of the wide earth are freighted to the market place. In curious contrast to this multiplex activity, the individual American by education and vocational training has become a specialist, so that his own contribution to his society generally lies narrowly within a limited field of endeavor. Perhaps it is this very restriction imposed upon individual effort that has raised over-all efficiency to the point where the individual can enjoy goods and services from the ends of the earth. To meet the expanding demand for goods and services, there exist in the United States more than three million miles of rural roads and city streets; over six hundred thousand miles of railroad; great harbors and a comprehensive system of waterways; air lanes reaching from any part of the nation to every port in the world. Maintaining—or developing—such a way of life and standard of living in any region or country of the world requires an efficient and well-integrated transportation system—one that (1) encompasses all modes of transport for people and goods and (2) becomes a direct and major part of over-all regional environmental design.

This incredible confusion of transport by land and sea and air is caught together in an equally incredible achievement of order; beneath it all is the hand and mind of the engineer. If the great achievement is to be sustained, if the present is not to be the end of progress, there must be continuing and expanding opportunities for engineers who have some notion of the whole pattern of transportation and who will each acquire that mastery of some specialized technology upon which the preservation of our culture depends. In this intricately conceived combination of many engineering functions the civil engineer plays an important role. It is the ambitious aim of the authors of this book to acquaint the student with the whole field of transportation in so far as civil engineering is involved, and at the same time to provide enough training in each branch of transportation engineering for an elementary understanding of the character of professional practice in that branch. At best this training cannot be more than a foundation for the further and more intensive study which is required of the professional engineer in these times.

A very large percentage of all civil engineers normally finds employment in the design, construction, or maintenance of transportation facilities. With this in mind, the civil-engineering student should familiarize himself with the diverse nature of the problems that characterize the several methods of transportation. Moreover, he should endeavor to acquire some idea of the opportunities offered by the different transportation agencies, both now and in the future. During the period of its major expansion one or another of these agencies may dominate the market for engineering services and thus, to some extent, the pattern of engineering education—as in the decades that witnessed the construction of the nation's railroads. Under such circumstances the man preparing himself for an engineering career should take care to form a balanced judgment of the over-all transportation picture. With this need in mind it is helpful to examine the distribution of traffic over the various types of carrier. This distribution is shown by the following table:

**Table 1  Estimated inland intercity traffic in the United States, in per cent***

| Carrier | Freight | | | | Passenger | | | |
|---|---|---|---|---|---|---|---|---|
| | 1916 (1) | 1939 (2) | 1950 (3) | 1965 (4) | 1916 (5) | 1939 (6) | 1950 (7) | 1965 (8) |
| Railways............. | 77.2 | 62.6 | 58.7 | 42.9 | 98.0 | 8.6 | 8.3 | 1.6 |
| Inland waterways....... | 18.4 | 17.9 | 16.2 | 15.6 | 2.0 | 0.5 | 0.3 | |
| Petroleum pipelines...... | 4.4 | 9.5 | 12.7 | 18.9 | | | | |
| Highways............. | .... | 10.0 | 12.4 | 22.5 | .... | 90.7 | 89.4 | 93.5 |
| Airways.............. | .... | .... | 0.03 | 0.12 | .... | 0.2 | 2.0 | 4.9 |

(1) 100 per cent represents 476 billion freight ton-miles in 1916.
(2) 100 per cent represents 537 billion freight ton-miles in 1939.
(3) 100 per cent represents 1,017 billion freight ton-miles in 1950.
(4) 100 per cent represents 1,644 billion freight ton-miles in 1965.
(5) 100 per cent represents 43 billion passenger-miles in 1916.
(6) 100 per cent represents 275 billion passenger-miles in 1939.
(7) 100 per cent represents 439 billion passenger-miles in 1950.
(8) 100 per cent represents 842 billion passenger-miles in 1965.

* Interstate Commerce Commission, Bureau of Transport Economics and Statistics.

Of course it must be considered that different types of transport differ widely in their need for engineers. Unfortunately, the sums paid for engineering services in the various fields of transportation are not conveniently recorded. As it stands, the above table does not present a fully satisfactory basis for evaluating the interests of the profession in these several types of carrier. Furthermore, coastwise and foreign trade were omitted from the table, mainly because they are not in direct competition with the inland carriers, and in any event the ton-mile would not be a satisfactory unit of measurement for such trade. Nevertheless, the development of seaports and coastal waterways does represent a major engineering activity. Some concept of its relative importance can be gained by comparing total tons of cargo—the amount handled at seaports is about double that carried on the inland waterways —but this comparison may be applied only to terminal facilities, as channel maintenance is relatively more important in the case of the inland waterways.

The years 1916 and 1939 were used in the table because data for those years were available and because they precede major wartime crises in transportation. They do not establish a trend which can be projected into the future. Any forecast from such premises would be unfair especially to air transportation, whose competitive position is already far stronger than in 1939. But it is important to realize also that each type of carrier enjoys possession of some field of service which it is likely to retain. Neither the ship nor the train is likely to lose the bulk of its freight business to its competitors in the foreseeable future. The loss of a substantial part of the rail passenger traffic to the airways has not been a crippling blow, for even before the advent of automobile competition the railways obtained 70 per cent of their gross operating revenue from freight traffic. Further losses in passenger and express traffic should be offset by an over-all expansion of all forms of transportation in pace with increases in the national income and anticipated improvements in world living standards. If one may be optimistic about the future progress of civilization, such optimism must include the transportation industry; for as men become more productive, they have both the means and the inclination to devote an increasingly larger proportion of an increasing income to the transportation of goods and persons. Commerce is a sensitive measure of material progress.

There are several factors which tend to confuse the engineer who tries to judge his opportunities for advancement in one or another branch of the transportation industry. An expanding industry is apt to require more construction work than one which has reached maturity. On the other hand, the rate of personal advancement is affected somewhat by the age of those filling responsible positions, at least in an organization where promotion must wait upon the retirement of incumbent superiors. Another factor is the relative merit of public as compared with private employment. The designers of highways, airports, and breakwaters are usually governmental employees, while the designers of railroads, pipe-lines, and docks are employed by private industry.

It is the purpose of this text to familiarize the student with the problems which characterize civil-engineering employment in the transportation field. This is not a treatise on national transportation policy. The development of economic policy in the integration of transportation agencies is left to more advanced and more specialized treatises.

# Fundamentals of
# Transportation Engineering

part one

# Roads and Pavements

# 1

# Foundation Soils

**1-1  Road and load**  The transportation of people or of goods must involve some provision for load support and some method of overcoming the inevitable resistance to motion.   As gravitational pull on transported objects can be sustained in various ways, land, sea, and air transportation are obvious possibilities.   Except for pipelines, overland transportation is keyed to use of the wheel.   Wheels rolling over unimproved earth surfaces encounter excessive resistance; and it is this fact which led to the development of railways and highways and which justifies the present study of road-building materials.   The nature of this rolling resistance is the factor which determines what properties are desirable in materials used for road construction.

A wheel load creates a depression in the supporting surface, because stress and strain are inseparable.   Figure 1-1 may serve to illustrate how the deformation of the earth surface is resisted by shear forces within the earth mass, and at the same time may suggest the thought that work was done in displacing soil particles against the resistance of internal friction.   As the vehicle advances, a constantly new body of soil is

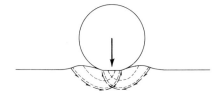

**Fig. 1-1** Incipient shear under wheel load.

similarly compressed, requiring continuous output of energy by the vehicle. The requisite force is transmitted to the soil by wheel thrust, and the power consumed is equal to the product of rolling resistance and the velocity of the vehicle.

The conditions of static equilibrium require that the vertical component of the earth reaction be precisely equal to the wheel load, regardless of the nature of the soil. Work is the product of force times distance; thus for a given wheel load it is only by the vertical displacement of the road surface under the wheel that power consumption is determined by the physical properties of the earth. It follows that greater tractive effort is required to travel over the more yielding surfaces. Soft soil should be replaced by firmer material or it should be covered with a substantial base course. In other words, the road surface should be hard.

The foregoing analysis ignores the effect of surface irregularity. Vehicles running over rough roads require additional tractive effort to provide energy for lifting wheel loads over projections in the roadway. Economical transportation, as well as personal comfort, is favored by smooth road surfaces.

The roadway reaction is not a vertical force. It does have a vertical component equal to the applied wheel load, but there is also a horizontal component to the reaction at a driving wheel, equal to the tractive effort, and at other wheels the horizontal component of the reaction is equal to the rolling resistance. The existence of a horizontal force at the point of tire contact necessitates shear resistance at that point. Even loose sands possess shear resistance, through their internal friction, but it cannot then exceed a specific proportion of the normal component of the reaction, perhaps 50 per cent of the wheel load. With modern vehicles the tractive effort can reach that magnitude, at least as an instantaneous value. In that case, the road surface is abraded, as anyone who has had the experience of digging his rear wheels into beach sand up to the hub caps will verify. Resistance to abrasion is provided by the cohesion of soil particles to each other, and this cohesion can be obtained by adding clay, bitumen, or portland cement to the granular aggregate. Insofar as the soil grains cohere, or stick together, they are able to resist tension as well as shear. It seems strange to speak of any tensile strength residual in a plastic clay or in a liquid asphalt, and in fact only in a dynamic sense does the notion have real validity. This sort of shear resistance is termed viscosity. Too often the dynamic nature of all transportation is over-looked, in a predilection for statics as a basis for design. Under moving wheel loads even a clay-bound gravel base course, or an oil mat, can act as an elastic slab over a soft subgrade. If the earth yields sufficiently under the wheel load the resulting curvature of the pavement or surfacing course throws the underside of the slab into tension. Of course, the relative thickness of the two layers is a determining factor. If the upper layer is very thick or if the earth is very stiff, there is little curvature and little or no tension. If the reverse is true, the tensile strength of the top layer may be exceeded, and the wheel will break through the surface crust. Over a weak soil the engineer must provide a strong elastic slab or a thick inelastic base.

In any event the engineer must attack his design problem reinforced with an adequate knowledge of soil properties and with a knowledge of those physical characteristics of his other constructional materials which will provide resistance to abrasion in the wearing surface, elastic stability in his slabs, and load-bearing capacity in his bases.

**1-2 Soil properties** In the course of road construction the surface of the ground is leveled off to a set elevation or grade, preliminary to laying the pavement. The soil below this grade is called the *subgrade*. The principal soil properties influencing subgrade performance are com-pressibility, stability, and permeability. Of these three, the first refers to the stress-strain relationship of the soil, the second is dependent upon shearing resistance, and the third concerns the rate of ground-water flow. Section 1-1 has shown the bearing of compressibility and stability upon road design; permeability is involved in such problems as drainage, frost damage, and spring breakup. The measurement of any of these three properties for a single sample involves considerable time, skill, and expense. When soil variations in a single test hole are multiplied by the soil changes between test holes, the cost of soil investigation by direct measurement becomes excessive.

The direct measurement of fundamental properties is a rational ap-proach to any problem, but in many fields it becomes an expensive luxury. The physician, for example, depends upon observation of symptoms for his diagnosis and prescription. Pulse, temperature, rash, pains—they are all given consideration because through long experience they have acquired statistical value in the identification of the malady. In highway engineering as well it has been found more practicable to use a few simple tests to identify the soil type in relation to its probable performance in the subgrade. The soil tests described in this chapter do not measure engineering properties directly, but they have proved to be valuable aids in the identification of soil types.

**1-3   Particle size**   The most apparent distinction to be made between soils is in the sizes of the constituent particles.   *Textural classification* is based upon grain size.   Of the various schemes that have been proposed to establish size limits for textural groups, highway engineers have most generally favored that originally proposed by the U.S. Bureau of Soils. In this system the term *silt* is reserved for the size range between 5 and 50 $\mu$.   A micron ($\mu$) is 0.001 mm.   All soil finer than silt is called *clay*. Soil coarser than silt, up to a diameter of 2 mm, is sand.   Specialists in soil mechanics tend to favor a roughly similar classification originating at Massachusetts Institute of Technology (MIT), because setting the maximum clay size at 2 $\mu$ is in better accord with the presence of the plasticity generally considered to be typical of clays.   This text will adhere to highway usage.

The grading of sands and gravels is determined by sieving.   The sieves have square openings formed by wire mesh.   The finest sieve in common use is the No. 200, whose openings have a width of 74 $\mu$.   Other methods than sieving must be used to establish the size of smaller soil particles.   Most of the usual methods are based upon measurement of the velocity with which they settle through water.   All particles of the same size move with the same velocity, other things being constant.   It is assumed that the grains are uniformly distributed throughout the suspension at the start of the test.   Thereafter the water above any specific level contains no grains larger than a determinable diameter, whereas immediately below that level soil particles of the computed size occupy the suspension with no change in concentration.   The relation between particle size and settling velocity is supplied by an equation originating with Stokes, which he derived for a solid sphere with a specific gravity of $G$ and a diameter of $D$ cm, settling through a liquid of specific gravity $G'$ and viscosity $n$.   The settling velocity in centimeters per second is

$$v = \frac{(G - G')D^2}{18n} \tag{1-1}$$

The coefficient of viscosity $n$ may be taken as 0.00001 g-sec/sq cm for water at 20°C.   Having computed the settling (or terminal) velocity for a grain of some specific size, it is possible to state the distance that the uppermost particle of that diameter will have settled in a given time. At that instant the last particle of the selected size will have passed the designated point, and the ratio of the density of the suspension at that point to the original density corresponds to the percentage of the total sample which is finer than the selected diameter.   It has become customary to use a hydrometer to measure the density of the suspension. The hydrometer test, or wet mechanical analysis, has been standardized by the American Society for Testing Materials (ASTM) and the American Association of State Highway Officials (AASHO). Under ordinary test conditions sand has settled past the hydrometer in about half a minute; silt, in about half an hour.

The results obtained from mechanical analysis are best shown as a graph, with grain size as abscissa and percentage finer than a given size as ordinate. Because of the relative importance of clay, grain size is plotted to a logarithmic scale. Highway engineers follow the American custom of plotting graphs in the first quadrant, but among other engineers there is a trend toward the use of the second quadrant for grain-size plots.

**1-4  Plasticity** Soils possess another property in such varying degree that it serves as a means of distinguishing between soil types, that is, the faculty of distortion without fracture or crumbling. Moist clay can be molded into stable shapes; clean dry sand is an aggregation of noncoherent separate grains. The plasticity which is characteristic of clays reflects, to some extent, the prevalence of flat flaky particles, which slide over one another without much interference; but even more, it results from the presence of adsorbed water films which coat the clay particles like a soft cement. To say that water *wets* a solid is to say that the water molecules immediately adjacent to the surface of the solid actually adhere to it with considerable force. Such water is not free; the force of gravity is too weak to remove it. But there is no sharp line of demarcation between the free water and the adsorbed film water; rather there is a zone of varying viscosity. Here two facts stand out: first, clay particles usually are separated by their films, without opportunity for physical interlocking; second, an interlacing network of continuous, contiguous, or overlapping films will create a coherent mass of particles without offering much resistance to remolding. The film exists because of its attraction to the surface of the solid; the quantity of film water is a function of the amount of surface area possessed by the solid particles. A gram of mineral will have more surface area if it is divided into a thousand particles than if it is left in a single piece; and the surface will be still greater if each of these grains is a thin flake instead of a sphere. A pound of corn flakes has a lot more surface area than a pound of raw corn kernels. To a considerable extent, plasticity is a function of grain size and of grain shape.

With increasing water content a plastic clay becomes softer as the average viscosity of its pore water is decreased by dilution. With the addition of enough water the clay is liquid. Conversely, stiffness results from a decrease in water content, until finally at some low water content the clay will crumble when remolded.    Between these two extremes the clay is in the *plastic state*.    The range in water content embraced in the plastic state is a measure of the power of the soil to attract film water. The water content at which the soil flows is called the *liquid limit*.    The water content at which the soil crumbles is called the *plastic limit*.    The numerical difference between these two values is called the *plasticity index*.    *Water content* is the ratio of the weight of the water in a soil sample to the weight of the dry soil, expressed in per cent.    Plasticity is undesirable in subgrade soils, although it will be noted later that in moderation it is a helpful property of binder material in untreated-gravel wearing surfaces.

**1-5  Capillarity and permeability**    *Free water* (or *gravitational water*) will drain from soil under the influence of gravity.    *Film water* is not free-draining.    Water also may resist gravitational drainage because it is held in the soil by capillarity.

If a narrow glass tube is held in a vertical position with its lower end immersed in a dish of water (Fig. 1-2), the water level in the tube will be above the water level in the dish.    A glance at a free-body diagram of the water column will show that the weight of the water must be balanced by the vertical component of surface tension at the meniscus.    Except for the film adsorbed by the walls of the tube, the column is made up of *capillary water* and is in the state of tension.    The walls of the glass tube are compressed by the corresponding *capillary pressure*.    If the tube is long enough to permit the water to rise to its maximum height, the meniscus will be tangent to the walls of the tube.    In that case the summation of vertical forces will place the weight of the water column equal to the product of surface tension and the circumference of the meniscus.    Let the internal diameter of the tube be $d$ cm, the height of capillary rise be $h_c$ cm, the unit weight of water be 1 g/cu cm, and take the surface tension

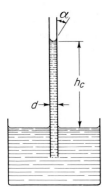

**Fig. 1-2**  Capillary rise in a tube.

of water at 0.08 g/cm.  Then

$$h_c \frac{\pi d^2}{4} 1 = 0.08\pi d$$

and

$$h_c = \frac{0.3}{d} \quad \text{cm} \tag{1-2}$$

In similar fashion capillary water may saturate the soil for a considerable height above the level of gravitational water, the water table. Moreover, when a saturated soil begins to dry by evaporation from its exposed surface, the immediate effect is the formation of a meniscus between each pair of particles.  The corresponding capillary pressure compresses the soil and causes shrinkage.  As the process continues the curvature of the meniscus increases until capillary pressure reaches its maximum value.  Evaporation beyond this point cannot produce further compression of the soil.  The water content at the point of maximum capillary compression is called the *shrinkage limit*.

Drainage lowers the water table by the removal of gravitational water. Water seeps through soil at velocities usually well below the critical value at which turbulence appears.  Taking a section normal to the direction of flow, the discharge through a cross-sectional area $A$ is given by Darcy's law:

$$Q = kiA \tag{1-3}$$

where $k$ is the *coefficient of permeability* for the soil and $i$ is the slope of the pressure gradient, or the head lost in unit distance.  The simplicity of this equation suggests that in principle a laboratory test to measure $k$ can be easily devised.  In actual practice some difficulties arise from entrained air, sampling, and temperature control.

**1-6  Laboratory tests**  In order to acquaint the student with basic engineering properties which lead to soil classification and determination of strength characteristics, the following laboratory tests are suggested:

| *Laboratory test* | *ASTM designation* |
|---|---|
| Liquid limit | D423-66 |
| Plastic limit | D424-59 (1965) |
| Plasticity index | D424-59 (1965) |
| Shrinkage limit | D427-61 |
| Mechanical analysis | D422-63 |
| Specific gravity | D854-58 (1965) |

A careful study of standard procedures should be made for the above tests, including the shrinkage tests, to gain knowledge of basic soil properties which will be significant in later discussion of soil engineering.

## QUESTIONS AND PROBLEMS

**1-1.** Define moisture content, liquid limit, plastic limit, shrinkage limit, and plasticity index.

**1-2.** A soil specimen weighs 30 g, oven-dry. When mixed with $7\frac{1}{2}$ cu cm of water the soil is at its plastic limit; the addition of another $10\frac{1}{2}$ cu cm of water brings the sample to the liquid limit. What is the plasticity index of the soil?

**1-3.** The shrinkage limit of a soil sample is 16.5 and the specific gravity of the solid grains is 2.65. If a sample of this soil weighs 50.2 g when wet and 40.5 g when dry, what percentage of the total volume is filled with solid material when wet? When dry?

**1-4.** Rework Prob. 1-3 using 45.5 g for the oven-dry weight instead of 40.5 g.

**1-5.** With what velocity in centimeters per second will a sand (specific gravity, 2.60) settle through water at 20°C? The diameter of the particle is 0.05 mm.

**1-6.** Compute the specific gravity of soil solids from the following data:

Volume of flask............................ 500.00 cu cm
Weight of flask............................ 98.70 g
Dry weight of soil......................... 80.00 g
Weight of flask + soil + water at 23°C....... 646.78 g
Weight of flask + water at 23°C............. 596.24 g
Density of water at 23°C.................... 0.9975

**1-7.** Determine the height to which water will rise in a capillary tube with an inside diameter of 0.01 mm, assuming the surface tension of water to be 0.075 g/cm.

**1-8.** A suspension with a volume of 1 liter is made by combining water (density = 1.0000) with 50 g (dry weight) of soil grains whose specific gravity is 2.65. Determine the specific gravity of the suspension.

## BIBLIOGRAPHY

Casagrande, A.: Research on the Atterberg Limits of Soil, *Public Roads*, **13**:121(1932).
The Formation, Distribution and Engineering Characteristics of Soil, *Purdue Univ. Research Ser.*, no. 87 (1943).
Hogentogler, C. A.: "Engineering Properties of Soil," McGraw-Hill Book Company, New York, 1937.
Wu, T. H.: "Soil Mechanics," Allyn and Bacon, Inc., Boston, 1966.
"Standard Specifications for Highway Materials and Methods of Sampling and Testing," American Association of State Highway Officials, Part I, Specifications, and Part II, Methods of Sampling and Testing, 1967.
Taylor, D. W.: "Fundamentals of Soil Mechanics," John Wiley & Sons, Inc., New York, 1948.

# 2

# Grading the Roadbed

**2-1  Preliminary considerations**  The purpose of grading in highway construction is to provide a stable roadbed to specified cross section (see Chap. 9) upon which the pavement structure is to be placed.  Often soil characteristics and drainage conditions will determine the location of a highway, and a proper balance of these factors with alignment and gradient will have to be made.  The parts of the highway cross section as referred to in subsequent discussion are shown in Fig. 2-1.

Among the first items of grading operations are clearing and grubbing of trees and stumps and removal of brush and debris from the area needed for cuts and fills.  Usually it will be desirable to leave within the right of way such natural growth of trees and shrubs as does not interfere with grading operations or drainage facilities.  Proper handling of such native growth requires careful study of roadside improvement afforded by its retention and effective marking of those trees and shrubs which are to be retained.

Topsoil should be removed and stockpiled for later use on finished cut and fill slopes.  Turfing and planting of finished slopes will add much

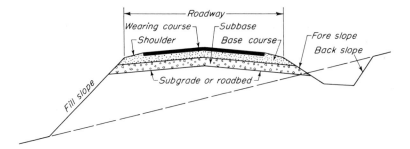

**Fig. 2-1** Highway cross section.

to appearance of the highway and will largely prevent erosion. It should be noted that native grasses and plants which are acclimated to the area are to be preferred over imported varieties. In many regions which are blessed with good topsoil and adequate rainfall, sod has been successfully used for shouldering material. In some cases the sod may be transplanted directly from other areas, but in general it is started from seed with a nurse crop of rye or other grain.

The grading contractor is usually paid for the quantity, in cubic yards, of excavation, and this payment includes reimbursement for the construction of embankments from said excavation. Included also in the item of excavation is a "free haul" of generally 500 or 1,000 ft, and any haul of excavation beyond this free-haul distance is paid for as a separate item of "overhaul."

**2-2    Basic concepts of soil engineering**    Excavation is usually classified as *common* and *rock*, but often the contract will call for *unclassified excavation* because of frequent difficulty in determining the dividing line between rock and common excavation. This is particularly true where numerous boulders of various sizes are to be expected and also where hard-soil strata such as clay pan or partially cemented hardpan may be encountered. In addition, items of *excavation for structures* and *borrow* will be included in the contract.

In general, a cubic yard of *solid-rock* excavation will make more than a cubic yard of embankment because void spaces will remain between the rock fragments as placed in the fill. Roughly, a *swell* of about 25 per cent can be expected, depending somewhat on the size range of chunks of rock. A mass of uniform-size chunks or particles will have a higher percentage of voids than one in which the chunks and particles are well graded from large to small.

Similarly, earth, or what is normally classified as common excavation, will shrink in the fill. In the construction of embankments, where the soil is compacted in 6- to 8-in. layers with present-day equipment, it is possible to attain a high degree of compaction in the fill.   It is not uncommon to experience a shrinkage of as much as 35 per cent, and seldom will the shrinkage be less than 10 per cent.   This is because the soil is compacted to a higher density in the fill than it had in its original state.   In other words, the voids or pore spaces in the soil have been reduced.   It is a common experience, upon having dug a post hole, to place the post in the hole and still be able to tamp all of the excavated soil into the hole around the post.   The solid soil particles are not compressed in any sense of the word, but the volume of voids or pore space between and around the soil grains has been reduced.

The amount of voids in a soil mass may be expressed in either of two ways: (1) as *porosity*, which is the percentage of voids based on the total volume of soil, and (2) as *void ratio*, which is the ratio of the volume of voids to the volume of solid soil grains, usually expressed as a decimal. The student should be thoroughly familiar with these expressions, and a better appreciation of their significance may be had by reference to Fig. 2-2a.   Suppose any given volume $V$ is exactly filled with $N$ balls, each having a diameter $d$.   The volume of solids $V_s$ is seen to be $N\pi d^3/6$, and the volume of voids $V_v$ is $Nd^3 - (N\pi d^3/6)$.   From these expressions it follows that, by definition, the *porosity* of this ideal mass is $V_v/V$ or 47.6 per cent.   In other words, nearly half of the total volume is void space.   Also, the *void ratio* is $V_v/V_s$ or 0.91.

It should be noted that the above values for porosity and void ratio are independent of the size of the spheres and will be true for any uniform-size spheres with this particular arrangement.   It can be shown that in a continuous mass with the balls arranged as in Fig. 2.2b, the void ratio is reduced to 0.74.   This amounts to a form of compaction by reorientation of particles and reduction of voids, and this principle applies in a general way to granular materials of which most soils are composed.

Density of the mass can also be increased by filling the large voids with smaller particles.   This can be seen by reference again to Fig. 2-2a, noting that numerous smaller spheres could be placed in the interstices, or void spaces of the spheres shown.   Obviously, porosity and void ratio will be reduced by this process.   A material, then, in which particles are

**Fig. 2-2**  Effect of compaction on void spaces.                    *(a)*                    *(b)*

well graded from coarse to fine can be most readily compacted to high density.

Particle-size gradation is best shown, for purposes of analysis, by plotting percentages of various sizes by weight against particle size, and a comparison of the plot against an ideal grading for maximum density will be helpful. Such an ideal grain-size curve for maximum density is shown in Fig. 2-3. It is plotted from Talbot's formula:

$$p = 100 \left(\frac{D}{M}\right)^m \tag{2-1}$$

in which $p$ is the desirable percentage by weight smaller than any particle size $D$ for a material whose maximum size is $M$. The best value of the exponent $m$ depends somewhat on characteristics of the soil grains—particularly angularity or roughness—but it is generally recognized that a soil is well graded if its gradation curve falls within the limits of $\frac{1}{4}$ and $\frac{2}{5}$ for $m$. The curve of Fig. 2-3 is plotted for $m = \frac{1}{3}$.

In its natural state soil contains more or less water in the pore spaces or voids of the soil mass. Below the ground-water table, the soil may

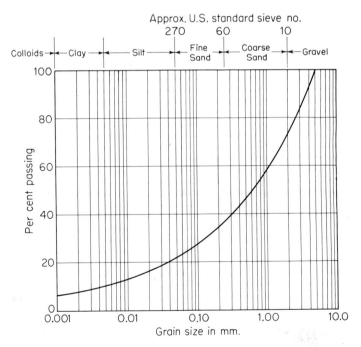

**Fig. 2-3** Talbot's curve. From $p = (D/M)^m \times 100$, where $m = \frac{1}{3}$ and $M = 5$ mm.

generally be assumed to be saturated. Above the ground-water table water may be held in the soil pores by capillarity in various stages of saturation. The degree of saturation is defined as the percentage of total voids filled with water. Air-dry soil either in its natural state or in the laboratory will contain *hygroscopic* moisture. This water is present as very thin films on the surface of the soil grains, and the amount will vary with temperature and humidity of the air. It is conventional to assume that oven drying of the soil to constant weight at 100 to 105°C will remove practically all the hygroscopic moisture.

Another way to visualize "soil solids," voids, and moisture content is illustrated in Fig. 2-4. The following formulas derived from these basic concepts will be found useful in many soil-engineering problems:

Total volume,

$$V = V_v + V_s = V_a + V_w + V_s \qquad (2\text{-}2)$$

Porosity,

$$n = \frac{V_v}{V} \times 100 \qquad \text{per cent} \qquad (2\text{-}3)$$

Void ratio,

$$e = \frac{V_v}{V_s} \qquad (2\text{-}4)$$

Degree of saturation,

$$S = \frac{V_w}{V_v} \times 100 \qquad \text{per cent} \qquad (2\text{-}5)$$

Also, it will be seen that

$$e = \frac{n}{100 - n} \qquad (2\text{-}6)$$

and

$$n = \frac{e}{1 + e} \times 100 \qquad (2\text{-}7)$$

*Specific gravity* may be defined as the ratio of the weight of a substance to the weight of an equal volume of water. *True* or *absolute specific gravity* of a soil is the specific gravity of the actual soil grains or solids. Several different minerals of varying absolute specific gravities may comprise a given soil mass, but its true or absolute specific gravity is determined as the average obtained from a representative soil sample. *Bulk* or *apparent specific gravity* is the ratio of the dry weight of soil to the weight of a volume of water equal to the total volume of soil (including voids).

*Moisture content* is defined to be the weight of water present in a soil expressed as a percentage of the weight of oven-dry soil. Thus, if a soil sample as taken from a test specimen weighs 126.0 g, and after oven drying to constant weight it weighs 105.0 g, the moisture content is $21.0/105.0 \times 100$, or 20 per cent.

If

$W_s$ = oven-dry weight of soil
$W_w$ = weight of water
$W$ = total weight of soil and water
$w$ = moisture content
$G$ = absolute specific gravity
$g$ = bulk specific gravity

then

$$w = \frac{W_w}{W_s} \times 100 = \frac{W - W_s}{W_s} \times 100 \tag{2-8}$$

$$V_s = \frac{W_s}{G \times 62.5} \tag{2-9}$$

$$V = \frac{W_s}{g \times 62.5} \tag{2-10}$$

$$W_s = \frac{W}{1 + w/100} \tag{2-11}$$

and

$$n = \frac{G - g}{G} \times 100 \tag{2-12}$$

The following example is given to illustrate the use of these basic concepts and definitions. An undisturbed sample of soil taken from a borrow pit is found to have a total volume of 0.317 cu ft and weighs 36.75 lb. The true specific gravity of the soil grains is 2.67 and the moisture content is 35 per cent. If it is expected that the borrow material will compact to a wet density of 130 pcf at a moisture content of 19 per cent in fill construction, what shrinkage factor should be used in computing quantities?

*Solution*  Since the shrinkage factor $F$ can be expressed as the ratio of the difference in volume of borrow excavation and volume of fill to volume of borrow excavation, it will be convenient to consider volume of solids as unity. Thus the question becomes: What total volume does 1 cu ft of soil solids occupy in the borrow pit and what total volume will it occupy in the fill, bearing in mind that reduction in total volume is accomplished by reduction in voids?

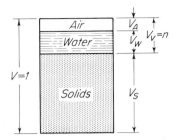

**Fig. 2-4** Volumetric relationships in terms of total volume.

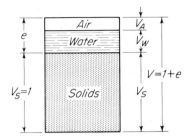

**Fig. 2-5** Volumetric relationships in terms of solid volume.

From Eq. (2-11) the weight of solids in the above sample is

$$\frac{36.75}{1.35} = 27.22 \text{ lb}$$

and from Eq. (2-9) the volume of solids is $27.22/(2.67 \times 62.5) = 0.163$ cu ft. $V_v$ is then 0.154 cu ft and the void ratio $e_1$ is 0.945. This means that 1 cu ft of soil solids will occupy $1 + e = 1.945$ cu ft in the borrow pit. Similarly, the void ratio $e_2$ of the compacted soil is found to be 0.528, or this same cubic foot of solid material will occupy 1.528 cu ft in the fill. The difference in volume is then seen to be the difference in void ratios or $e_1 - e_2$, and the shrinkage factor is determined to be $(e_1 - e_2)/(1 + e_1)$ $\times$ 100 or $0.417/1.945 \times 100 = 21.4$ per cent. The expression $(e_1 - e_2)/(1 + e_1)$ is quite useful in problems concerning compaction or consolidation of soils and is further exemplified in Fig. 2-5. With $V_s$ taken as unity, the total volume becomes $1 + e$.

**2-3  Soil compaction**  In the preceding example it is possible to determine the amount of air voids still present in the compacted soil. The absolute volume $V_s$ of soil grains present in 1 cu ft of soil is determined from Eqs. (2-11) and (2-9) to be 0.654 cu ft. The weight of water present is $0.19W_s$, or 20.8 lb, and its volume is 20.8/62.5, or 0.332 cu ft. These two volumes add up to 0.986 cu ft, leaving 0.014 cu ft of air voids, or 1.4 per cent air voids based on the total volume.

Theoretically it should be possible to compact the soil to zero air voids for any moisture content. This is not physically practicable or even possible at low moisture content owing to the frictional resistance and interlock of the soil grains which resist any practical amount of compactive effort so that some air space remains in the compacted mass. As the moisture content is increased, however, a certain amount of lubrication is furnished to the soil particles by the water and it will be possible to compact the soil to greater density with a given amount of compactive effort.  A point is then reached where further increase in moisture content will give less density under the same conditions of compaction because the soil will have reached a state of near saturation and further addition of water will act as an incompressible space filler to hold the soil grains apart.  Also, cohesive soils become plastic as moisture content is increased and tend to displace or flow around the compacting hammer or foot, thus minimizing the effectiveness of applied compactive effort.

In order to obtain soil densities in the laboratory as nearly compatible as possible with those obtained in the field during embankment construction, R. R. Proctor devised a method of laboratory compaction known as the Proctor density test, whereby the soil is compacted in a mold 4 in. in diameter and of $\frac{1}{30}$-cu-ft capacity in three equal layers.  Each layer is tamped 25 blows with a $5\frac{1}{2}$-lb hammer of specified size falling through a height of 1 ft.  The soil is struck off the level of the top of the mold, and the density is determined by weighing the sample.  This process is repeated for various moisture contents, and the results are plotted as the "wet-density" curve (Fig. 2-6).

The "dry-density" curve is then plotted as determined from Eq. (2-11).  The high point on the dry-density curve is indicated as the *optimum moisture content*.  It is the moisture content at which maximum density is attained by the prescribed method of compaction.

The modified Proctor and other methods have been devised, using much greater compactive energy in order to coordinate laboratory procedure with present-day construction methods which use heavier and more effective compaction equipment.  The higher density will be obtained, for the same soil, at a somewhat lower moisture content, as indicated by the broken-line curve of Fig. 2-6.  Increasingly greater

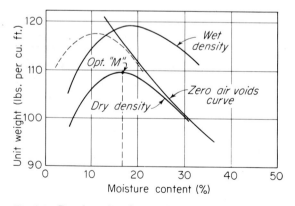

**Fig. 2-6**  Density related to moisture content.

compactive effort will tend to move the dry-density curve upward and
to the left.

The zero-air-voids curve is the dry-density curve for theoretically
perfect compaction of the soil to a state of complete saturation wherein
only soil and water are present. Taking total volume as unity, the curve
may be plotted from the fact that $V_s + V_w = 1$, at various values for $w$
and a specific value for $G$, as follows:

$$V_s + V_w = \frac{W_s}{G \times 62.5} + \frac{wW_s}{100 \times 62.5} = 1$$

from which

$$W_s = \frac{62.5G}{1 + wG/100} \tag{2-13}$$

The ultimate goal of such laboratory design procedures is the successful construction of embankments. A high degree of compaction is
desirable so that a stable roadbed will be assured and so that unequal
settlement will not impair a carefully constructed grade line and roadway
section. To this end it is generally required that the soil be compacted in
layers of 6 to 8 in. for most soils (sand embankments may be made in
layers up to 2 ft in thickness). Suitable rolling equipment, such as tamping rollers, wobbly-wheel rollers, or heavy pneumatic-tired rollers, should
be provided to compact each layer to 90 to 100 per cent of the density
established in the laboratory, depending on specification requirements.
It is important that soil moisture be kept within two or three percentage
points of optimum moisture content during this compaction process for
effective use of equipment.

Field density is generally measured by removing and weighing the
soil from a 4-in.-diameter test hole as deep as the layer being compacted.
A sample is taken for determination of moisture content. The volume
of soil is determined by measuring the quantity of dry uniform sand
required to fill the test hole. Water can also be used for this volumetric
test if a thin membrane of rubber or plastic is placed in the hole to prevent
the water from soaking into the soil. From these factors the dry density
and moisture content may be readily determined to check compliance
with compaction requirements.

**2-4  Soil strength**  Selective use of soils in connection with a grading
project can be made when an adequate evaluation of engineering properties of the various available soils has been made. Available soils will be
those taken from excavated portions of the roadway prism and those
obtainable from nearby borrow pits. Of primary importance is the load-
carrying capacity of each soil type or classification under the moisture

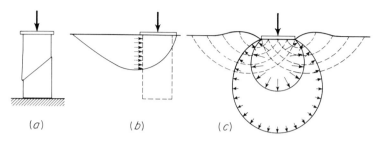

**Fig. 2-7**  Nature of subgrade stability.

conditions and degree of compaction which are likely to be experienced in actual field conditions during and after construction.

When a load is applied to a soil mass, compressive and shearing forces are mobilized within the mass, and as the load is increased the shearing strength of the soil is ultimately overcome.  If a test specimen of the soil is loaded in unconfined compression, one would expect failure to occur in a manner similar to that shown in Fig. 2-7a.  When this same soil specimen is confined within a continuous mass, lateral pressures are developed which tend to support lateral displacement and a larger portion of soil will have to be displaced, as illustrated in Fig. 2-7b.

This tendency for displacement will of course develop on an infinite number of such surfaces, and as the bearing plate is pushed into the surface of the soil, upward components of pressure tend to cause upward displacement of the soil mass, as illustrated in Fig. 2-7c, and it is conceivable that there exists a surface on which pressures normal to that surface are equal.  Such a surface is known as a *pressure bulb* or bulb of equal pressure and is illustrated in Fig. 2-7c.  It can be seen that these pressures are supported within the soil mass by shearing resistance developed between the soil grains or fragments.  This total shearing resistance is made up of cohesion, sliding friction, and mechanical interlock of soil particles.  Complete analyses of these stress factors are adequately covered in texts on soil mechanics.  The basic concepts are presented in the previous discussion for purposes of illustration.

**2-5  Pavement support**  The pavement structure, which is used here to include sub-base, base course, and wearing course (Fig. 2-1), serves in two ways to protect the subgrade material by virtue of its own weight so that upward displacement of the subgrade soil is largely prevented. First, the subgrade is materially strengthened by the confining influence of the protective cover.  It is not difficult to appreciate this strengthening influence if one considers the subgrade as consisting of somewhat plastic material.  Second, the subgrade is further protected by the fact that as a load is applied to the top of the pavement structure the pressure tends to spread out over a larger area as depth below the surface increases. Thus the intensity of vertical pressure on the subgrade will be considerably less than that under the load at the surface, depending on the depth of the pavement structure. In the case of a rigid pavement slab the pressure immediately under the slab is greatly reduced because of the beam strength of the slab. In the case of a flexible type of pavement the reduction in pressure may be assumed to be influenced alike by wearing course, base course, and sub-base materials inasmuch as these materials will be of a relatively coarse, granular nature such as reasonably well graded gravel and crushed rock.

For the reason that soil pressures and consequent shearing stresses due to applied surface loads decrease with increased depth, it follows that the weakest soils should be used in the lower layers of embankment construction and the most stable soils in the top layers. Also, the base course and wearing course are likely to be proportionately the largest cost items of highway construction, so it is highly desirable to build the subgrade to the highest practicable strength and stability. Further, it is desirable to build the subgrade to as *uniform* strength as possible, so that a uniform depth of base and wearing courses may be placed over the entire length of the grading project with perhaps only minor exceptions. This leads to the use of sub-base construction to bolster up weak sections of subgrade. This sub-base is considered part of the pavement structure. A stable material such as pit-run gravel available at reasonably low cost may be used for sub-base construction, the primary purpose being to bring the subgrade to a fair degree of uniformity in strength.

**2-6   Soil survey**   In general the load-carrying capacity of a soil after it has been saturated with water will be the determining factor in evaluating each soil type for use in the subgrade, and the horizons of the soil profile may vary considerably in this respect.

The top layer or A horizon (Fig. 5-5a) will generally contain a large amount of organic material resulting from vegetative growth. The presence of such plant residue in any appreciable amount tends to make the soil unsuitable for construction purposes. It is likely to have relatively large volume changes with changes in moisture content and therefore will shrink and swell detrimentally as ground moisture varies during wet and dry seasons of the year. Also, such soil exhibits elastic properties which make it difficult to compact because an appreciable rebound occurs upon the removal of applied loads. On the other hand, such soil is highly desirable for use in turfing of slopes and for general roadside-improvement planting. This layer may range from a few inches to 2 ft or more in thickness.

The second layer or B horizon of the soil profile will generally have characteristics of the underlying parent material, modified by infiltrated material such as water-soluble salts and minerals and the finer fractions of soil such as silt and clay brought in by leaching from the A horizon. In general it will have engineering properties of stability and strength which make it suitable for use as constructional material.

The C horizon, since it will not have been affected by surface activities, will have essentially the same characteristics and properties that it had in its original geological deposition.

The division between these layers of soil may not be clearly defined. Transition layers of considerable thickness may be sufficiently apparent that they should be designated in the log of soil borings, using subscripts such as $B_1$ and $B_2$. In the brief notation that describes each layer, the color, texture, and depth should be carefully noted because it is on these brief comments that the construction engineer will be able to recognize the soil types designated on soil profiles. A brief sketch of each test hole will provide the most convenient means of recording soil-survey data. Presence of the ground-water table within the limits of the boring should be noted. It is the elevation to which water rises in the test hole.

Economic considerations will often determine the completeness with which the soil survey is made. The nature of work to be done will also influence the extent of survey required; that is, if a highway is to be reconstructed in its present location, much information can be gained from previous soil studies, from performance characteristics of the present roadbed, and from analysis of soil profiles exposed on cut slopes, the principal requirement then being to investigate available borrow pits and supplementary borings in sections where cuts are to be deepened.

In the case of new location, however, a complete soil survey will be fully justified so that selective use can be made of available materials, and adequate design and construction procedures best suited to the various soil types and drainage conditions can be established. Generally, a test hole of each survey station will be adequate coverage for highway construction. Some intermediate holes may be required where marked changes in the soil profile occur. Holes should extend from 4 to 6 ft below any construction work; that is, from 4 to 6 ft below original ground line in fill sections of the roadway and below the final grade line in cut sections. In the case of bridge piers and abutments, specifications for the survey usually require that test holes be made to a depth equal to or somewhat greater than the largest horizontal dimension of the pier or abutment unless solid rock is reached.

A 4-in. post-hole auger, a $1\frac{1}{2}$-in.-diameter soil auger, and persistence will usually be adequate equipment for boring the test holes. Power equipment is available for this work and its use will often prove economical.  The excavated soil should be piled separately by type, and sufficient samples taken so that all variations in type are represented. Tightly woven sample bags should be provided for shipping the samples to the laboratory so that none of the fine material is lost.  If moisture-content determinations are to be made, airtight containers must be used.

Samples of soil taken from sites for structure footings, where the rate and amount of consolidation and consequent settlement of the structure need to be determined, should be taken as nearly as possible "undisturbed" so that the natural structural formation and moisture content are preserved for laboratory testing.  In general, however, for the area of roadway which is to be graded, and since all of the soil concerned in that operation will be more or less manipulated, disturbed samples are generally taken for laboratory testing.

Primarily, laboratory testing of the soil samples taken from the various horizons of the soil profiles leads to classification of each soil according to engineering properties, which in turn determine its suitability for use in construction and the specific treatment to be administered during construction.  Such classification also leads to prediction of behavior of each soil as subgrade material, based on past experience with similar soils under given conditions of moisture and frost, and, consequently, selection of the pavement structure required to protect it adequately.

**2-7  Soil classification**  A convenient and effective classification of highway subgrade soils, based on a few simple and routine tests, is the Revised Public Roads System, a modification by the Highway Research Board of the original Public Roads classification.  It includes a system of group indexing which practically eliminates overlapping of classifications.  The following descriptive paragraphs, Figs. 2-8 and 2-9, and Tables 2-1 and 2-2 have been taken from the report of the Committee on Classification of Materials for Subgrades and Granular Type Roads.[1]

*Group A-1*  The typical material of this group is a well-graded mixture of stone fragments or gravel, coarse sand, fine sand, and a nonplastic or feebly plastic soil binder.  However, this group includes also stone fragments, gravel, coarse sand, volcanic cinders, etc., without soil binder.

Subgroup A-1-a includes those materials consisting predominantly of stone fragments or gravel, either with or without a well-graded binder of fine material.

Subgroup A-1-b includes those materials consisting predominantly of coarse sand either with or without a well-graded soil binder.

*Group A-3*  The typical material of this group is fine beach sand or fine desert blow sand without silty or clay fines or with a very small

[1] *Highway Research Board Proc.*, **25** (1945).

amount of nonplastic silt. The group includes also stream-deposited mixtures of poorly graded fine sand and limited amounts of coarse sand and gravel.

*Group A-2* This group includes a wide variety of "granular" materials which are borderline between the materials falling in groups A-1 and A-3 and the silt-clay materials of groups A-4, A-5, A-6, and A-7. It includes all materials containing 35 per cent or less passing No. 200 sieve which cannot be classified as A-1 or A-3 because of fine content or plasticity, or both, in excess of the limitations for these groups.

Subgroups A-2-4 and A-2-5 include various granular materials containing 35 per cent or less passing No. 200 sieve and with a minus No. 40 portion having the characteristics of the A-4 and A-5 groups. They include such materials as gravel and coarse sand with silt content or plasticity index in excess of the limitations of group A-1 and fine sand with nonplastic silt content in excess of the limitations of group A-3.

**Table 2-1  Classification of highway subgrade materials**

| | Granular materials (35% or less passing No. 200) | | | Silt-clay materials (more than 35% passing No. 200) | | | |
|---|---|---|---|---|---|---|---|
| | Group A-1 | Group A-3* | Group A-2 | Group A-4 | Group A-5 | Group A-6 | Group A-7 |
| Sieve analysis, per cent passing: | | | | | | | |
| No. 10 | | | | | | | |
| No. 40 | 50 max | 51 min | | | | | |
| No. 200 | 25 max | 10 max | 35 max | 36 min | 36 min | 36 min | 36 min |
| Characteristics of fraction passing No. 40: | | | | | | | |
| Liquid limit | ...... | ...... | ...... | 40 max | 41 min | 40 max | 41 min |
| Plasticity index | 6 max | NP | ...... | 10 max | 10 max | 11 min | 11 min |
| Group index | ...... | ...... | 4 max | 8 max | 12 max | 16 max | 20 max |
| General rating as subgrade | Excellent to good | | | Fair to poor | | | |

Classification procedure: With required test data available, proceed from left to right on above chart, and correct group will be found by process of elimination. The first group from the left into which the test data will fit is the correct classification. (Note that all limiting test values are shown as whole numbers. If fractional numbers appear on test report, convert to nearest whole number for purposes of classification.)
* The placing of A-3 before A-2 is necessary in the left-to-right elimination process and does not indicate superiority of A-3 over A-2.

Subgroups A-2-6 and A-2-7 include materials similar to those described under subgroups A-2-4 and A-2-5, except that the fine portion contains plastic clay having the characteristics of the A-6 or A-7 group. The approximate combined effects of plasticity indexes in excess of 10 and percentages passing No. 200 sieve in excess of 15 is reflected by group-index values of 0 to 4. See group-index formula below.

Silt-clay materials—containing more than 35 per cent passing No. 200 sieve—may be listed as follows:

*Group A-4* The typical material of this group is a nonplastic or moderately plastic silty soil usually having 75 per cent or more passing No. 200 sieve. The group also includes mixtures of fine silty soil and up to 64 per cent of sand and gravel retained on No. 200 sieve. The group-index values range from 1 to 8, with increasing percentages of coarse material being reflected by decreasing group-index values.

*Group A-5* The typical material of this group is similar to that described under group A-4, except that it is usually of diatomaceous or micaceous character and may be highly elastic, as indicated by the high liquid limit. The group-index values range from 1 to 12, with increasing values indicating the combined effect of increasing liquid limits and decreasing percentages of coarse material.

*Group A-6* The typical material of this group is a plastic clay soil usually having 75 per cent or more passing No. 200 sieve. The group also includes mixtures of fine clayey soil and up to 64 per cent of sand and gravel retained on No. 200 sieve. Materials of this group usually have high volume change between wet and dry states. The group-index values range from 1 to 16, with increasing values indicating the combined effect of increasing plasticity indexes and decreasing percentages of coarse material.

*Group A-7* The typical material of this group is similar to that described under group A-6, except that it has the high liquid-limits characteristic of the A-5 group and may be elastic as well as subject to high volume change. The range of group-index values is 1 to 20, with increasing values indicating the combined effect of increasing liquid limits and plasticity indexes and decreasing percentages of coarse material.

Subgroup A-7-5 includes those materials with moderate plasticity indexes in relation to liquid limit and which may be highly elastic as well as subject to considerable volume change.

Subgroup A-7-6 includes those materials with high plasticity indexes in relation to liquid limit and which are subject to extremely high volume change.

**Table 2-2  Classification of highway subgrade materials (with suggested subgroups)**

| | Granular materials (35% or less passing No. 200) | | | | | | | Silt-clay materials (more than 35% passing No. 200) | | | |
|---|---|---|---|---|---|---|---|---|---|---|---|
| | Group A-1 | | Group A-3 | Group A-2 | | | | Group A-4 | Group A-5 | Group A-6 | Group A-7 (A-7-5, A-7-6) |
| | A-1-a | A-1-b | | A-2-4 | A-2-5 | A-2-6 | A-2-7 | | | | |
| Sieve analysis, per cent passing: | | | | | | | | | | | |
| No. 10............... | 50 max | | | | | | | | | | |
| No. 40............... | 30 max | 50 max | 51 min | | | | | | | | |
| No. 200.............. | 15 max | 25 max | 10 max | 35 max | 35 max | 35 max | 35 max | 36 min | 36 min | 36 min | 36 min |
| Characteristics of fraction passing No. 40: | | | | | | | | | | | |
| Liquid limit........ | ...... | | | 40 max | 41 min | 40 max | 41 min | 40 max | 41 min | 40 max | 41 min |
| Plasticity index..... | 6 max | | NP | 10 max | 10 max | 11 min | 11 min | 10 max | 10 max | 11 min | 11 min* |
| Group index†........ | 0 | | 0 | 0 | 0 | 4 max | 4 max | 8 max | 12 max | 16 max | 20 max |
| Usual types of significant constituent materials | Stone fragments, gravel, and sand | | Fine sand | Silty or clayey gravel and sand | | | | Silty soils | | Clayey soils | |
| General rating as subgrade | Excellent to good | | | | | | | Fair to poor | | | |

Classification procedure: With required test data available, proceed from left to right on above chart, and correct group will be found by process of elimination. The first group from the left into which the test data will fit is the correct classification.

*Plasticity index of A-7-5 subgroup is equal to or less than LL – 30. Plasticity index of A-7-6 subgroup is greater than LL – 30. (See Figs. 2-8 and 2-9.)

† See group-index formula and Figs. 2-8 and 2-9 for method of calculation. Group index should be shown in parentheses after group symbol, as A-2-6(3), A-4(5), A-6(12), A-7-5(17), etc.

The formula for the group index is as follows:

Group index $= 0.2a + 0.005ac + 0.01bd$

where $a =$ that portion of percentage passing No. 200 sieve greater than 35 and not exceeding 75, expressed as a positive whole number (1 to 40)

$b =$ that portion of percentage passing No. 200 sieve greater than 15 and not exceeding 55, expressed as a positive whole number (1 to 40)

$c =$ that portion of the numerical liquid limit greater than 40 and not exceeding 60, expressed as a positive whole number (1 to 20)

$d =$ that portion of the numerical plasticity index greater than 10 and not exceeding 30, expressed as a positive whole number (1 to 20)

The following are examples of calculations of the group index:

1. An A-6 material has 65 per cent passing No. 200 sieve, liquid limit of 32, and plasticity index of 13. The calculation is as follows:
   Let $a = 65 - 35 = 30$
   $b = 55 - 15 = 40$ (55 is substituted for 65 as critical range is 15 to 55)
   $c = 0$, since liquid limit is below 40
   $d = 13 - 10 = 3$

   Group index $= (0.2 \times 30) + (0.01 \times 40 \times 3) = 7.2$

   The group index should be recorded to nearest whole number, which is 7.

2. An A-7 material has 54 per cent passing No. 200 sieve, liquid limit of 62, and plasticity index of 33. The calculation is as follows:
   Let $a = 54 - 35 = 19$
   $b = 54 - 15 = 39$
   $c = 60 - 40 = 20$ (60 is substituted for 62 as critical range is 40 to 60)
   $d = 30 - 10 = 20$ (30 is substituted for 33 as critical range is 10 to 30)

   Group index $= (0.2 \times 19) + (0.005 \times 19 \times 20)$
   $$+ (0.01 \times 39 \times 20)$$
   $$= 13.5 \text{ (or 13)}$$

Charts for graphical determination of group index are shown in Figs. 2-8 and 2-9.

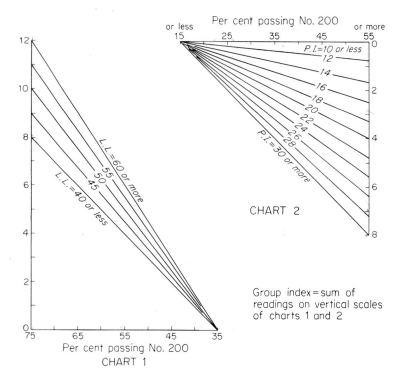

**Fig. 2-8**  Chart 1.                    **Fig. 2-9**  Chart 2.

**2-8  Related problems**  The engineer on a grading job is confronted
with many problems both in design and construction for which there are
no formulas or precise procedures to which he can turn for solution.  It is
not anticipated that all questions concerning problems presented by vast
areas of poor subgrade material, mud flows, landslides, rolling rocks,
and other problems such as back-slope stability and large areas with very
poor natural drainage can be adequately answered here.  Often, considering cost and the physical limitations of available construction equipment and materials, the only sensible answer will be to "go around" or
to relocate in such a way as to avoid situations which are admittedly not
impossible, but which are not economically feasible.  Such a way out,
however, is not always available, and each engineering problem must be
solved on the basis of its individual characteristics.  A brief summary
of some established practices based on successful past experience will
be helpful in acquiring acumen, good judgment, and skillful adaptation
of engineering principles to problems of design and construction.

**2-9  Subgrading**  Of paramount importance to the roadway itself is, of course, adequate foundation support.  Often the finally established grade line in a cut section of the roadway will not provide sufficient protective cover for the underlying soil.  The A horizon of the soil profile may be rich in humus for a considerable depth, or it may be composed of wind-deposited loess or silt and clay.  Many such materials are fairly stable when dry, but water from surface infiltration or from the ground-water table may be readily taken up by capillarity until an unstable state is reached.  In this condition such soils will often "pump" into the interstices of base-course material under the vibratory action of traffic and so impair or destroy the stability and supporting power of the entire pavement structure.

If the thickness of the layer is not more than about 5 ft, it will very likely be feasible to "subgrade" or remove the material to a more stable underlying soil.  It is notable, however, that subgrading and backfilling with better material will increase two pay items, namely, "common excavation" and "borrow."  For this reason a careful cost study should be made to determine whether subgrading can be economically carried down to more stable material, or a compromise be made by a combination of subgrading and raising the surface grade line so that sufficient depth of cover is provided to protect adequately the low-strength material.

**2-10  Filter layer**  When it is not feasible to excavate through the entire depth of a weak stratum, it is desirable to permit any accumulated free water to escape and at the same time prevent the silt or clay from "pumping" up into the voids of relatively coarse granular backfill.  A layer of filter sand may be used for this purpose, as illustrated in Fig. 2-10.

The layer of filter sand is usually made from 4 to 6 in. in thickness, depending on availability of suitable material, and it should extend to the ditches on both sides of the roadway.  In this way it serves as a lateral drain for any accumulated water which might otherwise be detrimental to both backfill and underlying soil.

In order to function properly as a filter material the sand must be fine enough to confine the foundation material and porous enough so that

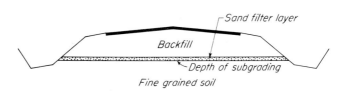

**Fig. 2-10**  Use of filter layer to control pumping.

water will not be held within it by capillarity. In general these conditions would be met by a sand passing a $\frac{3}{8}$-in. sieve and retained on the No. 40 sieve. It is important that the size of pore spaces be determined by the "sand" portion of the material; that is, the presence of some material coarser than the $\frac{3}{8}$-in. sieve will not be detrimental if it is dispersed through the sand in small enough proportion so that the net effect of its presence is to reduce rather than increase the porosity. The coarse portion of available filter material can usually be readily adjusted by screening out undesirable amounts of oversize material. Adjustments in grading of the finer fractions are not always so readily accomplished, and it is this finer material which will be effective in holding back the underlying soil while water percolates through the filter layer. If an appreciable amount of fine material such as fine sand, silt, and clay is present in the sand, it will generally be removed by washing, because screening is not practical on screens finer than about 8 meshes per inch. A cost comparison should be made between available commercial sands and processing of materials from other sources. A more complete discussion of filter sands is given in the section on airport engineering.

**2-11  Vertical sand drains**  Occasionally it is found necessary to construct an embankment on a water-bearing layer of clay or silt of such depth that it cannot be drained horizontally and its removal would not be economical. Such saturated soils, fed by underground water flow, must be drained if stability of the embankment is to be preserved. It is important, too, that consolidation of the foundation material be expedited so that unequal subsidence will not later impair embankment and pavement structure. In order to accomplish this, vertical sand drains are constructed by drilling holes 15 to 20 in. in diameter through the compressible stratum to firm soil and filling the holes with sand. Holes are spaced uniformly 8 to 20 ft center to center over the entire area to be drained. Since consolidation of water-bearing soil takes place by removal of the water, the rate of consolidation for a given embankment load is dependent on the permeability of the soil and the distance the water has to travel. For effective hastening of consolidation of foundation soils which are predominantly clay, the spacing of sand drains should be near the lower limit of the above range of spacing; for silt and fine sand layers, spacing of greater magnitude will still be effective.

The tops of the vertical sand drains should connect with a network of drain tile or a substantial layer of porous sand or gravel 18 to 30 in. in thickness at the ground surface to carry the water to the edge of the embankment.

Sand used for the vertical drains should comply with general requirements for filter sand previously discussed in the section on subgrading.

Consolidation of foundation soil under embankments by use of vertical drains can be further hastened by depositing an overload of several feet of additional embankment, which is later removed to bring the roadway to desired grade line.

**2-12   Displacement of peat or muck**   When an embankment is to be built over an area of peat or muck, it will generally be necessary to remove such material rather than to depend on consolidation.   Here, too, if the bed or layer is less than about 3 or 4 ft in thickness, it may be economically excavated and backfilled with more stable soil.   In the case of deeper beds the method of overloading the embankment has been successfully used.   The fill is built up to as much as 10 ft above the desired grade line, causing the peat or muck to flow outward from under the embankment. It should be recognized that this process of displacement might require several months to a year or more, and final adjustment of grade line and placement of pavement structure should not be attempted until elevation checks from established bench marks show that subsidence is complete to the point where further *uneven* settlement will not occur.

Blasting is often resorted to in order to expedite displacement of peat or muck so that roadway construction may be completed sooner. From 6 to 8 ft of the lower portion of the fill is placed on the area.   Then holes are drilled down through the fill to the muck and charges of dynamite are exploded so as to cause lateral displacement of the muck, allowing the embankment layer to settle down to firm soil.

**2-13   Low fills on marsh areas**   Careful investigation of foundation soils is always in order when there is any doubt as to the supporting power of the soil.   A marsh area or peat bog may have reached a stage of geological development where it has become reasonably firm and does not contain excessive amounts of water.   Existing vegetation or a mat of brush on the ground surface may sufficiently spread the load of a low embankment not exceeding 3 or 4 ft so that detrimental settlement will not occur. Engineers have estimated that such an embankment will be adequately supported on foundation material upon which a person can walk without making an appreciable indentation by virtue of his own weight.   Such a criterion might be helpful in judging whether an accurate determination of load-carrying capacity would be justified.   In making such a test one should be certain that firmness extends below the surface and that one's weight is not supported by surface tension of a well-established sod which may be virtually floating on saturated muck.

**2-14   Problems of sidehill construction**   In mountainous and very hilly country much of the roadway will be constructed along hillside slopes in order to avoid excessive grades, and many problems arise which must be handled in accordance with local characteristics of soil and moisture. In a general way, some principles of good practice are significant, because experience is the best guide to the solution of problems which do not lend themselves readily to the use of either rational or empirical formulas and where detailed investigations and studies are not economically justified.

Slides may occur in cuts or fills, but it is on long, relatively steep hillside slopes that control becomes difficult. Major factors which affect slides are steepness of slope, height of slope, and seepage. Two main tendencies develop when soil is placed on a steep slope, namely, sloughing or flow of material near the surface or sliding of a larger mass of soil because of failure in shear along a plane more or less parallel to the hillside slope, depending on any stratification or lamination present, or on a nearly spherical surface, as illustrated in Fig. 2-11.

The processes of weathering tend to loosen surface material and cause it to slough and drift down any slope greater than the angle of repose of the dry loose material. This applies to small particles and large boulders alike, with the result that ditches and other drainage structures become clogged on the cut side of the roadway and shoulders impaired on the fill side. Runoff from excessive rainfall further accelerates erosion of slopes and may set in motion relatively thick layers of accumulated loose surface soil as mud flows. Water which permeates into the soil may lubricate an underlying stratum of gumbo or clay pan to create a slip surface on which a slide may be mobilized; or it may create seepage pressures in the direction of ground-water flow which combine with the force of gravity to cause shear failure on a surface similar to that indicated on the cut side of Fig. 2-11. Infinite variety is given to the problem of

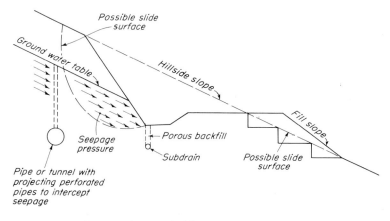

**Fig. 2-11** Roadway section on hillside.

control by the magnitude with which each contributory influence exists in a specific location.

Manifestly, it is desirable to keep the cut and fill slopes as flat as possible, but this desire must be balanced with the need to keep pay quantities within economical limits. Generally, it is better practice to risk some sliding and slope adjustment than to increase yardage to the point where a substantial factor of safety is provided against such events.

Further reference to Fig. 2-11 will show that back slope and fill slope will have to be considerably steeper than the long hillside slope if yardage is to be kept within reason. On the fill side the hillside slope should be benched so that the fill may be placed and effectively compacted in nearly horizontal layers and so that a slip surface will not develop on the interface between the original ground surface and the fill material. This will usually be accomplished by use of a crawler-type tractor with bull-dozer or angledozer attachment. The fill slope should not be steeper than $1\frac{1}{2}$ (horizontal) to 1 (vertical), and 3 to 4 ft of extra shoulder width should be provided to support guardrail posts adequately. A steeper fill slope, up to the angle of repose of the fill material, may have to be resorted to in certain cases, and if so, the shoulder should be made still wider to provide for anticipated natural slope adjustment during several years of weathering, recognizing, too, that considerable maintenance care will be necessary during this period to preserve the slope.

On the cut side of the roadway a wide ditch section, up to 8 or 10 ft, or more if possible, will be desirable to provide room for equipment to clean out the ditch and other drainage facilities that may exist and to provide a measure of protection to traffic against sliding material and rolling rocks which may proceed down the slope. This will, of course, increase excavation, but good design will allow some increase in this respect, depending largely on how critical are the adverse conditions of the soil mass.

In addition to providing a substantial ditch section on the cut side of the roadway, it may often be necessary to intercept ground-water flow more effectively, especially if the subgrade soil would be adversely affected by the near presence of such water. A subdrain of perforated or open joint pipe is often placed below the bottom of the ditch and back-filled with porous material so as to lower the water table at that point. If it is desired to relieve seepage pressure farther back in the soil mass to prevent sliding, a larger pipe or tunnel is sometimes placed with or without vertical leads, as illustrated in Fig. 2-11. Often such treatment will prevent a slide from starting or arrest the progress of one already started.

It would be overoptimistic indeed to expect that full control of a large mud flow or a large landslide can be realized. Such barriers as ordinary retaining walls or lines of piling will fail. It is better practice to attack the problem of drainage so as most effectively to mobilize and assist the restraining forces inherent within the soil mass itself—its shearing strength.   The cost of handling a large slide should be compared with alternate locations or a trestle span to support the roadway.

It is important to recognize the existence of a slide area in order to avoid some of the pitfalls that necessarily accompany construction on such areas.   Fence posts out of line or out of plumb with respect to the main line posts; telephone or power-line poles atilt or wires taut in some spans and sagging in others; trees growing at an angle with the vertical, or with a bow in the trunk, indicating that earlier movement of the soil tilted the tree and then it continued to grow straight up from that point; tension cracks in the surface layer of the soil or a depressed ledge near the top of the slide and slight ridges near the base are all indications that any substantial construction on that particular hillside would be costly to build and maintain because of the constant threat or actual movement of the soil mass upon which such construction must rest.

The benefits of vegetative growth upon a hillside slope of questionable stability should not be overlooked.   It serves in several ways to protect the slope; therefore any existing vegetation should be preserved and the possibility of making additional planting should be investigated. Trees, shrubs, and grass growth will largely prevent erosion, and, also, the larger roots build up a tensile strength in the soil mass which is additive to the shearing strength of the soil in resisting shallow sloughing.   In addition, transpiration of moisture tends to use excess moisture and thus establish more favorable moisture content for increased cohesion of plastic soils or soil layers.

**2-15   Frost heave**   One of the most difficult problems with which the highway engineer must cope is that of frost heave.   In northern areas where annual frost penetrates below the pavement structure the destructive action of freezing of soil water and subsequent thawing is often severe enough to weaken or break up the pavement structure, greatly reducing its service life and placing a strain on the maintenance budget.

Pavement heave over relatively large areas and more local breakups due to frost boils cannot be accounted for by the mere expansion of water as it freezes within the soil mass.   Water expands about 9 per cent as it freezes, and if a "quick-freeze" process were to occur in the moist soil mass, total vertical expansion would amount to less than an inch even though freezing were carried to a considerable depth.   The penetration of frost into the soil occurs naturally at a comparatively slow rate and usually with some alternate freezing and thawing as temperatures vary between night and day.   As a layer of water-bearing soil is frozen beneath the pavement it becomes essentially dry, and a dry surface of ice has a very high affinity for free water; that is, it will be recalled that water rises in a capillary tube only because the water is attracted to the surfaces of material from which the tube is made. The process of "wetting" continues up the inside of the tube until attraction of the water for the inside surface is balanced by the weight of water supported in the tube. A film of oil, or other substance for which water does not have great affinity, placed on the inner surface would greatly alter the height of rise and would actually result in a lowering of the water level within the tube below the level of free water in which the tube is placed, if water is repelled by the material on the surface of the tube. Because of its high affinity for a surface of ice, water would rise in a capillary tube made of ice to the full strength of its surface tension, and in the same way free water is attracted to the undersurface of the frozen layer of soil. This attracted water is then in turn frozen, as freezing temperature proceeds downward during cold weather. Field observations and laboratory tests have shown that layers or lenses of clear ice, several inches in thickness, are built up in this way if conditions are right.

For progressive growth of ice lenses there must be decreasing temperature so that accumulated free water is frozen on the surface of ice already formed. A rapid penetration of low temperature will of course freeze a layer of the composite soil mass between ice lenses. Such lenses form most readily when the heat of fusion of the ice as it forms on the lens is just absorbed by the frozen layer above.

The other essential is a supply of free water. Surface water may of course enter the subgrade through cracks in the pavement or from poorly constructed shoulders which tend to trap surface water; and any such accumulation would freeze during cold weather, causing more or less severe damage. But water for the formation of massive ice lenses which cause destructive pavement heave and frost boils is supplied by the soil itself through capillarity from the ground-water table. It is necessary that the pore spaces of the soil not only be small enough so that water will rise to a considerable height, but also that they be large enough so that water will rise at a fairly rapid rate. In the case of clay soil, the rate of penetration of the frost line into the soil would have to be very slow indeed to accommodate the formation of ice lenses, because water would be supplied to the freezing surface by capillary rise at a very slow rate. Experience has shown that fine sand and silt, particularly silt with a predominance of particle size of 0.02 mm, is most nearly ideal in respect to ability and rate of supplying water by capillarity for the formation of ice lenses under normal field conditions of soil freezing.

Pavements have been known to heave to a considerable extent without serious damage. This is particularly true of rigid pavements and in those cases where the heave has extended fairly uniformly over a consider-

able length of roadway. And if uniform heave could be assured, the
seriousness of the problem would be greatly reduced. This is not often
the case, however, because, in general, soil is not precisely uniform even
within a fairly restricted pavement area. Because of these variations in
soil characteristics and conditions, some cracking is likely to occur during
the time of freezing and formation of ice lenses. But it should be recognized that the real destruction comes during the time when frozen soil and
ice lenses are melting.

In general, frozen soil thaws out from above and below during warm
spring weather. During this period excessive amounts of water, freed
by the melting soil and ice lenses, may become trapped within the soil
between the relatively nonporous pavement surface and the still frozen
layer underneath, creating a supersaturated condition which may render
the soil virtually liquid and impotent. Support for the pavement structure may thus be reduced to practically zero, and this condition may prevail for a considerable length of time because the excess water cannot
readily evaporate through the pavement wearing course and will not
readily drain laterally into the ditches as such soils tend to hold water by
capillary attraction and thus prevent effective drainage.

Some solutions to the problem will of course present themselves.
The first of these is to prevent freezing. While this is not generally
practical, there may be instances where a network of heat conductors,
using water, steam, or electricity, is to be used for the prevention of
surface freezing at lower levels.

The silty soils which are susceptible to frost heave and subsequent
instability are easily recognized, and if moisture and frost conditions are
adverse, it is better practice to remove such material by subgrading and
backfilling to a depth equal to the average annual penetration of frost.
It is generally recognized that the pavement structure affords some
insulation against frost so that the average frost penetration may be
safely used. A chart showing average depth of annual frost is given in
the section on airport engineering.

Subdrainage to lower the water table will reduce the amount of
heave but will not prevent the nourishment of ice lenses by water supplied
by capillary action. Water will rise upward of 10 ft by capillarity in a
silty soil, depending on the fineness and degree of consolidation to which it
has been subjected. Raising the grade line will accomplish the same end,
and sometimes a combination of high grade and subdrainage will have
to be resorted to. Interception of lateral flow of ground water on slopes,
as illustrated in Fig. 2-11, will generally be economically justified on the
basis of effectiveness with respect to cost.

It is significant that a layer of fine-grained soil with a low rate of
capillary conductivity between the silt and the pavement structure will

prevent the formation of ice lenses.   A bituminous underseal to prevent
or retard capillary rise has also been successfully used.   Silt, because it is
easily worked and manipulated and is fairly stable when confined, lends
itself to bituminous treatment so that a layer of a few inches in thickness
of the foundation soil can be so treated by "mixed-in-place" operations
with blades, harrows, and rollers before backfill material or sub-base is
placed.   Such treatment also prevents pumping of the silt up into
coarser backfill and sub-base layers.

**2-16   Permafrost**   Permanently frozen ground, referred to as *permafrost*,
is found in vast areas of arctic and subarctic regions.   To some extent
this condition may be considered to be an underground continuation of
the polar ice cap, with the depth of the frozen layer generally decreasing
to the south and varying in thickness from several hundred feet to a few
feet.   Depth to the permafrost also becomes generally greater in latitudes
farther south.

However, much local variation exists with respect to depths because
of local variations in conditions of climate, topography, and soil.   There
is usually considerable difference, for instance, between depth to permafrost on a south slope as compared with a north slope in the same area,
other physical conditions being equal.   In inland regions of high altitude,
permafrost may be found in scattered areas as far south as 45 or 50°
latitude.   It is in regions such as Alaska and northern Canada, however,
that permafrost becomes an engineering problem on a significant scale.

The continued existence of permafrost within a few feet of the surface in subarctic regions is largely dependent on insulation provided by
the overlying soil mantle and vegetative cover.   Any change affecting
this insulation will alter the position of the permafrost table.   Experiments have shown, for instance, that placement of an earth fill or a building over permafrost provides additional insulation and the table will rise
to a higher position compatible with its insulation requirements in that
particular locality.   There is, of course, some seasonal variation in the
position of the permafrost table, and the effect of seasonal changes in
surface temperature varies with the nature of overlying soil, vegetative
cover, and snow cover.   Annual frost may or may not penetrate to the
permafrost table.

The effect of placing a roadway embankment over permafrost is
illustrated in Fig. 2-12.   It can be seen that the original overlying layer of
soil and part of the embankment may become frozen.   This may cause
considerable heave if the soil is susceptible to frost heave and if considerable moisture is present in the soil.   It can also be seen that such a rise
of the permafrost table can create a barrier to lateral flow of underground
water.   The detrimental or destructive effect of such impounding will be

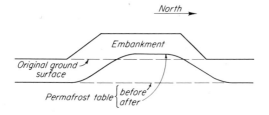

**Fig. 2-12** Effect of embankment on permafrost table.

determined by drainage characteristics and shearing strength of embankment and foundation soils, but direct interception of any natural groundwater flow should be avoided.

It is in the case of hillside construction that problems of permafrost become particularly significant. Reference to Fig. 2-13 will help to illustrate some of these problems and possible solutions. On the fill side a rise in the permafrost table is to be expected with attendant results, as previously discussed. On the cut side the permafrost table will be lowered. If the table is near the surface and the cut deep, excavation may be carried to the table, and then sufficient time provided for the permafrost to melt back to complete the cut section. This may require two or more summer seasons for completion of the work. Blasting may be needed to expedite matters.

Lowering of the permafrost table on the cut side is of course attended with release of water from the melting ice, and the seepage pressure caused by this water coupled with whatever underground water flow already exists may cause a mud flow into the excavated ditch section. This may require removal by maintenance crews with blade or power shovels for a period of 3 or 4 years in adverse cases before the permafrost table becomes stabilized in its new position. Such seepage also causes icing during winter months. It is not uncommon for ice to build up from slow seepage spreading in thin films over the ice already formed and in turn freezing until it crosses the entire roadway section. It is therefore

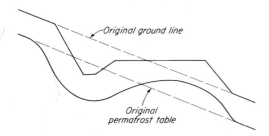

**Fig. 2-13** Alteration of permafrost table by hillside roadway section.

important to intercept natural ground-water flow on the slope above the cut and to maintain effective drainage in the ditches by providing adequate ditch grade and keeping them free of debris and slough from eroding back slopes.

In general, it is considered better practice in areas of permafrost to avoid cuts as much as possible and build the roadway in fill section. This greatly influences location, with respect to longitudinal gradients and to available borrow pits of acceptable granular materials. It also necessitates stripping of overburden from borrow pits a year or more in advance of construction to permit frozen material to thaw out in those areas where the permafrost table is close to the ground surface.

## QUESTIONS AND PROBLEMS

**2-1.** Determine the porosity, void ratio, and degree of saturation of the wet soil in Probs. 1-3 and 1-4.

**2-2.** A compacted soil with a wet density of 117.6 pcf has 97 per cent of its pore spaces filled with water at a moisture content of 29.5 per cent. What is the specific gravity of the soil solids? Determine the porosity and void ratio.

**2-3.** A clay soil has a void ratio of 0.53 at its liquid limit. The absolute specific gravity is 2.65, and the plastic limit is 14. What is the plasticity index?

**2-4.** An undisturbed sample of material taken from a borrow pit has a void ratio of 0.70. Results of the Proctor compaction test indicate that the material as compacted on the roadway will have a void ratio of 0.48. What shrinkage factor should be used in computing borrow and embankment quantities?

**2-5.** Determine a soil grading for which $n$ is 0.3 in Eq. (2-1), using 1-in. maximum size and the following sieve designations: $\frac{3}{4}$ in., $\frac{3}{8}$ in., No. 4, No. 10, No. 40, and No. 200.

**2-6.** Classify the following soils according to Highway Research Board subgrade groups and index numbers:

| Soil | Per cent passing | | | Liquid limit | Plasticity index |
|------|--------|--------|--------|------|------|
|      | No. 10 | No. 40 | No. 200 |      |      |
| 1 | 100 | 96 | 82 | 72 | 27 |
| 2 | 100 | 100 | 98 | 34 | 10 |
| 3 | 100 | 99 | 92 | 56 | 30 |
| 4 | 65 | 43 | 22 | 20 | 3 |
| 5 | 91 | 68 | 1 | 17 | NP |
| 6 | 96 | 52 | 26 | 30 | 11 |
| 7 | 100 | 97 | 89 | 55 | 36 |
| 8 | 86 | 52 | 24 | 17 | 4 |
| 9 | 83 | 78 | 4 | 20 | NP |
| 10 | 100 | 100 | 96 | 31 | NP |

**2-7.** A crushed rock has a specific gravity of 2.68.   In the loose state the porosity is 46 per cent and when compacted the porosity is 20 per cent.   Compute the dry unit weight in pounds per cubic foot for both the loose and compacted conditions, and the ratio of compacted volume to loose volume.

$A$                              $B$

## BIBLIOGRAPHY

Classification of Soils and Control Procedures Used in Construction of Embankments, *Public Roads*, **22**(12) (1942).

Classification of Soils and Subgrade Materials for Highway Construction Purposes, *Highway Research Board Proc.*, **25** (1945).

"Field Manual of Soil Engineering," rev. ed., Michigan State Highway Department, 1946.

Geophysical Methods and Statistical Soil Surveys in Highway Engineering, *Highway Research Record* 81, 6 Reports, Washington, 1965.

Proctor, R. R.: Fundamental Principles of Soil Compaction, *Eng. News-Record*, **3**(9):245 (1933).

"Standard Specifications for Highway Materials and Methods of Sampling and Testing," American Association of State Highway Officials, 1967.

Winn, H. F.: "Frost Action in Highway Subgrades and Bases," *Proc. Purdue Conf. on Soil Mech. and Its Appl.*, Purdue University, Lafayette, Ind., 1940.

# 3
# Stabilized Roads

**3-1 Good roads** The attainment of "good roads" has been a prominent goal in the planning of progressive communities throughout the country, particularly since the advent of the automobile. To meet these requirements, early programs of highway construction were primarily concerned with getting roads to everybody's door within a relatively short period of time. It is understandable that on so large a scale of new construction with limited funds, the roads were built to limited standards, sufficient to the needs of the time but with little margin for future increases in traffic. The colossal growth of automobile traffic coupled with heavier wheel loads on fast-moving trucks soon made the earlier roads inadequate, and a vast program of reconstruction has been under way in recent years to meet the demand for "better roads."

A question then arises as to what constitutes a good road—or an adequate road. Such a road should afford a smooth firm riding surface; it should be relatively dust-free; and its width, gradient, and alignment should be such as to afford safe and economical operation of traffic at reasonably high speeds. These conditions can be met in various ways, and it is certain that not all roads require a high-type pavement as a wearing course.

**3-2   Soil-stabilized roads**   The cost of construction and maintenance of a road can be economically justified primarily on the basis of saving in cost of vehicle operation to the user.   Many access roads, farm-to-market roads, and other light-traffic roads on which traffic volumes range from less than fifty vehicles in a day to three or four hundred vehicles per day have been built to serve adequately such traffic with some form of soil stabilization as a wearing course.   Then as traffic volumes increase beyond a few hundred vehicles per day, maintenance costs may become excessive, and it will be more economical to construct a higher type of surfacing or pavement with the previously constructed stabilized roadway providing a serviceable base course.

Basic ingredients for a stabilized soil are reasonably well graded granular material and a binder or cementing agent.   This combination is frequently found in natural sand-clay mixtures or it can be built up by mixing sand with a soil which has an overabundance of clay or by mixing clay with a sandy soil.

By itself, well-compacted sand would furnish adequate stability to static and slow-moving wheel loads.   Confined sand offers high resistance to displacement by virtue of the internal friction developed by interlocking sand grains.   It is important, then, that the sand be composed of hard durable particles and fragments preferably of somewhat angular shape.   Such stability against slow-moving and static loads is not sufficient, however, to the needs of a traffic surface because fast-moving tire contact tends to whip out aggregate particles even of considerable size.   This loose material gradually forms riffles under the action of traffic, and the resultant pounding of wheel and axle assemblies oscillating against the springing of fast-moving vehicles develops shallow transverse waves—"washboards"—across the roadway.   This phenomenon frequently occurs on gravel-surfaced roads where the fine material or binder has been whipped out of the surface by traffic and blown away as dust during dry weather.

Capillary tension of moisture films on aggregate particles, particularly the finer sand fractions, provides a cohesive or binding quality which makes a wet sandy surface more durable under traffic than a dry one.   Since it is generally not feasible to maintain a sandy surface in a constantly moist condition, a more substantial and permanent binder is necessary.   The most elementary of such binders is clay.

**3-3   Sand-clay roads**   As used here, clay is defined to be that portion of soil whose particles are smaller than 0.005 mm.   The finer fractions of clay—somewhat less than 0.001 mm—exhibit Brownian movement in water suspension and are known as colloids or colloidal clay, best described as a "gluey" material.

The adhesive and cohesive properties of clay are attributed largely to its ability to hold moisture in thin films on individual particle surfaces with exceedingly powerful molecular attraction which increases as the moisture films become minutely thin. Thus clay becomes hard and relatively stable at low moisture content. Its ability to retain moisture is much greater than that of materials, such as silt and sand, which have much less surface area. As the moisture content of clay increases, the adsorbed films of water become thick enough to provide lubrication between particles and the clay mass becomes plastic. With the addition of still more water the clay will become liquid, losing the adhesive and cohesive properties which had provided stability or resistance to displacement. By itself, then, a clay roadway surface would be expected to be slippery and form ruts under traffic when wet and to present a dust problem when the surface becomes very dry.

The proper combination of sand and clay provides all-weather stability and durability if adequately maintained. The sand or coarse material provides resistance to displacement, particularly when wet, provided that there is no excess of clay to disperse the coarser particles, with a resulting loss of interlock and surface friction. In this case, it is apparent that if sand grains are not in contact, the soil would adopt the properties of the filler. Consequently, a sand-clay road must be mostly sand. The clay provides durability to the extent that it prevents raveling of the surface under traffic, and it provides a tight surface which will shed water if proper crown is maintained.

Since particle-size gradation is a significant engineering property of soils, the divisions given in Table 3-1 are generally used to designate coarse aggregate, silt, and clay, and will be used here. Approximate sieve sizes are given for the coarser divisions. Sizes of particles smaller than

**Table 3-1  Particle-size designation**

| Size designation | Particle size, mm | U.S. Standard sieve | |
|---|---|---|---|
| | | Passing | Retained on |
| Gravel................. | Larger than 2.0 | ...... | No. 10 |
| Coarse sand............ | 2.0–0.25 | No. 10 | No. 60 |
| Fine sand.............. | 0.25–0.05 | No. 60 | No. 270 |
| Silt................... | 0.05–0.005 | | |
| Clay.................. | Finer than 0.005 | | |
| Colloidal clay.......... | Finer than 0.001 | | |

0.05 mm which include silt and clay are determined by the velocity with which the particles settle in a soil-water suspension.   Coarse aggregate is designated in the table as gravel.   (See also Fig. 2-3.)

For sand-clay surface courses—and base courses—it is desirable that from 10 to 25 per cent be coarse aggregate retained on the No. 10 sieve. As a general requirement for surface courses, all material should pass a 1½-in. sieve.   Larger particles tend to loosen and become dislodged from the surface under the action of traffic; also, as abrasion of the surface takes place, larger particles tend to project above the rest of the surface, creating a rough and noisy riding surface.   Considerable strength is afforded by the larger material, however, particularly if all the material is reasonably well graded from coarse to fine so that a high degree of compaction can be attained.

Of the portion passing the No. 10 sieve, the sand-clay portion, the gradings in Table 3-2 are recommended by AASHO.

Considerable range and variety of grading can be used, but for best results some attention should be given to prevailing moisture conditions which might be encountered in the field.   It is to be expected that grading 2 would be satisfactory under average conditions.   Grading 1, with a lower percentage of clay, is desirable in regions with a comparatively large amount of rainfall spread over a considerable portion of the year, so that the normal moisture content of the surface course might be expected to be rather high.   In comparatively arid regions a higher percentage of clay would be desirable and grading 3 could be effectively used.

It is further specified that the soil-mortar material passing the No. 40 sieve should have a liquid limit of not more than 30 and a plasticity index of not more than 10.   These specifications can be met by adjusting the proportion of clay in the mix according to the plasticity of the clay itself.   Rich clays are hard to mix.

Where natural material meeting these requirements is available, it is only necessary to spread it over the roadbed to the desired depth, com-

**Table 3-2   Suggested gradings for sand-clay roads***
Percentages by weight

|  | Grading 1 | Grading 2 | Grading 3 |
|---|---|---|---|
| Sand fraction: |  |  |  |
| Passing No. 10 sieve | 100 | 100 | 100 |
| Passing No. 40 sieve | 40–80 | 40–80 | 40–80 |
| Passing No. 60 sieve | 30–70 | 40–55 | 55–70 |
| Passing No. 270 sieve | 10–40 | 20–35 | 30–50 |
| Silt fraction, 0.05–0.005 mm | 3–20 | 0–15 | 10–20 |
| Clay fraction, finer than 0.005 mm | 7–20 | 9–18 | 15–25 |

* American Association of State Highway Officials.

pact it, and shape the surface to desired grade and section. Soil will generally compact to about two-thirds of its loose depth, so that 9 in. of loose soil will be needed to give a compacted depth of 6 in. Normally, a stabilized wearing course will be made from 4 to 8 in. in thickness, depending on the amount of anticipated commercial or truck traffic and the amount of protection deemed necessary to protect adequately the subgrade.

Compaction should be done with a sheeps-foot or other type of tamping roller so that the course is compacted from the bottom to the top. Control of moisture content during this operation is important, and it should be kept near optimum for the attainment of maximum density. If the material is considerably above optimum moisture content, but workable, then chances are that manipulation of the soil by such mixing operations as are necessary to obtain uniformity in the mixture will bring the moisture content to the desired amount. If the material is too dry, water is usually added by use of a distributor tank equipped with spray attachments while mixing operations are under way.

When stabilization is accomplished by the addition of either sand or clay to material already in place, the original surface is first scarified to a depth sufficient to furnish the required amount of in-place material for the desired depth of surface course to be built. Ordinarily it will be expedient to determine the proper combination of available materials to meet the grading requirements first, and then make such adjustments as necessary to keep within specified limits of liquid limit and plasticity index. These moisture tests should receive careful attention because detrimental shrinkage and swell with resultant cracking are likely to occur if the soil mortar has liquid limit and plasticity index of more than 30 and 10, respectively.

Mixing of materials is usually accomplished with blade graders, plows, and harrows to the satisfaction of the engineer that uniformity of texture and moisture have been attained. Some shaping of the cross section with blade graders usually accompanies compaction rolling. The final section should have a crown of about $\frac{1}{4}$ to $\frac{3}{8}$ in./ft to facilitate drainage of surface water.

Where stable shoulders are in place on the existing roadway, it may be desirable to construct the stabilized wearing course in a trench between the shoulders. Generally, it will be more expedient and effective to stabilize the entire roadway section from fore slope to fore slope. More effective drainage will be provided, and maintenance blading and dragging will not bring inferior material onto the riding surface. In either case, compaction should be carried across the full width of construction and should proceed from the outside in toward the center line so as to confine the material as much as possible during compaction.

Over-all width of roadway construction will generally be from 28 to 32 ft. This will provide for two 10- or 11-ft traffic lanes with a minimum of 4- or 5-ft shoulders. Traffic lanes will not be significant while the stabilized surface is functioning as a wearing course, and in fact it is better if traffic is well distributed over the roadway. Later improvement should be kept in mind, and adequate width for pavement mat and shoulders is a commendable provision.

**3-4 Gravel-surfaced roads** Engineering principles involved in the construction and maintenance of gravel-surfaced roadways are not far different from those discussed previously for sand-clay roads. In fact, a line of distinction between the two would be difficult to define. Pit-run gravel, screened gravel, or crushed gravel may be used for basic mineral aggregate, and the principles may be extended to include crushed rock and crushed slag. A suitable binder must be provided. Clay, rock screenings, slag screenings, or other cementitious material may be used. If sufficient binder is not present naturally or produced in crushing, it will have to be supplied from other sources.

Here again, particle-size gradation of composite material is of special significance. Some gradings as recommended by AASHO are given in Table 3-3.

It is further specified that liquid limit and plasticity index of that portion of material passing the No. 40 sieve should not exceed 25 and 6, respectively. This requirement is significant because excess binder may cause the surface to pack hard and form a crust which would be susceptible to cracking and pitting. Available materials and construction equipment will often dictate which grading is to be used. Attention should be given to attainment of uniform gradation from coarse to fine so that high density under compaction will be assured.

Considerable "processing" will usually be necessary to obtain a uniform mixture. This will generally be accomplished by the use of blade

**Table 3-3  Suggested gradings for gravel surfacing***

| Sieve designation | Percentage by weight passing indicated sieve sizes | | | |
|---|---|---|---|---|
|  | A | B | C | D |
| $1\frac{1}{2}$ in. .......... | 100 |  |  |  |
| 1 in. ............. | ..... | 100 |  |  |
| $\frac{3}{4}$ in. ............. | ..... | ..... | 100 |  |
| $\frac{1}{2}$ in. ............. | 45–75 | ..... | ..... | 100 |
| No. 4 ............ | 30–60 | 40–75 | 45–80 |  |
| No. 10 ............ | ..... | 25–60 | 25–60 | 25–60 |
| No. 200 ......... | 12– | 12– | 12 | 12 |

* American Association of State Highway Officials.

graders.  Moisture content is less critical than in the case of finer sand-clay gradings, and generally thorough mixing will require that the material be fairly moist to wet, depending on the nature and grading of aggregate used.  Some experimenting is usually necessary so as to coordinate physical conditions with construction equipment.  Uniformity in color and texture will usually be the best guide to determination of adequate mixing.

The trench section with built-up gravel or sod shoulders may be used; also, it is common practice to feather out the edges by reducing the thickness of wearing course from the center outward to the side slopes.  Best practice requires that the proper crown be built into the subgrade, so that a uniform thickness of wearing course can be provided, and the feathering of the wearing-course material can be accomplished across the width of the shoulders.

**3-5  Maintenance of clay-bound surfaces**  It should be recognized that constant maintenance will be required on these stabilized surfaces to smooth out surface irregularities as they appear.  If roughness and irregularities develop to any marked degree, the surface course may need to be scarified to a depth of several inches, or at least to the depth of the deepest depressions.  Adjustments by addition of clay or coarse material as needed and adequate mixing by harrowing and blading are then in order before the material is relaid.

In cases where failure is due to local weakness in the subgrade, it will be necessary to excavate deeper to remove inferior material or to improve subdrainage as needed.

Calcium chloride has played an important role in the maintenance and preservation of stabilized surfaces.  By virtue of its hygroscopic or deliquescent quality the salt absorbs moisture from the air, particularly during periods of high humidity, and thus maintains a favorable moist condition in the wearing-course material.  This preserves the adhesive and cohesive qualities of the binder, preventing raveling to a large extent and controlling dust.  The surface is usually loosened somewhat by light blading and the salt is applied by means of a suitable spreader, generally of the "drill" type, at a rate of from 1 to $1\frac{1}{2}$ lb/sq yd.  During dry summer weather two or three additional applications of about $\frac{1}{2}$ lb/sq yd may be needed, depending on rainfall conditions and the amount of traffic.  It is to be expected that considerable loss of salt by leaching will be experienced in sandy materials.  In case of predominance of clay on the surface, the salt would tend to produce slipperiness.

**3-6  Soil-cement roads**  For successful performance of soil-stabilized wearing courses under traffic, much depends on the quality of binder used, and it is only reasonable that a more substantial binder will produce better results. Portland cement has been successfully used for the purpose of binding together soil particles, thereby forming a hard stable mass with considerable resistance to abrasion.

The soil to be treated is scarified to a depth of 4 to 6 in. and usually of width comprising the traffic lanes within a trench formed by established shoulders. It is important to control depth of treatment, and gang plows will generally provide the best means for establishing depth after necessary scarifying has been done to break up hard material near the surface. The soil is then pulverized to break up lumps. If lumps are hard and do not break up easily, a smooth roller will prove effective, especially when the material is fairly dry. Thorough pulverizing of the soil is essential in order to promote an intimate mixture of the soil, cement, and water.

Any addition of borrow to improve grading should be spread uniformly over the loosened roadway material before mixing operations begin. Mixing is most effectively done by use of spring-tooth harrows, disk harrows, three- or four-bottom gang plows, and rotary tillers. Water is added during the mixing process, preferably with pressure distributors, so that good control of amount and coverage can be assured. It is important to have enough water in the mixture to hydrate the cement. Also, since a high degree of final compaction is essential, optimum moisture content for maximum density is highly desirable. Experience has shown that for sandy soil-cement mixtures the ideal moisture content for strength and durability is at optimum for maximum density. For silt and clay soil-cement mixtures, moisture content when cement is added should be two or three percentage points above optimum.

Portland cement is added to the mixture at a rate of from 6 to 12 per cent by weight of soil. The exact percentage to be used should be determined by laboratory tests, such as compressive strength of laboratory specimens fabricated so as to be representative of final field characteristics, wetting and drying tests, and, if significant, freezing and thawing tests. The cement should be spread evenly over the surface of the "processed" material. This is done by depositing sacks of cement at intervals which will give the required amount to the mixture. This means that actual spreading will have to be done by hand, however, and use of mechanical spreaders may generally prove more economical and efficient.

Harrows, cultivators, and gang plows are again put to use to incorporate the cement to a uniform mixture with the soil. Depth of plowing should be carefully controlled so as to bring up material from the bottom of the treated layer without increasing the depth of treatment. Mixing should be continued until uniform color and texture are obtained from top to bottom of the mix. Various traveling plants are also available for processing and mixing operations, which should be considered from the standpoint of construction costs and effectiveness.

Tamping rollers should be used for compacting the layer so that compaction is effectively applied from the bottom to the top of the layer. Upon completion of this main consolidation of material, the surface is usually lightly dragged with a spike-tooth harrow and finish rolling done with a self-propelled, smooth tandem roller so as to form a smooth surface.

It is not to be expected that cement-stabilized soil will have the strength and durability of even a low-grade concrete. In concrete, sufficient cement paste is furnished to coat each aggregate particle and form a nearly continuous mass having a relatively low porosity. In a cement-stabilized soil each particle of cement may hold several particles of soil in a cluster, forming a chainlike or clustered mass which one would expect to be friable as compared with concrete. Thus a cement-stabilized soil does not generally have a high resistance to abrasion, and its use as a wearing course should be limited to traffic volumes of around 500 vehicles per day. A surface coating of bituminous material followed with a blotter of sand or stone chips has proved to be necessary for establishing surface durability and also for improving water runoff characteristics.

Coarse-aggregate particles tend to loosen rather readily from the surface material under traffic, and for this reason a large amount of coarse material in the mixture is not desirable for soil-cement stabilization. Soils which are likely to prove successful in soil-cement stabilization will generally fall within the grading limits shown in Table 3-4. It should be acknowledged that finer soils than the limits of Table 3-4 have been successfully stabilized with cement, but a fairly high percentage of cement will usually be necessary, and field operations and control may prove more difficult, particularly in the case of presence of substantial amounts of clay.

Soil cement, like other forms of stabilized-soil wearing courses, affords an excellent base for later pavement construction provided its strength characteristics have been preserved during the interim.

**Table 3-4  Suggested soil gradings for soil-cement roads**

| Sieve size | Percentage passing |
|---|---|
| No. 4 | 80–100 |
| No. 10 | 50–100 |
| No. 40 | 25–90 |
| No. 80 | 10–60 |
| No. 200 | 5–20 |

## QUESTIONS AND PROBLEMS

**3-1.** Plot median values of the three gradings given in Table 3-2 and determine the corresponding values of $m$ in Eq. (2-1). Use the following logarithmic graph and note that $m$ is determined as the slope of plotted curves. [If $P$ represents the percentage finer than grain size $D$, then $m = (\log P_1 - \log P_2)/(\log D_1 - \log D_2)$.]

**3-2.** A crushed-rock base course 6 in. in compacted thickness is to be built on a roadway 28 ft in width, using the material and conditions of Prob. 2-7. The material is to be weighed loose in trucks as it is taken from a stockpile.

    *a.* Compute the number of tons of rock required per mile of roadway if the moisture content of the rock in the stockpile is 6.5 per cent.

    *b.* How many 5-yd truckloads of the crushed rock will be required per mile?

**3-3.** Aggregate $A$ has 85 per cent passing the $\frac{1}{2}$-in. sieve and aggregate $B$ has 30 per cent passing the $\frac{1}{2}$-in. sieve. What combination of the two will have 50 per cent passing the $\frac{1}{2}$-in. sieve?

**3-4.** A sample is to be graded for maximum density with the stipulation that 100 per cent shall pass the 1-in. sieve and 5 per cent shall pass the No. 200 sieve. What per cent of the sample should be retained on the No. 10 sieve?

**3-5.** It is desired to stabilize a gravel with 13 per cent passing the No. 40 sieve and containing no clay. The available binder soil has 90 per cent passing the No. 40 sieve, is 45 per cent clay, and has a plasticity index of 20. The mixture is to have a plasticity index of 8. Determine the proportioning of the mixture on the assumption that the plasticity index of a soil or mixture of soils is proportional to the clay content of the portion passing the No. 40 sieve. Note that the P-40 material in the mixture will have a clay content of 20 per cent. *Ans.* 9.6 parts binder soil to 100 parts gravel by weight.

## BIBLIOGRAPHY

Brown, V. J., et al.: Soil Stabilization, *Roads and Streets*, August, 1938.
Cotton, Miles D.: Soil-cement Mixtures for Roads, *Highway Research Board Proc.*, **18**(2):314–321 (1937).
Frost, Physical Properties and Stabilization, *Highway Research Record* 128, 7 Reports, Washington, 1966.
Hennes, R. G.: How to Design Stabilized Soil Mixtures, *Eng. News-Record*, **130**(20): 761–762 (1943).
Hogentogler, C. A.: Soil Road Surfaces, *Public Roads*, **15**(12) (1935).
Hough, K. B., Jr.: Practical Application of Soil Mechanics: A Symposium, *Trans. ASCE*, **103**:1414–1431 (1948).
Influence of Stabilizers on Properties of Soils and Soil-aggregate Mixtures, *Highway Research Board Bull.* 282, Washington, 1961.
Stabilized Soils: Mix Design and Properties, *Highway Research Record* 198, 6 Reports, Washington, 1967.

# 4

# Bituminous Materials:
# Properties and Basic Uses

In the attainment of better roads the pavement structure will represent a major portion of the cost of construction. Sound engineering practice dictates that the pavement structure be geared to traffic demands which are likely to prevail during the service life of the roadway. In the preceding chapter subgrade and surfacing requirements were discussed for traffic volumes up to roughly 400 to 500 vehicles per day.

A discussion of the many phases of highway economics will not be attempted here, but it should be recognized that many communities and regions of the country which are not blessed with natural resources for great economic growth are not likely to experience greatly increasing demands on local highways. Consequently, many low-cost roads with only a light surface treatment will, with adequate maintenance, serve such traffic needs as can reasonably be expected in the foreseeable future.

On the other hand, it can be expected that traffic volumes will continue to increase within and between growing population centers, with the result that good roads will have to be improved and the better roads will become main arterial highways.  It can be seen that such development lends itself to stage construction and that the application of sound engineering practices to initial stages of construction will pay off to the extent that these roadways can be utilized for later improvements.

Between the extremes of soil-stabilized roads for low traffic volumes and high-type pavement construction serving tens of thousands of vehicles per day, there is an enormous variety of service demand to be provided for with intermediate types of construction.  And it is in this large field of road construction that bituminous products such as asphalts and tars are widely used to bind together the aggregate particles of which the pavement structure is built.

**4-1  Asphalt**  According to the definition given by ASTM, asphalts are "black to dark brown solid or semi-solid cementitious materials which gradually liquefy when heated, in which the predominating constituents are bitumens, all of which occur in the solid or semi-solid form in nature or are obtained by refining petroleum, or which are combinations of the bitumens mentioned with each other or with petroleum or derivatives. thereof."  And bitumens are "mixtures of hydrocarbons of natural or pyrogenous origin, or combinations of both, frequently accompanied by their non-metallic derivatives, which may be gaseous, liquid, semi-solid or solid, and which are completely soluble in carbon disulfide."

Petroleum, then, consists principally of a mixture of bitumens which in turn are hydrocarbons of the methane series.  The student may recall that the basic chemical formula for this series is $C_nH_{2n+2}$.  The first of the series is methane (marsh gas), $CH_4$; then ethane, $C_2H_6$, . . . ; octane, $C_8H_{18}$, . . . ; etc.

Petroleum may be separated into various commercial products by fractional distillation because the boiling point of the various components increases as the molecular weight increases.  See Fig. 4-1.  Each commercial product represents a considerable range of molecular structure and there is, of course, overlapping of products in this respect.  It is helpful, however, to visualize approximately the order and distillation temperatures for various products, and for this purpose some of the chief products of petroleum are listed in Table 4-1.

The compounds at the top of the series, beginning with methane, are gases at ordinary room temperature and pressure.  The lighter fractions shown in Table 4-1 are in general the solvents and motor fuels.  It should be noted that no fine line of distinction separates the various products. Each is a mixture or blend of several compounds, and only the main compound of each is indicated.

The middle fractions are mainly the furnace oils and diesel fuel oils. They are distilled under prolonged high temperature upward of 700°F.

Table 4-1  Petroleum products in order of molecular weight

| Product | Main chemical composition | Approx. bp, °F |
|---|---|---|
| Petroleum ether.............. | $C_5H_{12}$—$C_6H_{14}$ | 100–160 |
| Naphtha.................... | $C_6H_{14}$—$C_7H_{16}$ | 160–195 |
| Gasoline................... | $C_7H_{16}$—$C_8H_{18}$ | 195–250 |
| Benzene.................... | $C_8H_{18}$—$C_9H_{20}$ | 250–300 |
| Kerosene................... | $C_{10}H_{22}$—$C_{16}H_{34}$ | 300–600 |
| Middle fractions............. | | |
| Lubricating oil.............. | | |
| Heavy greases............... | | |
| Paraffin and/or asphalt....... | | |

When this temperature is maintained with a pressure of 4 or 5 atmospheres the chemical structure of these fractions breaks down—"cracks"—into lighter compounds of the gasoline range.

Petroleums found in the United States may be divided into three main groups according to their basic residues. These are (1) the paraffin-base petroleum of the Eastern states, as from the fields of Pennsylvania, Ohio, and southwestern New York, (2) mixed-base petroleum of mid-continent fields such as many of those of Oklahoma and northern Texas (these produce both asphaltic and greasy residues), and (3) the asphaltic-base petroleum of southern Texas, California, and Mexico which yields most of the asphalt used in road building.

Market demands for the various products of petroleum distillation have great influence on the amount and degree of fractional distillation, the amount of "cracking" of heavier molecules into lighter fractions, and the amount of steam refining and purification of heavier residual oils which is to be done. Percentages of asphalt in crude oils vary greatly. "Liquid asphalts" will contain varying amounts of volatile and nonvolatile oils, whereas "asphalt cements" are semisolids.

Liquid asphalts are classified into three groups in accordance with the rate of curing or "setting" when mixed with aggregate and used in roadway construction. They are (1) the slow-curing group, designated as SC, (2) the medium-curing group, MC, and (3) the rapid-curing group, RC. Curing or setting of liquid asphalts is dependent primarily on loss of volatile oils. Subsequent hardening is affected by other influences of weathering, particularly oxidation, which tends ultimately to make the asphalt somewhat nonductile or brittle.

The slow-curing (SC) materials may be either crude or partially refined petroleums with high percentages of asphalt, or they may be blends of two or more asphaltic residues or crude oils. Residual asphalts contained in these materials may vary over a considerable range of consistency but will generally be very soft because of the presence of low-volatility oils, mostly in the range of the middle fractions or fuel oils. More or less permanent softness will prevail in these materials to the extent that they contain oils of the lubricating range.

Medium-curing (MC) liquid asphalts are produced by fluxing heavy asphaltic residue with kerosene or the light fuel oils. These materials will more definitely cure out to the basic asphaltic residue in a reasonably short time, and yet this process will allow considerable time for mixing and manipulation of cold-application procedures or for penetration in priming or penetration treatments. They are adapted to a wide variety of uses for both hot and cold applications in road construction, as will be seen in later discussion.

Rapid-curing (RC) liquid asphalts are produced by fluxing heavy asphaltic residue with naphtha or gasoline. This assures rapid curing or quick setting where this feature is particularly desirable in roadway construction.

MC and RC liquid asphalts are referred to as cutbacks because they are produced by fluxing or cutting back asphalt residue with volatile oils. The amount of volatile oil added to the basic asphalt determines the consistency or viscosity of the product. Numbers 30 to 3000 are used in connection with RC, MC, and SC designation to indicate a range of kinematic viscosity, with the larger numbers indicating the more viscous materials. For example, an MC-800 liquid asphalt is "heavier" or more viscous than an MC-70. Furol viscosity is the measure of resistance to flow of a bituminous material and is usually stated as the time in seconds required for a given amount of material to flow through a given orifice at a specified temperature. Kinematic viscosity is the dynamic shear resistance of liquids, measured in centistokes. Note that the number in the grade designation signifies the lower limit of the kinematic viscosity for the grade (Table 4-2). The upper viscosity limit is twice the lower limit. For example, MC-70 indicates an MC liquid asphalt having a kinematic viscosity within the range of 70 to 140 centistokes at 140°F.

Asphalt cements are semisolid asphalt residues. The degree of solidity varies over a wide range from very soft to hard. They may be

**Table 4-2  Liquid asphalts and asphalt cements**

|  |  | Increasing viscosity ⟶ |  |  |  |
|---|---|---|---|---|---|
| Increasing volatility | RC | | 70 | 250 | 800 | 3000 |
| | MC | 30 | 70 | 250 | 800 | 3000 |
| | SC | | 70 | 250 | 800 | 3000 |

AC 200–300, 120–150, 85–100, 60–70, 40–50

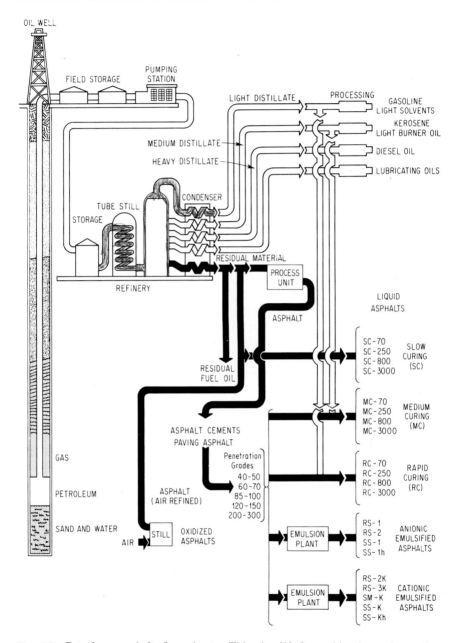

**Fig. 4-1**  Petroleum-asphalt flow chart.  This simplified graphic chart shows the interrelationships of petroleum products, with gasoline, oil, and asphalt flowing from the same oil well.  (*Courtesy of The Asphalt Institute.*)

considered to be a continuation of the SC liquid asphalts; that is, the difference between an SC-3000 and a very soft asphalt cement may be insignificant. The viscosity of asphalt cements is expressed as the distance in hundredths of a centimeter which a standard needle vertically penetrates a sample under known conditions of loading, time, and temperature. Thus a designation 85–100 pen. AC denotes an asphalt cement of 85 to 100 penetration. That is, the standard needle under load of 100 g would penetrate the material a distance of from 0.85 to 1.00 cm in 5 sec at 77°F. A bituminous material is considered to be liquid if it has a penetration of more than 350 at 77°F under a load of 50 g applied for 1 sec, and semisolid if not more than 350 under these conditions.

Some of the more common identification tests for asphalt are schematically depicted in Fig. 4-2. Normal penetration (Fig. 4-2a) has been briefly described in previous paragraphs. Viscosity (Fig. 4-2b) is tested at various temperatures because of the wide range of viscosities of liquid asphalts at any given temperature. Those most commonly used are 77, 122, 140, and 180°F.

Ductility, as illustrated in Fig. 4-2c, is expressed as the distance in centimeters that a standard sample of asphalt cement with a least cross-section area of 1 sq cm will stretch before breaking. The test is made at a temperature of 77°F and the rate of pull is 5 cm/min. The test is important because high ductility is a desirable property for asphalt cements which are to be used in roadway wearing courses where flexibility and toughness are needed to resist cracking. High ductility in an asphalt cement also gives reasonable assurance that the material will exhibit not only good cohesion within itself but also good adhesion to aggregate surfaces.

Determination of the flash point of an asphaltic product is important because it is the critical temperature above which precautions should be taken to prevent fire when the material is heated for transfer to and from storage and for hot applications on roadway construction or maintenance. It is determined (Fig. 4-2d) as the temperature at which evolved vapors while heating temporarily ignite or flash when a very small flame is passed near the surface of the material. The flash point of a mixture of hydrocarbons is not likely to be that of the lighter fractions; that is, an RC cutback which is a mixture of asphaltic residue and gasoline will not have a flash point equal to that of gasoline, but it will be considerably higher because vaporization of the lighter, more volatile material is held back by the heavier hydrocarbons of the mixture. The flash point of straight residuals will vary directly with the distillation temperature.

The distillation test as illustrated in Fig. 4-2e is applied to liquid bituminous products to determine amount and characteristics of both distillate and residue. The percentages of distillate by volume are

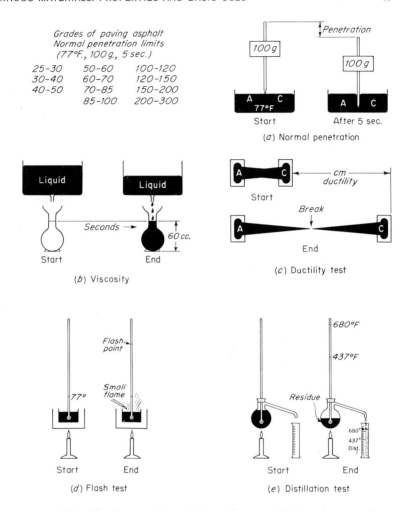

Grades of paving asphalt
Normal penetration limits
(77°F., 100 g., 5 sec.)

| 25-30 | 50-60 | 100-120 |
| 30-40 | 60-70 | 120-150 |
| 40-50 | 70-85 | 150-200 |
|       | 85-100 | 200-300 |

(a) Normal penetration

(b) Viscosity

(c) Ductility test

(d) Flash test

(e) Distillation test

**Fig. 4-2**  Identification tests for asphalt products.   (a) Normal penetration;
(b) viscosity; (c) ductility test; (d) flash test; (e) distillation test.   (*Courtesy
of The Asphalt Institute.*)

recorded during the test at various temperatures.   For liquid asphalts
the following temperatures are commonly used: 190°C (374°F), 225°C
(437°F), 260°C (500°F), and 360°C (680°F).   The residue is usually subjected to a consistency test, and if it is a semisolid so that a penetration
test can be made, it is also tested for ductility.   If the residue is too soft
for a penetration test, the material may be safely classified a slow-curing
road oil because the cutbacks are generally produced from basic asphalt
residues of about 85 to 100 penetration, although the next higher and

lower ranges may also be used.   The material is then a cutback product
if a normal penetration test can be run; and if one-fourth or more of the
total distillate is accumulated at 225°C (437°F), it may be classified as
rapid-curing, while if less than one-fourth is recorded at this temperature
the material should be classified as a medium-curing product.

Asphalt cements, particularly those of 85 to 100 and higher normal
penetration, are also cut back or "liquefied" with water.   These products
are called asphalt *emulsions*, in which minute droplets of asphalt are held
in suspension in water.   The amount of asphalt varies from about 55 to
70 per cent of the total emulsion.   Emulsifiers such as soap, glue, starch,
and colloidal clay are used to prevent coalescence of the asphalt particles.
The emulsifier is adsorbed on the surfaces of the asphalt particles, and
the amount and character of emulsifier used largely determine the rate
at which the emulsion "breaks" or "sets" when it comes in contact with
materials such as mineral aggregates.   Emulsions are classified as (1)
rapid-setting, RS—these are generally used for penetration treatments
where little or no mixing or manipulation is to be done; (2) medium-
setting, MS—this type will generally be used with coarse aggregate and
macadam or broken-stone type of grading where mixing operations can
be effectively accomplished rather quickly; (3) slow-setting, SS—these
are generally required for fine aggregates and dense gradings and for soil
stabilization where mixing and spreading is likely to be comparatively
slow.   See Table 4-7 for specifications for emulsified asphalts.

Beds or lakes of native asphalts are found in many places on or near
the earth's surface.   Perhaps the best known of these is the deposit in
the form of a lake covering about 100 acres and more than 100 ft deep on
the island of Trinidad off the north coast of Venezuela.   It is fed from
beneath from an underground source which maintains the lake within a
few feet of its original level in spite of many decades of asphalt production
on a commercial scale.   Refining of the asphalt consists mainly of
removal of foreign matter and impurities.   Even in the refined state the
bitumen content is somewhat less than 60 per cent.   It contains a large
amount of fine mineral matter on the order of colloidal clay, much of
which is not removed for road-construction purposes.   It is fluxed with
petroleum residues which increase bitumen content and improve consistency for general use.

Another deposit of native asphalt known as Bermudez is found on
the north coast of Venezuela.   The bitumen content of refined Bermudez
asphalt is about 95 per cent.

In this country a hard and brittle native asphalt called gilsonite is
found in Colorado and Utah.   It is practically pure bitumen, but production is limited because of the cost of mining.   It is often used to enrich

asphalts which are low in bitumen content.   A similar native material called grahamite is found in Oklahoma.

Rock asphalts are deposits of limestone or sandstone naturally impregnated with bitumen.   In the United States deposits are found mainly in California, Kentucky, Oklahoma, and Texas.   Production for road construction consists of quarrying and crushing.   The bitumen content varies from about 5 to 15 per cent.   Adjustment by addition of either bitumen or mineral aggregate will usually have to be made.

**4-2   Tars**   Tars are defined by ASTM as "black to dark brown bituminous condensates which yield substantial qualities of pitch when partially evaporated or fractionally distilled, and which are produced by destructive distillation of organic material, such as coal, oil, lignite, peat and wood."   Pitch, then, is the basic cementitious residue which furnishes the desirable qualities for binding together aggregate particles in road work.   Like asphalt it gradually liquefies when heated, and it can be brought to any desired consistency by control of initial distillation or by fluxing with tar distillates, thus producing a cutback product similar to asphalt cutbacks.   Tars for road purposes are further classified by ASTM in accordance with manufacture as follows:

*Gas-house tar*, produced as a by-product during destructive distillation of
     bituminous coal in the manufacture of coal gas
*Coke-oven tar*, produced as a by-product during destructive distillation of
     bituminous coal in the manufacture of coke
*Water-gas tar*, produced by cracking petroleum (middle fractions) at high
     temperatures in the process of carbureting water gas

The bitumen content of tars is generally low as compared with petroleum asphalts, ranging in refined road tars from 75 to 95 per cent. Notable among impurities are water and free carbon.   Most specifications require, however, that the amount of water present shall be less than 2 per cent.

Important laboratory tests for identification and determination of selective use are viscosity, bitumen content, specific gravity, distillation, and determination of softening point of the residue.   During the distillation test fractions are usually recorded at 170, 270, and 300°C.   This divides the material into *light oils*, the distillate up to 170°C; *middle oils*, the distillate which comes over between 170 and 270°C—these are the creosote oils used in fungicides, insecticides, and in the preservation of wood; *heavy oils*, the distillate between 270 and 300°C—these contain the crystalline anthracene which is the basis for many artificial dyes.

The softening point is a relative measure of the consistency of the residual binder that remains in the road after volatile fractions have evaporated. The standard ASTM test, known as the ring-and-ball method, is schematically illustrated in Fig. 4-3.

Road tars are classified according to consistency into 12 grades and designated as RT-1 to RT-12. Two cutbacks are recognized; designations for these are RTCB-5 and RTCB-6.

Many of the significant properties, specifications, and test designations for bituminous materials in general are summarized in Tables 4-3 to 4-8. Suggested temperatures for use are given for guidance of the user and are not part of the specifications.

Before going ahead with a discussion of the many ways in which bituminous materials are used in highway construction and maintenance, it will be helpful to emphasize some basic definitions and considerations which must be weighed in determining which available product shall be used for a given purpose. To this end, reference is made to Fig. 4-4.

**4-3  Prime coat**  When a mixture of bituminous material and mineral aggregate is to be applied to a surface, that surface is usually given a light application of liquid bituminous material to bind together surface particles and to furnish a bond or continuation between the applied bituminous mat and the foundation surface. When such treatment is applied to a relatively porous, granular surface such as compacted gravel or crushed rock where it is desirable to attain considerable penetration of the bituminous material, it is referred to as a *prime coat*. Inasmuch as it is desirable to obtain considerable depth of penetration, it follows that a reasonably slow-curing product of low viscosity should be used.

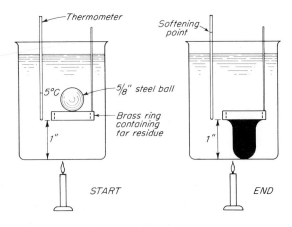

**Fig. 4-3**  Ring-and-ball test.

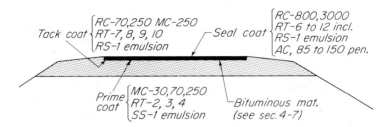

**Fig. 4-4**  Illustration of some basic highway uses of bituminous products.

Engineers have found that the light grades of slow-curing asphaltic road oils tend to separate fractionally when used for priming, particularly if the surface treated is tight or dense-graded; that is, lighter fractions penetrate into interstices of surface aggregate while the heavier fractions tend to remain as a coating or film over the top of the treated surface.  The medium-curing road oils which are more stable solutions in this respect are for this reason generally used for priming.  The light grades may be applied cold during warm summer weather, while MC-250 is usually heated to about 180°F.  Of the tars, the low-viscosity grade of straight-run tars are usually used.  Cutback tars do not give desired uniformity of treatment in priming applications and are not generally used for this purpose.

The rate at which asphalt emulsions "break" is largely dependent on mixing operations, and since priming and other penetration treatments do not require such manipulation, the rapid-setting emulsion is most satisfactory to use.

The amount of bituminous material to be used may vary from about 0.2 to 0.4 gal/sq yd of surface, depending on grading and density of aggregate or soil to be primed.  Accurate control of application may be realized with the use of modern pressure distributors (Fig. 4-5) with spray bars which can be adjusted for width of coverage by opening or closing nozzles at the ends of the bar.  Automatic control of tank pressure assures uniform flow through the nozzles, and rate of application is finally controlled by rate of travel, which is accurately measured in feet per minute by a tachometer.  A slight excess of material sometimes occurs because of variations in surface texture, and a blotter of sand or stone chips is spread to take up any excess bitumen and make the surface safe for further construction operations and for traffic.

**4-4  Tack coat**  In cases where some type of bituminous mat is to be placed over old portland-cement concrete pavements, old brick pavements, previously treated bituminous surfaces, and others where no

**Fig. 4-5**   Pressure distributor in operation.

appreciable penetration will be experienced, the initial surface treatment
of a light application of liquid bituminous material is referred to as a
*tack coat.*   The main purpose is to provide a thorough bond between the
two courses.   Since in general no delaying action is required, a rapid-
curing cutback asphalt or the heavier grades of tar are generally used
so that construction of the overlying mat may proceed as quickly as
possible.   If the old pavement is somewhat porous or the mortar joints of
old brick pavements are cracked and porous, it may be desirable to use
MC-250 or RT-5 and 6.   From 0.08 to 0.12 gal/sq yd of surface will
usually provide sufficient bitumen for bonding unless appreciable penetration into the old surface occurs.

**4-5   Seal coat**   A seal coat is an application of bituminous material
followed by a cover of sand or stone chips applied to a new or old pavement surface to:

1. Provide a waterproof surface to improve surface drainage and afford
   protection to the underlying mat against weathering and the abrasive action of traffic.
2. Improve visibility by providing a bright aggregate surface supplied by
   the sand or stone chips.
3. Improve skid resistance.   This is accomplished by using angular sand
   particles and stone chips for seal-coat cover, embedded in a firm
   bituminous membrane over both new and old bituminous mats or
   surface treatments.

## Table 4-3  Specifications for asphalt cements*

| Characteristics | AASHO test method | ASTM test method | Grades | | | | |
|---|---|---|---|---|---|---|---|
| | | | 40–50† | 60–70 | 85–100 | 120–150 | 200–300 |
| Penetration, 77°F, 100 g, 5 sec... | T49 | D5 | | | | | |
| Viscosity at 275°F: | | | | | | | |
| Saybolt furol, SSF.......... | .... | E102 | 120+ | 100+ | 85+ | 70+ | 50+ |
| Kinematic, centistokes....... | .... | D2170 | 240+ | 200+ | 170+ | 140+ | 100+ |
| Flash point (Cleveland open cup), °F....................... | T48 | D92 | 450+ | 450+ | 450+ | 425+ | 350+ |
| Thin-film oven test............. | T179 | | | | | | |
| Penetration after test, 77°F, 100 g, 5 sec, per cent of original...... | T49 | D5 | 55+ | 52+ | 47+ | 42+ | 37+ |
| Ductility: | | | | | | | |
| At 77°F, cm................. | T51 | D113 | 100+ | 100+ | 100+ | 60+ | |
| At 60°F, cm................. | .... | ...... | ....... | ....... | ....... | ....... | 60+ |
| Solubility in carbon tetrachloride, per cent.................... | T44‡ | D4‡ | 99.5+ | 99.5+ | 99.5+ | 99.5+ | 99.5+ |
| General requirements.......... | .... | ...... | The asphalt shall be prepared by the refining of petroleum. It shall be uniform in character and shall not foam when heated to 350°F. | | | | |

* Asphalt Institute.
† Except that carbon tetrachloride is used instead of carbon disulfide as solvent, Method No. 1 in AASHO Method T44 or Procedure No. 1 in ASTM Method D4.
‡ Also special and industrial uses.

## Table 4-4  Specifications for rapid-curing (RC) liquid asphalts*

| Characteristics | AASHO test method | ASTM test method | Grades | | | |
|---|---|---|---|---|---|---|
| | | | RC-70 | RC-250 | RC-800 | RC-3000 |
| Flash point (open tag.), °F...... | T79 | D1310 | ......... | 80 min | 80 min | 80 min |
| Kinematic viscosity at 140°F, centistokes†................. | .... | D445 | 70–140 | 250–500 | 800–1600 | 3,000–6,000 |
| Distillation—distillate (per cent of total distillate to 680°F): | | | | | | |
| To 374°F................... | .... | ...... | 10 min | | | |
| To 437°F................... | T78 | D402 | 50 min | 35 min | 15 min | |
| To 500°F................... | .... | ...... | 70 min | 60 min | 45 min | 25 min |
| To 600°F................... | .... | ...... | 85 min | 80 min | 75 min | 70 min |
| Residue from distillation to 680°F, volume per cent by difference.................... | .... | ...... | 55 min | 65 min | 75 min | 80 min |
| Tests on residue from distillation: | | | | | | |
| Penetration, 77°F, 100 g, 5 sec | T49 | D5 | 80–120 | 80–120 | 80–120 | 80–120 |
| Ductility, 77°F, cm.......... | T51 | D113 | 100 min | 100 min | 100 min | 100 min |
| Solubility in carbon tetrachloride, per cent.............. | T44‡ | D4‡ | 99.5 min | 99.5 min | 99.5 min | 99.5 min |
| Water, per cent............... | T55 | D95 | 0.2 max | 0.2 max | 0.2 max | 0.2 max |

Note: When the Heptane-Xylene Equivalent Test is specified by the consumer, a negative test with 35 per cent xylene after 1 hr will be required, AASHO Method T102.
* Asphalt Institute.
† As an alternate, Saybolt furol viscosities may be specified as shown in Table 4-10.
‡ Except that carbon tetrachloride or trichloroethylene is used instead of carbon disulfide as solvent, Method No. 1 in AASHO Method T44 or Procedure No. 1 in ASTM Method D4.

On new bituminous mat construction it is important to recognize that inasmuch as the mat has been adequately designed with respect to bitumen content, no appreciable penetration or "priming" of the mat should occur in seal-coating operations. Such priming could lead to an overasphalted condition in the mat which would result in undue softening, especially during hot weather. It is highly desirable, rather, to produce a firm membrane over the surface into which sand or stone chips may be substantially embedded.

It is desirable, then, to use a rapid-curing product with a fairly hard—say, 85 to 120 normal penetration—residue. Softening point of tar residues for this purpose should preferably be in the range of from 50°C (122°F) to 70°C (158°F) because it is not uncommon for pavement temperatures to reach this range during hot summer weather. Asphalt cements are also used for seal-coat applications.

Dried-out and weathered bituminous surfaces are often rejuvenated by resealing and the addition of cover material. In these cases new

**Table 4-5  Specifications for medium-curing (MC) liquid asphalts***

| Characteristics | AASHO test method | ASTM test method | Grades | | | |
|---|---|---|---|---|---|---|
| | | | MC-70 | MC-250 | MC-800 | MC-3000 |
| Flash point (open tag.), °F† | T79 | D1310 | 100 min | 150 min | 150 min | 150 min |
| Kinematic viscosity at 140°F, centistokes‡ | .... | D445 | 70–140 | 250–500 | 800–1,600 | 3,000–6,000 |
| Distillation—distillate (per cent of total distillate to 680°F): | | | | | | |
| To 437°F | .... | ...... | 20 max | 0–10 | | |
| To 500°F | T78 | D402 | 20–60 | 15–55 | 35 max | 15 max |
| To 600°F | .... | ...... | 65–90 | 60–87 | 45–80 | 15–75 |
| Residue from distillation to 680°F, volume per cent by difference | .... | ...... | 55 min | 67 min | 75 min | 80 min |
| Tests on residue from distillation: | | | | | | |
| Penetration, 77°F, 100 g, 5 sec | T49 | D5 | 120–250 | 120–250 | 120–250 | 120–250 |
| Ductility, 77°F, cm§ | T51 | D113 | 100 min | 100 min | 100 min | 100 min |
| Solubility in carbon tetrachloride, per cent | T44¶ | D4¶ | 99.5 min | 99.5 min | 99.5 min | 99.5 min |
| Water, per cent | T55 | D95 | 0.2 max | 0.2 max | 0.2 max | 0.2 max |

Note: When the Heptane-Xylene Equivalent Test is specified by the consumer, a negative test with 35 per cent xylene after 1 hr will be required, AASHO Method T102.
* Asphalt Institute.
† Flash point by Cleveland open cup may be used for products having a flash point greater than 175°F.
‡ As an alternate, Saybolt furol viscosities may be specified as shown in Table 4-10.
§ If penetration of residue is more than 200 and its ductility at 77°F is less than 100, the material will be acceptable if its ductility at 60°F is 100+.
¶ Except that carbon tetrachloride or trichloroethylene is used instead of carbon disulfide as solvent. Method No. 1 in AASHO Method T44 or Procedure No. 1 in ASTM Method D4.

**Table 4-6  Specifications for slow-curing (SC) liquid asphalts***

| Characteristics | AASHO test method | ASTM test method | Grades | | | |
|---|---|---|---|---|---|---|
| | | | SC-70 | SC-250 | SC-800 | SC-3000 |
| Flash point (Cleveland open cup), °F | T48 | D92 | 150 min | 175 min | 200 min | 225 min |
| Kinematic viscosity at 140°F, centistokes† | .... | D445 | 70–140 | 250–500 | 800–1,600 | 3,000–6,000 |
| Distillation: | | | | | | |
| Total distillate to 680°F, per cent by volume | T78 | D402 | 10–30 | 4–20 | 2–12 | 5 max |
| Float test on distillation residue at 122°F, sec | T50 | D139 | 20–100 | 25–110 | 50–140 | 75–200 |
| Asphalt residue of 100 penetration, per cent | T56 | D243 | 50 min | 60 min | 70 min | 80 min |
| Ductility of 100 penetration asphalt residue at 77°F, cm | T51 | D113 | 100 min | 100 min | 100 min | 100 min |
| Solubility in carbon tetrachloride, per cent | T44‡ | D4‡ | 99.5 min | 99.5 min | 99.5 min | 99.5 min |
| Water, per cent | T55 | D95 | 0.5 max | 0.5 max | 0.5 max | 0.5 max |

Note: When the Heptane-Xylene Equivalent Test is specified by the consumer, a negative test with 35 per cent xylene after 1 hour will be required, AASHO Method T102.
* Asphalt Institute.
† As an alternate, Saybolt furol viscosities may be specified as shown in Table 4-10.
‡ Except that carbon tetrachloride or trichloroethylene is used instead of carbon disulfide as solvent, Method No. 1 in AASHO Method T44 or Procedure No. 1 in ASTM Method D 4.

"life" (flexibility and resistance to cracking) can be given to material near the surface of the old mat by permitting a small amount of penetration or priming. The heavier grades of MC liquid asphalt, or tars in the range of RT-6 to RT-8, have in this respect been successfully used for seal-coat application.

Application temperatures (for spraying) for the different bituminous products used in seal-coat construction are given in Tables 4-3 to 4-8.

Amounts of bituminous materials and cover are interrelated. Table 4-9 gives a range of working values from which it will be seen that larger-size cover aggregate requires more bitumen; and it should also be emphasized that larger cover aggregate requires in general a bituminous product with a heavier, harder residue. This latter requirement stems from the fact that larger particles are more readily loosened from the roadway surface by fast-moving traffic. In general the amount of bituminous material required will be about 0.01 gal/sq yd for each pound of cover aggregate per square yard.

**4-6  Surface treatment**  In order to extend the service of soil-stabilized and gravel-surfaced roads from 400 to 500 vehicles per day to meet the

**Table 4-7a   Specifications for anionic emulsified asphalts**

| Characteristics | AASHO test method | ASTM test method | Rapid setting | | Slow setting | |
|---|---|---|---|---|---|---|
| | | | RS-1 | RS-2 | SS-1 | SS-1h |
| Tests on emulsion: | | | | | | |
| Furol viscosity at 77°F, sec............ | .... | ..... | 20–100 | ....... | 20–100 | 20–100 |
| Furol viscosity at 122°F, sec........... | .... | ..... | ....... | 75–400 | | |
| Residue from distillation, per cent by weight............................. | .... | ..... | 57+ | 62+ | 57+ | 57+ |
| Settlement, 5 days, per cent difference... | T59 | D244 | 3– | 3– | 3– | 3– |
| Demulsibility: | | | | | | |
| 35 ml of 0.02 N CaCl₂, per cent...... | .... | ..... | 60+ | 50+ | | |
| 50 ml of 0.10 N CaCl₂, per cent...... | .... | ..... | | | | |
| Sieve test (retained on No. 20), per cent. | .... | ..... | 0.10– | 0.10– | 0.10– | 0.10– |
| Cement mixing test, per cent.......... | .... | ..... | ....... | ....... | 2.0– | 2.0– |
| Tests on residue: | | | | | | |
| Penetration, 77°F, 100 g, 5 sec......... | T49 | D5 | 100–200 | 100–200 | 100–200 | 40–90 |
| Solubility in carbon tetrachloride, per cent............................ | T44* | D4* | 97.5+ | 97.5+ | 97.5+ | 97.5+ |
| Ductility, 77°F, cm.................. | T51 | D113 | 40+ | 40+ | 40+ | 40+ |

* Except that carbon tetrachloride is used instead of carbon disulfide as solvent.   Method No. 1 in AASHO Method T44 or Procedure No. 1 in ASTM Method D4.

traffic need of 800 to about 1,500 vehicles per day, a bituminous treatment of the traffic surface to provide a dust-free and reasonably durable surface will usually be adequate.   Basic procedures for such treatment are the application of a prime coat, followed by a seal coat with aggregate cover. Such treatment does not add appreciably to the thickness of wearing course—less than 1 in.—and is therefore referred to as *surface treatment*.

Much of the successful performance of surface treatment depends on a well-compacted, stable roadway surface, free of dust pockets and brought to true grade and section by careful blading.   Gravel roads and well-graded crushed-rock roads compacted by traffic under proper maintenance are well adapted to surface treatment.   Light blading and sweeping with a power-driven rotary broom will usually prepare such a surface for priming.   The surface should be dry, but some dampness in the material just beneath the surface may be helpful rather than objectionable.   The situation is somewhat analogous to the familiar fact that spilled water or other liquid is more readily wiped up with a damp rag than one which is bone-dry.

It is important that ground temperature as well as air temperature be warm.   If ground temperature is below 40 to 45°F just beneath the

**Table 4-7b  Specifications for cationic emulsified asphalts**

| Characteristics | AASHO test method | ASTM test method | Rapid setting | | Medium setting | Slow setting | |
|---|---|---|---|---|---|---|---|
| | | | RS-2K | RS-3K | SM-K | SS-K | SS-Kh |
| Tests on emulsion: | | | | | | | |
| Furol viscosity at 77°F, sec.... | T59 | D244 | ...... | ....... | ....... | 20–100 | 20–100 |
| Furol viscosity at 122°F, sec... | T59 | D244 | 20–100 | 100–400 | 50–500 | | |
| Residue from distillation, per cent by weight............ | T59 | D244 | 60+ | 65+ | 60+ | 57+ | 57+ |
| Settlement, 7 days, per cent difference................ | T59 | D244 | 3− | 3− | 3− | 3− | 3− |
| Sieve test (retained on No. 20), per cent................. | T59* | D244* | 0.10− | 0.10− | 0.10− | 0.10− | 0.10− |
| Aggregate coating-water resistance test............... | .... | D244 | | | | | |
| Dry aggregate (job), per cent coated................. | .... | ...... | ....... | ....... | 80+ | | |
| Wet aggregate (job), per cent coated................. | .... | ...... | ....... | ....... | 60+ | | |
| Cement mixing test, per cent.. | T59 | D244 | ....... | ....... | ....... | 2− | 2− |
| Particle charge test.......... | .... | ...... | Positive | Positive | Positive | | |
| pH....................... | .... | E70 | ....... | ....... | ....... | 6.7− | 6.7− |
| Oil distillate, per cent by volume.................... | T59 | D244 | 5− | 5− | 20− | | |
| Tests on residue: | | | | | | | |
| Penetration, 77°F, 100 g, 5 sec | T49 | D5 | 100–250 | 100–250 | 100–250 | 100–200 | 40–90 |
| Solubility in carbon tetrachloride, per cent............. | T44† | D4† | 97.0+ | 97.0+ | 97.0+ | 97.0+ | 97.0+ |
| Ductility, 77°F, cm.......... | T51 | D113 | 40+ | 40+ | 40+ | 40+ | 40+ |

* Except that distilled water is used instead of sodium oleate solution.
† Except that carbon tetrachloride is used instead of carbon disulphide as solvent, Method No. 1 in AASHO Method T44 or Procedure No. 1 in ASTM Method D4.
Note: (a) "K" in grade designations signifies cationic type.
    (b) In medium setting grade "SM" indicates sand mixing grade.

surface, the thin film of applied bitumen will quickly lose heat and tend to congeal on the surface rather than penetrate. Most specifications require that no bituminous work be done when the air temperature is below 60°F. It is also usually specified, particularly in Northern states, that bituminous work be confined to certain months of the year, roughly, May to October in the North Central states.

Application of prime coat and blotter sand, of liquid bituminous material or asphalt cement followed by cover aggregate, is done in the order previously outlined for prime coat and seal coat. When cover aggregate up to $\frac{3}{4}$-in. size is used, a second "shot" of bituminous material and a second application of finer stone cover is usually desirable so that

# Table 4-8 Specifications for tar for use in road construction

| | Grade | | | | | | | | | | | | | | AASHO testing method |
|---|---|---|---|---|---|---|---|---|---|---|---|---|---|---|---|
| | RT-1 | RT-2 | RT-3 | RT-4 | RT-5 | RT-6 | RT-7 | RT-8 | RT-9 | RT-10 | RT-11 | RT-12 | RTCB-5 | RTCB-6 | |
| Consistency: | | | | | | | | | | | | | | | |
| Engler spec. vis. at 40°C | 5–8 | | | | | | | | | | | | | | T-54 |
| Engler spec. vis. at 50°C | | 8–13 | | | | | | | | | | | | | T-54 |
| Float test at 32°C | | | 13–22 | 22–35 | 17–26 | 26–40 | | | | | | | 17–26 | 26–40 | T-50 |
| Float test at 50°C | | | | | | | 50–80 | 80–120 | 120–200 | 75–100 | 100–150 | 150–220 | | | T-50 |
| Spec. grav. at 25°C/25°C | 1.08+ | 1.08+ | 1.09+ | 1.09+ | 1.10+ | 1.10+ | 1.12+ | 1.14+ | 1.14+ | 1.15+ | 1.16+ | 1.16+ | 1.09+ | 1.09+ | T-43 |
| Total bitumen, per cent by weight | 88+ | 88+ | 88+ | 88+ | 83+ | 83+ | 78+ | 78+ | 78+ | 75+ | 75+ | 75+ | 80+ | 80+ | T-44 |
| Water, per cent by volume | 2.0– | 2.0– | 2.0– | 2.0– | 1.5– | 1.5– | 1.0– | 0 | 0 | 0 | 0 | 0 | 1.0– | 1.0– | T-55 |
| Distillation, per cent by weight: | | | | | | | | | | | | | | | |
| To 170°C | 7.0– | 7.0– | 7.0– | 5.0– | 5.0– | 5.0– | 3.0– | 1.0– | 1.0– | 1.0– | 1.0– | 1.0– | 2.0–8.0 | 2.0–8.0 | |
| To 200°C | | | | | | | | | | | | | 5.0+ | 5.0+ | |
| To 235°C | 35.0– | 35.0– | 30.0– | 30.0– | 25.0– | 25.0– | 20.0– | 15.0– | 15.0– | 10.0– | 10.0– | 10.0– | 8.0–18.0 | 8.0–18.0 | |
| To 270°C | 45.0– | 45.0– | 40.0– | 35.0– | 35.0– | 35.0– | 30.0– | 25.0– | 25.0– | 20.0– | 20.0– | 20.0– | | | |
| To 300°C | | | | | | | | | | | | | 35.0– | 35.0– | |
| Softening point of distillation residue, °C | 30–60 | 30–60 | 35–65 | 35–70 | 35–70 | 35–70 | 35–70 | 35–70 | 35–70 | 40–70 | 40–70 | 40–70 | 40–70 | 40–70 | T-53 |
| Suggested temperatures for application, °F | 60–125 | 60–125 | 80–150 | 80–150 | 80–150 | 80–150 | 150–225 | 150–225 | 150–225 | 175–250 | 175–250 | 175–250 | 60–120 | 60–120 | |

**Table 4-9  Materials for seal coat**

| Size of cover aggregate | Gallons of bitumen per square yard | Pounds of aggregate per square yard |
|---|---|---|
| $\frac{3}{4}$ in. to $\frac{3}{8}$ in............ | 0.40–0.50 | 45–50 |
| $\frac{5}{8}$ in. to No. 4............ | 0.35–0.45 | 35–45 |
| $\frac{1}{2}$ in. to No. 4............ | 0.25–0.30 | 25–30 |
| $\frac{1}{2}$ in. to No. 10.......... | 0.20–0.25 | 20–25 |
| No. 4 to No. 10.......... | 0.15–0.20 | 15–20 |
| No. 10 minus............ | 0.10–0.15 | 10–15 |

the larger particles are firmly held by the bituminous binder and by the interlocked or "keyed" finer particles.

Surface treatment of sand-clay and other stabilized-soil roads follows essentially the same procedure as for gravel and crushed-rock surfaces.  In the case of finer material, however, the lightest grades of priming liquid should be used, and a period of several days allowed, preferably without traffic, for the prime coat to penetrate into the soil.  It is well to recognize that very plastic soil will not respond to such treatment without considerable manipulation and mixing; therefore the liquid limit

**Table 4-10  Alternate viscosity specifications for liquid asphalts* †**

| Characteristics | AASHO test method | ASTM test method | Grades | | | |
|---|---|---|---|---|---|---|
| | | | RC-, MC-, or SC-70 | RC-, MC-, or SC-250 | RC-, MC-, or SC-800 | RC-, MC-, or SC-3000 |
| Furol viscosity at 140°F, sec..... | T72 | D88 | 35–70 | 125–250 | 400–800 | 1,500–3,000 |
| Or: | | | | | | |
| Furol viscosity at 122°F, sec | ... | ... | 60–120 | | | |
| Furol viscosity at 140°F, sec | T72 | D88 | ...... | 125–250 | | |
| Furol viscosity at 180°F, sec | ... | ... | ...... | ....... | 100–200 | 300–600 |

* Asphalt Institute.
† As an alternate for kinematic viscosity limits included in specifications on Tables 4-4 to 4-6, furol viscosity limits may be used.  These furol viscosity limits may be specified, as shown above, either at a constant temperature (140°F) or at three temperatures.  This alternate is provided in recognition of the transition period required in changing from the Saybolt furol test to the more fundamental kinematic viscosity test.

## Table 4-11 Principal uses of bituminous products*

| Class of work | 40-50 | 60-70 | 85-100 | 120-150 | 200-300 | RC 70 | RC 250 | RC 800 | RC 3000 | MC 70 | MC 250 | MC 800 | MC 3000 | SC 70 | SC 250 | SC 800 | SC 3000 | RS-1 | RS-2 | SS-1 | SS-1h | RS-2K | RS-3K | SM-K | SS-K | SS-Kh |
|---|---|---|---|---|---|---|---|---|---|---|---|---|---|---|---|---|---|---|---|---|---|---|---|---|---|---|
| *(Paving asphalts)* | | | | | | *(Liquid asphalts†: RC)* | | | | *(MC)* | | | | *(SC)* | | | | *(Anionic)* | | | | *(Cationic)* | | | | |
| Dust palliative | | | | | | | | | | | | | | x | | | | | | x‡ | | | | | x‡ | |
| Tack coat: | | | | | | x | | | | | | | | | | | | x | | x‡ | x‡ | | | | x | x |
| Prime coat: Tightly bonded surface | | | | | | x | | | | x | | | | x | | | | | | | | | | | | |
| Open surface | | | | | | | x | | | | x | | | | | | | | | | | | | | | |
| Seal coats and surface treatments: Fog seal, light application without cover | | | | | | | | | | | | | | | | | | x | | x‡ | x‡ | | | | | x‡ |
| Sand seal, light application with sand cover | | | | | | | x | | | x | | | | | | | | x | | | | x | | | | |
| Chip seal | | | | x | | | | x | x | | | | x | | | | | | x | | | x | x | | | |
| Slurry seal | | | | | | | | | | | | | | | | | | | | x | x | | | | x | x |
| Penetration treatment | | | | x | | | | | | | | | | | | x | | x | | | | | | | | |
| Penetration macadam | | | x | x | | | | | x | | | | | | | | | x | | | | x | | | | |
| Mixed-in-place: Open-graded aggregate, low on Nos. 8 and 200 | | | | | | | x | x | | | x | x | | | | | | | | | | | | x | | |
| Clean sand, 100% ± pass No. 4, 0-10% pass No. 200 | | | | | | x | | | | x | x | | | x | | | | | | x | x | | | | | x |

Graded aggregate, up to 15% No. 200

Sandy soil, to 20% pass No. 200

Plant mix, cold laid:
Graded aggregate, 4–10% pass No. 200

Patching mix:
Immediate use
Stockpile

Asphalt concrete:
Industrial floors

Parking areas

Highways, good quality, rough-textured aggregate

Highways, borderline quality aggregate

Airports, good quality, rough-textured aggregate

Curbs

Open-graded surface course, $\frac{3}{8}$-in. max, 0–2% No. 200

* Asphalt Institute.
† In northern areas where rate of curing is slower, a shift from MC to RC or from SC to MC may be desirable. For very warm climates, a shift to next heavier grade may be warranted.
‡ Diluted with water.

and plasticity index of the soil mortar should be low.  The plasticity index should not exceed 6.

Dust palliatives which are light applications of low-viscosity, slow-curing liquid bituminous materials are not included here in the category of surface treatment, but are included rather in maintenance expediencies for laying the dust, as the name implies, on roadways and city streets.

Thus the expression *prime and seal* describes what is, substantially, surface treatment without the intervening bituminous mat.

**4-7  Bituminous mats**  As used here *bituminous mat* is intended, for purposes of distinction and clarity, to describe such application or construction which increases the thickness of the wearing course 1 in. or more.  Such terms as *armor coat, road mix,* and *plant mix* are used to describe bituminous mats, using liquid bituminous materials, which may be classified as intermediate types of wearing courses and will be taken up in the next chapter, Intermediate Types of Pavement Structure. Such terms as *sheet asphalt, penetration macadam,* and *bituminous concrete* describe carefully designed and controlled operations using asphalt cements to construct bituminous mats for more severe conditions of highway use.  These will be discussed in Chap. 6, High-type Bituminous Pavements.

A summary of principal uses of various bituminous products is given in Table 4-11.  Reference will be made to this table in subsequent discussion of various types of bituminous wearing courses.

**4-8  Laboratory tests**  The following identification tests are suggested for laboratory work in order that the student may become familiar with basic properties of bituminous road-building materials:

| Test | ASTM designation | AASHO designation |
|------|------------------|-------------------|
| Penetration | D5-65 | T49-53 |
| Ductility | D113-44 (1961) | T51-44 |
| Furol viscosity | D88-56 | T72-57 |
| Distillation | D402-55 (1961) | T78-60 |
| Flash point | D92-66 | T48-60 |
| Softening point | D36-66T | T53-42 |

**QUESTIONS AND PROBLEMS**

**4-1.** Test results for six bituminous products are given in the table below.  Identify them with the proper letter and number designations.

| Bituminous product | Distillate in per cent of total distillate | | | Normal penetration of residue (0.01 cm) | Water per cent by volume | Softening point of residue at 300°C (°C) |
|---|---|---|---|---|---|---|
| | To 437°F | To 518°F | To 680°F | | | |
| 1 | 8 | .. | 100 | 127 | 0 | |
| 2 | 31 | .. | 100 | 96 | 0 | |
| 3 | .. | .. | 37 | Intermediate | 0.4 | |
| 4 | .. | .. | .... | 95 | 43 | |
| 5 | .. | .. | ... | 68 | 0 | |
| 6 | .. | 38 | ... | ... | 2 | 67 |

**4-2.** What specific properties are desirable in a bituminous product to be used for:

    *a.* Priming a well-compacted, dense-graded roadway surface?

    *b.* A tack coat on old brick pavement?

    *c.* A seal coat using cover aggregate of $\frac{1}{2}$ in. to No. 10 size? What products would you recommend for use in each case, and what should be the approximate temperature of application?

**4-3.** A gravel roadway surface is to be primed with 0.3 gal/sq yd of MC-250 at 180°F. Using the full length of spray bar, 12 ft of roadway width will be covered. With a constant tank pressure of 45 psi under above conditions, the discharge from the nozzles will be 160 gal/min.

    *a.* At what speed, as indicated by the tachometer gauge in feet per minute, should the operator run the truck?

    *b.* The tank holds 1,100 gal, but in order to ensure control of discharge the operator will cut off the spray when the gauge reads 100 gal. Over what length of roadway should he have traveled to spread the 1,000 gal at the specified rate?

## BIBLIOGRAPHY

"The Asphalt Handbook," The Asphalt Institute, 1965.

"Asphalts, Paving and Liquid," The Asphalt Institute, 1966.

Bauer, Edward E.: "Highway Materials," McGraw-Hill Book Company, New York, 1932.

Bituminous Materials, *Highway Research Record* 134, 6 Reports, Washington, 1966.

"Highway Materials, Specifications and Tests," vols. 1 and 2, American Association of State Highway Officials, 1961.

# 5

# Intermediate Types
# of Pavement Structure

**5-1  Pavement types**  Pavement mats or wearing courses are classified into two general types: flexible and rigid.  Flexible pavement includes all types of bituminous mats and in general has little "beam" strength: it does not distribute load over the subgrade by its flexural resistance but depends upon the shear strength of base and surfacing.  Rigid pavement includes plain and reinforced portland-cement concrete pavement slabs which act like any elastic floor slab except that the positions of the distributed and concentrated forces are reversed from their conventional positions in a building frame.

Flexible-pavement mats or wearing courses may be further classified into (1) intermediate types, those in which liquid bituminous materials are used as the binder, and (2) high type of flexible pavement using asphalt cements and the heaviest grades of tars.  The intermediate types of flexible pavement will be discussed in this chapter.

**5-2  Load distribution**  When a load is applied to the surface of a flexible mat it will conform to the deflected shape of supporting base course and subgrade soils.  The mat will, to some extent, "bridge" rapidly moving wheel loads over relatively weak spots in underlying materials in the same way that a person will be supported on thin ice—if he keeps moving.  But slow-moving truckloads will soon find the weak spots and meticulously outline them for the attention of maintenance forces.  A slight rut with longitudinal cracking of the mat will usually develop first, then as traffic moves aside to avoid the depression or rutting, lateral cracks also form and the entire weak spot or area becomes mapped in "alligator" pattern. The depression, coupled with cracking, readily admits surface water to aggravate underlying weakness until major surgery is required.  Removal by excavation of poor materials, whether it be in the base course or in the subgrade soil, and backfilling with stable granular material are essential. Lateral drains to side ditches or other parts of the drainage system may be necessary.

Because of the interdependence of flexible mats, base courses, and subgrade in supporting applied loads, it will be convenient to treat the mat or wearing course, the base course, and the sub-base—in other words, that portion of road construction above the subgrade—as a unit which will be referred to as the pavement structure.  And before proceeding with a discussion of types of mats, consideration will be given to thickness of pavement structure with respect to the load-carrying capacity or supporting power of the subgrade.

It will be recalled from Chap. 2, Grading the Roadbed, that a substantial depth of cover with stable material serves to protect the subgrade against applied wheel loads.  The surfacing and base act together as a cover or surcharge which restricts the subgrade soil from escaping from under the load, thus reducing that part of the vertical deformation which would result from lateral yielding (Fig. 2-7).  This fact is significant only in the case of wheel loads which, although they produce a relatively high unit pressure on the surface, are confined to a small area.  If this same intensity of pressure were to be applied over a large area of the roadway surface, a far different situation would arise in which a landslide might develop, particularly in the case of a high embankment.  It is with local stresses in the subgrade that the following discussion is concerned.

A further influence of a substantial cover on subgrade soil is to spread the wheel load over a larger area and so reduce intensity of pressure.  Some appreciation of this phenomenon may be gained by reference to Fig. 5-1.  Consider a depth $d$ of sand with a load $P$ applied on a circular bearing plate of radius $a$ on the surface of the sand, as illustrated in Fig. 5-1$a$.  If the enclosing sides were removed, much of the surrounding sand would readily fall away, but because of internal friction and interlock of sand grains created by the applied pressure, considerable effort would be required to remove the sand beyond a conical surface, as shown in Fig. 5-1$b$.

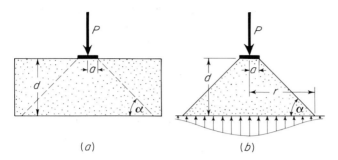

**Fig. 5-1**  Load distribution through granular material.

By virtue of such internal resistance the sand spreads the load over an area which increases with the depth, so that intensity of pressure is reduced.    The supporting pressure on the bottom of the sand layer is not uniform, however, but varies from a maximum directly under the load to zero at some indefinite radius $r$.    Because granular materials such as sand and other soils are not homogeneous and isotropic, it becomes desirable to adopt empirical procedures for determination of required thickness of cover adequately to reduce intensity of pressure on a given subgrade for a given wheel load.

One such method assumes that vertical pressure due to the load $P$ of Fig. 5-1 is distributed uniformly over the base of a cone whose base angle $\alpha$ is 45 deg.    Within the limits of normal thicknesses of flexible-pavement structure the method has proven to be of practical value.    On this assumption it will be seen that intensity of pressure is reduced from $P/\pi a^2$ directly under the bearing plate to $P/\pi r^2$ at the bottom of the layer. And if $\alpha = 45$ deg, then $r = d + a$.

Equating the load $P$ to the total vertical supporting pressure at the bottom of the layer—neglecting the weight of the cover material—the following formula is available:

$$P = p\pi(a + d)^2 \tag{5-1}$$

in which $p$ is the average pressure on the subgrade caused by a wheel load $P$ acting through base course and sub-base material of depth $d$. Thus, if an allowable unit pressure $p$ can be determined for a particular subgrade soil, the required thickness of cover can be readily determined for the maximum truck wheel load that is likely to be experienced.

Magnitude of wheel loads or axle loads (two-wheel loads) is restricted by state laws.    These restrictions are usually made more severe during spring months or prolonged wet periods when subgrade soils may become unstable because of excessive moisture.    Tire pressure is also generally limited to 80 or 90 psi, so that subgrade pressures do not reach the high intensities that would accompany highly concentrated loads.

Wheel loads exceeding about 4,000 lb are carried on dual tires, and tandem duals are used for carrying larger loads. Spacing of tandem duals is also controlled by law because "total load" becomes more critical than "load intensity" when proper limitations are exceeded. It is quite obvious, for instance, that a continuous series of closely spaced tandem duals could cause shear failure, creating a slide in an embankment or graded section of roadway, even though allowable pressure intensity were not exceeded.

Individual tire-contact areas on a smooth surface are somewhat elliptical, and there is some variation in contact pressure over the area because of the stiffness of tire walls. For purposes of subgrade-pressure analysis, however, these factors are not particularly significant, and the pressure is assumed to be uniform over the contact area. Also, the contact area is reduced to an equivalent circular area for both single and dual tires. Thus a wheel load of 8,000 lb with a tire pressure of 80 psi will have a contact area of 100 sq in., and the radius $a$ of an equivalent circular area will be about 5.6 in. Then from preceding considerations the proper depth of total flexible-pavement structure can be determined if the bearing capacity of the subgrade soil in pounds per square inch is known.

The bearing value of a subgrade soil is determined as the load intensity required to produce a small deflection of a bearing plate, usually 0.1 in. Loading may be applied directly to the compacted subgrade in the field with the use of loaded trucks, jacks, and strain gauges. It is important that the diameter of the bearing plate used be reasonably near the diameter of the area upon which the design load will be supported by the subgrade. Bearing plates ranging from 10 to 30 in. in diameter are commonly used. Bearing values of subgrade soils may vary from 5 to 50 psi.

Westergaard introduced a modulus of subgrade reaction designated as $k$ to denote subgrade strength. It is the load in pounds per square inch required per inch of soil deflection. Values commonly range from 50 to 800. Thus a subgrade with a modulus $k = 150$ requires 15 psi to compress it 0.1 in.

Figure 5-2 presents in scalar form a summary of approximate values for subgrade evaluation which may prove helpful in estimation or judgment as to what might be expected from different soil groups. The modified Bureau of Public Roads (BPR) soil-classification groups are shown at the top of the figure in accordance with total required thicknesses of protective cover or flexible-pavement structure as recommended by the Highway Research Board (HRB) for an assumed axle load of 18,000 lb, using granular stabilized base courses and 2 in. of bituminous wearing course (Table 5-1).

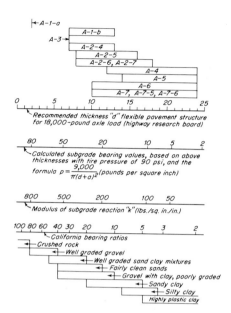

**Fig. 5-2** Approximate correlation of subgrade strength evaluations.

On the basis of these recommended total thicknesses, subgrade bearing values have been calculated as indicated by the second scale from the top. These values were arrived at by using the 45-deg-cone method of pressure distribution previously discussed, with an assumed tire pressure of 90 psi, and the principle that the total load $P$ (in this case 9,000 lb) is supported by a unit subgrade pressure $p$ on an area $\pi(a + d)^2$, where $a$ is the radius of tire-contact area and $d$ is the depth from the surface of the wearing course to the subgrade.

There is not at present final agreement as to standard methods of measurement of subgrade moduli, but for purposes of discussion here and later in the design of rigid pavements, the modulus of subgrade reaction is taken to be 10 times the bearing value of the subgrade for 0.1-in. deflection. These values are indicated on the third scale of Fig. 5-2.

It will be seen from the figure that only soil groups A-1-a and A-3 are definitely evaluated as to flexible-pavement requirements. The other groups are spread over a considerable range of strength characteristics, primarily because of such physical influences and construction factors as moisture content as affected by good or poor drainage, frost action, degree of compaction as compared with maximum density at optimum moisture content, and the amount of subgrade stabilization done with the use of such materials as calcium chloride, portland cement, and bituminous products. The use of granular materials so far as possible, attainment of optimum density during construction, and maintenance of uniform

Table 5-1  Recommended thicknesses of highway-pavement courses for 18,000-lb axle loads

Thickness, in.

| BPR soil classification: | A-1-a | A-1-b | A-3 | A-2-4 | A-2-5 | A-2-6 | A-2-7 | A-4 | A-5 | A-6 | A-7 A-7-5 A-7-6 |
|---|---|---|---|---|---|---|---|---|---|---|---|
| Wearing course | 2 | 2 | 2 | 2 | 2 | 2 | 2 | 2 | 2 | 2 | 2 |
| Base course | 0 | 5 | 5 | 5 | $6^a$ | $6^a$ | $6^a$ | $8^a$ | $8^a$ | 8 | $8^a$ |
| Sub-base | $0^b$ | 0-6 | $0^c$ | $0-6^c$ | $0-8^b$ | $0-10^{d,e}$ | $0-10^{d,e}$ | $2-14^{f,e}$ | $4-14^{f,e}$ | $0-14^{d,e}$ | $0-14^{d,e}$ |
| Total thickness | 2 | 7-13 | 7 | 7-13 | 8-16 | 8-18 | 8-18 | 12-24 | 14-24 | 10-24 | 10-24 |

[a] Soil-cement bases on A-2-5, A-2-6, and A-2-7 soils may be made 5 in. thick, and those on A-4, A-5, A-7-5, and A-7-6 soils may be 6 in. thick.

[b] No sub-base is required over A-1-a and A-2-5 soils where frost action is absent and the water table is low.  In areas subject to severe frost action or where the water table is near the surface, the maximum thickness should be used.

[c] In the case of fine-grained A-2-4 and A-3 soils, it is often necessary to modify the upper layer of the subgrade by admixture to a depth of several inches with binder soil, stone screenings, or bituminous materials to produce a stable surface upon which to place the granular base course.

[d] On A-6 and A-7 soils no sub-base is required where the water table is permanently below the elevation from which moisture may reach the base by capillarity.  The maximum thickness is required where the water table is high and for A-6 soils where frost conditions are severe.

[e] Soils of groups A-2-6, A-2-7, A-4, A-5, A-6, A-7-5, and A-7-6 are highly capillary and rapidly lose supporting power when saturated. When pavements are placed in cuts over such soils, the maximum thickness of sub-base should be employed.

[f] On A-4 and A-5 soils a sub-base course of the minimum thickness specified, consisting of stone screenings or similar material, is desirable to produce a firm support upon which to place the overlying granular course.  The maximum thickness is required only in areas subject to severe frost or where the water table is close to the surface.

79

moisture content by adequate drainage facilities are factors of roadway construction which greatly influence establishment and preservation of subgrade strength, primarily by improving and preserving shearing strength of the soil.

**5-3  California Bearing Ratios**  An empirical design method for flexible pavements which is widely used with various modifications by highway and airport organizations throughout the country is based on the California Bearing Ratio (CBR) test.   The test establishes a standard grading of crushed rock compacted under specified conditions as a standard or basis for measurement of bearing power of any soil.   The standard curve of penetration versus load intensity for the crushed rock is given in Fig. 5-3.

When a subgrade soil is to be compared with the standard, it is compacted to modified AASHO density in a 6-in.-diameter mold and put to soak for a period of 4 days with a surcharge weight approximately equal to the weight of pavement structure to be placed over it.   The amount of swell during the soaking period is recorded as a percentage of original height of specimen (this is also volumetric increase in a confining mold), and swell of more than 2 or 3 per cent is critical and indicates the need for

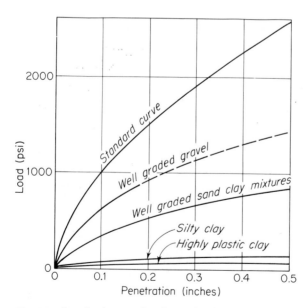

**Fig. 5-3**  Standard curve for determination of CBR and some average values for other soil types.

stabilization, extra cover, or special drainage in areas where moisture or frost conditions are adverse.

After soaking, the specimens are tested for strength (in the molds with surcharge weight in place) by application of load on a bearing surface of 3 sq in. at the rate of 0.05 in./min. Load readings are taken at 0.025, 0.05, 0.1, 0.2, 0.3, 0.4, and 0.5 in. penetration. Stress values at 0.1 and 0.2 in. penetration are compared with those on the standard curve for crushed rock, which are 1,000 and 1,500 psi, respectively. The load intensity on the soil for 0.1 in. penetration, expressed as a percentage of that for the crushed rock (1,000 psi), is the CBR for the soil. The CBR as determined from the bearing values at 0.2 in. penetration should be within one or two percentage points of that for 0.1 in. penetration or the test should be rerun; and if a consistently higher value of CBR is obtained for 0.2 in. penetration, it should be recorded as the CBR of the soil.

CBR values are usually determined at the moisture which is anticipated in the field. Some average values for different types of soil are shown by the curves of Fig. 5-3, and approximate ranges for various general soil types are shown on the lower scale of Fig. 5-2.

It is significant that pressure intensity to produce 0.1 in. deflection in a given soil under standard procedure with a bearing plate of 3 sq in. is much greater than the intensity of pressure required to produce an equal deflection in the same soil when applied directly to the subgrade with a bearing plate of perhaps 100 times the area. It serves to reemphasize the statement previously made that "total load" has as much significance as "unit pressure" in determining design limitations and legal restrictions to be placed on wheel loads.

The CBR method has been widely used by the U.S. Corps of Engineers and various highway and airport organizations for design of flexible pavements by coordination of CBR values with field requirements of protective cover. Past experience, performance records of existing roads, accelerated road tests, and direct field measurements of soil strengths have resulted in design curves and charts which correlate laboratory determinations of CBR with anticipated field conditions. Such influences as drainage conditions and severity of frost have brought about various modifications in the application of the test to design procedures in various states and regions. The "number" as well as the weight of commercial trucks using the roadway per day is also of significance in modifying thickness requirements. A typical basic design chart is given in Fig. 5-4.

Use of the chart is best illustrated by an example. It is required to establish a grade line and determine thickness of flexible-pavement structure for a roadway on which a maximum wheel load of 12,000 lb is anticipated in the near future. Data from the soil survey indicate three

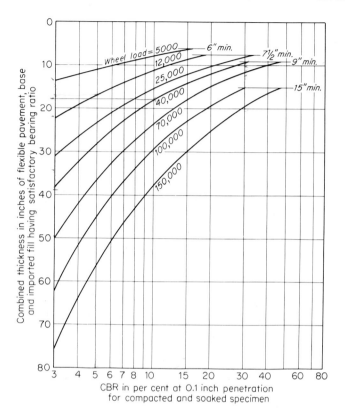

**Fig. 5-4** Design chart for CBR method of design of flexible pavements for highways and airport runways. Increase runway design thickness by 10 per cent for airport taxiways, aprons, and warm-up pads.

definite horizons of soil as illustrated in Fig. 5-5a, and laboratory tests have been made on sufficient samples to determine CBR values for the different types of soil as indicated.

From the design chart of Fig. 5-4 it will be seen that a soil with a CBR of 7 requires a minimum of 14 in. of cover with material of higher bearing strength. This means that the B-horizon soil may be used to within 14 in. of the top of the pavement or final grade line; provided, of course, that construction procedures will assure optimum density so that the maximum strength of the soil will be realized and that either frost will have no adverse effect on the soil or annual frost does not on the average penetrate below 14 in. If the number of trucks with wheel loads larger than about 4,000 lb expected on the roadway exceeds 100 per day, the thickness will usually be increased somewhat over the minimum because significant stresses below the design maximum should be given consideration. Service records of existing roads nearby or under similar physical conditions and proved local practices should be studied for possible modification of design thicknesses of protective cover as indicated by general design curves.

Attention should always be given to the transition zone between cut and fill so that weak materials are not left in place directly beneath the pavement. Subgrading and backfilling with more stable material should be done, as indicated in Fig. 5-5b. The soil from horizon C may be used for this purpose. Since a minimum of $7\frac{1}{2}$ in. of flexible-pavement structure is indicated by the design chart for a 12,000-lb wheel load, the soil with a CBR of 25 may be used up to within that distance from the surface of the wearing course. Thus a lift of $6\frac{1}{2}$ in. of this soil will be utilized in the fill section, where it is to be placed on soil with a CBR of 7.

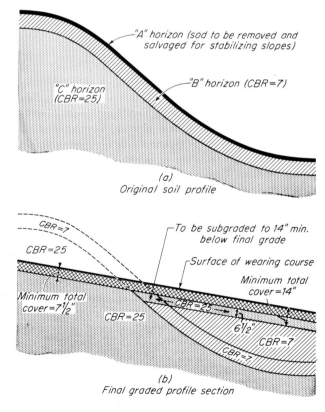

(a)
Original soil profile

(b)
Final graded profile section

Fig. 5-5

While the balancing of cut and fill quantities is of economic importance in a highway grading project, it can be seen that establishment of final grade lines should be influenced by the nature of the soils which are to create such a balance.  In the illustrated example, for instance, cutting deeper into the C horizon would pay off in making available more of the better material for use on the project to the extent that overhaul does not become excessive.  It is generally better practice to build the subgrade to uniform strength over considerable length of roadway so that the pavement structure may be built to uniform depth.  This requires close control of grading operations so that the best selective use is made of available materials.  Once the various soil horizons have been identified and evaluated, they will generally be fairly easily recognized in the field by color and texture.  Therefore it is important that soil surveys include such information in detail, as well as depth of various layers of soil.

**5-4  Hveem-Carmany method**  This method of design is currently used by highway departments in several states.  The primary soil strength criterion is determined as the resistance value $R$ by use of the Hveem Stabilometer, a closed triaxial cell.  This is a test in which the vertical load on the test specimen is increased while lateral expansion is restrained. Because lateral expansion is restricted, the lateral pressure builds up as the sample deforms.  The magnitude of the restraining pressure is evidence with which to judge the intrinsic stability of bituminous paving mixes and bases, as well as that of soil and gravel.  The $R$ value is used in conjunction with a traffic index to determine a required thickness of cover, expressed as a gravel equivalent.  (See scales $A$, $B$, and $C$ of the nomograph of Fig. 5-6).  The gravel equivalent is the thickness, in inches, of base course and surfacing required for minimal tensile strength materials such as untreated gravel or crushed rock.

A secondary strength criterion is the cohesiometer value $C$ which measures the relative flexural strength of the paving mat or treated base material.  In this test the flexural strength of a disk-shaped test specimen is determined much as in the case of tests for the modulus of rupture of portland-cement concrete.  A variation of this method, standardized by the California Department of Highways in 1964, replaces the cohesiometer value on scale $D$ (Fig. 5-6) by a "gravel equipment factor" which is based on the cohesiometer test and road experience.  Values of $C$ above 100 serve to reduce the total thickness of cover from that indicated by the gravel equivalent, thus taking advantage of the load-spreading capability of paving or surfacing materials with inherent tensile strength.  For instance a well-designed and -constructed asphaltic concrete paving mat may be expected to have a cohesiometer value of about

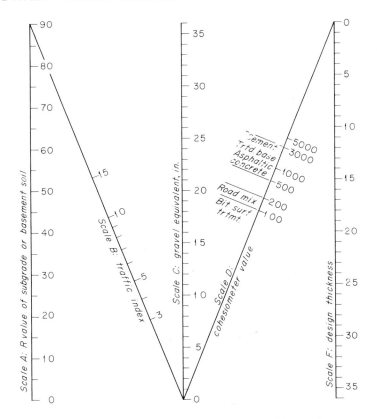

**Fig. 5-6** Thickness design chart for base course and/or paving mat using Hveem Stabilometer method.

500 to 1,000 and to reduce the total thickness of pavement structure accordingly. Continuation of the nomograph of Fig. 5-6 through scale $D$ to scale $E$ gives the adjusted design thickness.

The traffic index measures the severity of vehicular traffic, particularly as affected by the number of repetitions of the heavier wheel loads of commercial trucks and buses. It can be seen, then, that implications of the design wheel load are inherent in the numerical value of traffic index to be used for a given traffic demand. An estimated correlation of traffic index with design wheel load and general category of traffic use is given in Table 5-2.

Another important feature of this method of thickness design is a determination of the swell pressure developed by the compacted soil as it is subjected to soaking in water. The swell pressure may vary, in general, from zero for predominantly granular material to possibly 2 psi for fine-

**Table 5-2  Estimated correlation of traffic index with design wheel load**

| Traffic index | Design* wheel load, lb | Traffic severity | General category of traffic use |
|---|---|---|---|
| 5 | . . . . . . | Very light | Property access roads and streets |
| 6 | 5,000 | Light | Feeder roads and streets |
| 7 | 9,000 | Medium | County and city neighborhood arterials |
| 8 | 12,000 | Heavy | Secondary state highways |
| 9 | 16,000 | Very heavy | Primary state highways |
| 10 | 20,000 | Severe | Truck routes and truck lanes |
| 15 | 30,000 | . . . . . . . . . . . | Logging roads |

* One-half of single-axle load or one-half of tandem-axle load.

grained soils.   The design procedure requires that the minimum thickness of cover be sufficient to prevent the swell from taking place, assuming the cover material to have a unit weight of 130 pcf.   On this basis, the required minimum cover thickness $t$ in inches is

$$t = 13.3p \tag{5-2}$$

where $p$ is the swell pressure of the soil in psi.   The cover thickness so determined will govern if this is greater than the thickness determined from the design chart of Fig. 5-6.

**5-5   The AASHO design procedure**   It will be profitable to consider yet another design procedure—one which provides for consideration of economic balance of thickness design in the various segments of a layered system.   Findings of the road test of the American Association of State Highway Officials in northern Illinois, completed in the early sixties, are pointing the way toward balanced thickness design which is geared to traffic requirements.

The AASHO and other road tests have established equivalency as between tandem-axle loads and single-axle loads.   Basically a 32-kip tandem-axle load is equivalent to an 18-kip single-axle load so far as general reaction of pavement structure and basement soil is concerned. Furthermore, the relationship holds for greater or lesser loads so that a tandem-axle load times 0.57 gives the single-axle load equivalent, designated by $L$ in kips.

Also, a number of repetitions of a given single-axle load are equivalent to a different number of 18-kip single-axle loads.   This relationship is established by a load factor $F_L$ as follows

$$F_L = 10^{0.118(L-18)} \tag{5-3}$$

Then if $W_L$ is the number of applications of single-axle load $L$, the equivalent number of 18-kip single-axle load repetitions is

$$W_{18} = \Sigma(W_L \times F_L) \qquad (5\text{-}4)$$

These relationship are easily established by use of the chart in Fig. 5-7. For instance, 100 repetitions of 30-kip tandem-axle loads are

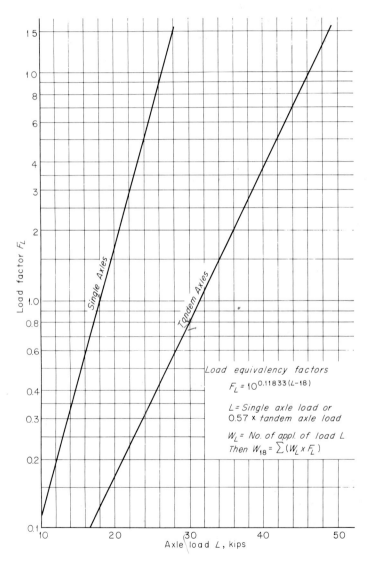

**Fig. 5-7**   Load equivalency factors.   (*From Asphalt Institute Manual, MS-1, September, 1963.*)

equivalent to $(100 \times 0.74)$ or 74 repetitions of 18-kip single-axle loads. Highway department loadometer or truck-weight data may thus be reduced to $W_{18}$. In this determination, axle loads less than 10 kips may generally be neglected so far as "destructive equivalency" is concerned.

The thickness design procedure is best illustrated by use of the chart in Fig. 5-8. Suppose that it has been previously determined that the subgrade or basement soil has a CBR value of 10. The soil-support value $S$ is determined to be 6 from the scale at the top of the chart. With this value of $S$ and an estimated value of, say, 200 for $W_{18}$, the daily

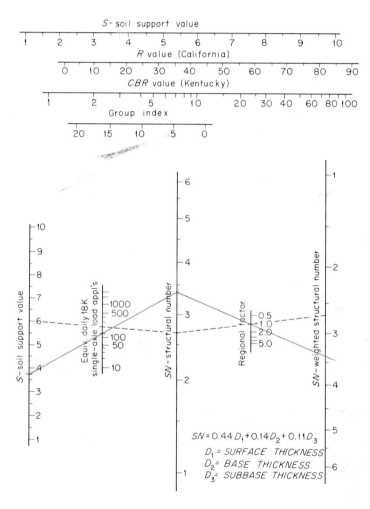

**Fig. 5-8** AASHO design chart, flexible pavements (20-year traffic analysis).

equivalent 18-kip single-axle load applications, the nomograph in the lower portion of the chart indicates a required structural number of 2.7, based on the climate and soil conditions of the AASHO Road Test.

Further modification of the structural number requirement is made in accordance with variation of climate and soil conditions from those experienced at the test road location.   This is done by means of a regional factor, the determination of which requires considerable judgment in evaluation of performance records of similarly constructed roads under comparable traffic use in the region in question.   Considerable research and study is presently under way to determine such regional factors. Assuming a regional factor of 1, in the example, the resulting weighted structural number SN is 2.7.

The structural requirement can be met by a variety of different combinations of $D_1$, $D_2$, and $D_3$ as indicated by the equation in Fig. 5-8.   The coefficients for $D_1$, $D_2$, and $D_3$ suggest the relative structural value of the layers of the pavement structure.   These coefficients, as presently established in the formula, may of course be subject to future change as research indicates, but will serve here to illustrate the usefulness of this design procedure.   Some of the possible solutions to the example problem are as follows:

1. Conceivably the structural requirement could be supplied by $D_3$, with $D_1$ and $D_2$ equal to zero.   The thickness requirement would be 2.7 divided 0.11 or about 25 in.   Sub-base material is generally gravel or select borrow.   The decision to do this would depend on availability (and in-place cost) of a well-graded gravel of excellent quality.   Surface treatment (prime and seal) would provide the wearing surface.   Then later, as traffic increases and/or as construction funds become available, this earlier construction would provide an excellent base for a paving mat.
2. A 2-in. paving mat $D_1$ may be combined with a base course $D_2$ (generally crushed rock or crushed gravel) of minimal thickness, say $3\frac{1}{2}$ in., and a sub-base layer $D_3$.   Then

$$2.7 = 0.44 \times 2 + 0.14 \times 3.5 + 0.11D_3$$

from which $D_3$ = 12 in. of sub-base material.   Note that the resulting total thickness is reduced from 25 in. in (1) to $17\frac{1}{2}$ in.   This could be important where acceptable material is in short supply at, or reasonably near, the construction site.
3. Properly designed and constructed bituminous-treated or portland-cement–treated base courses may be included in $D_1$ surface thickness.   That is, it may be decided to consider a 2-in. asphalt paving

mat placed on a portland-cement–treated base.   Then

$2.7 = 0.44D_1$

from which $D_1 = 6+$, say $6\frac{1}{2}$ in., with thickness of treated base of $4\frac{1}{2}$ in.

Cost comparison of possible alternatives would then determine the most economical design.

**5-6   The group-index method of flexible-pavement design**   The group index of a subgrade soil (see BPR soil classification, Chap. 2) may often serve as a valuable guide in determination of thickness of flexible-pavement structure.   Under average conditions of good drainage and thorough compaction, the load-bearing values of a subgrade soil may be taken as an inverse function of its group index; that is, a group index of 0 indicates a high bearing value and a group index of 20 indicates a very low bearing value.

D. J. Steele describes[1] application of thickness requirements.   A summary of recommendations for sub-base thickness for various group indices is given in the following table:

| Group index | Description | Thickness of sub-base |
|---|---|---|
| 0–1 | Good subgrades | No sub-base is necessary |
| 2–4 | Fair subgrades | 4 in. |
| 5–9 | Poor subgrades | 8 in. |
| 10–20 | Very poor subgrades | 12 in. |

These thicknesses of sub-base are additive to those of surface-course and base-course requirement to protect subgrade material adequately.

**5-7   Base courses**   The base course, because of its vulnerable position directly under a relatively thin wearing course, is a very important structural part of flexible pavements.   Its purpose is to provide a uniform and nonyielding support for the wearing course and to transfer and distribute traffic wheel loads upon the subgrade.   The material used must not only offer resistance to displacement, but should be wear-resistant.   Experience has shown that base-course materials are subject to considerable wear and deterioration because of the "kneading" action of traffic which is carried through the protecting flexible mat to a signifi-

[1] *Highway Research Board Proc.*, **25**:388 (1945).

cant extent, especially if a large proportion of traffic wheel loads are heavy truckloads.

Base-course material, then, should be composed of hard, strong, and durable particles relatively free from clay lumps, soft fragments, friable or laminated pieces, and organic matter. Gravel, crushed gravel, and crushed rock are the materials generally used. Specifications usually require that the percentage of wear as determined by the standard Los Angeles Rattler Test be not more than 50. In this test 5,000 g of a specified grading of the coarse portion of the aggregate is placed in a cylindrical steel drum along with a charge of six to twelve $1\frac{7}{8}$-in.-diameter steel spheres, depending on the grading. The drum is rotated at about 30 rpm for 500 revolutions, the aggregate and charge of steel spheres falling through the diameter of the drum (2 ft) each revolution because of a 4-in. shelf running the length of the drum on the inside. The percentage of material passing a No. 12 sieve after this treatment is recorded as the percentage of wear.

Base-course materials, which usually represent a substantial part of pavement cost, should be used to the extent that qualities of high stability and resistance to wear are needed and utilized. A distinction is made between base course and sub-base or subgrade, but the distinction between sub-base and subgrade is not so well defined nor is it particularly significant. Sub-base material is placed to bring the subgrade to fairly uniform strength characteristics so that base-course thickness may be reduced to a uniform and most effective value. Availability of good borrow material will usually dictate what is to be used for sub-base construction. It is important that it have greater stability and bearing power than the subgrade soil it is to protect, but quality of material, stability, and wear resistance such as required in the base course would be inefficiently used in the sub-base. Thickness of wearing course, base course, and sub-base on subgrades of soil types of the BPR classification for 18,000-lb axle loads as recommended by HRB are given in Table 5-1 for granular stabilized materials such as gravel, crushed gravel, crushed rock, water-bound macadam, and similar materials.

Base-course gradings recommended by AASHO are given in Table 5-3. The wide variety of gradings is the result of successful use of many different combinations under various physical conditions of climate, traffic, and subgrade soils and the need for making the best possible use of locally available materials. In many localities the best deposits of mineral aggregates have been used, especially those on public lands available for public use. Consequently materials must often be purchased locally or shipped in or inferior materials improved by admixing or processing, such as washing, screening out excess fine material, and crushing excessive oversize material.

Recent studies have shown that traffic accidents are definitely reduced by the provision of substantial shoulders (5 to 10 ft, depending on traffic volumes) so that stalled vehicles can get out of traffic lanes. Substantial shoulders also strengthen the entire roadway structure by virtue of the confining influence that is afforded by them. A good foundation for the shoulders is provided by extending the base course clear across the roadway. Stabilization of the surface of the shoulders is generally done by admixing clay in the top 3 or 4 in. of depth, by light surface treatment of bituminous materials, or by establishment of 4- or 5-in. sod cover in areas where climate and soil are conducive to growth of grass. Another advantage of extending granular base material across the entire roadway is that it provides a continuous lateral drain to the side ditches for any accumulated water which might occur on the surface of the subgrade,

**Table 5-3   Base-course gradations**

| Sieve designation | Percentages by weight passing square-mesh sieves (AASHO T27) | | | |
|---|---|---|---|---|
|  | Grading A | Grading B | Grading C | Grading D |
| Class I (coarse-graded): |  |  |  |  |
| 2 in. | 100 |  |  |  |
| 2 in. |  | 100 |  |  |
| $1\frac{1}{2}$ in. |  |  | 100 |  |
| 1 in. | 35–65 | 50–80 |  | 100 |
| No. 4 | 10–30 | 15–35 | 20–40 | 25–45 |
| Less than No. 200 |  |  | 10 | 10 |
| Class II (intermediate-graded): |  |  |  |  |
| 3 in. | 100 |  |  |  |
| 2 in. |  | 100 |  |  |
| $1\frac{1}{2}$ in. |  |  | 100 |  |
| 1 in. | 45–75 | 40–80 |  | 100 |
| No. 4 | 15–45 | 20–50 | 25–55 | 30–60 |
| No. 10 |  |  |  | 20–50 |
| Less than No. 200 | 10 | 12 | 12 | 12 |
| Class III (stabilized): |  |  |  |  |
| 3 in. | 100 |  |  |  |
| 2 in. | 65–100 | 100 |  |  |
| $1\frac{1}{2}$ in. |  | 70–100 | 100 |  |
| 1 in. | 45–75 | 55–85 | 70–100 | 100 |
| $\frac{3}{4}$ in. |  | 50–80 | 60–90 | 70–100 |
| $\frac{3}{8}$ in. | 30–60 | 40–70 | 45–75 | 50–80 |
| No. 4 | 25–50 | 30–60 | 30–60 | 35–65 |
| No. 10 | 20–40 | 20–50 | 20–50 | 25–50 |
| No. 40 | 10–25 | 10–30 | 10–30 | 15–30 |
| No. 200 | 3–10 | 5–15 | 5–15 | 5–15 |

because generally base-course gradings provide sufficient porosity for this, even in a well-compacted state.

Macadam is essentially broken stone. The name implies a coarse, open grading with little fine material. Common usage, however, has variously construed the meaning with the result that the distinction between macadam and crushed rock is not significant. Limestones and dolomites are among the most widely used road-building rock. They have less toughness, hardness, and resistance to wear than do the trap rocks, but they have a high cementing value. Crushed basalt is also in common use. While still harder rocks would be excellent road metal, there is a point at which the cost of rock crushing becomes excessive. Broken or crushed limestone, processed with water with the use of blade graders and compacted on the roadway, forms a well-bound, partially cemented mass that provides a high degree of stability as base-course material. Often during the "processing" or mixing, enough water is used so that the material forms a slurry in front of the blades and it is virtually washed into place. The presence of some fine material and a low percentage (3 to 5 per cent) of dust is advantageous in improving cementing value and forming a well-knit layer. The cementing action is also improved by adding sodium chloride or calcium chloride. This type of base-course construction is called water-bound macadam. It is more specifically water-bound crushed limestone, but for obvious phonetic reasons the former name will undoubtedly prevail.

Large-size particles, up to 2 or 3 in., as indicated by Table 5-3, are commonly used in base-course construction, because any tendency for them to loosen and become dislodged is prevented by the overlying wearing course. Stability is improved by the presence of angular particles, and therefore the larger particles from a material pit are usually crushed.

The most suitable gradings of mineral aggregates for base-course construction vary widely, but perhaps the greatest influence is availability of suitable material in the vicinity of the project. It is significant, however, that it is largely the coarse material and not the fines which carries the load and resists wear. Consequently the coarse particles should not be separated by fines in excess of the amount necessary to bind them together. This applies in particular to the dense or stabilized gradings.

The soil-stabilized roadways discussed in Chap. 3 very often become base courses for flexible mats. If the old surface is badly rutted or pitted it is usually scarified to the depth of any surface irregularities, and some crushed rock or stable material is added. The surface is then brought to true section by blading and rolling before the new wearing course is laid. The extent to which repairs are made and the old surface "built up" is of course dependent on the condition of the old roadway and on the nature and magnitude of anticipated traffic that the new surface will be expected to carry. The high plasticity of these old roads becomes undesirable when the surfacing is converted to a base, because of the excessive water content which builds up when evaporation is restricted by the oil-mat cover.

Base-course materials, particularly the intermediate and dense gradings, in general are usually subjected to considerable processing or mixing with blade graders on the roadway before they are laid. This ensures uniform distribution of fines so that full advantage is taken of such stabilizing or cementing qualities as the fine material may possess. For effective processing, water is also added, so that the material describes a circular or rolling motion ahead of the curved blade, and does not merely slide or shove. Good blade operators are critical of the moisture content. The moisture also affords some "lubrication" of aggregate particles so that compaction and interlock are effectively accomplished when the material is spread and bladed to final section and rolled.

**5-8  Bituminous mats**  A flexible mat or wearing course is an intimate mixture of mineral aggregate and bituminous binder. Thorough mixing is necessary to ensure complete film coverage on all aggregate particles. Many minierials have a greater preference or affinity for water than for bituminous material. Such aggregates are referred to as *hydrophilic* in character. Generally quartzite and granite show comparatively poor adhesion with bituminous binders. Consequently, when hydrophilic materials are used considerable care should be exercised to obtain a thorough coating of aggregate particles. A somewhat thicker film should be provided so that moisture is effectively prevented from reaching the aggregate surfaces, where it would cause displacement or *stripping* of the binder from aggregate particles. There are on the market commercial additives which, when mixed with asphalt in small proportions, tend to improve adhesive qualities and, to some extent, prevent stripping. It is desirable that aggregates be dry to ensure effective coating with bituminous materials. To this end specifications usually require that the moisture content of mineral aggregates be not more than 1 or 2 per cent at the time of mixing with bituminous materials.

Mineral aggregates of basic or alkaline nature such as limestone may attract or have affinity for bitumens. These are described as *hydrophobic*. The bitumen-retaining power of an aggregate is particularly important where moisture conditions are adverse. Aggregates should be clean and free from clay coatings which prevent effective adhesion of bitumen and mineral surfaces.

A wide variety of aggregate gradings are used for bituminous mats; some of those in common use are listed in Table 5-4. It is notable that the maximum size of aggregate is 1 in. A maximum size of $\frac{3}{4}$ in. is also

**Table 5-4   Aggregate gradings for bituminous mats***

| Sieve size | Percentages by weight passing | | | | |
|---|---|---|---|---|---|
| | A | B | C | D | E |
| 1 in. | 100 | 100 | 100 | 100 | 100 |
| $\frac{3}{4}$ in. | 75–100 | 75–100 | 85–100 | 85–100 | 85–100 |
| No. 4 | 35–45 | 40–60 | 45–65 | 50–70 | 60–95 |
| No. 10 | 20–35 | 25–45 | 30–50 | 35–55 | 45–80 |
| No. 200 | 2–7 | 3–8 | 5–10 | 5–12 | 5–15 |

* In connection with the given gradings, AASHO further specifies that the aggregate should be so graded that at least 10 per cent of the total will pass the No. 4 sieve and be retained on the No. 10 sieve.

frequently specified. Larger-size particles are not desirable because bituminous binders are not sufficiently strong to prevent their loosening and becoming dislodged under the action of high-speed traffic. They would also tend to produce a rough and noisy riding surface. It is desirable, however, that a fairly large proportion of the coarse material be angular in shape so as to improve internal friction and mechanical interlock of particles. To this end specifications commonly require that not less than 40 per cent of particles retained on the No. 4 sieve shall have at least one fracture face. Consequently, oversize material from a gravel pit is usually crushed instead of being screened out. Requirements for base-course material apply with even greater significance to wearing-course aggregates where stability and wear resistance are especially important. Manifestly the best available materials should be reserved for the wearing course. Percentage of wear as determined by the Los Angeles Rattler Test should not exceed 40.

The portion of aggregate passing the No. 200 sieve is called *filler*. It is desirable that filler have the properties of the parent mineral and should therefore be silt rather than clay. Clay is generally objectionable because it tends to coat coarser particles, interfering with proper adhesion of bitumen, and because most clays experience volume changes with variation in moisture content which may produce detrimental swell and shrinkage within the mat. Natural silt deposits, stone dust from crushing operations—limestone dust is particularly desirable—and portland cement are commonly used as fillers to supplement the portion naturally present or produced by crushing in the aggregate to be used.

The most obvious effect of filler is implied by its name. The dust occupies void space, reducing the amount of bitumen required to make the mat waterproof. The effect of filler upon stability is less clearly understood, but it is generally credited with contributing to the toughness of the paving mix. The film of asphalt in direct contact with a mineral surface has lost much of its mobility by reason of that adhesion. Any increase in surface area of the aggregate means a corresponding increase in the proportion of the total asphalt content which has lost some of its liquid properties. A pound of filler has far more surface area than a pound of coarse aggregate and consequently produces more rigidity in the asphaltic fraction of the mat.

Within limits, then, mineral filler is desirable in bituminous mixtures. The amount is basically controlled by properly proportioning filler and asphalt. A filler-asphalt ratio of 1:1 by weight is recommended where rock dust or mineral material of similar unit weight is used for filler.

Before proceeding with a discussion of various types of bituminous mats, some generalizations can be made with respect to amounts and grades of liquid bituminous materials to be used. Frequently during stage construction of roads to meet growing traffic demands the item of salvage and reworkability of a bituminous mat is important. A summary of working temperatures and reworking-time limits of mixtures using liquid asphalts, as taken from the Asphalt Institute design manual, is given in Table 5-5. Minimum working temperatures indicate that such salvage and reworking be done during hot summer weather when pavement temperatures are high.

Other factors to be given some consideration in selecting the type and grade of bituminous material to be used are climate, type of work—whether penetration treatment, road mix, or plant mix—and available equipment. Comparative cost of products as between road tars and liquid asphalts will also be of significance.

Thickness of mat will generally vary from 1 to 2 or 3 in. It should be emphasized that the mat is primarily a wearing course and stability must be built into the pavement structure from the subgrade up in order to give proper support to the mat. It is not good economy to build a thick mat to offset deficiencies of stability in base course or subgrade. For highway purposes the range of thickness of wearing course will generally be from 1 to 2 in., with 2 in. most commonly used for intermediate types of pavement structure for traffic volumes which are not excessive. When a bituminous mat is to be built upon a sand base which needs a considerable degree of confinement to develop stability, a thick mat of 3 to 5 in. may be placed. But in general it is better practice to stabilize the sandy base by admixing clay binder, by use of portland cement, or by treating it with a bituminous binder to provide cohesion in lieu of confinement.

**5-9 Asphalt content** The quantity of bituminous material to be used may be estimated by empirical formulas which have proved to be reason-

**Table 5-5 Suggested temperatures for use of asphalts***

| Grade of paving asphalt | Distributor spraying temperature, °F | | Pugmill mixing temperature of aggregates, °F | |
|---|---|---|---|---|
| | min | max | min | max |
| 40– 50 penetration | ...† | ...† | 300 | 350 |
| 60– 70 penetration | 285 | 350 | 275 | 325 |
| 85–100 penetration | 285 | 350 | 275 | 325 |
| 120–150 penetration | 285 | 350 | 275 | 325 |
| 200–300 penetration | 260 | 325 | 225 | 275 |

| Grade of liquid asphalt | Distributor spraying temperature, °F | | Pugmill mixing temperature of aggregates for MC and SC liquid asphalts, °F | |
|---|---|---|---|---|
| | min | max | min | max |
| RC, MC, SC grade 70 | 120 | 180 | 95 | 140 |
| RC, MC, SC grade 250 | 165 | 220 | 135 | 175 |
| RC, MC, SC grade 800 | 200 | 255 | 165 | 205 |
| RC, MC, SC grade 3000 | 235 | 290 | 200 | 240 |

| Grade of asphalt emulsions | Distributor spraying temperature, °F | | Pugmill mixing temperature of aggregate, °F | |
|---|---|---|---|---|
| | min | max | min | max |
| RS-1, RS-2K | 75 | 130 | (Not used | |
| RS-2, RS-3K | 110 | 160 | for mixing) | |
| SS-1, SS-K | 75 | 130 | 50 | 130 |
| SS-1h, SS-Kh | 75 | 130 | 50 | 130 |
| SM-K | 100 | 160 | 60 | 140 |

* "Asphalts—Paving and Liquid," The Asphalt Institute, 1965.
† Seldom used for spraying.
Note: Stockpile mixes should be mixed at lower temperatures. Pugmill mixing temperatures for open graded mixes should be between 225 and 250°F.
Caution: The purpose of the above table is to indicate temperature ranges necessary to provide proper asphalt viscosity for spraying and mixing applications for the grades of asphalts shown. It must be recognized, however, that temperature ranges indicated by this table generally are above the minimum flash point for the RC, MC, and SC liquid asphaltic materials as specified by The Asphalt Institute and other agencies. In fact, some of these liquid asphalts will "flash" at temperatures below these indicated ranges. Accordingly, suitable safety precautions are mandatory at all times when handling these liquid asphalts. These safety precautions include, but are not limited to, the following:

1. Do not permit open flame or sparks of any kind close to these materials except in heating kettles, mixers, distributors, or other equipment properly designed and approved for handling and applying them.
2. Do not use an open flame to inspect or examine drums, tank cars, or other containers in which these materials have been stored.
3. Properly vent and ground all vehicles transporting these materials.
4. Permit only experienced personnel to supervise the handling of these materials.
5. Comply with all applicable intra- and interstate commerce requirements.

ably reliable or by laboratory tests. One of the earliest formulas to be offered was that of McKesson and Frickstad in 1927. The original was modified somewhat to conform with field experience to

$$p = 0.02a + 0.045b + 0.18c \tag{5-5}$$

where $p$ is percentage by weight of liquid asphalt required (this may be considered to be the percentage by weight of total mixture) and $a$, $b$, and $c$ are the percentages of aggregate retained on the No. 10 sieve, passing the No. 10 sieve and retained on the No. 200 sieve, and passing the No. 200 sieve, respectively. Thus an estimated percentage of bituminous material required for grading C of Table 5-3, using median values, is

$$p = (0.02 \times 60) + (0.045 \times 32.5) + (0.18 \times 7.5)$$
$$= 1.20 + 1.46 + 1.35 = 4.01 \text{ per cent, say 4 per cent}$$

*Asphalt percentage* or *percentage of bitumen* is usually expressed as a percentage by weight of total mixture. The percentage of bitumen based on weight of aggregate is commonly referred to as *oil ratio*, being the ratio of oil to aggregate.

The empirical formula for determination of estimated per cent of RC, MC, and SC asphalt content as presented in the Asphalt Institute "Mixed-in-Place" (Road-Mix) Manual, 1965, is

$$p = 0.02a + 0.07b + 0.15c + 0.20d$$

where $p$ = per cent[1] of asphalt material by weight of dry aggregate
$a$ = per cent[1] of mineral aggregate retained on No. 50 sieve
$b$ = per cent[1] of mineral aggregate passing No. 50 and retained on No. 100 sieve
$c$ = per cent[1] of mineral aggregate passing No. 100 and retained on No. 200 sieve
$d$ = per cent[1] of mineral aggregate passing the No. 200 sieve
Absorptive aggregates—such as slag, limerock, vesicular lava and coral—will require additional asphalt. The formula for determination of estimated per cent asphalt emulsion requirement is

$$P = 0.05A + 0.1B + 0.5C$$

where $P$ = per cent[1] by weight asphalt emulsion, based on weight of graded mineral aggregate
$A$ = per cent[1] of mineral aggregate retained on No. 8 sieve
$B$ = per cent[1] of mineral aggregate passing No. 8 sieve, and retained on No. 200 sieve
$C$ = per cent[1] of mineral aggregate passing No. 200 sieve

[1] Expressed as a whole number.

It can be seen that such formulas are at best a rough estimate of asphalt content to be used. However, use of the McKesson-Frickstad formula results in mixtures which are generally somewhat "lean" or slightly deficient in bitumen, and the chances are that a satisfactory asphalt content for grading C of Table 5-4 would be closer to 5 per cent than 4 per cent, depending somewhat on the nature of aggregates used, particularly the specific gravity and surface texture of aggregate particles and the amount of fractured or crushed material.

The quantity of bitumen used should be somewhat less than that required to fill the voids of the compacted mineral aggregate. Completely filling the voids with bitumen would tend to separate aggregate particles when the bitumen expands during hot weather, thus impairing stability of the mixture on the roadway, with the result that the mat would soften and become subject to rutting and "shoving." The phenomenon of shoving is frequently seen at approaches to railroad crossings or at intersections where heavy trucks and buses are brought to a stop. The horizontal thrust created by such braking of heavy vehicles causes softened mat material to shove or flow horizontally and form ridges or washboards. Tractive effort of heavy vehicles while accelerating or climbing steep grades may also produce shoving in the mat. Rutting and surface "bleeding" of the mat are also the result of too much bitumen in the mixture. The surface should retain a "mosaic" or "pebbly" texture with some aggregate surface exposed so as to preserve stability and skid resistance. When surplus bitumen is brought to the surface by "bleeding" or squeezing out, a sleek surface is formed which may become dangerously slippery when wet.

Such "overasphalted" conditions may be brought about by expansion of bitumen during hot weather or it may be brought about over a period of time by densification of the mat under the kneading action of traffic wheel loads. This densification or compaction after construction results in a reduction of voids in the mineral aggregate which may ultimately produce a squeeze play on the bitumen, with attendant impairment of stability.

It is important, then, that the mat be compacted by effective rolling to a high degree of density during construction and that some porosity or void space be retained in the final compacted mat. In dense and intermediate aggregate gradings it is desirable to fill about 75 to 80 per cent of voids in the aggregate at maximum compaction with bitumen. For open or macadam gradings a smaller percentage of aggregate voids will normally be filled.

The use of too little asphalt will of course reduce durability, with the result that raveling or cracking may occur. Also excessive water absorption by the porous mat would expose it to the likelihood of stripping and the destructive action of freezing and thawing. Free circulation of air in porous mats also increases oxidation of films of the bituminous binder. Such weathering or oxidation hardens the bitumen, making it brittle and subject to excessive wear and cracking. The length of service life of a bituminous mat depends upon the retention of its initial flexibility.

Some stability, then, must generally be sacrificed for preservation of durability and watertightness, and a proper balance is attained in design when a porosity of about 3 to 6 per cent is present in the mat as compacted on the roadway. A minimum porosity of about 3 per cent will provide room for expansion of bitumen during hot weather and for some densification under traffic after construction. A maximum of 6 per cent porosity will prevent excessive absorption of free surface water if proper crown (about $\frac{1}{4}$ in./ft, increased slightly on steeper grades) in the cross section and good surface drainage are otherwise provided for and maintained.

In order to obtain a reasonably accurate estimate of voids in the mineral aggregate as it will be finally compacted in the wearing course, it is necessary to determine the absolute specific gravity and apparent specific gravity of a well-compacted laboratory sample. The specific gravities of coarse aggregate (retained on the No. 4 sieve) and fine aggregate (passing the No. 4 sieve) are usually determined separately by standard procedures. Then the absolute specific gravity or theoretical maximum density of the composite aggregate (assuming no air voids) is given by

$$G = \frac{100}{P_c/G_c + P_f/G_f} \tag{5-6}$$

where $G$ = absolute specific gravity of composite aggregate

$P_c$, $P_f$ = percentages of coarse and fine materials by weight, respectively

$G_c$, $G_f$ = specific gravities of coarse and fine materials, respectively, as they are to be used in the mix

The bulk specific gravity of the compacted aggregate may be determined by carefully compacting a representative sample of the aggregate into a container of known volume and weighing it. An accurate volume measure of from $\frac{1}{4}$ to 1 cu ft, depending on the maximum size of the aggregate, should be used. The material should be tamped into the container in thin layers—only slightly thicker than the maximum size of aggregate—with a weighted, flat tamping surface. The 10-lb hammer used for the modified AASHO density test is effective for this purpose. Care should be used so as not to break or crush aggregate particles. Since the material is dry, some kneading action should be applied to the surface to overcome internal friction and obtain the most favorable orientation of aggregate particles for maximum density. This can be done by a circular motion of the hammer handle while the flat impact surface is pressed down on the material. Vibration is also effective in obtaining maximum density where equipment for effectively applying it is available. Each layer should be tamped and kneaded to a firm unyielding surface. The top surface is best "struck off" by using a section of round rod or pipe and rolling the material to an even surface, flush with the top of the container rather than scraping off excess material. The Triaxial Institute kneading compactor is ideally adapted to this type of aggregate compaction and its use is highly recommended. A more complete discussion of design procedure for bituminous mixtures is given in the following chapter.

If the bulk specific gravity of the aggregate so determined is represented by $g$ and the porosity—percentage of voids based on total volume—by $n$, then

$$n = \frac{100(G - g)}{G} \tag{5-7}$$

Considering, then, 100 cu cm of total volume of mixture and assuming 75 per cent of compacted aggregate voids to be filled with bitumen, the volume of bitumen to be used is seen to be $0.75n$, and the absolute volume of aggregate is $100 - n$, leaving a final porosity in the mixture of $0.25n$. By weight the proportions will be $0.75n \times G_a$ of bitumen to $(100 - n)G$ of mineral aggregate, where $G_a$ is the specific gravity of the bituminous binder to be used. From these proportions by weight the oil ratio or the bitumen content by weight of total mix is readily determined. Such determinations are of value for purposes of estimating quantities to be included in the construction contract and for guidance in getting construction operations under way. The shortcoming is, of course, in the lack of coordination of laboratory compaction with field compaction under heavy rollers. There is also some notable difference in the densities obtained in the field, with the use of a smooth iron roller, for instance, as compared with pneumatic-tired rollers, and as between weights of rollers and the number of passes used in compacting the mat. Consequently, a close check should be kept on bulk specific gravity of the mat as portions are completed. This may be effectively done in the field from samples taken directly from the finished pavement by determining weight in air $(W_a)$ and weight when suspended in water $(W_w)$; then

$$g = \frac{W_a}{W_a - W_w} \tag{5-8}$$

For dense mixtures this may be done directly with the sample as taken from the wearing course if the weight in water is determined quickly before any appreciable absorption can take place. If appreciable absorption is likely, the sample should be coated with paraffin and the above formula corrected for the weight and volume of paraffin used.

Frequent laboratory checks should also be made on specific gravities of aggregates used. Specific gravities of liquid asphalts at 60°F may be determined for design purposes from the average unit weights given in Table 5-6. As given in the table, equal average weights are assumed for any one grade of any type—RC, MC, or SC. Average unit weights of asphalt cements are also tabulated. The standard temperature for determination of weights and volumes of bituminous materials for purposes of payment and design is 60°F. Conversion tables are available for determining precise weights and volumes of various products for any given application temperature used in construction. Approximate coefficients of thermal expansion are given in the last column of the table.

**5-10  Armor coat**  The distinctive feature of an armor coat is that its thickness is only slightly more, if any, than the maximum size of aggregate used. Consequently, the thickness of a wearing course referred to as an armor coat is about 1 in. It was first used in California and is now used in many states. It is just within the range of types included here as bituminous mats but in fact is often referred to as a heavy-duty surface treatment.

After the base material has been brought to uniform grade and properly crowned section and swept clean of excessive dust and loose material with a rotary broom, the surface is primed as described in Chap. 3. The prime coat should be allowed to penetrate, preferably without traffic, for several days (usually 4 or 5 days). The precise amount of liquid bituminous material to be used for the prime coat varies so widely that it is

**Table 5-6  Unit weights of asphaltic products**

| Asphaltic product | Average gallon per ton at 60°F | Average pound per gallon | Approximate coefficient of expansion at 60°F |
|---|---|---|---|
| Grade: | | | |
| 30............ | 251 | 7.97 | 0.00040 |
| 70............ | 248 | 8.07 | 0.00050 |
| 250.......... | 245 | 8.16 | 0.00060 |
| 800.......... | 243 | 8.23 | 0.00070 |
| 3000......... | 241 | 8.30 | 0.00080 |
| Penetration: | | | |
| 200–300....... | 239 | 8.36 | 0.00085 |
| 100–200....... | 237 | 8.44 | 0.00090 |
| 40–100........ | 235 | 8.51 | 0.0010 |

often advisable to test a small area of the roadway by applying—at the application temperature to be used in construction—a measured quantity with a sprinkling can so that only a slight excess of liquid remains on the surface after about 6 hr. This small area is then covered with heavy building paper so that it does not get a second application from the pressure distributor. Experienced operators, however, will usually be able to estimate the quantity to be used so that only slight adjustment in rate of application has to be made after the first trial run.

Depth of penetration to be expected from the prime-coat application—and in general for most penetration treatments at specified application temperatures—is about $\frac{1}{2}$ to $\frac{3}{4}$ in. for dense base materials, $\frac{3}{4}$ to $1\frac{1}{4}$ in. for intermediate aggregate gradings, and 1 to 3 in. for open or macadam aggregate gradings.

The first application of heavier liquid bituminous material—light grades of asphalt cements are sometimes used—is placed on the primed surface at about 0.15 to 0.2 gal/sq yd depending somewhat on the consistency and application temperature of product used. Obviously it should not flow to any extent on the roadway surface. This application is immediately covered with clean crushed rock—preferably of the trap-rock group—passing the $\frac{3}{4}$-in. sieve and retained on the $\frac{1}{2}$-in. sieve, spread uniformly on the roadway at the rate of 55 to 65 lb/sq yd. The aggregate is then smoothed with a broom drag and/or blade and rolled with a power roller.

A second application of bitumen is made at the rate of about 0.4 gal/sq yd and followed with about 24 lb/sq yd of stone chips passing the $\frac{1}{2}$-in. sieve and retained on the No. 8 sieve. Broom drag and roller are then used to work the chips into the voids of the first layer of stone and thoroughly key them to form a well-knit layer. Traffic will often bring up excess bituminous material in places. This should be blotted up immediately with stone chips or coarse sand.

When larger-size aggregate is used, it is usually applied in three applications according to size. After the first application of about 0.25 gal/sq yd of bitumen, crushed stone ranging in size from $1\frac{1}{2}$ to 1 in. is spread on the roadway so that a continuous one-stone coverage is made. A second application of bitumen at the rate of about $\frac{1}{2}$ gal/sq yd is then made and followed by keystone sized from $\frac{3}{4}$ to $\frac{1}{4}$ in. which is broom-dragged and rolled to a smooth well-knit layer. A surface coat, which is essentially a seal coat, is then applied, using about 0.3 gal/sq yd of bitumen and about 25 lb/sq yd of clean stone screenings or coarse, sharp sand passing the $\frac{3}{8}$-in. sieve and retained on the No. 8 sieve.

While the armor coat may be considered to be a heavy-duty surface treatment, it will also be seen that it is not far removed from penetration macadam of the high type of bituminous-pavement classification.

**5-11  Road-mixed bituminous mats**  A road-mixed or mixed-in-place bituminous mat is characterized by the fact that mineral aggregates and bituminous materials are mixed directly on the roadway base on which it is to be laid.  Compacted thickness of mat is generally about $2\frac{1}{2}$ in.

Dense gradings with maximum size of aggregate at $\frac{1}{2}$ or $\frac{3}{4}$ in., often referred to as fine-crushed gradings, have been most successfully used (Table 5-4).

Graded gravel and crushed aggregates generally compact to about 0.7 of loose depth on the roadway.  Thus the loose volume of mineral aggregate as it is contained in trucks and deposited on the roadway should be sufficient to cover the wearing-course area to a loose depth of about $3\frac{1}{2}$ in. in order to make a $2\frac{1}{2}$-in. mat.  The percentage of voids (porosity) of compacted fine-crushed aggregates generally used in road-mixed bituminous mats will range from about 15 to 20.  Optimum oil ratios of about $4\frac{1}{2}$ per cent are common.

Preparation of the base course is usually done considerably in advance of constructing the mat so that the surface is firm and well established.  Final shaping of base material and priming are often done during one construction season, and mat construction the next.  The primed base thus becomes the surface upon which mixing operations for mat construction take place.  Care must be exercised during mixing so as not to gouge into the base with equipment.

Some blading of aggregate after it has been deposited on the roadway is necessary to establish uniformity of grading and of quantity over the length of the project.  This is especially true when the aggregate grading from a particular source has to be "built up" by admixture of fine or coarse material from another source.  The mixing may be done centrally as the material is loaded into trucks, but any spottiness as to distribution of particle sizes should be eliminated by manipulation on the roadway with the use of blade graders.  The aggregate is then brought to a windrow along one side of the roadway until mixing operations are started.  At this time the windrow is checked for uniformity in size by measuring the cross section at regular intervals of 50 or 100 ft in order to ensure a uniform thickness of mat.

Mixing operations are started by spreading the windrow of aggregate uniformly over the entire width of proposed construction.  Air temperature should preferably be 60°F or above, and the aggregate should be dry.  The liquid bituminous material is sprayed over the aggregate with the use of a pressure distributor in increments of about $\frac{1}{2}$ gal/sq yd over the full width.  Application temperatures generally range from 175 to 225°F.  Each increment or application of bitumen is followed by tractor-drawn disk harrows and/or spring-tooth harrows to accomplish the initial mixing.  Rotary tillers are also used for this purpose.

After the final application of bitumen has been so mixed with the aggregate, the mixture is again bladed into a windrow and the rest of the mixing is done with blade graders.  Two or three are usually needed, as it is not uncommon to require as many as 30 or more bladings to complete the mixing.  Uniformity of texture and color is the best guide to a complete mixing job.  Sometimes there is a tendency for the bitumen to ball up with the fine material, especially if some segregation of fine material was present in the aggregate when mixing operations were started.  It may also occur if the bitumen is applied in too large increments.  These balls should be thoroughly broken up and distributed before blading is stopped.  All particles of aggregate should be coated with bitumen, and experienced inspectors insist that the mix should be "alive."  The test for this is that on any steeply inclined surface such as the side of a windrow the material should appear to be "crawling" downward.  If such a surface appears inert or "dead" and densely black, it is generally an indication of too much bitumen.  Also, a quantity squeezed in the hand should leave only a brown stain, not a black smudge.  These are only indications, however, and a sample quickly compacted in the field laboratory and checked for density and void content will determine more reliably whether the mix is acceptable.  Some adjustment by addition of more bitumen or more aggregate is frequently necessary in the first portion of the mat to be laid.

Before the mat is laid, provision is often made for edge thickening by making a cut about 2 in. deep at the edge of the wearing course and tapering to zero about 2 ft in from the edge with a blade, casting this portion of base material outward, where it becomes part of the shouldering material.  This reinforces the edges of the mat, where cracking and breaking away might otherwise occur because of feather edges where faulty shaping of base or uneven mat thickness may exist.  Care must be exercised to obtain uniformity of mat thickness on superelevated curves.

The mixed material is spread to accurate section over the roadway and compacted by rolling with 7- or 8-ton power rollers.  Rolling should begin at the edges and proceed toward the center on the first passes so as to confine the material as much as possible.  The process should continue to the satisfaction of the engineer that effective compaction has been attained.  Samples taken directly from the completed mat should be tested for density and void content so that minor adjustments in aggregate grading or bitumen content can be made intelligently as the work progresses.

It is advisable to mix extra material on the roadway and stockpile it for patching and general maintenance.  It will be expedient to make such a stockpile at the beginning of the project with the first mix made.

This will afford an opportunity to lay and compact a small area of mat for inspection and testing before mat construction proper begins.

Base courses and old gravel or fine-crushed aggregate roads may be stabilized or treated by road-mixing operations with bituminous materials. The old surface is scarified to the desired depth and thoroughly pulverized and mixed with such additional aggregate as needed to bring the original aggregate to desired grading. Then operations proceed as outlined above. The resulting mat may constitute a new wearing course, or in the case where coarser materials are involved the operations may constitute bituminous treatment of base course upon which a wearing course is to be built. In the latter case—bituminous-treated base course—considerably less bitumen is usually used because increased stability is the desired result, and durability is not a significant factor.

A seal coat may or may not be placed on a newly constructed road-mix mat. This may be taken over as a maintenance item so that at intervals of about 2 years, or as needed, a seal coat of bitumen and rock chips is applied to furnish a new wearing surface with improved skid resistance and improved visibility and to protect the mat by waterproofing the surface. The aim of seal coating should be to form a new top surface integral with the mat but forming a distinct layer of quick-setting or rapid-curing bitumen and cover aggregate. Any appreciable penetration or "priming" of the mat by seal-coat application should not be tolerated because it tends to fill the voids in the mat which were carefully preserved initially for specific reasons of balanced design. The only justification for allowing some penetration or for making a bituminous seal without aggregate cover is in the case where the existing mat might be subject to surface raveling because it is too dry or lean, whether from deterioration or from defects in its initial construction.

**5-12 Traveling plants** A variation on road-mixed bituminous mats is made with the use of the traveling plant. The mineral aggregate is picked up by the machine from the prepared windrow and elevated to a hopper from which it is meted out into the mixing chamber. As the aggregate enters the mixer it is sprayed with bitumen at a predetermined and adjustable rate. The mixer is generally of the pugmill type. It is composed essentially of stubby blades or paddles projecting from two horizontal shafts which rotate in opposite directions. As the sprayed aggregate is worked through the length of the mixer from one end of the shafts to the other, it is buoyed up and "cut" by the paddles. This churning action provides effective mixing and coating of aggregate particles. The rate at which the mix moves through the mixer is governed by the adjustable pitch of the blades and slope of the shafts. Thus a continuous flow of materials is provided from aggregate windrow in front of the traveling plant to a windrow of bituminous mixture behind. The windrow of mixture is then spread with blades. As the mixture leaves the traveling plant it may also be dropped directly into the hopper of a paver which follows behind, spreading and finishing the mat except for final rolling.

**5-13  Plant-mixed bituminous mats**  Stationary and semiportable plants for production of bituminous mixtures have steadily increased in use. Better control of production resulting in better quality and uniformity of product has been the greatest advantage in favor of these plants. Heating and drying of aggregates at the plant eliminate dependence on the weather for this operation. Aggregates are separated according to size into three separate hoppers so that coarse, intermediate, and fine aggregates are fed through separate control gates into the mixer. This makes possible a high degree of accuracy in the control of aggregate grading. A dust collector at the drier returns filler material to the "fines" hopper so that this material is not blown away. Excess material from any bin or hopper overflows into a truck or waste pile. Bituminous materials are sprayed on the aggregate as it enters the pugmill, as in the traveling plant. The mixture is then hauled to the roadway in trucks and deposited into the hopper of the finishing machine or paver, which spreads it and strikes it off to proper depth with a vibrating screed, covering one-half the roadway (10 to 12 ft) at a time. The depth of material behind the paver will compact about 25 per cent, depending somewhat on the aggregate grading used.

Aggregate gradings commonly used parallel those given in Table 5-3, and bitumen content should be consistent with balanced design in respect to stability and durability, as discussed earlier in this chapter. When the lighter grades of liquid asphalts containing considerable solvent or distillate are used unheated in plant mixes, rolling or compaction should be deferred from 24 to 48 hr to allow solvents to evaporate, otherwise the surface sealing which is sometimes created by rolling may entrap these distillates, thus greatly delaying curing of the mixture. For this reason there is considerable preference for the use of heavier grades of bituminous materials applied hot to warm, dry aggregates at the plant.

Many engineers feel that a seal coat is not immediately necessary on a well-designed bituminous mat. The low void content and the mosaic or pebbly surface texture of a well-constructed mat provide desirable roadway characteristics until surface-aggregate particles become worn smooth by traffic and surface bitumen becomes weathered to the extent that durability is impaired. When these events take place it is time to apply a seal coat to provide a new wearing surface.

## QUESTIONS AND PROBLEMS

**5-1.** The maximum tire pressure for truck operation in a certain state is limited by law to 90 psi. Assuming this pressure to be uniformly distributed over the area of tire contact on the roadway, find the radius $a$ of the equivalent circular contact area for a 12,000-lb wheel load.

**5-2.** For the wheel load of Prob. 5-1, assume that subgrade pressure is not to exceed 20 psi. Find the required thickness of flexible-pavement structure, according to the principle of the cone of pressure distribution.

**5-3.** Upon completion of grading operations a subgrade was tested for bearing capacity by loading large bearing plates. It was found that a load of 5,000 lb produced a deflection of 0.1 in. under a plate 18 in. in diameter. Based on this load, what is the modulus of subgrade reaction according to Westergaard?

**5-4.** Laboratory tests indicate that a certain subgrade soil has a CBR of 5. What thickness of flexible-pavement structure should be specified for a 12,000-lb wheel load? See Fig. 5-4.

**5-5.** By improving the underdrainage, the CBR of the soil of Prob. 5-4 may be doubled. Under this condition, what total thickness of flexible-pavement structure would be satisfactory?

**5-6.** An embankment is to be constructed so that it will be safe for 40,000-lb wheel loads. It is desired to make the best possible use of soil which has a CBR of 3. Assume that this soil will be covered with a layer of sandy clay with a CBR of 10 and a layer of pit-run gravel with a CBR of 30. The base course is to be constructed of well-graded crushed rock and topped with a 2-in. bituminous mat. Find the required thickness of each layer and sketch the soil-and-pavement profile.

**5-7.** *a.* Compute the volume in cubic yards of crushed-rock base-course material for Prob. 5-6 per mile of roadway 24 ft wide, if the ratio of compacted thickness to loose thickness is 0.67.

*b.* If the void ratio of the loose material is 0.875, what is the void ratio of the compacted base course?

*c.* The specific gravity of the crushed rock is 2.65. How many tons (dry weight) of rock are required per mile of roadway?

**5-8.** A bituminous mat is to be constructed using the upper limits of grading C of Table 5-3. This aggregate mixture is composed of 92 per cent crushed rock (sp gr = 2.63) and 8 per cent limestone dust (sp gr = 2.72). Determine the theoretical maximum density of the combined aggregate. Note that this is also the absolute specific gravity of the aggregate mixture.

**5-9.** The compacted aggregate of Prob. 5-8 has a porosity of 20 per cent. Compute the ratio, by weight, of asphalt cement (sp gr = 1.035) to aggregate so that the compacted mixture will have a porosity of 5 per cent. What is the asphalt content expressed as a percentage by weight of total mix?

**5-10.** *a.* What is the apparent or bulk specific gravity of the bituminous mixture of Prob. 5-8?

*b.* Compute the number of tons of the mix required per mile if the mat is to be constructed in two 10-ft lanes, with a compacted thickness of 2 in.

**5-11.** A truck route is to be built over a silty-clay subgrade soil with an $R$ value of 28, and an expansion pressure of 1.8 psi, for a maximum design wheel load of 20,000 lb. There is available a select gravel borrow with an $R$ value of 70. A crushed-rock base course will be used, with a 2-in. road-mix asphalt mat which has an estimated cohesiometer value of 200.  Find the required thickness of each layer and sketch the pavement and soil profile down to the subgrade.

**5-12.** *a.* Determine the weighted structural number of the subgrade soil of Prob. 5-11 for equivalent 18-kip single-axle loads of 150 repetitions per day, using a regional factor of 1.0.

   *b.* Based on estimated in-place unit costs currently experienced in your locality for sub-base, base course, and surface course, determine the most economical thickness for this layered system in accordance with suggested procedure given in Sec. 5-5.

## BIBLIOGRAPHY

AASHO Road Test Results Applied to Pavement Design in Illinois, *Highway Research Record* 90, Washington, 1965.

"The Asphalt Handbook," The Asphalt Institute, 1965.

Bituminous Construction Operations, *Highway Research Board Bull.* 280, 1961.

Design of Flexible Pavements Using the Triaxial Compression Test, *Highway Research Board Bull.* 8, 1947.

Flexible Pavement Design Studies, *Highway Research Board Bull.* 269, 1960.

Hubbard, Provost: Flexible Pavement Reaction under Field Load Bearing Tests, *Asphalt Inst. Res. Ser.*, no. 9, 1943.

"Manual on Design and Construction of Asphaltic Roads and Streets," The Asphalt Institute, 1963.

Performance and Tests of Asphaltic Concrete, *Highway Research Board Bull.* 234, 1959.

Surface Treatments, Bituminous Mixtures and Pavements, *Highway Research Record* 104, 9 Reports, Washington, 1965.

Symposium on Compaction of Earthwork and Granular Bases, *Highway Research Record* 177, 17 Reports, Washington, 1967.

Symposium on Mineral Fillers for Bituminous Mixtures, *Highway Research Board Bull.* 329, 1962.

# 6

# High-type Bituminous Pavements

Sheet asphalt is a wearing course, usually constructed about $1\frac{1}{2}$ in. thick, composed of sand, filler, and asphalt cement. It is common practice to place it on a binder course of about the same thickness.

**6-1  The binder course**  The purpose of the binder course is to provide a uniform transition from the thin mat to the base on which new construction is placed. Old worn pavements are often rejuvenated by construction of a sheet-asphalt wearing course. The binder course should not be considered primarily as a leveling course, but it does serve to iron out minor surface roughness, old joints, and cracks. Primarily, a good binder course will provide a firm and uniform cushion for the wearing course; it will assure a firm but somewhat flexible bond with the old surface; and it will provide substantial resistance to shoving, rutting, or creeping of the wearing course.

Table 6-1 Typical aggregate gradings, binder course

| Sieve sizes | Percentages by weight passing | |
| --- | --- | --- |
| | A | B |
| 1 in. | 95–100 | 100 |
| $\frac{3}{4}$ in. | ...... | 95–100 |
| $\frac{3}{8}$ in. | 40–80 | |
| No. 4 | ...... | 20–40 |
| No. 10 | 15–30 | 15–35 |

Typical aggregate gradings for the binder course are given in Table 6-1. Crushed rock and broken slag are materials most commonly used. Asphalt-cement content generally ranges from about 4 to 7 per cent by weight of dry aggregate, the lower percentages being used for coarser gradings. Construction operations are essentially those of intermediate plant-mix wearing courses except that mixing temperatures are much higher, consistent with the use of asphalt cements.

When a sheet-asphalt wearing course is applied to a bituminous-treated base course recently constructed, the binder course is not always needed and is sometimes omitted. In either case, the surface upon which sheet asphalt is laid should be true to section and longitudinal grade. A template and a 10-ft straightedge is used for testing trueness, and deviations of more than $\frac{1}{4}$ in. are corrected.

**6-2 Character of the surface course** Sheet asphalt provides a very dense, smooth, waterproof surface. A properly constructed mat is highly durable and has considerable resiliency. It presents a traffic surface which is comparatively noiseless and easy to clean. These properties make it particularly well adapted for use on city streets.

Some disadvantages are (1) less stability than other well-controlled hot plant mixes using well-graded but coarser aggregates, resulting in some tendency to "shove" under applied longitudinal thrust of braking or tractive effort of heavy vehicles; (2) a greater tendency to become slippery when wet, for which reason it should not be used on grades steeper than about 5 per cent; (3) the need for precise control of construction operations, both in mixing and laying, making it somewhat more expensive than comparable types of wearing course.

Sheet asphalt is highly serviceable. Its properties of resilience and durability are in fact improved and preserved by continuous traffic, whereas checking and cracking may develop on a surface which is little used. Its greatest acceptance has been for use on busy city streets.

**Table 6-2   Typical aggregate gradings, surface course**

| Passing | Retained on | Percentages by weight |
|---------|-------------|----------------------|
| No. 10............. | ........ | 95–100 |
| No. 10............. | No. 40 | 15–50 |
| No. 40............. | No. 80 | 30–60 |
| No. 80............. | No. 200 | 18–40 |
| No. 200........... | ........ | 0–5 |

**6-3   Sand**   The sand should consist of fairly sharp or angular, hard, durable grains.   Grading may vary within considerable range but practically all should pass the No. 10 sieve.   Grading limits are given in Table 6-2.   These are indicative of common practice.

It will be noted in Table 6-2 that three main size divisions control the grading limits.   Even with the considerable range of percentages permitted in each of the three size groups, it will often be necessary to mix or blend sands from two or more available sources to meet these requirements.   Use of the triaxial diagram as illustrated by Fig. 6-1 will be of help in determining proper proportions.   Use of the diagram is based on

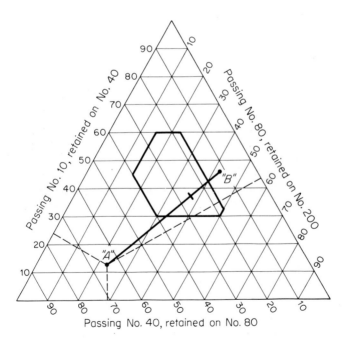

**Fig. 6-1**   Triaxial diagram.

the geometric proposition that the sum of perpendiculars to the sides, from any point within an equilateral triangle, is constant and equal to the altitude of the triangle.   Thus if the altitude of the triangle is taken as 100, each perpendicular distance from any point to a side will represent a percentage portion of the total of the three divisions chosen, each side representing the zero line for one of the three divisions and the opposite apex 100 per cent for that division.   In this case each of the three sides is used as the zero line for one of the size groups or divisions of sand as given in Table 6-1.

Use of the diagram is illustrated by an example.   Let it be required to determine the proportions of sand A and sand B to use in a blend for a sheet-asphalt mixture.   Actual gradings for the separate sands are:

|  | Sand A, per cent | Sand B, per cent |
|---|---|---|
| P10-R40.......... | 23 | 42 |
| P40-R80.......... | 12 | 46 |
| P80–R200......... | 65 | 12 |

The grading of sand A is represented by point $A$ in Fig. 6-1, each of the dotted lines representing a percentage division of the material.   Similarly, the grading of sand B is represented by point $B$ in the diagram. The grading limits for sheet asphalt from Table 6-2 are represented by the heavy lines forming the closed polygon within the diagram.   Any point on or within the periphery of the polygon represents an acceptable grading.   Also, any point on the line connecting $A$ and $B$ represents a possible blend of A and B from 100 per cent A to 100 per cent B.   It will be seen that a blend composed of about 75 per cent of sand B and 25 per cent of sand A will provide a satisfactory grading as indicated.   Some considerable range of proportions may be used, but a grading near the middle of the acceptable range would be used unless supply of one or the other of the sands is limited.   A blend of three or more sands is possible by combining two at a time.

**6-4   Filler**   Filler is material passing the 200-mesh sieve.   It is used primarily as a void filler.   The addition of filler results in a reduction in the optimum asphalt content.   Optimum asphalt content is considered to be the amount required to produce a porosity of 2 to 5 per cent in the finally compacted mixture in the wearing course.

Limestone dust and portland cement have been widely used as fillers. Other stone dusts and diatomaceous earth have also been successfully used. Amount used varies from about 12 to 20 per cent of the total aggregate by weight. According to AASHO, the combined sand and filler should conform approximately to the following grading:

| Passing | Retained on | Percentages by weight |
|---------|-------------|----------------------|
| No. 10.......... | ....... | 95–100 |
| No. 10.......... | No. 40 | 10–40 |
| No. 40.......... | No. 80 | 20–50 |
| No. 80.......... | No. 200 | 15–35 |
| No. 200......... | ....... | 12–20 |

The influence of filler in a sheet-asphalt mixture is to increase stability and toughness. While these properties are highly desirable in the wearing course, the quality of workability must be considered. In order to prevent the formation of a wavy surface it is essential to obtain a uniform spread of the mix on the roadway. This required hand raking, in the earlier days of sheet-asphalt construction, and filler in excess of about 20 per cent made such raking difficult. The use of modern spreading and finishing machines with the auger spreader and vibrating screed finisher has overcome the physical difficulties of raking, but a tendency for "tough" mixes to pull apart or "drag" under the screed still limits the use of filler to a maximum of around 20 per cent of the total aggregate.

**6-5  Asphalt cement**  Successful performance of the sheet-asphalt mat under traffic will depend to a considerable degree on the consistency of asphalt cement used. In cold climates mats containing harder asphalts may tend to crack. This tendency is overcome somewhat by the kneading action of heavy traffic, which maintains integrity and flexibility by a more or less continuous rolling process. In warm climates softening of the cement during prolonged hot weather may contribute to rutting and shoving in the mat. Asphalt cements of penetration grades 50 to 60, 60 to 70, and 70 to 85 are commonly used. Table 6-3 gives recommended consistencies in accordance with general traffic and temperature conditions.

The amount of asphalt cement used should be consistent with the well-established principle of good design that the finally compacted mixture on the roadway should have a porosity of 2 to 5 per cent. Ordinarily the optimum amount will be within the range of about 8 to 12 per cent of the weight of the aggregate. Careful determination of the optimum asphalt content is important, because if there is a deficiency the sheet

**Table 6-3   Recommended normal penetration of asphalt cements for sheet asphalt**

| Traffic | Climate | |
|---|---|---|
| | Hot | Cold |
| Heavy............ | 40–50 | 60–70 |
| Moderate......... | 60–70 | 85–100 |
| Light............ | 85–100 | 120–150 |

asphalt will lack durability and be subject to cracking and raveling even though the mix is stable so far as resistance to displacement under applied load is concerned.   An excess of asphalt may reduce stability to the point where rutting and shoving are imminent while durability remains high.   Good design will provide a proper balance between stability and durability, and it is significant that a variation of half a percentage point in asphalt content either side of optimum will normally be critical.

Any local experience with sheet-asphalt design and construction should be carefully studied.   Past performance may prove to be an excellent guide to best practice, provided any changed conditions such as increased traffic are taken into consideration.

When it is feasible to do so, a test batch should be made up, spread, and compacted with rollers similar to those to be used in actual construction.   Samples may then be taken from this small test plot for determination of bulk specific gravity from which, with the calculated theoretical maximum density, the actual void content or porosity may be determined, in accordance with the procedure discussed in the previous chapter on intermediate types of pavement structure.   Such a test procedure should follow considerable laboratory study with fabricated specimens so that repeated field tests can be avoided.   Any cost connected with such direct field study on a small scale will generally be insignificant when it is considered that a costly type of pavement is to be constructed and that the use of asphalt cements is not conducive to reworking of the mix.

In preparation for preconstruction field testing, or in lieu of such testing, laboratory samples are made, using the particular aggregate grading adopted for the construction job and using different asphalt percentages ranging from about 8 to 12 per cent.   Each sample is made in a 2-in.-diameter mold, using enough mixture to make a compacted height of 1 to $1\frac{1}{2}$ in., making all the samples as nearly uniform in height as possible. Both aggregates and asphalt cement are heated to about 300 to 325°F before mixing. Mixing may be done by hand with a large spoon or by use of small mechanical mixers. Compaction in the mold is accomplished by the application of a static load, usually 3,000 psi, applied for 1 min. The samples are extracted from the molds and, after cooling, are weighed in air and weighed submerged in water to determine the bulk specific gravity obtained for each asphalt content. The asphalt content selected will usually be that yielding a porosity of between 2 and 5 per cent. Field compaction with the use of 8- to 12-ton rollers may give greater or less density than that indicated by laboratory procedure, depending, mainly, on aggregate grading; therefore a close follow-up of tests on samples taken directly from the finished pavement as construction progresses, or from small test plots, should be made.

**6-6 Mixing** Sheet-asphalt mixtures are made in stationary or semi-portable plants. The sand supply usually governs the location of the plant. Close control of proportioning and mixing generally favors use of the batch-type pugmill mixer rather than the continuous flow type for these mixes. The sand is heated to 325 to 400°F and combined with the filler, which is usually not heated. The asphalt cement is heated to 275 to 350°F. The materials are weighed separately into uniform batches, conveyed to the pugmill, and mixed for about 1 min or until all particles are uniformly coated and the mixture is homogeneous. Prolonged mixing should be avoided, because aeration of the hot asphalt cement in the mixer may cause considerable oxidation, with resultant hardening or brittleness. The temperature of the mixture as it is dumped from the mixer into trucks should be between 250 and 325°F.

**6-7 Spreading and finishing** The hot mixture is conveyed to the job and immediately spread with the use of a mechanical finishing or paving machine in a uniform layer to produce the required compacted thickness. The tonnage or weight of mix required for any unit of area is readily determined from previously determined bulk specific gravity of the compacted mix.

**6-8 Compaction** While it is still hot, the mixture is rolled with a 10- or 12-ton roller, beginning at the outside edges of the laid strip with the first passes and proceeding inward toward the center of the strip by overlapping about half the width of the roller on each pass. Diagonal rolling in both directions usually is required in addition to the initial longitudinal rolling. Some hand tamping is usually necessary around curbs and manholes.

Specifications usually require a finished surface smoothness to $\frac{1}{8}$- or $\frac{1}{4}$-in. deviation from a cross-section template or from a 10-ft straightedge placed parallel to the center line.

**BITUMINOUS CONCRETE**

In general, bituminous concrete includes a considerable variety of mixtures using a wide range of aggregate grading, relatively small amounts of filler, and asphalt cement or the heavier grades of refined tar. Two general types are recognized, namely, (1) hot bituminous concrete, made at a central plant on the job under close supervision and control, conveyed to the roadway and laid hot, and (2) cold bituminous concrete, made at a central plant under close supervision and control, shipped to the user and laid cold.

Cold-laid bituminous concretes are made and sold under various trade names. Of those using asphalt cement, Amiesite is the oldest and best known. The aggregate is first treated with a liquefier such as gasoline or kerosene, which keeps the mix workable until it is laid. The treated aggregate is then mixed with hot asphalt cement, loaded into cars or trucks, and shipped to the job. Cutback tars are used in making cold-laid mixtures such as Tarvia, Tarmac, and Slagmac, the last using crushed-slag aggregate.

Principles of good design and construction procedures are essentially the same for both types of mixtures. They are generally laid in two courses—a binder course and a surface course. Total thickness of pavement will usually range from 2 to 3 in. They are laid on stable bases of gravel, crushed rock, soil cement, portland-cement concrete, bituminous-treated bases, or other suitable construction. The binder course is frequently omitted when the so-called "black base" or bituminous-treated base course is used.

**6-9  Aggregates** Hard, tough, nonporous rocks of the trap-rock group are preferred for bituminous concretes. Crushed rock is generally specified for the coarse fractions. Clean sharp sand is generally used to build up intermediate and fine fractions. Thus, high-quality crushed gravel is acceptable aggregate. Fillers similar to those used in sheet asphalt are generally included, but in much smaller amounts. Aggregate gradings specified by AASHO for hot bituminous concrete are given in Table 6-4. Gradings containing sizes of 1 in. and over are used for bituminous-treated bases and binder courses. Maximum size of aggregate for the surface course is usually specified as $\frac{3}{4}$ in. For the gradings of Table 6-4 it is further specified that the fraction retained between any two consecutive sieves be not less than 4 per cent of the total. Percentage of wear is frequently limited to a maximum of 25 as determined by the Los Angeles Rattler Test.      40

**6-10  Bitumen content** The optimum bitumen content for a pavement mixture is that which will afford a proper balance of stability and durability. The best presently available criterion for attaining and main-

**Table 6-4   Aggregate gradings for bituminous concrete**

| Sieve size | Percentages by weight passing | | | |
|---|---|---|---|---|
| | A | B | C | D |
| 2 in. | 100 | | | |
| $1\frac{1}{2}$ in. | 95–100 | 100 | | |
| 1 in. | 75–100 | 95–100 | | |
| $\frac{3}{4}$ in. | 60–90 | 80–100 | 100 | |
| $\frac{1}{2}$ in. | . . . . . . | . . . . . . | 85–100 | 100 |
| $\frac{3}{8}$ in. | 35–65 | 45–80 | 75–100 | 90–100 |
| No. 4 | 25–50 | 28–60 | 50–85 | 70–100 |
| No. 10 | 20–40 | 20–45 | 30–75 | 60–90 |
| No. 40 | 10–30 | 10–32 | 15–40 | 30–70 |
| No. 80 | 5–20 | 8–20 | 8–30 | 10–40 |
| No. 200 | 1–8 | 3–8 | 5–10 | 5–12 |

taining such a balance is the preservation of a small amount of porousness or porosity in the mixture as finally compacted on the roadway. This principle has been previously discussed in the chapter on Intermediate Types of Pavement Structure and in this chapter under the topic of Sheet Asphalt. As a further consideration of this criterion, it is desirable to provide watertightness in the wearing course so that free water from rainfall does not enter the mat appreciably but will readily drain off on the crown slopes. At the same time, engineers generally feel that a pavement should be able to "breathe"; that is, it should be possible for water vapor to escape through the mat, especially in case of considerable vapor pressure from below the mat during hot weather. Cases of blistering or oozing of bitumen to the surface because of the combination of excess bitumen and vapor pressure are not uncommon. Such surface accumulation of excess bitumen will tend to make the surface slippery, especially when wet. Free circulation of air in the mat, however, is not desirable because oxidation of the bitumen makes it hard and brittle. In addition tot hese factors there is the likelihood that some further densification of the mat will occur under traffic subsequent to final compaction during construction. It is desirable, then, so to proportion the mixture that the bituminous concrete mat, on completion of construction, will have a porosity of 4 to 7 per cent. Experience has shown that this provision is consistent with preservation of most desirable properties of the bituminous mat.

**6-11   Laboratory tests**   A fairly reliable test for obtaining an estimate of the optimum amount of bitumen for a given aggregate grading is made by determination of its centrifuge kerosene equivalent (CKE). In this test a 100-g representative sample of that portion of the aggregate passing the No. 4 sieve is placed (dry) in a centrifuge cup fitted with a disk of filter paper. The bottom of the cup is placed in kerosene until the aggregate becomes saturated. The sample is then centrifuged for 2 min at a force of 400 times gravity. The CKE is the amount of kerosene retained, expressed as a percentage of the dry weight of aggregate. If the specific gravity of the aggregate is appreciably different from 2.65, a correction is applied to the CKE according to the scale at the bottom of Fig. 6-2. Using this corrected CKE and the percentage of aggregate passing the No. 4 sieve, the indicated oil ratio (ratio of bitumen to aggregate by weight) is obtained from the alignment chart of Fig. 6-2.

Laboratory-test samples, using the aggregate grading adopted for construction and bitumen contents within a range above and below that indicated by the CKE test, are made for the purpose of determining optimum stability and cohesion in accordance with standard procedures and apparatus such as the open-cell triaxial test, the Hveem Stabilometer and Cohesiometer (or an equivalent measure), the Marshall apparatus, and the Hubbard field test. Standard procedures and equipment for

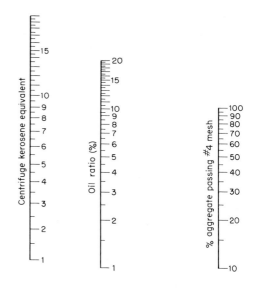

**Fig. 6-2** Chart for converting CKE to oil ratio.

these and other tests are readily available in engineering literature. Tests for water absorption and swell, and determination of porosity of the mixtures will usually supplement stability tests, so that a careful study can be made of all factors contributing to successful performance of each mixture as a wearing course.

Care must be exercised in the fabrication of specimens to obtain orientation of particles and reduction in voids, which are consistent with high densities obtainable in the field by the kneading action of heavy rollers and subsequent traffic. Most fabrication procedures include provision for effective compaction, but it should be recognized that a close follow-up of tests on samples taken from the mixer and from the finished pavement as construction progresses is essential for proper control.

**6-12 Mixing, spreading, and compaction** The close control of temperatures, aggregate grading, bitumen content, and mixing time necessary for construction of high-type bituminous concrete pavements is readily attainable in modern mixing plants of either continuous-flow or batch type. Many features of the mixing plant have been discussed under plant-mix type of intermediate surfacing in Chap. 5. Mixing, spreading, and compaction are essentially the same for bituminous concrete as for sheet asphalt.

**6-13 Seal coat** The regular seal coat is not immediately necessary on a well designed and constructed bituminous concrete mat. This is especially true for the dense-aggregate gradings. Frequently a light spread of dry or bitumen-coated stone chips sized from $\frac{3}{8}$ in. to No. 8 sieve, firmly rolled into the surface to form a tightly knit surface texture, is desirable when the coarser and more open aggregate gradings are used in the mat.

**PENETRATION MACADAM**

Penetration macadam, sometimes called bituminous macadam, designates a wearing course constructed by placing and compacting a layer of coarse broken stone, applying hot bituminous material which is absorbed by penetration, then spreading and working in sufficient choker stone or keystone to fill the voids of the first course and form an effective aggregate interlock, following each such spread with an application of hot bitumen and a final application of stone chips. The thickness of the course is determined by the maximum size of stone used in the first layer so that it is essentially one stone thick, generally $2\frac{1}{2}$ to 3 in. In basic principle it is not different from the armor coat discussed in the preceding chapter. It is indeed a heavy-duty type of surface course, suitable for the heaviest traffic volumes and wheel loads. Economic justification for its use demands that it be placed on a sound, unyielding base.

**6-14 Base course** As in other types of construction, it is desirable that the base course extend across the entire roadway including the shoulders. It should be placed on stable granular material which extends below annual frost penetration and in any case deep enough adequately to protect underlying soil. In other words, the subgrade or sub-base must be of unquestionable quality. Under these conditions the base course is generally made 4 in. in thickness. It is usually made of relatively coarse broken stone ranging from about 3 to 4 in. maximum size to 1 in., rolled to a firm surface with a 10- or 12-ton roller, then choked with bonding fines passing the ½-in. sieve and 70 to 90 per cent passing the No. 4 sieve. The fines are thoroughly drag-broomed and rolled into the interstices of the coarse layer; then excess material is removed with rotary brooms, leaving the larger stones of the first layer projecting slightly above surrounding fines. This provides an interlock with the wearing course, which effectively prevents shoving or bunching.

**Table 6-5 Penetrated base course, using asphalt cement or tar***

| Operation | Bitumen, gal | Aggregate, lb | |
| --- | --- | --- | --- |
| | | Coarse | Key |
| 1st spreading | .... | 285 | |
| 1st application | 1.85 | | |
| 2d spreading | .... | ... | 30 |
| 2d application | 0.30 | | |

Aggregate Grading

| Sieve designation | Percentages by weight | |
| --- | --- | --- |
| | Coarse | Key |
| $3\frac{1}{2}$ in. | 100 | |
| 3 in. | 90–100 | |
| 2 in. | 0–15 | |
| $\frac{3}{4}$ in. | ..... | 100 |
| $\frac{1}{2}$ in. | ..... | 90–100 |
| $\frac{3}{8}$ in. | ..... | 40–75 |
| No. 4 | ..... | 0–15 |
| No. 8 | ..... | 0–5 |

* American Association of State Highway Officials.

When the base course is treated by penetration of hot bituminous material like that used in the wearing course, a somewhat lesser thickness of about 3 to $3\frac{1}{2}$ in. is used.   Construction procedure is similar to that described for the wearing course in subsequent paragraphs.   Tables 6-5 and 6-6 give the sequence of operations, materials required per square yard, and aggregate gradings for penetration macadam construction of base course and surface course taken from AASHO recommended construction practices.

**6-15  Construction of the wearing course**   The first layer of coarse stone is spread uniformly on the roadway from moving trucks by controlling the opening of endgates.   The stone should not be dumped in piles and then spread, because considerable segregation of sizes of stone is bound to

**Table 6-6  Surface course, using asphalt cement or tar***

| Operation | Bitumen, gal | Aggregate, lb | | |
|---|---|---|---|---|
| | | Coarse | Key | Chips |
| 1st spreading......... | .... | 270 | | |
| 2d application......... | 1.50 | | | |
| 2d spreading......... | .... | | 30 | |
| 2d application......... | 0.50 | | | |
| 3d spreading......... | .... | | 25 | |
| 3d application......... | 0.30 | | | |
| 4th spreading......... | .... | | | 15 |

Aggregate Grading

| Sieve designation | Percentage by weight passing | | |
|---|---|---|---|
| | Coarse | Key | Chips |
| 3 in. .......... | 100 | | |
| $2\frac{1}{2}$ in. .......... | 90–100 | | |
| 2 in. .......... | 35–70 | | |
| $1\frac{1}{2}$ in. .......... | 0–15 | | |
| 1 in. .......... | ...... | 100 | |
| $\frac{3}{4}$ in. .......... | ...... | 90–100 | |
| $\frac{1}{2}$ in. .......... | ...... | ...... | 100 |
| $\frac{3}{8}$ in. .......... | ...... | 20–55 | 90–100 |
| No. 4.......... | ...... | 0–10 | 10–30 |
| No. 8.......... | ...... | 0–5 | 0–8 |

* American Association of State Highway Officials.

occur.   Some blading or dragging will be done to make a uniform layer over the area to be occupied by the wearing course.   It is then rolled to a firm, unyielding spread with the use of 10- or 12-ton rollers.

The first application of bitumen may range from 1.5 to 2 gal/sq yd. The figures in the preceding tables for sequence of operations and quantities of materials are good average values but may vary somewhat in accordance with the nature of materials used.   The bitumen is applied with a pressure distributor at a temperature of about 325 to 375°F.   The stone should be thoroughly coated, but an excess of bitumen should not appear at the bottom of the layer.

Keystone is then worked into the layer by thorough drag brooming and rolling to form an effective interlock of aggregate particles.   This will usually require two spreads of keystone, each followed by an application of hot bitumen.

The last application of bitumen and stone chips is essentially the seal coat, which furnishes the actual wearing surface.   Usually a slight excess of chips is spread, rolled, and broomed in the same manner as the keystone, and the surplus chips swept from the surface.   The use of wobbly-wheel, pneumatic-tired rollers is effective in firmly setting the chips so that they are not readily whipped out by traffic.

## QUESTIONS AND PROBLEMS

**6-1.** The gradings of two sands, suitable for sheet-asphalt construction, are as follows:

|  | Sand A, per cent | Sand B, per cent |
|---|---|---|
| P10-R40......... | 15 | 35 |
| P40-R80......... | 40 | 55 |
| P80-R200........ | 45 | 10 |

Select a blend to conform to the limits shown in Fig. 6-1.

**6-2.** A sheet-asphalt mixture is to be made using the following percentages by weight of total mix:

*Per Cent*

Sand (sp gr = 2.67).....................  78
Filler (sp gr = 2.71).....................  12
Asphalt cement (sp gr = 1.01)...........  10

*a.* Find the theoretical maximum density of the mixture.

*b.* A compacted test specimen weighing 1,140 g in air was found to weigh 645 g when suspended in water.   What is the porosity of the compacted specimen?

**6-3.** An asphaltic concrete is made up of the following materials, proportioned by weight as shown:

*Per Cent*

Crushed stone (sp gr = 2.67)............. 45
Sand (sp gr = 2.70).................... 44
Stone dust (sp gr = 2.60)............... 7
Asphalt cement (sp gr = 0.98)............ 4

     *a.* Compute pounds of each ingredient per 1,500-lb batch.
     *b.* Compute yield in cubic feet per batch, allowing 3 per cent of total volume for unfilled voids.

**6-4.** A plant mix is composed of materials combined in the following proportions by weight:

*Per Cent*

Crushed stone (sp gr = 2.70)............. 50
Sand (sp gr = 2.65).................... 39
Limestone dust (sp gr = 2.70)............ 5
Asphalt cement (sp gr = 1.01)............ 6

The compacted mixture has a specific gravity of 2.36 after rolling.
     *a.* Determine the porosity of the compacted mixture.
     *b.* Compute the weight of a 2-in.-thick surface course in pounds per square yard.

**BIBLIOGRAPHY**

"The Asphalt Handbook," The Asphalt Institute, 1965.
Bituminous Concrete Construction, *Highway Research Record* 132, 5 Reports, Washington, 1966.
Bituminous Concrete Mixes, *Highway Research Record* 158, 6 Reports, Washington, 1967.
Design, Performance and Surface Properties of Pavement, *Highway Research Record* 189, 9 Reports, Washington, 1967.
Flexible Pavement Design, *Highway Research Record* 71, 10 Reports, Washington, 1965.

# 7

# Portland-cement
# Concrete Pavement

**7-1 Advantages of use** The modern portland-cement concrete pavement leaves little to be desired as a traffic surface and as a pavement structure. Within a relatively small depth it distributes wheel loads upon the subgrade, and bridges minor inequities in subgrade support by virtue of flexural strength and load-transfer capacity in shear. Maintenance costs are comparatively low. Service records indicate a present-day life expectancy of 30 to 40 years. When it finally does fail or become inadequate, it usually has a high salvage value as a base for a new wearing course. Tractive or rolling resistance is low and nighttime visibility is high.

**7-2 Disadvantages of use** Initial cost is high, limiting its use, along with other high types of pavements, to relatively high traffic volumes and to built-up municipal areas where costs can be distributed effectively. Time consumed in construction plus a curing period of several days causes considerable delay in opening the road to traffic. The need for providing and maintaining joints is a disadvantage over other high-type pavements. Glare from reflected sunlight is objectionable.

**Table 7-1  Typical grading, coarse aggregate**

| Sieve size | Percentages passing |
|---|---|
| 2 in. | 100 |
| $1\frac{1}{2}$ in. | 70–95 |
| 1 in. | 50–80 |
| $\frac{1}{2}$ in. | 10–30 |
| No. 4 | 0–5 |

**Table 7-2  Typical grading, fine aggregate**

| Sieve size | Percentages passing |
|---|---|
| $\frac{3}{8}$ in. | 100 |
| No. 4 | 95–100 |
| No. 8 | 70–90 |
| No. 16 | 45–75 |
| No. 50 | 5–20 |
| No. 100 | 0–5 |

**7-3  Coarse aggregate**  Coarse aggregate consists of crushed stone or gravel having clean, hard, durable pieces, free from friable, thin, or laminated pieces, clay lumps, coal, and other deleterious substances. Air-cooled blast-furnace slag is also sometimes used. Maximum percentage of wear as determined by the Los Angeles abrasion machine is commonly specified at 25 to 40. Aggregate gradings may vary considerably to meet local conditions, and some adjustments will frequently be made on the job to comply with construction methods and requirements for workability. In general, coarse-aggregate gradings for pavement construction will range within the limits given in Table 7-1.

**7-4  Fine aggregate**  The division between coarse aggregate and fine aggregate or sand for concrete work is the No. 4 sieve. Specifications for fine aggregate usually require that the sum of percentages of shale, coal, clay lumps, soft fragments, and other deleterious substances shall not exceed 3 per cent by weight and that it shall be free from injurious amounts of organic impurities as indicated by the standard colorimetric test. In addition, it is usually required that mortar made with the sand develop a strength between 90 to 100 per cent of that for standard Ottawa sand mortar. The strength of pavement concrete in tension is important, and much depends on the use of high-quality sand.

Grading requirements for fine aggregate usually conform to size limits given in Table 7-2.

**7-5  Cement paste**  Portland cement combines with water in a chemical reaction called hydration. The resulting paste hardens over a considerable period of time to form a solid mass comparable in strength to mineral aggregates previously discussed. Adhesive qualities of the paste form a strong bond with clean mineral surfaces to bind them firmly together. Thus when each particle of aggregate in a mixture of aggregate and cement paste is thinly coated with the paste, a new, somewhat heterogeneous, rocklike structure called concrete is formed. It is desirable that a mass so formed be continuous or "solid" to the extent that no air voids will appear in a dried sample other than the finely dispersed voids created by evaporation of excess mixing water beyond that required for hydration of the cement or by controlled air entrainment during mixing. Since the cement paste is the costliest ingredient in concrete, it is important not only to use it to best advantage with respect to amount but also to make it as strong and durable as is consistent with economical design.

**7-6  Water-cement ratio**  The best quality of cement paste—and consequently, concrete which it produces—from the standpoint of strength and durability is made when only enough water is used completely to combine with or hydrate the cement. The minimum requirement for this purpose is approximately $3\frac{1}{2}$ gal of water per sack of cement (a sack of cement contains 94 lb and occupies 1 cu ft, loose volume). However, in order to produce a workable concrete it is necessary that the cement paste be somewhat "liquid." But as more water is used, that is, as the water-cement ratio is increased, the concrete or mortar produced is weaker.

Also, the strength of concrete increases with time or curing period after mixing and placing. These facts are shown graphically in Fig. 7-1.

Water-cement ratio may also be expressed as a ratio by weight, usually as a decimal. For instance, 5 gal of water per sack of cement is equivalent to a water-cement ratio of 0.444 by weight, assuming 1 gal of

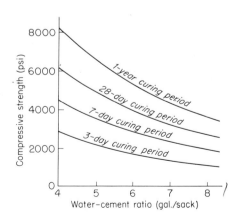

**Fig. 7-1**  Chart illustrating variation of concrete strength with time of curing.

water weighs 8.345 lb.   In practice, water-cement ratios of between 5 and 6 gal per sack of cement are common; 6 gal per sack is most common for pavement construction.   In determining the amount of water to be added to a mix to produce a specified water-cement ratio, allowance must be made for water contained in the sand or coarse aggregate.   Many aggregates are somewhat porous.   Any deficiency or excess of water in the aggregate beyond that required for a saturated, surface-dry condition is adjusted by correction to the amount of mixing water added.

**7-7  Proportioning materials**   The ideal conditions within a concrete mixture from the standpoint of strength and economic use of materials would prevail if materials were so proportioned that pore spaces or voids of the coarse aggregate were just filled with fine aggregate and if sufficient cement paste were furnished to coat all aggregate particles and fill voids in the fine aggregate.   The volume of concrete so produced would then be the volume of coarse aggregate used.   Obviously, such a mix would not only be difficult to produce, but placement in forms, creating effective bond with reinforcing steel, and finishing to a true surface would be impractical, if not impossible.   Workability then is essential for proper mixing, handling, placement, and finishing.

An excess of fine material will overcome harshness by separating coarse-aggregate particles so that interlock and friction between the larger pieces are minimized.   In turn, an excess of cement paste will serve to "lubricate" all aggregate particles so that the concrete will "flow" into proper placement with only a reasonable amount of encouragement from spading, rodding, tamping, and mechanical vibration.   This, of course, works to the detriment of economy, because fine aggregate greatly increases the surface area to be coated with cement paste and because of the extra paste required to increase plasticity of the mix.

It is necessary, then, to establish proper limits of plasticity or con-sistency that will assure workability and still maintain a desirably low water-cement ratio and a reasonably low cement content per cubic yard of concrete produced.   Such a measure of consistency is made possible by the slump test, which is illustrated in Fig. 7-2.   Freshly mixed con-crete is placed by specified procedure in a standard mold, in the form of

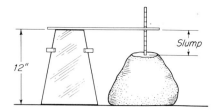

Fig. 7-2   The slump test.

the frustum of a cone, 12 in. in height. The *slump* of the mixture is measured as the reduction in height of the concrete when the mold is withdrawn. A slump of 3 or 4 in. will normally indicate proper control of consistency for pavement construction.

Specifications usually require that concrete for pavement shall have a minimum compressive strength in 28 days of 3,000 to 3,500 psi. With these two factors—compressive strength and slump—in mind, then, and using coarse and fine aggregates representative of those to be used in construction, trial batches are made up in the laboratory with various proportions of coarse and fine aggregates and with a limited range of water-cement ratios, usually between 5 and 6 gal per sack.

In making the slump test and in making the specimens for determining compressive strength, the qualities contributing to workability as affected by additional sand and a larger proportion of cement paste will be readily recognized. From a study of the results of the series of tests the most economical proportions consistent with required strength and workability can then be made.

Proportioning of cement and aggregates at modern mixing plants and on paving jobs is done by weight. As previously discussed, the moisture content of aggregates greater or less than that required to produce a saturated, surface-dry condition must be taken into account in proportioning aggregates and in adding mixing water. Proportions of cement and aggregate are commonly expressed by a numerical sequence such as 1:2.50:3.75 by weight, where the numbers represent cement, surface-dry sand, and surface-dry coarse aggregate, respectively. Cement is generally procured by the sack of 94 lb, or by the barrel, which is 4 sacks or 376 lb.

**7-8  Yield**  The yield of concrete is usually expressed in terms of one sack of cement. It is readily determined by summation of absolute volumes when specific gravities of ingredients are known. The method is best illustrated by use of an example. Let it be required to determine the volume of concrete produced from one bag of cement, using the proportions indicated above, viz., 1:2.50:3.75 by weight, with a water-cement ratio of 6 gal per sack. The specific gravities of cement, sand, and coarse aggregate are, respectively, 3.12, 2.65, and 2.63. Note also that the proportions will be 94:235:352.5 by weight.

The following tabulation of absolute volumes will illustrate the solution for

Cement

$$\frac{94}{3.12 \times 62.5} = 0.48 \text{ cu ft}$$

Sand

$$\frac{235}{2.65 \times 62.5} = 1.42 \text{ cu ft}$$

Coarse aggregate

$$\frac{352.5}{2.63 \times 62.5} = 2.15 \text{ cu ft}$$

Water

$$\frac{6 \times 8.345}{62.5} = 0.80 \text{ cu ft}$$

Total concrete per sack of cement = 4.85 cu ft.

The quantities of materials required per cubic yard of concrete are readily determined as follows for

Cement

$$\frac{27}{4.85} = 5.58 \text{ sacks}$$

Sand

$$5.58 \times 235 = 1,310 \text{ lb}$$

Coarse aggregate

$$5.58 \times 352.5 = 1,960 \text{ lb}$$

Water

$$5.58 \times 6 = 33.5 \text{ gal}$$

Quantities of materials required for the job are then obtainable. It should be emphasized again that these are quantities for surface-dry aggregates, and a close check on moisture content of aggregates will be maintained at the batching plant so that adjustment of weights is made as any significant change in moisture content occurs.

**7-9  Pavement stress**  Modern design of concrete mixtures and universal use of pneumatic tires have practically eliminated any need for consideration of surface wear or abrasion by traffic on portland-cement concrete pavements. Durability, then, as it has been discussed in other types of wearing courses, may be largely taken for granted here. It is obvious, too, that stability or resistance to displacement of material within the pavement itself, while a major factor in flexible pavements, is not significant for rigid pavement. Of particular significance in the design of a rigid pavement slab are those influences which produce tension cracks.

Among the influences which produce tension in a concrete slab are bending or deflection of the slab under wheel loads, warping of the slab due to a difference in temperature of the top and bottom of the slab, warping of the slab due to a difference in moisture content at the top and bottom of the slab, and direct tension produced as contraction of the slab during falling temperature or while drying out is resisted by sliding friction between the slab and subgrade. Of these factors, the first—tension produced by flexure or bending of the slab due to wheel loads—is the stress used as the basis for thickness design. The basic formula for determination of thickness was first suggested by Goldbeck and later developed by Older, Sheets, Westergaard, et al.

**7-10  The Older formula**  Complete steel reinforcement of concrete slabs is generally conceded to be economically impractical. Consequently, thickness design is based on the strength of concrete in tension. The most vulnerable or weakest part of a slab with respect to support of wheel loads is near a corner. The most critical conditions of stress will occur if it is assumed that the wheel load $W$ is applied directly at the corner and that there is no subgrade support at the corner. The slab corner then acts as a cantilever beam. The weakest plane of the section will be the shortest plane; this will make an angle of 45 deg with each side of the corner, as indicated in Fig. 7-3, and a break may occur at some distance $x$ from the corner. The stress produced in the extreme fiber at the plane of weakness is determined by use of the flexure formula $S = Mc/I$, where $S$ is unit stress in psi, the moment $M = Wx$, the distance $c$ from the neutral axis of the section to the extreme fiber $= d/2$, and the moment of inertia $I = 2xd^3/12$. From these relations it develops that

$$S = \frac{12Wxd}{4xd^3} = \frac{3W}{d^2} \qquad (7\text{-}1)$$

The cantilever beam, then, is a constant-strength beam, and a break may occur at any distance $x$ from the corner over which "no subgrade support" might occur.

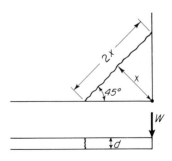

**Fig. 7-3**  Corner break in a rigid slab.

The conditions of no subgrade support and concentration of the wheel load at a point on the corner are drastic assumptions which have been variously modified in practice for application to Eq. (7-1). Most early modifications of the formula were made by arbitrary variations of the constant 3 to values ranging from 1.5 to 2.5, and in this form were generally referred to as the *Older formula*.

**7-11   The Sheets formulas**   Empirical relations for evaluation of subgrade support were developed by Sheets for use in the basic-thickness design Eq. (7-1). Based on the modulus of subgrade reaction $k$ discussed in Sec. 5-2, his formulas take the form

$$S = \frac{1.92Wc}{d^2} \qquad \text{for protected corners} \tag{7-2}$$

and

$$S = \frac{2.4Wc}{d^2} \qquad \text{for unprotected corners} \tag{7-3}$$

A protected corner is one which is provided with a load-transfer device, such as dowels, supporting sills, or aggregate interlock, for connecting two adjacent corners so that each is aided in carrying the load (Fig. 7-13). For protected corners the stress is 0.8 that of the stress for an unprotected corner under the same load. The variable $c$ in the formulas is determined according to the value of $k$ from Fig. 7-4.

The coefficients in Eqs. (7-2) and (7-3) include an impact factor of about 1.2. $W$ is taken as the static wheel load for use in the formulas.

An approximate evaluation of $k$ in accordance with CBR values may be obtained from Fig. 7-5.

**7-12   The Westergaard formulas**   Modern roadbed construction methods with respect to optimum density, uniformity of compaction across the entire roadway, provision of substantial shoulders, and effective drainage make possible a more rational approach to the problem of subgrade

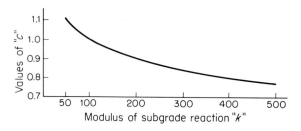

**Fig. 7-4**   Curve for determination of $c$ in Sheets' formula.

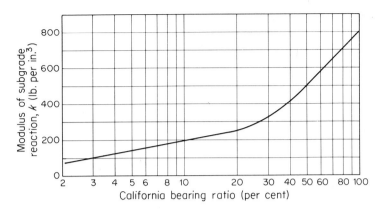

**Fig. 7-5**   Approximate relation of modulus of subgrade reaction $k$ to CBR.   (*Courtesy of Portland Cement Association.*)

support for a pavement slab.   The slab functions as an elastic plate supported on continuous but yielding support.   Thus deflection of the slab due to applied wheel load is accompanied by an equal deformation of the subgrade.   Consequently, net deflection is directly determined by the stiffness of the slab and stiffness of the subgrade, and so becomes a function of the relative stiffness of the slab to that of the subgrade. According to Westergaard, this relationship—called the radius of relative stiffness—takes the form of a linear dimension expressed by the formula

$$l = \sqrt[4]{\frac{Ed^3}{12(1 - \mu^2)k}}$$

where $l$ = radius of relative stiffness, in.
   $E$ = modulus of elasticity of the concrete, psi
   $d$ = slab thickness, in.
   $\mu$ = Poisson's ratio for the concrete
   $k$ = subgrade modulus, lb/cu in.

   Values of $l$ are given in Table 7-3 for various values of $k$ and $d$, with $E = 4,000,000$ psi and $\mu = 0.15$.

   The radius $a$ of wheel-load distribution on a slab-corner loading, as

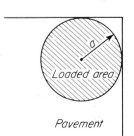

**Fig. 7-6**   Corner loading.

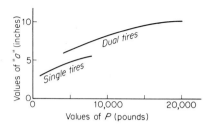

**Fig. 7-7** Curves for determination of $a$ in the Westergaard formulas.

illustrated in Fig. 7-6, may be determined for various wheel loads from the curves of Fig. 7-7.

Upward warping of the slab corner due to differential temperature and moisture content in top and bottom of the slab reduces subgrade support with consequent increased stress in the concrete due to a wheel load on the corner. For this reason the derived Westergaard formulas for stress have been modified to semiempirical formulas which have given results in close agreement with field-measured stresses under a considerable range of field conditions. These modified Westergaard formulas in terms of factors previously discussed are

$$S = \frac{3.36P}{d^2}\left(1 - \frac{\sqrt{a/l}}{0.925 + 0.22a/l}\right) \tag{7-4}$$

for protected corners, and

$$S = \frac{4.2P}{d^2}\left(1 - \frac{\sqrt{a/l}}{0.925 + 0.22a/l}\right) \tag{7-5}$$

for unprotected corners, where $P$ is the static wheel load multiplied by an impact factor, usually 1.2. In terms previously described, $P = 1.2W$.

The required thickness of slab may then be determined by use of Eqs. (7-2) to (7-5) when loading conditions and allowable stress are known.

**Table 7-3  Radii of relative stiffness $l$ in inches**

| Subgrade modulus $k$ | $d = 6$ | $d = 7$ | $d = 8$ | $d = 9$ | $d = 10$ | $d = 12$ |
|---|---|---|---|---|---|---|
| 50 | 34.8 | 39.1 | 43.2 | 47.2 | 51.1 | 58.6 |
| 100 | 29.3 | 32.9 | 36.3 | 39.7 | 43.0 | 49.2 |
| 200 | 24.6 | 27.6 | 30.6 | 33.4 | 36.1 | 41.4 |
| 300 | 22.3 | 25.0 | 27.6 | 30.2 | 32.7 | 37.4 |
| 400 | 20.7 | 23.2 | 25.7 | 28.1 | 30.4 | 34.8 |
| 500 | 19.6 | 22.0 | 24.3 | 26.6 | 28.7 | 33.0 |

**7-13  Allowable design stress**   For pavement design the flexural strength of concrete in tension is determined by use of the flexure formula

$$S = \frac{Mc}{I}$$

the terms of which are described in Sec. 7-10.   The flexural test is made on beams usually 6 in. square in cross section and 30 in. long, cured 7, 14, or 28 days and loaded at the third points in accordance with standard procedure given by ASTM designation C-78-57.   The strength indicated by the flexure formula when the beam breaks under the standard conditions of loading is taken to be the flexural strength of the concrete.   It will be recognized that this is not a true stress because the proportional limit of the concrete has been exceeded when it breaks.   For this reason the flexural strength of the concrete so determined is usually referred to as the "modulus of rupture."

The modulus of rupture or flexural strength of concrete varies with respect to water-cement ratio and curing time in much the same pattern as compressive strength (Fig. 7-1).   Although the relationship between the two is not definite or constant, the modulus or rupture will commonly be around 20 per cent of the compressive strength.   Consequently the modulus of rupture may range from 600 to 800 psi for pavement concrete.

In the past, one-half of the 28-day modulus of rupture was frequently used as the working stress $S$ upon which thickness design was based.   This provided a *factor of safety* of 2, or a *stress ratio* of 0.5, based on the modulus of rupture.   Experience has shown that this often provides a somewhat greater margin of safety than is justified in the best interests of economy.   A better design procedure makes full use of available *fatigue resistance* of the concrete in accordance with the number of load-stress repetitions anticipated for a given traffic condition during the estimated service life of the pavement.

**7-14  Design procedure based on fatigue resistance of concrete**   Comprehensive laboratory research and other studies of fatigue behavior of concrete subjected to repetitions of stress have shown that when the repeated stress does not exceed 50 per cent of the ultimate strength, the concrete will withstand an unlimited number of stress repetitions without failure.   When the repeated stress materially exceeds 50 per cent of the ultimate strength (stress ratio of 0.5 to 1.0), the number of stress repetitions to cause failure decreases as the stress ratio increases.   This relationship between allowable load-stress repetitions and stress ratio is given in Table 7-4.

Design charts for determination of load stresses are given in Figs. 7-8 and 7-9 for single-axle loads and tandem-axle loads, respectively.   These

**Table 7-4  Stress ratios and allowable load repetitions***

| Stress† ratio | Allowable repetition | Stress† ratio | Allowable repetition |
|---|---|---|---|
| 0.51‡ | 400,000 | 0.69 | 2,500 |
| 0.52 | 300,000 | 0.70 | 2,000 |
| 0.53 | 240,000 | 0.71 | 1,500 |
| 0.54 | 180,000 | 0.72 | 1,100 |
| 0.55 | 130,000 | 0.73 | 850 |
| 0.56 | 100,000 | 0.74 | 650 |
| 0.57 | 75,000 | 0.75 | 490 |
| 0.58 | 57,000 | 0.76 | 360 |
| 0.59 | 42,000 | 0.77 | 270 |
| 0.60 | 32,000 | 0.78 | 210 |
| 0.61 | 24,000 | 0.79 | 160 |
| 0.62 | 18,000 | 0.80 | 120 |
| 0.63 | 14,000 | 0.81 | 90 |
| 0.64 | 11,000 | 0.82 | 70 |
| 0.65 | 8,000 | 0.83 | 50 |
| 0.66 | 6,000 | 0.84 | 40 |
| 0.67 | 4,500 | 0.85 | 30 |
| 0.68 | 3,500 | | |

* Portland Cement Association.
† Load stress divided by modulus of rupture.
‡ Unlimited repetitions for stress ratios of 0.50 or less.

design charts, developed by the Portland Cement Association, are based on many considerations, such as the theoretical studies of pavement-slab behavior by Westergaard and others, the findings of full-scale road tests with controlled test traffic, and actual performance records of street and highway pavements subjected to normal mixed traffic.

In the past, a load-impact factor was used for thickness design as described in Secs. 7-11 and 7-12. Road-test findings and actual field-deflection measurements have developed substantial evidence to show that the axle loads of a moving truck cause smaller pavement stresses than the stresses produced when the truck is stopped. Consequently the impact factors formerly used for pavement-thickness design can be more accurately described as load safety factors. The following load safety factors (LSF) are recommended by the Portland Cement Association:

1. For interstate and other multilane projects where there will be uninterrupted traffic flow and high volumes of truck traffic, LSF = 1.2
2. For highways and arterial streets where there will be moderate volumes of truck traffic, LSF = 1.1

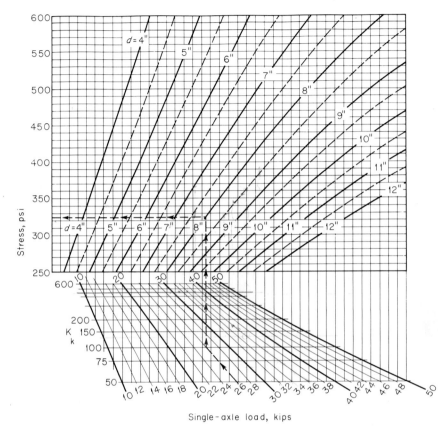

**Fig. 7-8**   Design chart for single-axle loads.   (*Courtesy of Portland Cement Association.*)

3. For highways, residential streets, and other streets that will carry
   small volumes of truck traffic, LSF = 1.0

   Procedure for use of the design charts for pavement-thickness design
is exemplified in Fig. 7-10.   From highway department traffic counts and
loadometer data, the present average daily number of repetitions of a
given single- or tandem-axle load may be determined.   The expected
number of such repetitions during the service life of the pavement is then
readily estimated with the use of a projection factor which reflects probable increase in traffic volume for the particular route in question.   With
present-day design and construction practices, the service life may be
taken as 40 years.   The expected number of repetitions of particular
axle loads are tabulated in column 6 of Fig. 7-10 as illustrated.   Use of
the design charts and Table 7-4 will then give the percentage of available

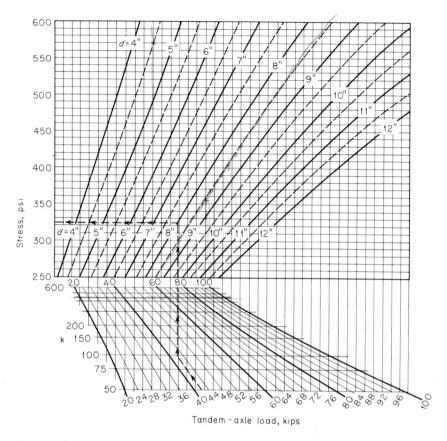

**Fig. 7-9**  Design chart for tandem-axle loads.  (*Courtesy of Portland Cement Association.*)

fatigue resistance used by each category of axle load.  This is recorded in column 7.  The total of column 7 is the total fatigue resistance used by the anticipated traffic in 40 years, using a 9-in. pavement thickness.  The procedure is then repeated using a lesser trial thickness—say 8 in., etc.— until a design thickness is determined which will utilize 100 to 125 per cent of the available fatigue resistance.

The $k$ value (modulus of subgrade reaction) of the subgrade or basement soil will have an appreciable influence on the required pavement-slab thickness.  For soils of low $k$ value (say 100 or less), it may be economical to use an untreated or cement-treated sub-base on which the slab is then placed.  Tables 7-5a and b present useful data for making such economic comparisons where sub-base material, generally gravel, is available near the construction site.

## CALCULATION OF CONCRETE PAVEMENT THICKNESS
(Use with Single & Tandem Axle Design Charts)

Project _DESIGN ONE - A_

Type _Rural Interstate-Rolling Terrain_ No. of Lanes _4_

Subgrade k _100_ pci., Subbase _4-in. Granular Untreated_

Combined k _130_ pci., Load Safety Factor _1.2_ (L.S.F.)

### PROCEDURE

1. Fill in Col. 1,2 and 6, listing axle loads in decreasing order.
2. Assume 1st trial depth. Use 1/2-in. increments.
3. Analyze 1st trial depth by completing columns 3,4,5 and 7.
4. Analyze other trial depths, varying M.R.[*], slab depth and subbase type.[**]

| 1 | 2 | 3 | 4 | 5 | 6 | 7 |
|---|---|---|---|---|---|---|
| Axle Loads | Axle Loads X/.2L.S.F. | Stress | Stress Ratios | Allowable Repetitions | Expected Repetitions | Fatigue Resistance Used[***] |
| kips | kips | psi | | No. | No. | percent |

Trial depth _9.0_ in.   M.R.[*] _650_ psi   k _130_ pci

X PLSF

**SINGLE AXLES**

| 1 | 2 | 3 | 4 | 5 | 6 | 7 |
|---|---|---|---|---|---|---|
| 30 | 36.0 | 340 | .52 | 300,000 | 3100 | 1 |
| 28 | 33.6 | 325 | .50 | Unlimited | 3100 | 0 |
| 26 | 31.2 | | <.50 | " | 6200 | 0 |
| 24 | 28.8 | | " | " | 163,200 | 0 |
| 22 | 26.4 | | " | " | 639,740 | 0 |
| | | | | | | |
| | | | | | | |
| | | | | | | |
| | | | | | | |

X H LSF

**TANDEM AXLES**

| 1 | 2 | 3 | 4 | 5 | 6 | 7 |
|---|---|---|---|---|---|---|
| 54 | 64.8 | 382 | .59 | 42,000 | 3100 | 7 |
| 52 | 62.4 | 368 | .57 | 75,000 | 3100 | 4 |
| 50 | 60.0 | 358 | .55 | 130,000 | 30,360 | 23 |
| 48 | 57.6 | 348 | .54 | 180,000 | 30,360 | 17 |
| 46 | 55.2 | .333 | .51 | 400,000 | 48,140 | 12 |
| 44 | 52.8 | 318 | <.50 | Unlimited | 150,470 | 0 |
| 42 | 50.4 | | " | " | 171,360 | 0 |
| 40 | 48.0 | | " | " | 248,060 | 0 |
| | | | | | | |
| | | | | | | |
| | | | | | | |

_Total = 64_

[*] M. R. Modulus of Rupture for 3rd pt. loading.

[**] Cement-treated subbases result in greatly increased combined k values.

[***] Total fatigue resistance used should not exceed about 125 percent.

100 - 125

**Fig. 7-10** Design thickness calculation. (*Courtesy of Portland Cement Association.*)

**Table 7-5a  Effect of untreated sub-base on** $k$ **values, pci***

| Subgrade $k$ value | Sub-base $k$ value | | | |
|---|---|---|---|---|
| | 4 in. | 6 in. | 9 in. | 12 in. |
| 50 | 65 | 75 | 85 | 110 |
| 100 | 130 | 140 | 160 | 190 |
| 200 | 220 | 230 | 270 | 320 |
| 300 | 320 | 330 | 370 | 430 |

**Table 7-5b  Design** $k$ **values for cement-treated sub-bases***

Subgrade $k$ value—approximately 100 pci

| Thickness, in. | $k$ value, pci |
|---|---|
| 4 | 300 |
| 5 | 450 |
| 6 | 550 |
| 7 | 600 |

* Portland Cement Association.

It is interesting to note that there is an equivalency between the single-axle load and the tandem-axle load which is at least implied in the two design charts of Figs. 7-8 and 7-9. Note for instance that for a stress of 300 psi in a 7-in. slab on a soil with a $k$ value of 100, a 32-kip tandem-axle load is equivalent to a single-axle load of a little more than 19 kips, indicating an equivalency ratio of about 0.6. This agrees reasonably well with the ratio of 0.57 as discussed in Sec. 5-5.

**7-15  Purpose of pavement joints**  It is axiomatic that a continuous pavement slab will crack. Pavements are exposed to severe temperature and moisture conditions which in themselves may combine to produce excessive stresses sufficient to crack the slab, primarily because of warping produced by temperature and moisture differences between the top and bottom of the slab. The slab is thus obliged to raise its own weight off the subgrade, and when a dimension is of excessive length, the weight cannot be supported by the relatively thin section. Direct tension in the slab due to more or less uniform contraction also develops cracks. Compression failures and buckling or "blowups" also occur because of expansion of the slab. Experience has shown that a continuous concrete-pavement slab will crack under these adverse influences combined with

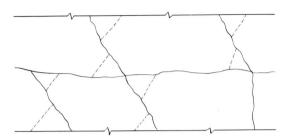

**Fig. 7-11**   Pavement cracks.

wheel-load stresses at fairly regular intervals of 12 to 18 ft, depending on the degree of uniformity of contributing factors, particularly subgrade support.

When a continuous slab is constructed across two traffic lanes, a more or less irregular longitudinal crack is likely to develop as heavy wheel loads occur simultaneously in both lanes at a cross section of the roadway, thus creating severe bending moments, particularly at such times when even a slight warping of the pavement is present or when transverse contraction of the slab is taking place.   Also, many of the transverse cracks which develop at more or less regular intervals across the pavement are likely to make acute angles with the free edges and the longitudinal crack (see solid-line cracks of Fig. 7-11).   The acute-angle corners then become weak sections which break further under heavy wheel loads, producing cracks, as indicated by the broken lines of the figure.

To prevent such progressive breakup of the pavement and for control of expansion and contraction of the slab, longitudinal and transverse joints are placed.

**7-16   Longitudinal joints**   Three types of longitudinal joints are shown in Fig. 7-12.   They may properly be considered as "hinge" joints, dividing the pavement into 10-, 11-, or 12-ft lanes.   Tie bars are used to hold the slabs together as alternate expansion and contraction tend to cause separation or opening up of the joint, thus admitting water and sediment from the roadway surface.   Size and spacing of the tie bars are determined so that they will develop sufficient strength to overcome frictional resistance of the pavement slab on the subgrade at such times when contraction of the slab is taking place.   Deformed steel reinforcing bars $\frac{3}{8}$ to $\frac{5}{8}$ in. in size are used for this purpose.   Length of bars and spacing are illustrated by the following example.

A concrete pavement 24 ft wide and 6 in. thick is to be provided with a center longitudinal joint using $\frac{1}{2}$-in.-square tie bars.   The unit weight of the concrete is taken as 144 pcf.   The coefficient of friction of the slab on

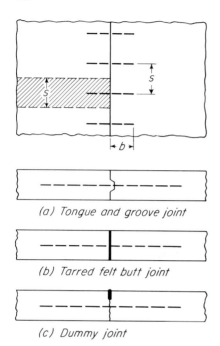

(a) Tongue and groove joint

(b) Tarred felt butt joint

(c) Dummy joint                    **Fig. 7-12**  Longitudinal joints.

the subgrade may vary from 1 to 2.5; a value of 2 is commonly used for
design purposes.  Assuming an allowable working stress in tension
for the steel bars at 20,000 psi, the strength of one bar is 5,000 lb.  Then

$$2 \times 144 \times \frac{6}{12} \times 12 \times \frac{S}{12} = 5,000$$

where $S$ is the spacing of bars, in inches, from which $S = 35$ in. (see
shaded area of Fig. 7-12).

The length of the tie bar is determined by its strength in bond with
the concrete.  Assuming an allowable working stress in bond of 120 psi,
the dimension $b$ in Fig. 7-12 (one-half the required length) is made sufficient to develop the full strength of the bar in tension.  Since the perimeter of the bar is 2 in., then

$$2b \times 120 = 5,000$$

from which $b = 21$ in., making the required length 42 in.

When one lane at a time is poured, either the tongue-and-groove
type of joint or the butt joint is generally used.  The joint surface of the
previously constructed lane is painted with tar or asphalt, or a bituminous-treated felt strip is provided to break the concrete bond and yet
form a watertight joint.

When two lanes are poured at once, the dummy joint is usually made. A plane of weakness is created by making a longitudinal groove, usually about 2 to 3 in. deep, cut by a tool attached to the finishing machine. The groove is later filled with bituminous material to within about $\frac{1}{4}$ in. from the top of the slab before the pavement is opened to traffic.

Generally not more than three lanes are tied together with tie bars because longitudinal cracking may occur at the ends of the bars. One or more properly sealed free longitudinal joints are provided when the pavement is more than three lanes in width.

**7-17 Transverse expansion joints** A transverse joint, usually about $\frac{3}{4}$ in. in width, is provided at intervals to relieve temperature expansion and increase in length due to moisture absorption. A joint filler will be required to make the joint watertight. Premolded fillers of such materials as felt, cork, or wood fiber impregnated with bituminous material, and rubber have been successfully used. Poured joints, usually of heavy grades of bituminous materials, sometimes with fibrous or resilient materials in suspension, are also used. Details of the joint are shown in Fig. 7-13.

Engineers are not in agreement as to proper spacing of expansion joints. State specifications in this regard may vary from 50 ft to several hundred feet. Recent trends have been toward greatly increased spacings of 200 to 500 ft.

Load-transfer devices, usually in the form of steel dowels (Fig. 7-13), are used to prevent vertical displacement of the slab on one side of the joint with respect to the adjoining slab. The dowels are designed to transfer 50 per cent of the design wheel load across the joint. Recent experience has shown that the length of the dowels should be from 12 to 15 in., so that the dowel is embedded 6 or 7 in. into the slab on either side of the joint. At least one end of each dowel must be greased, painted, or provided with a sleeve to prevent bonding with the concrete so that the joint may function to relieve expansion of the pavement. One end must, also, be provided with an expansion cap—a short metal sleeve, cork, or similar compressible material. Steel reinforcing bars and standard steel pipe are commonly used as dowels. Load-transfer capacity of various dowels as commonly used are given in Table 7-6 for joint openings of $\frac{1}{2}$ to 1 in.

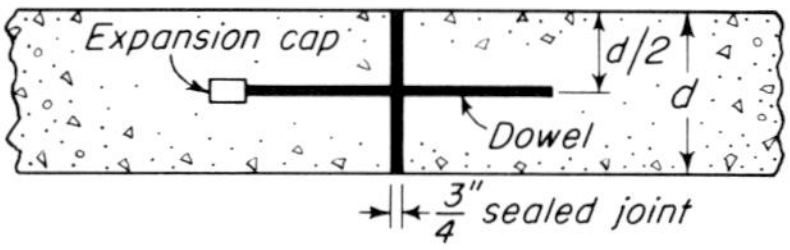

**Fig. 7-13** Expansion joint.

**Table 7-6   Load-transfer capacity of dowels**

| | Load-transfer capacity, lb | | | Recommended length, in. |
|---|---|---|---|---|
| | $\frac{1}{2}$-in. joint | $\frac{3}{4}$-in. joint | 1-in. joint | |
| $\frac{3}{4}$-in. bar............... | 1,200 | 1,100 | 1,000 | 12 |
| $\frac{3}{4}$-in. pipe............... | 1,800 | 1,800 | 1,600 | 12 |
| 1-in. bar............... | 2,200 | 2,100 | 2,000 | 13 |
| 1-in. pipe............... | 2,800 | 2,700 | 2,600 | 13 |
| $1\frac{1}{4}$-in. bar............... | 3,400 | 3,300 | 3,200 | 14 |
| $1\frac{1}{4}$-in. pipe............... | 4,400 | 4,200 | 4,000 | 14 |
| $1\frac{1}{2}$-in. bar............... | 5,000 | 4,800 | 4,600 | 15 |
| $1\frac{1}{2}$-in. pipe............... | 5,600 | 5,400 | 5,200 | 15 |

The number of dowels which are operative in transferring 50 per cent of the design wheel load across the joint depends on the amount of deflection of the slab, which in turn is a function of the radius of relative stiffness $l$ (Table 7-3). Friberg[1] describes a method for dowel spacing in which a dowel directly under the wheel load is assumed to develop its full load-transfer capacity. The effectiveness of neighboring dowels is directly proportional to their distances from the load and is zero for a dowel placed a distance equal to $1.8l$ from the load. The most severe condition will prevail at a corner, as illustrated in Fig. 7-14. The first dowel will generally be placed a minimum distance of 6 in. in from the outside edge of the pavement. With the wheel load $W$ placed on one side of the joint directly over it, this dowel will transfer its full capacity,

[1] Friberg, Bengt F., Design of Dowels in Transverse Joints, *Trans. ASCE,* **105** (1940).

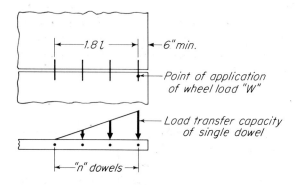

**Fig. 7-14**   Spacing of dowels.

as determined from Table 7-6.   The average load transfer of the $n$ dowels placed in the distance $1.8l$ is then the full load-transfer capacity of a single dowel divided by 2, and $n$ times this average must equal $\frac{1}{2}W$.   It follows that the required number of dowels to be placed by equal spacing in the distance $1.8l$ is $W/$(load-transfer capacity of one dowel), and the required spacing is given by $1.8l/(n-1)$, in inches.   This spacing is then carried across the entire joint.   The size of dowel should be selected so that the spacing will be about 12 to 15 in. for effectiveness and for convenience in placing.

**7-18  Contraction joints**  Contraction joints, placed primarily to control cracking due to warping stresses and tension produced by contraction of the slab, are usually of the dummy type in which a groove about $\frac{1}{4}$ in. wide and 2 in. deep is placed in the fresh concrete at right angles with pavement edges with a template or other grooving device.   This creates a plane of weakness along which cracks may be expected to develop.   The groove is then sealed, usually with bituminous materials or mixtures, before the pavement is opened to traffic.   Specifications usually require that they be placed at intervals of from 12 to 20 ft apart.   Best practice indicates that if contraction joints are placed at intervals of 15 to 18 ft or less, no dowels are necessary.   This closer spacing assures that the crack will not open excessively, because there are more cracks to take up any contraction of the slab which may take place.   Thus load transfer across the crack in the dummy joint is provided by aggregate interlock still inherent in the slightly opened crack.   For dummy-joint spacing of 20 ft or more, dowels should be provided as for the transverse expansion joint. Dummy joints are comparatively easy to place with a grooving tool or template as the fresh concrete is finished.   Sawed joints, which are generally preferred, are usually cut to a minimum depth of one-sixth of the slab thickness for the transverse contraction joints, and one-fourth of the slab thickness for the longitudinal joint to ensure control of cracking. They should be cut after the initial set of the concrete has taken place, early enough to control cracking but late enough to prevent raveling by the saw.   Best practice indicates that spacing of 15 ft or less should be used.   Among the advantages to be gained by closer spacing of dummy joints are:

1. Warping stresses are greatly reduced.
2. Crack opening is reduced, thus preserving the load-transfer capacity of the slab without the use of dowels.
3. The numerous cracks so formed provide some relief for expansion of the pavement, thus increasing the spacing required for transverse expansion joints, which are far costlier to make.

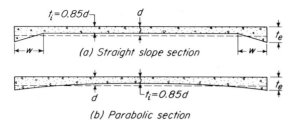

**Fig. 7-15** Pavement cross section.

**7-19  Edge thickening of concrete pavement**  Design of the pavement
slab is made on the basis of uniform thickness, and present practice
appears to be inclined to favor construction of such a slab.  There is,
however, some inherent weakness in a slab edge which favors extra thickening of the edge beyond the thickness $d$ determined in design and accompanying this increase with a reduction of interior thickness, as illustrated
in Fig. 7-15.  Sheets, in 1933, developed edge-thickening requirements
for the straight-slope thickened-edge section, as illustrated in part $a$ of
the figure and for the parabolic section illustrated in part $b$.  For the
straight-slope section when the thickening is increased uniformly through
a slope width $w$ of 2 ft, his development leads to the recommendation that
the interior thickness $t_i$ may be reduced to $0.85d$, and the edge thickness $t_e$
increased to $1.275d$ or $1.5t_i$.  These recommendations, with some modifications in different states, have been widely used.  The saving in concrete
per mile of pavement resulting from such a more nearly balanced design
is significant and may be readily computed from the foregoing relations.
It should be realized, however, that the extra cost of roadbed shaping for
the thickened edge and the need for larger forms tend to offset the saving
in concrete, so that from the standpoint of over-all cost per mile the
thickened-edge section has no great advantage over the section of uniform
thickness.  There is added argument also for the use of uniform cross
section in that modern 11- and 12-ft lanes minimize traffic use of the
pavement edges.  Further, the strength of the slab against a corner break
is not increased by edge thickening and reduction of interior thickness in
proportions that will result in the use of less concrete.

**7-20  Reinforcement of concrete slabs**  It is not generally considered
economically practical to provide steel as true reinforcement in a pavement slab.  However, when greater spacing of contraction joints is used,
and in cases of adverse moisture conditions and frost action affecting
stability of the subgrade, it is often advisable to reinforce the slab with
welded-wire mesh or bar mats, not for the purpose of preventing cracks,

but to prevent them from opening up. Local experience will largely determine whether or not such reinforcement should be used.

In general, the welded-wire fabric is more effective than the bar mats because of the closer network of steel. The reinforcement is placed about 2 in. below the surface of the pavement, and it should not cross any free joint. Details of typical horizontal arrangement of a bar mat are shown in Fig. 7-16.

To determine the required area $A$ of steel per foot of one slab dimension, the steel should be strong enough in tension to "drag" one-half of the other dimension of slab against the resisting friction of the subgrade, in the same way that tie bars are designed for the longitudinal joint. Thus if $l$ is the length of the slab between transverse joints, $W$ the weight of concrete in pounds per square foot of surface area, $f$ the allowable tensile stress of steel in pounds per square inch, and the coefficient of friction between the slab and subgrade is assumed to be 2, then the required cross-section area $A$ of longitudinal steel in square inches per foot of slab width is given by $A = lW/f$. The required area of transverse steel per foot of length of slab may be determined in the same manner. Bar sizes most commonly used are $\frac{1}{4}$-in. and $\frac{3}{8}$-in. round or square reinforcing rods. The bars should preferably be welded at intersections. Spacing of longitudinal wire in welded-wire-mesh fabric is usually 6 in., and spacing of transverse wires from 6 to 12 in., depending on slab dimensions. Size or gauge of wires is then determined by the required cross-section area of steel required per foot of slab dimension.

Extra reinforcing bars are sometimes placed along edges of the slab between joints to prevent opening of edge cracks. This may be done whether or not mesh reinforcement is used.

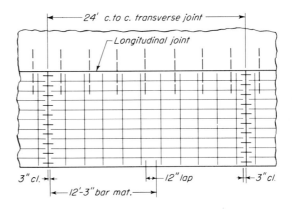

**Fig. 7-16**  Bar-mat reinforcing.

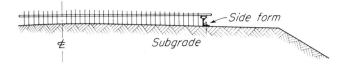

**Fig. 7-17**  Use of metal template to check accuracy of roadbed shaping.

**7-21   Roadbed preparation**   Accurate shaping of the roadbed to crown section and grade preparatory to laying concrete pavement is essential for precise control of slab depth.   Concrete-pavement crown is usually not more than $\frac{1}{10}$ to $\frac{1}{8}$ in./ft.   This crown is built into the subgrade surface so that the subgrade becomes the datum surface for effective depth control.   When the parabolic pavement section or edge thickening is used, only careful shaping of the subgrade to conform to the required shape will ensure a uniform pavement section.

Side forms are set true to line and grade and are then used as rails to support planers and templates as they are moved along the roadbed. Figure 7-17 illustrates the use of a template with heavy adjustable spikes on a rigid frame to check accuracy of roadbed shaping as blade graders and planers operate to trim the surface to true section between the forms. Obviously, the forms must be set and staked firmly because spreading and finishing machines will also use the forms as rails during paving operations.   The importance of thorough compaction of the subgrade clear across the roadway again becomes significant, not only to give firm support to pavement forms but also to give firm support to pavement edges throughout the service life of the pavement.   The subgrade is usually left a little high during grading construction, and excess material is removed by planing or blading so as to avoid filling in low spots with loose material.   Such excess material, if of good quality, may be placed outside the forms to build up the shoulders.   As in other types of pavement construction, best practice dictates that uniformity of subgrade support is essential to successful pavement performance.   Alternate hard and soft spots or areas in the subgrade are bound to contribute heavily to cracking of a rigid pavement.

The subgrade should be moist when the pavement is laid so that mixing water is not absorbed from the fresh concrete.   It will usually be sufficient to sprinkle fine-grained soils such as heavy clay or gumbo immediately ahead of the paver, but more porous soils will usually be wet down thoroughly the preceding day or evening so that no appreciable amount of water will escape from the fresh concrete during the critical early curing period.   Considerable care must be exercised in those cases where subgrade soil is subject to "swell" when it becomes moist or wet.

Such volume changes can cause severe damage to "green" concrete while it is in the curing stage. Many states require that such subgrades be covered with building paper before the pavement is placed so that the underlying soil will not be subjected to moisture from the fresh concrete.

**7-22 Mixing and placing concrete** Dry materials for concrete are proportioned at central batching plants in accordance with design requirements and hauled to the job in motortrucks holding from one to five batches. Each batch is contained separately in the truck box by substantial dividers so that one batch at a time may be deposited in the paver "skip" on the job. The skip, in the form of a large scoop, raises the batch of dry materials into the mixing drum.

The paver is a self-propelled machine usually mounted on caterpillar traction and equipped with gas or diesel engines or electric motors for moving and operating the machine directly on the subgrade. It is equipped with accurate control devices for measuring water into the mixer and automatically regulating time of mixing for each batch.

In general, increased time of mixing improves the quality of concrete, but practical considerations limit the time to about 1 min. States vary considerably as to the required minimum mixing time—commonly between 45 and 90 sec.

Experience has shown that for proper control and efficient operation it is desirable to limit the capacity of the mixing drum to maximum batches of about $1\frac{1}{4}$ cu yd. The capacity of the paver is designated by the size of batch in cubic feet that it is designed to handle. This number is followed by the letter E to complete the designation. Thus a 21-E paver is designed to handle batches producing 21 cu ft of mixed concrete. Common sizes may vary from 21-E to 34-E.

Many of the new pavers are equipped with dual mixing drums. Each batch is mixed one-half the required mixing time in one drum and then transferred by continued mixing action to the second drum, where mixing is completed. With this provision the capacity of the paver is doubled.

It is important that the freshly mixed concrete be deposited on the subgrade as nearly as possible in the position which it is to retain, so as to avoid excessive shoveling and raking which might result in segregation and nonuniform texture. To accomplish uniform spreading, the concrete is emptied from the mixing drum into a bottom-dump bucket, which moves out on a boom. The boom can be swung in a horizontal plane over a considerable area of subgrade. It is fixed in position, and as the bucket is run out on the boom, the concrete is dumped so that it is spread over considerable length of run, making a uniform depth of spread on the subgrade.

When bar-mat or wire-mesh reinforcement is to be used in the slab, one-half the depth of the slab is poured and struck off with a separate plate or template attached by hinges to the finishing machine. The template may then be raised or lowered accordingly as final finishing or striking off for placement of reinforcement is to be done.

The finishing machine operates on the forms which serve as rails. It has a screed which can be set to strike off the concrete slightly above the finish grade. This first screed is followed by a second screed or tamper. Both operate with a combined transverse and longitudinal motion to strike off and compact the concrete. The machine may also be equipped with a finishing belt. Some machines also have vibrating elements to compact effectively the concrete. A second pass is usually made with the machine to provide the required true surface. Some handwork to tamp the concrete next to the forms and to true up the surface with strike boards, floats, and similar tools is usually required. A broom finish is frequently given to the surface to improve traction by providing slight transverse grooves to the concrete surface.

The slip-form paver replaces the machines of earlier conventional paving trains, except the mixer, and eliminates the use of side forms. The machine spreads the concrete over the subgrade, vibrates and tamps it, strikes it off, and shapes it to the desired crown and thickness. Attached to the paver are mechanical transverse reciprocating rubber belt and burlap for final finish.

The concrete for slip-form paving should be uniform in consistency, and the slump should not exceed $1\frac{1}{2}$ in. Air entrainment (generally 4 to 6 per cent) will protect the pavement against the effects of freezing and thawing and will provide good workability of the concrete at the specified slump.

**7-23 Curing** During a curing period—usually from 7 to 21 days— the concrete must be protected from undue loss of water. This may be accomplished in several ways. The effectiveness of the method used will necessarily be dependent on climatic factors such as hot dry winds, rainfall, and humidity.

Frequently the initial curing—the first day—is accomplished by placing wet burlap on the slab as soon as the concrete is set sufficiently so that it is not easily marred. The burlap is kept wet by sprinkling. The following day, or sooner if the concrete has hardened sufficiently, the burlap is replaced by less costly material, such as earth from the roadway shoulders or straw which is kept moist for the rest of the curing period. Earth dikes may also be built around sections of pavement slab and the enclosures flooded with water. This method is called ponding. When a pressurized water system can be tapped, as in municipal areas, a sprinkling system may be set up to keep the pavement surface moist.   Common modern methods of curing concrete merely prevent loss of moisture by sealing the surface.   This may be done by means of waterproof paper, plastic sheets, or liquid membrane-forming compounds.

## QUESTIONS AND PROBLEMS

**7-1.** On a paving job the following materials are to be used in proportions 1:2.2:2.3:1.2 by weight with a water-cement ratio of 0.8 cu ft per sack: cement, sp gr = 3.05; sand, sp gr = 2.67; gravel A, sp gr = 2.60; and gravel B, sp gr = 2.65.

  *a.* Compute the yield for a 7-sack mixer, in cubic yards per batch.

  *b.* Determine quantities of materials required per mile of pavement using 8-in. uniform thickness and 24-ft width.

  *c.* Allowing $1\frac{1}{2}$ min per batch, find the length of pavement that can be laid in an 8-hr day.

  *d.* The sand has a loose weight of 115 pcf and a moisture content of 12 per cent based on surface-dry weight.   How many cubic yards of sand will be needed per mile of pavement?

  *e.* Adjust the proportioning of materials to take into account the moisture in the sand.

**7-2.** Determine thickness (uniform) of a concrete pavement to carry an 18,000-lb axle load on dual tires using Sheets' formula for doweled joints with modulus of rupture of 700 psi and modulus of subgrade reaction of 100 psi per inch.   Assume a factor of safety of 2.

**7-3.** Using the modified Westergaard formula, find the maximum tensile stress at the top of a slab corner for the pavement designed in Prob. 7-2.

**7-4.** Make a balanced thickened-edge design for the pavement of Prob. 7-2 in accordance with Sec. 7-19 and determine the saving in concrete in cubic yards per mile for two 10-ft lanes.

**7-5.** Complete the thickness design for the traffic data given in Fig. 7-10 to the nearest $\frac{1}{2}$ in. using 100 to 125 per cent of available fatigue resistance.   Assume a $k$ value of 100 for the subgrade soil and make the thickness designs for the following conditions:

  *a.* No sub-base
  *b.* 4 in. of granular untreated sub-base
  *c.* 5 in. of cement-treated sub-base

**7-6.** A pavement slab 48 ft wide and 8 in. thick is laid in three strips, each 16 ft wide. The coefficient of friction between pavement and subgrade is 1.40, allowable tensile stress in steel is 20,000 psi, and allowable bond stress between concrete and steel is 110 psi.   Determine the length and spacing of $\frac{5}{8}$-in. round tie bars.

**7-7.** Design size and spacing of dowels across a $\frac{3}{4}$-in. expansion joint.   Slab is 8-in. uniform thickness.   Dowels are to transfer 50 per cent of a 10,800-lb load (9,000 + impact).   Subgrade modulus is 200 psi per inch.   $E$ is 4,000,000 psi.   Poisson's ratio is assumed to be 0.15.

**7-8.** Determine lateral and longitudinal spacing of $\frac{1}{4}$-in. square bars in a bar mat for the pavement of Prob. 7-6 if transverse joints are placed at 18-ft intervals.

**7-9.** Estimate the water required per day for the paving operation described in Prob. 7-1.   Express results in gallons and include water required for ponding.   Contractor works a 5-day week.   Evaporation and other losses amount to $\frac{1}{2}$ in. in 24 hr.   A 7-day curing period is specified.

**7-10.** A concrete pavement is designed to carry a 12,000-lb wheel load with a factor of safety of 1.8.   From Table 7-4, how many repetitions of a 15,000-lb wheel load will cause the slab to fail?

## BIBLIOGRAPHY

Aggregates for Concrete, *Highway Research Record* 124, 1966.
Portland Cement Association, "Concrete Pavement Inspector's Manual," 1959.
Portland Cement Association, "Design and Control of Concrete Mixtures," 1960.
Design, Performance and Surface Properties of Pavements, *Highway Research Record* 131, 12 Reports, 1966.
Oglesby and Hewes, "Highway Engineering," John Wiley & Sons, Inc., New York, 1963.
Report of the Committee on Concrete Pavement Design, *Am. Road Builders Assoc. Tech. Bull.* 121, 1947.
Slip-form Paving in the United States, *Am. Road Builders Assoc. Tech. Bull.* 263, 1967.
A Survey of Pumping in Illinois, *Highway Research Board Research Rept.* 1D, 1948.
Symposium on Slip-form Paving, *Highway Research Record* 98, 1965.
"Thickness Design for Concrete Pavements," Portland Cement Association, 1966.
Use of Air-entraining Concrete in Pavements and Bridges, *Highway Research Board Bull.* 13, 1946.

# 8
# Drainage

**8-1 General considerations** Drainage is admittedly among the most important problems in the field of design and construction of transportation facilities. It is not discussed here in the order of its importance but has rather been deferred to this late chapter because the significance of good drainage is not fully appreciated until the effects of moisture on subgrade soil properties, on base course and ballast materials, on flexible mats, and on rigid pavements are better understood.

The solution of the drainage problem begins with location, whether it be for a highway, an airport, or a railroad. The adverse effects of excessive moisture on foundation soils with resultant detrimental influences on superstructure should be fully appreciated by the locating engineer and should be an ever-present guide to his judgment, so that he may balance these considerations with good alignment and permissible gradients. Many situations involving expensive control of underground water can often be avoided, and strategic crossing of natural streams may often dictate precise-location control points. Two main divisions of drainage will be discussed here. They are (1) surface drainage, which is

153

concerned with runoff water from rain and melting snow, and (2) interception and control of underground water, which may flow laterally under the influence of gravity or rise vertically by capillarity to soften foundation soils and impair superstructure.

**8-2  Surface drainage**  On straight portions of the roadway a crown is built into the cross section so that water will readily drain off into the side ditches.  Normally the transverse slope provided is small and is often in the form of a curve, such as the parabola.  In general, for efficient removal of surface water a slope of $\frac{1}{4}$ in./ft or slightly more is provided on stabilized soil and gravel surfaces.  This is equivalent to a transverse grade of about 2 per cent.  On bituminous surfaces, slopes of $\frac{1}{8}$ to $\frac{1}{4}$ in./ft are common, and on portland-cement concrete pavements transverse slopes seldom exceed $\frac{1}{10}$ to $\frac{1}{8}$ in./ft.  The transverse grade or slope is increased somewhat on steep longitudinal grades so that surface water drains to side ditches before it travels excessively long distances along the roadway.  This becomes more critical on soil-stabilized wearing courses and on shoulders.  For this reason transverse slopes on shoulders are generally made about 4 or 5 per cent and should not be appreciably less than the steeper longitudinal grade.  Superelevation on curves will usually provide for surface drainage to the inside ditch or the inside slope on fill sections.  The outside shoulders, however, are sloped toward the outside ditch.

It is desirable to keep fill slopes and fore slopes in cut sections as flat as is economically possible for reasons of safety and also to minimize erosion due to surface runoff from rainfall.  Slopes of 4 (horizontal) to 1 (vertical) have become standard for main arterial highways, with 3 to 1 and 2 to 1 common for lower-cost roads.  When slopes steeper than those specified for the particular highway in question must be resorted to, as in the case of high embankments, a guardrail is usually provided for safety; but it should be recognized that maintenance of steeper slopes to combat erosion becomes a significant item of cost, depending on the nature of the soil and the convenience with which sodding or other stabilizing methods may be employed.

In cut sections it will often be necessary to protect the back slope on one side where there is any appreciable drainage down to the back slope from higher ground above the cut.  This diversion ditch, as illustrated in Fig. 8-1, should carry runoff water on an easy gradient to the stream channel.  It is important in diversion ditches and roadside ditches as well that scouring be prevented.  To this end the velocity of the water should be kept below about 3 ft/sec in soils which are likely to erode or scour.  Since velocity of water in a channel or conduit is a function of depth of flow as well as slope, it is often necessary to use a wide-bottom

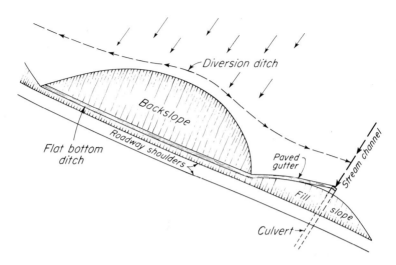

**Fig. 8-1**  Elements of surface drainage.

ditch instead of the V ditch where a considerable quantity of water has to be carried on a grade. Good turf is an excellent water-velocity retarder, and its use in flat-bottom ditches will often prove an effective means of erosion control for handling intermittent flow of runoff to natural stream channels. In general, it is desirable to provide a minimum grade of 0.5 per cent in ditches for effective drainage so that water will not stand in puddles beside the roadway to any appreciable degree. To meet minimum and maximum requirements of flow in side ditches, separate ditch grades independent of the roadway centerline grade frequently must be constructed.

Paved ditches or gutters and metal or wood flumes are used where excessive grades for flow of water cannot otherwise be avoided. Such occasions may arise in the case of conveying water from the ditches in the transition zone between cut and fill to the stream channel (Fig. 8-1), from roadway surface to ditch over steep slopes of erodible soil, or from culvert-pipe outlet to channel when the culvert must be placed in the fill above the channel flow line.

Small dams, called ditch checks, made of wood, stone, concrete, or sod are often used in ditches on steep longitudinal grades to check or retard flow.

**8-3  Drainage surveys**  Drainage structures are provided to carry natural channel flow across the right of way with as little alteration or disturbance of normal courses as possible. Chief among the problems which confront the engineer is the determination of the required size of structure. The waterway opening should be large enough to discharge runoff from the drainage area or watershed without impounding water to the extent that the roadway may be topped or the embankment softened by prolonged submergence.

For larger streams the high-water mark on either bank will largely determine both height and span of a bridge, since no lateral constriction of the natural channel can safely be made. The likelihood of ice jams and the amount of floating logs and debris during high water will be important factors in determining minimum clearance of structure above high water and the number and type—if any—of center piers used. Amount and kind of navigation to be provided for will be important to proper design of larger bridges.

The high-water mark on the banks or the floodplain is usually indicated by signs of erosion or scour and deposition of debris during high water. Reliable information can also frequently be obtained from local residents who have observed characteristics of the stream during flood stages over a period of many years. Size and other details of existing structures upstream and downstream from the present site should also be included in the drainage survey for any structure.

The drainage survey will also include information regarding foundation soils which are to support bridge piers and abutments or inverts and floors of culverts. As a general rule, for bridges the test holes should be carried to a depth of about $1\frac{1}{2}$ times the estimated greatest horizontal dimension of the pier or abutment unless solid rock is reached at lesser depth. If the presence of an underlying compressible stratum is suspected, the subsurface exploration should be carried to a sufficient depth to permit analysis of settlement due to consolidation. A detailed log of borings is made in the permanent survey notes, and appropriate samples are taken from various soil layers for laboratory testing. For pipe culverts and box culverts, borings 6 to 10 ft deep will usually reveal any conditions which might contribute to excessive or uneven subsidence.

A plan and profile of the flow line in the vicinity of the roadway for each existing channel is important not only to determine flow characteristics of the channel but also to determine the proper placement of the structure, whether it be large or small.

The size in acres of small drainage areas may be estimated. Larger drainage areas are measured by field survey or, more conveniently, determined from available maps of the U.S. Geological Survey, or other contour maps, or from aerial photographs of the area. Photogrammetric contour maps of the area may also be available or obtainable, thus greatly reducing the amount of field work.

Characteristics of each drainage area with respect to general slope—

such as flat, rolling, hilly, and mountainous—nature of vegetation, and land use should be noted so that runoff characteristics can be estimated with reasonable accuracy.

**8-4   Types of structures**   Drainage structures provided for passage of water from one side of the road to the other are divided into two general classes, major and minor.   Major structures are those having a total clear opening between end walls or abutments of more than 20 ft measured along the center line of the roadway.   Minor structures are those having a total clear opening of 20 ft or less.

Major structures are generally bridges but may include multiple-span culverts.   Minor structures include small bridges, concrete box and arch culverts, masonry arches, treated-timber culverts, and pipe culverts. Pipe culverts may include vitrified-clay sewer pipe, corrugated-cast-iron pipe, cast-iron water pipe, reinforced-concrete culvert pipe, and corrugated-sheet-metal pipe.   Reinforced-concrete and corrugated-sheet-metal pipes are most commonly used.

**8-5   Area of waterway opening**   The larger major structures properly require more extensive and specialized treatment with respect to span and vertical clearance requirements, anchorage, and foundation problems than it is feasible to include here.   The same care in determination of high-water scour and drift and foundation conditions applies, however, to small bridges.   Minor structures will often constitute all of the cross-road drainage requirements of a road-improvement project; and while it is not desirable to provide large factors of safety with respect to adequacy in case of the worst flood conditions which might occur, provision should be made for runoff conditions which might occur during a reasonably normal life expectancy for the structure.   This will provide for such adverse influences as clogging of the structure due to floating debris, rolling boulders on steep rocky channels, and heavy rainfall combined with melting snow and ice on frozen or partially frozen ground in the watershed.

Factors which affect the amount of runoff from rainfall are intensity of rainfall and duration of storm, porosity and absorptive qualities of soil and other surfaces in the watershed, slope and surface roughness, amount and nature of vegetative cover, and the distance from the remotest part of the drainage area to the structure inlet.   A rational approach to determination of required size of opening or waterway will take all these factors into account, and it becomes apparent that because of some indeterminate and uncontrollable factors, a completely rational procedure is not available.

Use of the well-known rational formula $Q = CIA$ is described in the

section on airports in this text.   This method is particularly well suited
to problems of surface drainage for airports and city streets.   Use of
empirical formulas such as the Talbot formula and the Burkli-Ziegler
formula, which are given below, has become accepted practice for general
highway and railroad work where runoff conditions may often vary considerably over a construction project of even a few miles.   They are
applicable in the general range of minor structures.   While it can be seen
that all the influencing factors are not represented by variables in the
formulas, their reliability has been well established through many years
of successful use.

Talbot's formula is

$$a = c \sqrt[4]{A^3} \qquad\qquad\qquad (8\text{-}1)$$

where $a$ = area of the waterway opening, sq ft
$c$ = coefficient depending on topography
$A$ = drainage area, acres

The constant $c$ is usually selected in accordance with experience in a
particular locality.   Average values are recommended as follows:

$c = \frac{2}{3}$ to 1 for mountainous or very hilly country
$c = \frac{1}{3}$ to $\frac{1}{2}$ for terrain which is gently rolling to hilly
$c = \frac{1}{5}$ to $\frac{1}{3}$ for flat to gently rolling country

The Burkli-Ziegler formula is

$$Q = cRA \sqrt[4]{\frac{S}{A}} \qquad\qquad\qquad (8\text{-}2)$$

where $Q$ = quantity of water reaching the structure, cfs
$c$ = coefficient depending on the character of surface drained
$R$ = average rainfall, in./hr, during the heaviest rainfall considered, as determined from available weather records and
generally based on the most severe storm which occurs once
in 10 years
$S$ = average slope of the ground comprising the drainage area,
ft per 1,000 ft
$A$ = drainage area, acres

Commonly recommended average values for $c$ are as follows:

$c = 0.75$ for downtown districts where areas are mainly paved streets,
sidewalks, and roof area
$c = 0.65$ for residential areas with paved streets
$c = 0.40$ to $0.55$ for sparsely built-up areas or those having large yards
with turf, shrubs, and trees
$c = 0.25$ to $0.30$ for farming country and rural areas

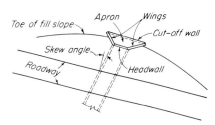

**Fig. 8-2** Protection of culvert ends.

Locally established values for $c$ and $R$ should be used where available.

Having arrived at a value of $Q$, the size of drainage structure is commonly determined from Manning's formula, use of which is described in the chapter on drainage of airports.

Minimum diameter of pipe culverts for cross-road drainage is generally 24 in. Smaller pipes tend to become clogged too readily and they are difficult to clean. For driveways and farm and field entrances crossing the side ditches, 18-in.-diameter pipe is a common minimum.

A small bridge or a large box culvert will often be made use of as a cattle pass in crossing from one side of the road to the other. Where request is made for such use it is desirable to provide minimum vertical clearance of about 7 or 8 ft.

Headwalls and wings (Fig. 8-2) are provided at each end of box culverts to retain ground slopes and embankment slopes in their proper positions. They should extend high enough above the ground slopes and fill slopes to prevent material from sloughing into the channel at the inlet or outlet of the structure. A floor or "apron" is placed on the area between the wings, and a cutoff wall extends across the free end of the apron to prevent the water from undercutting the structure. Wing and headwall footings also extend well into the ground. For best appearance the headwalls are usually placed parallel to the roadway tangent, regardless of culvert skew angle (Fig. 8-2).

Wings should be so placed as to provide the best transition from the approach-channel cross section to the culvert cross section.

**8-6  Laying pipe culverts**  It is important that a pipe culvert have stable and uniform support throughout its entire length, so that localized stresses will not crack or deform the pipe and a uniform flow line or invert grade is established and maintained. Soft or unstable soil should be removed from the channel throughout the length of the culvert, and a width of at least the outside dimension of the pipe or pipes to be placed and backfilled with stable material thoroughly tamped into place. Where solid rock ledges are encountered, excavation should be made to a depth of about 4 in. below the final elevation of the pipe bottom and backfilled with sand thoroughly tamped.

When the drainage channel lies in stable, uniform soil, a trench is dug and carefully shaped to conform to the shape of the bottom of the culvert. The pipe should be laid on a straight longitudinal axis, and it is also desirable to maintain a minimum invert grade of 1 or 2 per cent where possible without disturbing the original channel at inlet and outlet unduly. Consequently the shaped bedding trench will usually vary in depth.

When the height of embankment above the top of the pipe is 15 ft or more, concrete cradles are used to support the pipe unless the foundation soil is exceptionally stable and unyielding. Extra-strength reinforced-concrete pipe or cast-iron pipe are frequently specified where such fills are to be built.

**8-7   Backfilling**   It is imperative that the embankment material be thoroughly tamped with hand or mechanical tampers around the haunches of the pipe. Large stones and boulders should not be placed in the fill in contact with the culvert. Tamping should be carried to the top of the culvert because it is difficult to work in close to the structure with com-paction equipment. Rollers and heavy tractors operated along the sides of the pipe culvert, combined with tamping, will assure lateral support for the pipe and will minimize settlement of embankment mate-rial around the pipe.

When multiple pipes are placed, a clear distance between pipes of one-half the diameter of pipes used and not less than 1 ft is commonly specified in order to allow room for adequate tamping of fill material between the pipes.

**8-8   Minimum depth of cover over pipe culverts**   The final grade should be so adjusted that the subgrade will be at least one-half the diameter of the pipe and not less than 2 ft above the top of any pipe culvert. When heavy equipment is to operate over the pipe during construction, a minimum of 3 ft of cover should be provided and cut down later to proper subgrade elevation before pavement structure is placed.

**8-9   Subsurface drainage**   It is not possible to treat grading operations entirely apart from drainage. Many of the problems connected with grading the roadbed are intimately associated with those of drainage, particularly those concerned with ground-water control. For this reason, some discussion of subsurface drainage was included in various topics treated in Chap. 2, Grading the Roadbed. Only a summary, with some extension of basic principles, will be presented here.

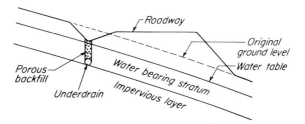

**Fig. 8-3**  Interception of ground water.

Problems of ground-water control may be divided into two main divisions: (1) those concerned with ground-water flow, where the water table is on a slope or grade, and (2) those concerned with a water table which is level, or nearly so, where fluctuation in elevation of the water table is largely dependent on wet and dry seasons.

The first of these is illustrated by Fig. 8-3. It is important to recognize that when any considerable fill is placed on the original ground surface on part of the cross section, this additional overburden will consolidate the soil of the water-bearing stratum. Such consolidation reduces porosity and to that extent chokes off the normal flow in the stratum. The water table then rises on the side above the fill and may start springs or seepage in the ditch or along the back slope. The real danger, however, is that the water table may also rise in the fill, softening the material and creating hydrostatic and seepage pressures which may mobilize a slide comprising part or all of the embankment and possibly a substantial portion of original foundation soil, depending, of course, on permeability and shearing-resistance characteristics of both embankment and foundation soils. A complete analysis of the problem is available to the engineer through the application of principles of soil mechanics. Not only is the treatment of such a problem beyond the scope of this text, but it is in relatively few instances that the cost of a detailed investigation is justified.

The placement of an underdrain with porous backfill to intercept ground-water flow will generally provide adequate insurance against adverse effects by lowering rather than raising the level of the water table in the vicinity of the ditch and roadway. It is desirable to intercept completely the water-bearing stratum, but this is not always feasible, nor is it necessary so long as any choking off of natural flow is prevented.

The second type of subsurface drainage problem is illustrated by Fig. 8-4. It is often desirable to lower the water table so that a seasonal rise of the table will not reach overlying fill material. This can be done effectively with the use of underdrains and porous backfill if the soil

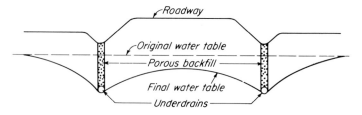

**Fig. 8-4**   Local depression of ground-water table.

comprising the water-bearing stratum is drainable.   However, if the water-bearing soil has high capillarity, such underdrains will be ineffective, except insofar as consolidation of the fine-grained soil is hastened by the presence of the drains.

## QUESTIONS AND PROBLEMS

**8-1.** *a.* Compute the required area of waterway opening for a culvert to drain 1 sq mile of hilly terrain, using Talbot's formula.

*b.* What size (inside diameter) of pipe culvert should be used?

**8-2.** Using the Burkli-Ziegler formula, find the quantity of water reaching a storm sewer from 160 acres of residential area with paved streets.   The average slope of the watershed is 0.020, and the heaviest rainfall considered is 0.8 in. in 20 min.

**8-3.** A diversion ditch is to be constructed as a 90-deg V ditch to intercept the runoff from 40 acres of farming country for which the average slope is 0.016.   The heaviest rate of rainfall considered is 1.5 in./hr.

*a.* Find the quantity of water reaching the ditch, using the Burkli-Ziegler formula.

*b.* If the maximum depth of flow in the ditch is limited to 21 in., find the necessary slope of the ditch to handle this quantity of flow, using Manning's formula with $n = 0.035$.

*c.* What is the mean velocity of flow in the ditch?

## BIBLIOGRAPHY

Buried Circular Conduits and Behavior of Foundations, *Highway Research Record* 145, 1966.

Culverts and Storm Drains, *Highway Research Record* 116, 1966.

Erickson, E. L.: The Economics of Design and Construction of Bridges for Local Roads, *Proc. Ill. Conf. on Highway Eng.*, University of Illinois, 1949.

Greenman, R. L.: The Use and Treatment of Granular Backfill, *Mich. Eng. Expt. Sta. Bull.* 107, 1948.

Highway Drainage and Scour Studies, *Highway Research Record* 123, 1966.

Izzard, C. F.: Design of Roadside Drainage Channels, *Highway Research Board Roadside Development Rept.* **1942,** 47–63.

"Standard Specifications for Highway Bridges," American Association of State Highway Officials, 1965.

"Subsurface Drainage Investigation," U.S. Army, Corps of Engineers, 1945–1946.

# 9

# Public Roads and Streets in the United States

**9-1 The interstate system** Strictly speaking, there is no Federal system of highways in the United States. Only relatively small mileages in national parks, national forests, and Indian reservations are built and maintained by the Federal government. There does exist, however, a national system of interstate and defense highways, designated by the Federal-Aid Highway Act of 1944 to comprise not more than 40,000 miles in extent, built to high design standards consistent with traffic needs of various sections of the system and located so as to connect principal metropolitan areas, cities, and industrial centers and to serve the national defense.

The present authorized total extent of this interstate system is 41,000 miles, all of which has been designated. General location of the system is shown in Fig. 9-1. Although it consists of only a little more than 1 per cent of the country's total road and street mileage, it is estimated that it will ultimately carry 20 per cent of the total traffic. It may well be considered as the main arterial system or basic framework of the nation's roads.

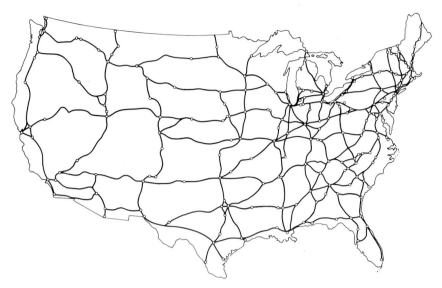

**Fig. 9-1**  National system of interstate and defense highways.  (*Source: U.S. Bureau of Public Roads.*)

Following is the status of improvement on the interstate system as of September 30, 1965, as reported by the Bureau of Public Roads:

|                                                    | *Miles* |
|----------------------------------------------------|--------:|
| Preliminary status or not yet in progress............ | 3,276 |
| Engineering and right of way in progress........... | 11,536 |
| Under construction............................... | 6,202 |
| Open to traffic.................................. | 19,986 |
| Total......................................... | 41,000 |

**9-2  State highway systems**  In general, state governments, acting through their state highway departments, have the responsibility for development of primary and secondary road systems within state boundaries.  The state primary systems include the portion of the national interstate highway system lying within the borders of each state.  Assisting the states in this development is the U.S. Bureau of Public Roads. The principal function of the Bureau is administration of the Federal-aid program.  It also cooperates with other government agencies in constructing roads in national parks, forests, and reservations and conducts comprehensive research programs in all phases of highway work.

The state primary and secondary systems are interconnected with networks of county roads which serve mainly as "feeder" roads from

smaller towns and communities to main arteries of travel and accommodate industry-to-market and other traffic of local or sectional interest
within the state. Such county road systems comprise, in extent, more
than half of the nation's roads. In some states—Delaware, North
Carolina, Virginia, and West Virginia—county roads are under supervision of the state governments, but generally, the responsibility for
development of county roads rests with county governments, and state
supervision is confined to state systems, which include roads having
state-wide and interstate traffic interest.

Local roads and city streets, then, complete the network which
makes possible door-to-door automotive travel from any place to every-
where. Some appreciation of this vast and intricate network of roads
may be gained from the tabulation of existing mileages in the United
States at the beginning of the year 1965. These mileages are shown in
Table 9-1.

**Table 9-1   System mileages**

*Thousands of miles*

| | |
|---|---:|
| State primary systems (including urban extensions)................. | 460 |
| State secondary systems....................................... | 283 |
| County roads................................................. | 1,741 |
| Town, township, and other local roads........................... | 604 |
| Local city streets............................................. | 429 |
| Roads under Federal control in national parks, forests, and reservations | 127 |
| Total................................................... | 3,644 |

**9-3   Serviceability**   Of the total 3,644,000 miles of roads and streets
in the United States as of Jan. 1, 1965, 2,730,000 miles were surfaced, and
914,000 were nonsurfaced. Of the nonsurfaced roads and streets, about
half were graded and drained, and half were primitive and unimproved.
Many of the nonsurfaced roads are remote-access roads and post roads
which have very little traffic use but which must exist for the accommodation of the few who must use them.

The level of service to be provided for remote access to individual
farms, mines, privately owned forests, and other enterprises of private
industry is largely determined by whatever such individual enterprise
can afford, and this may often be a primitive or unimproved road. As
communities develop and more property owners can share the costs of
construction and maintenance, the graded and drained roadway may
become economically justifiable or feasible. Where relatively few property owners share the cost of several miles of improvement, it can be seen
that cost must be kept at a minimum. Grade lines will generally follow
natural ground slopes, and surfacing, if any, will consist of some form

**Table 9-2 Total road and street mileage in the United States, 1964**

Classified by system and type of surface[a]

| System | Nonsurfaced mileage[b] | | | Surfaced mileage[c] | | | | | Total existing mileage |
|---|---|---|---|---|---|---|---|---|---|
| | A B | C | Total | D E | F G-1 H-1 | G-2 H-2 I | J | Total surfaced mileage | |
| Rural mileage: | | | | | | | | | |
| Under state control: | | | | | | | | | |
| State primary systems | 640 | 3,508 | 4,148 | 16,538 | 130,657 | 213,288 | 46,445 | 406,928 | 411,076 |
| Secondary roads under state control: | | | | | | | | | |
| State secondary systems[d] | 3,059 | 1,015 | 4,074 | 17,072 | 56,107 | 32,775 | 1,969 | 107,923 | 111,997 |
| County roads under state control[e] | 12,064 | 8,274 | 20,338 | 52,954 | 48,227 | 14,191 | 474 | 115,846 | 136,184 |
| Subtotal | 15,763 | 12,797 | 28,560 | 86,564 | 234,991 | 260,254 | 48,888 | 630,697 | 659,257 |
| State parks, forests, and reservations, etc.[f] | 1,559 | 6,596 | 8,155 | 7,945 | 1,145 | 2,654 | 2,218 | 13,962 | 22,117 |
| Total | 17,322 | 19,393 | 36,715 | 94,509 | 236,136 | 262,908 | 51,106 | 644,659 | 681,374 |
| Under local control: | | | | | | | | | |
| County roads | 261,098 | 292,098 | 553,196 | 818,750 | 268,663 | 91,300 | 8,687 | 1,187,400 | 1,740,596 |
| Town and township roads | 71,826 | 67,263 | 139,089 | 289,139 | 64,812 | 41,030 | 2,015 | 396,996 | 536,085 |
| Other local roads | 52,972 | 8,264 | 61,236 | 5,006 | 1,056 | 110 | 34 | 6,206 | 67,442 |
| Total | 385,896 | 367,625 | 753,521 | 1,112,895 | 334,531 | 132,440 | 10,736 | 1,590,602 | 2,344,123 |
| Under federal control: | | | | | | | | | |
| National parks, forests, reservations, etc.[f] | 43,514 | 46,941 | 90,455 | 29,844 | 3,615 | 3,064 | 102 | 36,625 | 127,080 |
| Total rural mileage | 446,732 | 433,959 | 880,691 | 1,237,248 | 574,282 | 398,412 | 61,944 | 2,271,886 | 3,152,577 |
| Municipal mileage: | | | | | | | | | |
| Under state control: | | | | | | | | | |
| Extensions of state primary systems | 8 | 52 | 60 | 163 | 5,638 | 31,722 | 11,504 | 49,027 | 49,087 |
| Extensions of secondary roads under state control[d,e] | 217 | 59 | 276 | 362 | 6,197 | 5,244 | 959 | 12,762 | 13,038 |
| Total | 225 | 111 | 336 | 525 | 11,835 | 36,966 | 12,463 | 61,789 | 62,125 |
| Under local control: | | | | | | | | | |
| Local city streets | 7,447 | 25,259 | 32,706 | 85,643 | 140,289 | 135,136 | 35,593 | 396,661 | 429,367 |
| Total municipal mileage | 7,672 | 25,370 | 33,042 | 86,168 | 152,124 | 172,102 | 48,056 | 458,450 | 491,492 |
| Total rural and municipal mileage in the United States | 454,404 | 459,329 | 913,733 | 1,323,416 | 726,406 | 570,514 | 110,000 | 2,730,336 | 3,644,069 |

[a] U.S. Bureau of Public Roads: "Highway Statistics," 1964.

[b] Nonsurfaced includes A and B, primitive and unimproved, and C, graded and drained roads.

[c] Surface types indicated by symbols in these columns are as follows: D, soil-surfaced; E, slag, gravel, or stone; F, bituminous surface treated; G-1, mixed bituminous, and H-1, bituminous penetration having a combined thickness of surface and base less than 7 in. and/or low load-bearing capacity; G-2, mixed bituminous, and H-2, bituminous penetration having a combined thickness of surface and base 7 in. or more and/or a high load-bearing capacity with or without portland cement concrete base; I, bituminous concrete and sheet asphalt with or without portland-cement concrete base; and J, portland-cement concrete with or without bituminous wearing surface less than 1 in. in compacted thickness. Segregation of G and H surfaces according to thickness and load-bearing capacity is not uniform for all states. Where no segregation was reported for them, the mileage was classified as G-1 and H-1.

[d] Includes mileage designated as farm-to-market in Louisiana and as state aid in Maine.

[e] Includes mileage of county roads under state control in all counties of Delaware, North Carolina, and West Virginia; eight counties in Alabama; all but two counties in Virginia; and some county mileage in Nevada.

[f] State and national park, forest, reservation, toll, and other roads that are not a part of the state system.

# Table 9-3  Estimated motor-vehicle travel in the United States and related data, calendar year, 1964*†

| Item | Passenger vehicles | | | | | Trucks and combinations | All motor vehicles |
|---|---|---|---|---|---|---|---|
| | Passenger cars‡ | Buses | | | All passenger vehicles | | |
| | | Commercial | School and nonrevenue | All buses | | | |
| Motor-vehicle travel, million vehicle-miles: | | | | | | | |
| Main rural roads.......... | 246,850 | 908 | 674 | 1,582 | 248,432 | 68,180 | 316,612 |
| Local rural roads.......... | 94,853 | 181 | 743 | 924 | 95,777 | 25,416 | 121,193 |
| All rural roads.......... | 341,703 | 1,089 | 1,417 | 2,506 | 344,209 | 93,596 | 437,805 |
| Urban streets.......... | 345,432 | 1,803 | 307 | 2,110 | 347,542 | 56,562 | 404,104 |
| Total travel.......... | 687,135 | 2,892 | 1,724 | 4,616 | 691,751 | 150,158 | 841,909 |
| Number of vehicles registered, thousands.. | 72,970 | 82.3 | 223.1 | 305.4 | 73,276 | 14,019 | 87,295 |
| Average miles traveled per vehicle........ | 9,417 | 35,140 | 7,727 | 15,115 | 9,440 | 10,711 | 9,644 |
| Fuel consumed, million gallons........... | 47,924 | 619 | 241 | 860 | 48,784 | 19,117 | 67,901 |
| Average fuel consumption per vehicle, gal. | 657 | 7,521 | 1,080 | 2,816 | 666 | 1,364 | 778 |
| Average miles traveled per gallon of fuel consumed................... | 14.34 | 4.67 | 7.15 | 5.37 | 14.18 | 7.85 | 12.40 |

* U.S. Bureau of Public Roads: "Highway Statistics," 1964.
† For the 50 states and District of Columbia.
‡ Includes taxicabs; also 985,445 motorcycles, which are estimated to account for 0.4 per cent of the total travel.

of soil stabilization or light surfacing of pit-run gravel, with perhaps a light surface treatment of bituminous material for waterproofing and control of dust where a reasonable amount of traffic warrants the expense.

On the other hand, in the case of city streets and alleys the cost of improvement may be divided among many property owners as residential areas become built-up, so that street improvement may be made with the use of high-type pavement far beyond the actual traffic needs. The justification for the additional expenditure is that a relatively "permanent" pavement will add to the continuing valuation of abutting property, with very little depreciation and little or no maintenance cost.

Classification of road and street mileage by system and type of surface as of the end of calendar year 1964 is given in Table 9-2. Improving and maintaining each mile of the entire network to a level of serviceability adequate for the traffic which is to use it is a continuing process rather than an ultimate goal. In general, the level of service demand raises with ever-increasing vehicular traffic volumes and higher operating speeds. Some characteristics of motor-vehicle travel in the United States are given in Table 9-3.

Evaluation of performance or serviceability of pavements by determining a Present Serviceability Index was developed in 1961 at the AASHO Road Test in Illinois. Many states have adopted this procedure in evaluating pavement performance as a measure of condition or sufficiency ratings. The Present Serviceability Index or PSI may be computed from the following or similar expression:

$$PSI = 12.54 - 4.49 \log R - 0.01 \sqrt{C + P} - 1.38D^2 \qquad (9\text{-}1)$$

where $R$ = roughness, in./mile
$C$ = pronounced cracking, sq ft/1,000 sq ft
$P$ = patching, sq ft/1,000 sq ft
$D$ = average rut depth in both wheel paths (depths of depression under straight edge 4 ft long), in.

General condition categories corresponding to various PSI ranges are as follows:

| PSI range | Typical values within range | | | | Condition |
|---|---|---|---|---|---|
| | $R$ | $C + P$ | $D$ | PSI | |
| 4–5 (max) | 65 | 0 | 0 | 4.3 | Very good |
| 3–4 | 100 | 0 | 0.25 | 3.5 | Good |
| 2–3 | 125 | 100 | 0.50 | 2.7 | Fair |
| 1–2 | 150 | 400 | 0.75 | 1.8 | Poor |
| 0–1 | 200 | 1000 | 1.00 | 0.5 | Very poor |

Note that a PSI of about 5 is maximum for a newly constructed pavement. That is, the specification limit for surface "roughness" tolerance in construction may be $\frac{1}{8}$ to $\frac{1}{16}$ in. under a straight edge 10 ft long. For example, this may be, say, $\frac{1}{10}$ in. in 10 ft, or 52.8 in. per mile. The log of 52.8 is 1.723, and the roughness factor in the formula is

$$4.49 \times 1.723 = 7.74$$

Then, since $P$, $C$, and $R$ would be zero for the new pavement, the PSI is $12.54 - 7.74 = 4.8$. Surface roughness is usually determined with the use of a roughometer or profilometer which measures and records surface variations as it travels over the surface.

**Table 9-4    Minimum design standards for secondary and feeder roads**

| Design control | Average annual daily traffic volume | | | | | |
|---|---|---|---|---|---|---|
| | Under 100 | | 100–400 | | 400–1,000 | |
| | Mini-mum | Desir-able | Mini-mum | Desir-able | Mini-mum | Desir-able |
| Design speed, mph: | | | | | | |
| Flat topography.............. | 40 | ..... | 45 | 55 | 50 | 60 |
| Rolling topography.......... | 30 | ..... | 35 | 45 | 40 | 50 |
| Mountainous topography.... | 20 | ..... | 25 | 35 | 30 | 40 |
| Sharpest curve, deg: | | | | | | |
| Flat topography.............. | 14 | ..... | 11 | 7 | 9 | 6 |
| Rolling topography.......... | 25 | ..... | 18 | 11 | 14 | 9 |
| Mountainous topography.... | 56 | ..... | 36 | 18 | 25 | 14 |
| Maximum gradient, per cent: | | | | | | |
| Flat topography.............. | 8 | 5 | 8 | 5 | 7 | 5 |
| Rolling topography.......... | 12 | 7 | 10 | 7 | 8 | 6 |
| Mountainous topography.... | 15 | 10 | 12 | 9 | 10 | 7 |
| Nonpassing sight distance, ft (defined in Sec. 9-9): | | | | | | |
| Flat topography.............. | ..... | ..... | 315 | 375 | 350 | 415 |
| Rolling topography.......... | ...... | ..... | 240 | 315 | 275 | 350 |
| Mountainous topography.... | ...... | ..... | 165 | 240 | 200 | 275 |
| Width of surfacing or pavement, ft........................ | 12 (if any) | ..... | 16 | 20 | 18 | 20 |
| Width of roadbed............. | 20 | ..... | 24 | 28 | 26 | 30 |
| New bridges: | | | | | | |
| Clear width, ft.............. | 14 | 20 | 22 | 24 | 24* | |
| Design load, AASHO........ | H-10 | H-15 | H-15 | ... | H-15 | |
| Width of right of way, ft....... | 40† | ..... | 40† | 80 | 50 | 80 |

* Minimum of 24 ft, or 4 ft more than approach-pavement width.
† Minimum of 40 ft, or as required for construction.

**Table 9-5  Average bid prices of major items on federal-aid primary highway construction contracts awarded, calendar year, 1964**[a]

| Major item[b] | | Total bid quantity reported, thousands | Weighted average unit price, dollars |
|---|---|---|---|
| Roadway excavation | | | |
| Borrow | | 289,446 cu yd | $ 0.66 |
| Common | | 270,984 cu yd | 0.46 |
| Unclassified | | 673,180 cu yd | 0.59 |
| Solid rock | | 16,873 cu yd | 1.29 |
| Steel: | | | |
| Pavement reinforcement | | 407,131 lb | 0.101 |
| Structural reinforcement | | 1,035,039 lb[c] | 0.112 |
| Structural steel | | 1,084,548 lb | 0.193 |
| Steel H piling | | 242,163 lb | 0.123 |
| Prestressing steel | | 20,003 lb | 0.576[d] |
| | **Weighted average thickness** | | |
| Bases: | | | |
| Gravel and clay gravel | 8.96 in. | 145,192 sq yd | 0.71 |
| Macadam or stone | 9.46 in. | 91,391 sq yd | 1.03 |
| Bituminous concrete | 5.22 in. | 42,922 sq yd | 1.31 |
| Portland-cement concrete | 8.39 in. | 2,815 sq yd | 5.00[e] |
| Surfaces: | | | |
| Bituminous surface treatment | 0.54 in. | 37,466 sq yd | 0.25 |
| Bituminous road mix | 3.94 in. | 811 sq yd | 0.94 |
| Bituminous plant mix medium | 3.35 in. | 51,723 sq yd | 1.02 |
| Bituminous penetration | 1.85 in. | 1,007 sq yd | 0.60 |
| Bituminous concrete | 2.99 in. | 101,616 sq yd | 1.09 |
| Portland-cement concrete | 9.02 in. | 73,887 sq yd | 4.24[e] |
| | **Diameter** | | |
| Pipe: | | | |
| Clay | 6.00 in. | 3,417 lin ft | 1.51 |
| Reinforced concrete | 24.00 in. | 1,107 lin ft | 7.20 |
| Corrugated steel | 24.00 in. | 492 lin ft | 7.03 |
| Structural concrete: | | | |
| Superstructures | | 2,717 cu yd | 61.34[f] |
| Substructures | | 2,529 cu yd | 55.16[f] |
| Foundations and footings | | 408 cu yd | 49.31[f] |
| Prestressed concrete | | 222 cu yd | 71.46[g] |

[a] U.S. Bureau of Public Roads: "Highway Statistics," 1964.
[b] Total cost of major items is 70 per cent of total contract cost.
[c] Includes 19,830,000 lb in prestressed concrete.
[d] Cost of prestressing operations included.  (Estimated.)
[e] Excludes costs of reinforcement and joints.
[f] Reinforcement cost excluded.
[g] Excludes costs of reinforcing and prestressing steel and cost of prestressing operations.  (Estimated.)

**9-4  Design standards**  The American Association of State Highway Officials adopted, in 1945, design standards for construction and reconstruction of secondary and feeder roads to be used for minimum design under normal conditions, but not necessarily for exceptional cases for which lower values will have to be determined separately.  The standards were not specifically intended for county roads as a whole, but they serve as an excellent guide to planning and design of any roads where traffic volumes will justify their use.  The annual average daily traffic volume is used as the basis for the standards, as shown in Table 9-4.  It is desirable that the standards be used for traffic volumes which allow for foreseeable future increases in traffic.  The design peak-hour traffic density is assumed to be approximately 10 per cent of the annual average daily traffic.

Design requirements for rural roads and urban extensions of the state primary systems, including the national interstate system, present an extremely varied pattern in fitting the needs of a wide range of vehicular traffic volumes to a great variety of topography, land use, and other environmental considerations.  These requirements involve many varying provisions for right of way, grade separation, sight distances, median strips, and other factors related to efficient—and safe—operation of vehicular traffic at relatively high speeds.  The American Association of State Highway Officials has assembled design requirements in publications on policy referred to in the bibliography at the end of this chapter.

Greater emphasis is being placed on safety in highway transportation.  Recent national legislation requires greater attention to safety from standpoints of vehicle and driver as well as roadway design.  Another important aspect of highway design receiving national attention is roadside development and beautification.

**9-5  Construction costs**  Because roads are built over a wide range of topographic and other physical conditions, to serve the needs of a considerable variety of vehicular traffic, construction costs per mile—or per project—must generally be determined from unit costs for at least the major items.  A useful tabulation for such estimating purposes is given in Table 9-5 for major items of Federal-aid primary highway construction.  Note that these major items will generally constitute about 70 per cent of the total contract cost.

### BIBLIOGRAPHY

Cunningham, H. E.: Analysis of Short Count Methods in Measuring Traffic Flow, *Highway Research Board Proc.*, **25**:329–342 (1945).

Fairbanks, H. S.: Objects and Methods of State-wide Highway Planning Surveys, *American Highways*, January, 1937.

Forecasting Models and Economic Impact of Highways, *Highway Research Record* 149, 1966.

Geometric Aspects of Highways, *Highway Research Record* 162, 1967.

Highway Finance and Benefits, *Highway Research Record* 138, 1966.

Highway Safety, *Highway Research Record* 163, 1967.

"Highway Statistics," U.S. Bureau of Public Roads, 1966.

Pavement Design and Evaluation, *Highway Research Record* 121, 1966.

"A Policy on Arterial Highways in Urban Areas," American Association of State Highway Officials, 1957.

"A Policy on Geometric Design of Rural Highways," American Association of State Highway Officials, 1965.

Roadside Development, *Highway Research Record* 161, 1967.

Road User and Vehicle Characteristics, *Highway Research Record* 159, 1967.

Shelton, W. A.: Methods of Estimating Highway Traffic Volume, *Highway Research Board Proc.*, **16**:239–252 (1936).

Tentative Skid-resistance Requirements for Main Rural Highways, National Cooperative Highway Research Program Report 37, 1967.

Transportation System Evaluation, *Highway Research Record* 148, 1966.

# 10

# General Maintenance
# of Roads and Streets

**10-1 The maintenance budget** Figures compiled by the Highway Research Board show that nearly 3 billion dollars was spent in 1963 for maintenance on all roads and streets in the United States. Thus a large portion of funds available for highway purposes is set aside for maintenance—protection of the highway structure and keeping it in condition for service to traffic.

Of the maintenance dollar, about 45 cents is spent on care and repair of the roadway surfaces; about 25 cents on shoulders, roadside, and drainage; 10 cents on control of snow and ice; 5 cents on bridge maintenance; and 15 cents on traffic services—signs, markings, and lighting.

In most states, the state primary and secondary Federal-aid highways, including urban extensions, are maintained by state forces and state-owned equipment, and county and local roads and streets by county and local governments. Many states, however, contract with counties and the larger cities for maintenance of state highways within the jurisdiction of each. This latter arrangement provides greater economy by elimination of duplicated supervision, equipment, and shops.

Since maintenance requirements for roadway surfaces vary greatly with the type of surface, each general type will be discussed briefly.

**10-2 Earth, soil-stabilized, and gravel surfaces** These surfaces, in which the binder is silt or clay, are kept smooth by repeated dragging or blading. The road drag, consisting of two or more blades attached to a rigid frame, is pulled with a tractor or truck. For heavier work—reshaping and establishing a crown—the motor-driven blade grader is generally used.

Blading and dragging are most effectively done as soon as practicable after a rain, when the surface material is moist. Surface material is dragged or bladed from the edges toward the center so that a crown of $\frac{1}{4}$ to $\frac{1}{2}$ in./ft is maintained.

During dry weather much of the fine material—binder—is whipped out by traffic and blown from the roadway material in the form of clouds of dust. This dust is a hazard to traffic and its loss from the surface material causes loosening and raveling of coarser particles. Dust palliatives such as calcium chloride or bituminous materials are used to promote safety and to preserve the "fines" as binder in the roadway material.

Calcium chloride is usually applied at the rate of $\frac{1}{2}$ to $1\frac{1}{2}$ lb/sq yd. Two or more applications may be required during the summer season depending on the climate and the speed and volume of traffic.

Bituminous materials—light grades of liquid asphalt and tars, and diluted emulsions—are usually applied with a pressure distributor at a rate of 0.1 to 0.25 gal/sq yd.

When a considerable amount of binder is lost from the roadway surface material it is usually replaced by scarifying a few inches of the surface material and adding clay or other binder, mixing, and relaying. This may constitute a "betterment" or reconstruction of the roadway rather than maintenance. Administrative policy will usually determine a proper division of maintenance and construction for accounting purposes.

**10-3 Bituminous surfaces** Maintenance of bituminous surfaces consists mainly of patching, filling cracks, and resealing. Rutting, shoving, and raveling of bituminous mats may occur in areas where there is a concentration of heavy bus and truck traffic. Considerable areas of mat may need to be replaced with more stable mixtures or they may be reworked if slow- or medium-curing liquid asphalts or light-grade tars were used in the initial construction. Where surface failure is due to base-course and subgrade weakness, the weak material should be removed and backfilled, usually by tamping by hand or by use of pneumatic tampers with more stable materials. Sometimes improvement of underdrainage is necessary. Stockpiles of bituminous mixture using MC-250, MC-800, SC-250, or SC-800 should be placed at convenient points along the roadway during construction of bituminous surfaces. These mixtures, if placed in fairly large piles of several tons, will remain workable for several seasons for patching.

Patching may also be done by penetration treatment of aggregate which has been compacted into place, using coarse sand or stone chips or blotter aggregate and rolling to proper grade.

Random cracks which often develop in bituminous surfaces should be filled with bituminous material to seal the opening against surface water which might otherwise enter base-course and subgrade materials to weaken them. This is usually done by hand, pouring with cold liquid bituminous materials such as RC-800 or heated asphalt cements or tars.

Resealing of bituminous surfaces is done by the application of a seal coat using bituminous material and blotter aggregate. Purposes and procedure for seal-coat applications are described in earlier chapters.

**10-4 Portland-cement concrete surfaces**  Filling and sealing of joints and cracks, to prevent seepage of water into the subgrade, constitutes the main item of maintenance of concrete pavements. Loose material should be removed from the joint or crack before it is sealed. Cleaning is usually done with compressed air from portable compressors. The filler material—commonly air-blown asphalt cement with low susceptibility to temperature change—is placed by means of hand-pouring pots. Prepared mixtures of powdered rubber and asphalt are also used for this purpose. The work should be done in the summer when temperature is high.

Scaled and spalled areas should be thoroughly cleaned of all loose material and treated with an application of liquid asphalt such as RC-3 or light tar. Coarse sand or finely crushed stone chips are then applied. If necessary, a second application of bitumen and blotter aggregate may be applied to build up the patch to grade. Deeper spalled and abraded areas may require a light bituminous mat and seal-coat application.

Areas of concrete pavement which have broken up should be replaced with concrete, first correcting any adverse subgrade conditions which may have directly or indirectly caused the failure.

Open joints and cracks and faulty drainage may often lead to "pumping," whereby deflection of the pavement due to heavy wheel loads forces accumulated water from beneath the slab through open joints or cracks or out from the pavement edge. The water carries fine material with it from the subgrade, and gradually a considerable area of slab may be impaired by such undermining. In these cases and where the pavement may have settled unduly because of differential settlement of the subgrade, it may be restored to proper elevation and grade by means of "mud jacking." A mixture of water and fine soil or soil plus cement is pumped through a hole in the concrete slab to fill any void space under the slab and to raise it to grade. Liquid asphalts such as SC-70, RC-70, and MC-70 are also used in place of water to "liquefy" the mixture for ease in pumping it, under pressure, to raise the slab.

**10-5  Shoulders**  Adequate shoulders, properly maintained, are necessary to development of the full capacity of the roadway surface. For purposes of emergency traffic use, shoulders require maintenance similar to that for the roadway surfaces. They should be kept smooth and free from holes. Erosion of fill slopes during heavy rains should be repaired as soon as possible so that the full width of shoulder is preserved for safe emergency use. The same general treatment is required on road approaches.

Gravel and earth shoulders are kept smooth by blading or dragging. The transverse shoulder slope is usually maintained at about 5 per cent to provide good surface drainage. This slope is usually reduced somewhat on the steeper longitudinal grades.

Turf shoulders require mowing, cleaning, and weed control. They are kept smooth by occasional rolling. Deep ruts and holes may need to be corrected by blading and the turf reestablished by reseeding or resodding and fertilizing.

Bituminous-treated and paved shoulders require maintenance similar to that required for roadway surfaces of similar types.

**10-6  The roadside**  Mowing of grass, weeds, and brush on the roadside within the limits of the right of way is necessary to improve sight distances and to provide a better appearance along the roadway. The work is done at intervals as needed during the growing season. Grass clippings may be left as mulch, but brush and weeds should be raked and burned.

Weed control is often a major problem. Weeds should be cut before going to seed. Small areas are sometimes burned *in situ* by direct flame. Special chemicals, such as 2,4 dichlorophenoxyacetic acid, are widely used at present for weed control by directly spraying the plants.

Seeding and sodding of slopes and steeper ditch grades are important maintenance operations for control of erosion. Strips of sod will usually be used instead of seeding on the steeper slopes. Planting of bushes and vines and similar ground cover may be required on steep slopes where seeding or sodding is not practical. For most regions it is important that native grasses and plants be used.

Special roadside areas—parks and turnouts—should be maintained so as to meet approved aesthetic and health standards. Many state highway organizations have roadside improvement and forestry departments that have charge of planting, trimming, spraying, and fertilizing of turf, trees, and shrubs in roadside and park areas.

Responsibility for the removal of fallen branches, landslides, rocks, and other debris rests heavily with the highway maintenance departments for the protection of the traveling public.

**10-7 The drainage system**  Preserving adequate roadway crown and shoulder slope, as previously discussed, will do much toward promoting good surface drainage. In addition to this, ditches should be kept clean and culverts checked regularly to see that they are free from obstructions. Inlets and outlets should be kept open and free from debris, and the entire culvert cleaned and repaired when necessary.

Periodic inspection of bridges—superstructure, substructure, and the stream bed—is made to determine what maintenance work is necessary. Inspection should be complete enough to discover any undermining of footings and damage to or deterioration of substructure by corrosion, erosion, floating debris, or attack by organisms.

Frequent cleaning and painting of all surfaces of steel superstructures is essential for protection from the elements; bridge floors must be repaired and replaced; damage caused by traffic accidents should be repaired promptly and damaged members replaced; and expansion elements—bridge seats, rockers, and rollers—must function properly.

**10-8 Snow and ice removal and control**  In a large section of the country various types of snowplows, large trucks, and power graders must be kept in working order for removal of snow from the roadway. Good organization is required for this work because removal must usually follow closely in the wake of a storm.

Snow fences, consisting of fabricated wood slats which may be rolled into bundles when not in use, are placed on steel posts parallel to the roadway about 80 to 140 ft from the center line, in areas where drifting of snow occurs, and on the side of the roadway from which the wind blows during storms, to drift or pile up the snow before it reaches the roadway. Natural snow fences or windbreaks may be made in critical areas by planting rows of trees and shrubs parallel to the roadway where right of way or property easement may be procured.

Ice-control abrasives such as coarse sand and cinders should be placed in stockpiles or containers at strategic points along the roadway where icing of the surface is likely to occur. Such materials are also quickly hauled in trucks from central storage or pits to places where they are needed. Calcium chloride is frequently mixed with the abrasive at the rate of 50 to 100 lb/cu yd of abrasive when the stockpiling is done.

Maintenance crews must frequently operate in alternating shifts during emergencies. Increasing use is being made of both one-way and two-way radio communications for reporting of adverse conditions; rapid dispatching of equipment, personnel, materials, and supplies; and for general control of operations.

**10-9  Traffic-control devices**  The maintenance department usually has charge of making and painting traffic signs, their installation, repair, and replacement to promote safe and efficient motor-vehicle operation. Pavement markings must be renewed periodically because of the wear and abrasion they receive from traffic.  Most departments have mechanized equipment for application of the painted strips.  In recent years wide use has been made of small glass beads mixed in the paint or added to the painted surface to reflectorize the strip for greater night visibility.

**BIBLIOGRAPHY**

Pavement and Bridge Maintenance, *Highway Research Record* 146, 1966.
Jorgensen, Roy, et al.: "Virginia Maintenance Study, 1963–1966," Final Report, Virginia Department of Highways, 1966.
"A Policy on Maintenance of Roadsides," American Association of State Highway Officials, 1948.
"A Policy on Maintenance of Roadway Surfaces," American Association of State Highway Officials, 1948.
"A Policy on Maintenance of Shoulders, Road Approaches and Sidewalks," American Association of State Highway Officials, 1948.
Radzikowski, H. A.: Mechanized Maintenance of Low Cost Roads, *Proc. Am. Road Builders Assoc.*, 1950.
Surface Drainage of Highways, *Highway Research Board Rept.* 6B, 1948.
A Survey of Pumping in Illinois, *Highway Research Board Rept.* 1D, 1948.

# part two

# Airport Engineering

# 11
# Civil Aviation
# in the United States

**11-1 Introduction** In the United States over 9,000 airports and airfields serve the aeronautical needs of more than 90,000 active U.S.-registered civil aircraft. These airports and airfields range in size from 20 to 10,000 acres of land. Some provide facilities for general aviation, scheduled air carriers, and the military, while others are designed to serve only general civil aviation. Roughly 30 per cent of the facilities are paved, and about the same percentage are lighted. Approximately 60 per cent of the airports in the nation are privately owned.

Federal regulation of civil aviation is carried out by the Federal Aviation Administration (FAA), created by the Federal Aviation Act of 1958 to meet the current and future needs of American aviation. The FAA is directed by an administrator appointed by the President and confirmed by the Senate. In meeting the needs of American aviation the FAA develops the major policies necessary to guide its long-range growth, modernizes the air-traffic-control system, establishes in a single authority the essential management functions to support the common needs of civil and military operations, and provide for the most effective and efficient use of the airspace over the United States.

Table 11-1   Airports\* on record with FAA, by state and other area, Dec. 31, 1957–1965†

| State or other area | 1957 | 1958 | 1959 | 1960 | 1961 | 1962 | 1963 | 1964 | 1965 |
|---|---|---|---|---|---|---|---|---|---|
| Total................... | 6,412 | 6,018 | 6,426 | 6,881 | 7,715 | 8,084 | 8,814 | 9,490 | 9,566 |
| United States........... | 6,395 | 6,001 | 6,409 | 6,865 | 7,695 | 8,062 | 8,788 | 9,463 | 9,542 |
| Alabama.............. | 62 | 68 | 77 | 76 | 85 | 90 | 100 | 116 | 113 |
| Alaska............... | 281 | 297 | 344 | 366 | 373 | 431 | 519 | 549 | 547 |
| Arizona.............. | 151 | 106 | 111 | 118 | 133 | 143 | 171 | 185 | 183 |
| Arkansas............. | 79 | 78 | 86 | 94 | 97 | 101 | 104 | 119 | 122 |
| California............ | 369 | 395 | 394 | 401 | 448 | 505 | 574 | 627 | 624 |
| Colorado............. | 95 | 81 | 83 | 91 | 98 | 124 | 158 | 163 | 164 |
| Connecticut.......... | 28 | 28 | 30 | 40 | 82 | 89 | 91 | 88 | 85 |
| Delaware............. | 19 | 20 | 18 | 17 | 17 | 16 | 18 | 19 | 19 |
| District of Columbia... | 1 | 1 | 1 | 1 | 2 | 4 | 3 | 5 | 4 |
| Florida.............. | 137 | 140 | 151 | 152 | 168 | 173 | 218 | 252 | 266 |
| Georgia.............. | 95 | 94 | 107 | 110 | 121 | 122 | 153 | 161 | 158 |
| Hawaii.............. | 16 | 16 | 35 | 39 | 38 | 38 | 42 | 46 | 46 |
| Idaho............... | 164 | 152 | 152 | 160 | 172 | 172 | 173 | 178 | 164 |
| Illinois.............. | 269 | 241 | 270 | 267 | 272 | 277 | 302 | 319 | 323 |
| Indiana.............. | 124 | 128 | 133 | 137 | 140 | 144 | 152 | 159 | 150 |
| Iowa................ | 158 | 150 | 169 | 179 | 190 | 193 | 203 | 212 | 213 |
| Kansas.............. | 191 | 171 | 185 | 214 | 220 | 219 | 236 | 260 | 263 |
| Kentucky............ | 48 | 33 | 32 | 35 | 40 | 51 | 55 | 59 | 60 |
| Louisiana............ | 75 | 69 | 82 | 150 | 155 | 155 | 161 | 193 | 186 |
| Maine............... | 110 | 86 | 83 | 75 | 133 | 134 | 137 | 139 | 137 |
| Maryland............ | 47 | 41 | 44 | 50 | 63 | 70 | 72 | 76 | 79 |
| Massachusetts........ | 87 | 80 | 80 | 74 | 109 | 109 | 111 | 124 | 118 |
| Michigan............ | 194 | 187 | 197 | 215 | 216 | 226 | 226 | 235 | 241 |
| Minnesota........... | 199 | 189 | 210 | 215 | 244 | 246 | 245 | 240 | 238 |
| Mississippi........... | 74 | 73 | 76 | 77 | 79 | 86 | 105 | 147 | 149 |
| Missouri............. | 114 | 109 | 128 | 158 | 186 | 191 | 210 | 231 | 248 |
| Montana............. | 151 | 137 | 142 | 158 | 170 | 168 | 175 | 186 | 185 |
| Nebraska............ | 171 | 171 | 168 | 181 | 179 | 180 | 193 | 195 | 269 |
| Nevada.............. | 57 | 50 | 50 | 52 | 64 | 68 | 66 | 76 | 73 |
| New Hampshire....... | 32 | 28 | 24 | 24 | 33 | 34 | 35 | 37 | 41 |
| New Jersey........... | 104 | 93 | 88 | 97 | 112 | 118 | 119 | 130 | 126 |
| New Mexico.......... | 100 | 81 | 83 | 90 | 100 | 102 | 110 | 114 | 113 |
| New York............ | 266 | 216 | 225 | 243 | 305 | 308 | 312 | 321 | 328 |
| North Carolina....... | 101 | 99 | 105 | 103 | 113 | 127 | 142 | 167 | 169 |
| North Dakota......... | 165 | 137 | 155 | 161 | 162 | 163 | 169 | 170 | 172 |
| Ohio................ | 271 | 245 | 263 | 340 | 359 | 371 | 377 | 397 | 401 |
| Oklahoma............ | 117 | 112 | 125 | 130 | 137 | 143 | 163 | 195 | 195 |
| Oregon.............. | 116 | 128 | 135 | 148 | 156 | 158 | 165 | 187 | 181 |
| Pennsylvania......... | 278 | 250 | 261 | 258 | 427 | 450 | 473 | 481 | 463 |
| Rhode Island......... | 9 | 10 | 10 | 10 | 13 | 13 | 13 | 14 | 11 |
| South Carolina....... | 61 | 64 | 69 | 71 | 78 | 85 | 90 | 99 | 95 |
| South Dakota........ | 70 | 69 | 69 | 77 | 83 | 87 | 89 | 98 | 102 |
| Tennessee............ | 48 | 47 | 54 | 58 | 66 | 71 | 105 | 110 | 107 |

**Table 11-1   Airports\* on record with FAA, by state and other area, Dec. 31, 1957–1965†**
**(continued)**

| State or other area | 1957 | 1958 | 1959 | 1960 | 1961 | 1962 | 1963 | 1964 | 1965 |
|---|---|---|---|---|---|---|---|---|---|
| Texas................ | 477 | 464 | 505 | 543 | 588 | 621 | 705 | 812 | 846 |
| Utah................ | 53 | 56 | 57 | 59 | 66 | 69 | 70 | 69 | 68 |
| Vermont............. | 23 | 23 | 24 | 26 | 28 | 29 | 32 | 38 | 38 |
| Virginia............. | 100 | 87 | 87 | 90 | 94 | 97 | 105 | 110 | 113 |
| Washington........... | 179 | 165 | 172 | 168 | 189 | 197 | 212 | 215 | 203 |
| West Virginia......... | 50 | 43 | 44 | 46 | 53 | 52 | 53 | 53 | 54 |
| Wisconsin............ | 150 | 141 | 161 | 161 | 170 | 173 | 192 | 202 | 203 |
| Wyoming............. | 59 | 52 | 55 | 60 | 69 | 69 | 84 | 85 | 86 |
| Outside United States.... | 17 | 17 | 17 | 16 | 20 | 22 | 26 | 27 | 24 |
| Puerto Rico.......... | 9 | 9 | 9 | 8 | 12 | 13 | 18 | 20 | 17 |
| Virgin Islands........ | 2 | 2 | 2 | 2 | 2 | 3 | 3 | 2 | 2 |
| South Pacific Islands‡.. | 6 | 6 | 6 | 6 | 6 | 6 | 5 | 5 | 5 |

\* Includes seaplane bases, heliports, and military fields having joint civil-military use.
† Office of Management Services, FAA.
‡ American Samoa, Canton, Guam, Wake.

Federal regulation of civil aviation is also administered by the Civil
Aeronautics Board (CAB), an independent five-man panel.   The Board
issues certificates permitting persons to engage in air transportation as a
business and is concerned with the economic regulation of air carriers.
It investigates and determines the probable cause of aircraft accidents
and compiles accident reports and statistics.

**11-2   Airports**   At the end of 1964, there were 9,490 airports and airfields
on record with the Federal Aviation Administration.   Of these facilities,
2,773 had runway lighting equipment and 2,620 had paved runways.
The increasing number of airports providing runway lengths of 7,000 ft
and over reflects the continued growth of turbine-powered aircraft traffic.
   Airports, forming a network of the national aviation system, are
classified by FAA into two major categories: those for air-carrier operations and those for general aviation.   Air-carrier airports serve primarily the needs of scheduled air carriers, but they also accommodate
a large percentage of all the other segments of civil aviation.   General
aviation airports serve all civil aircraft other than civil air carriers,
catering to a wide variety of aviation activity including air taxi services,
business transportation, and many agricultural and industrial pursuits.
FAA records for 1962 show that the general aviation fleet, over 80,000
aircraft, flew over four times as many hours and over twice as many
miles as the domestic scheduled airlines.

Table 11-2   Eligible aircraft per 1,000 square miles and per 10,000 population by state, Dec. 31, 1965*

| State or other area | Aircraft per 1,000 square miles | Aircraft per 10,000 population | Total eligible aircraft | State area (square miles) | Estimated 1964 population (000) |
|---|---|---|---|---|---|
| Total | 27.0 | 5.0 | 97,741 | 3,618,779 | 196,428 |
| United States | 27.0 | 5.0 | 97,459 | 3,615,211 | 193,816 |
| Alabama | 23.2 | 3.4 | 1,201 | 51,609 | 3,462 |
| Alaska | 2.9 | 67.2 | 1,702 | 586,400 | 253 |
| Arizona | 12.8 | 9.0 | 1,462 | 113,909 | 1,608 |
| Arkansas | 26.3 | 7.1 | 1,397 | 53,104 | 1,960 |
| California | 82.6 | 7.0 | 13,108 | 158,693 | 18,602 |
| Colorado | 13.9 | 7.4 | 1,458 | 104,247 | 1,969 |
| Connecticut | 138.8 | 2.4 | 694 | 5,009 | 2,832 |
| Delaware | 156.0 | 6.3 | 320 | 2,057 | 505 |
| District of Columbia | ..... | .... | 599 | 69 | 803 |
| Florida | 61.9 | 6.2 | 3,630 | 58,560 | 5,805 |
| Georgia | 28.4 | 3.8 | 1,676 | 58,876 | 4,357 |
| Hawaii | 26.3 | 2.3 | 169 | 6,424 | 711 |
| Idaho | 11.1 | 13.4 | 929 | 83,557 | 692 |
| Illinois | 79.1 | 4.1 | 4,482 | 56,400 | 10,644 |
| Indiana | 70.5 | 5.2 | 2,562 | 36,291 | 4,885 |
| Iowa | 33.9 | 6.9 | 1,911 | 56,290 | 2,760 |
| Kansas | 33.8 | 12.4 | 2,782 | 82,264 | 2,234 |
| Kentucky | 17.4 | 2.2 | 704 | 40,395 | 3,179 |
| Louisiana | 31.5 | 4.3 | 1,530 | 48,523 | 3,534 |
| Maine | 13.1 | 4.4 | 438 | 33,215 | 993 |
| Maryland | 90.0 | 2.7 | 952 | 10,577 | 3,519 |
| Massachusetts | 151.8 | 2.3 | 1,253 | 8,257 | 5,348 |
| Michigan | 64.3 | 4.5 | 3,744 | 58,216 | 8,218 |
| Minnesota | 30.2 | 7.1 | 2,540 | 84,068 | 3,554 |
| Mississippi | 23.2 | 4.7 | 1,111 | 47,716 | 2,321 |
| Missouri | 35.4 | 5.4 | 2,472 | 69,686 | 4,497 |
| Montana | 8.1 | 16.9 | 1,199 | 147,138 | 706 |
| Nebraska | 17.6 | 9.2 | 1,363 | 77,227 | 1,477 |
| Nevada | 6.6 | 16.7 | 736 | 110,540 | 440 |
| New Hampshire | 28.2 | 3.9 | 263 | 9,304 | 669 |
| New Jersey | 264.8 | 3.0 | 2,074 | 7,836 | 6,774 |
| New Mexico | 8.9 | 10.5 | 1,089 | 121,666 | 1,029 |
| New York | 87.2 | 2.3 | 4,368 | 49,576 | 18,073 |
| North Carolina | 31.4 | 3.3 | 1,660 | 52,712 | 4,914 |
| North Dakota | 10.8 | 11.7 | 764 | 70,665 | 652 |
| Ohio | 100.8 | 4.0 | 4,148 | 41,222 | 10,245 |
| Oklahoma | 30.4 | 8.5 | 2,132 | 69,919 | 2,482 |
| Oregon | 19.8 | 10.1 | 1,925 | 96,981 | 1,899 |
| Pennsylvania | 67.3 | 2.6 | 3,048 | 45,333 | 11,520 |

Table 11-2  Eligible aircraft per 1,000 square miles and per 10,000 population by state,
Dec. 31, 1965* (continued)

| State or other area | Aircraft per 1,000 square miles | Aircraft per 10,000 population | Total eligible aircraft | State area (square miles) | Estimated 1964 population (000) |
|---|---|---|---|---|---|
| Rhode Island | 110.7 | 1.4 | 134 | 1,214 | 920 |
| South Carolina | 22.2 | 2.7 | 692 | 31,055 | 2,542 |
| South Dakota | 10.0 | 11.0 | 775 | 77,047 | 703 |
| Tennessee | 29.9 | 3.2 | 1,266 | 42,244 | 3,845 |
| Texas | 29.8 | 7.5 | 7,971 | 267,339 | 10,551 |
| Utah | 6.8 | 5.8 | 583 | 84,916 | 990 |
| Vermont | 17.9 | 4.3 | 172 | 9,609 | 397 |
| Virginia | 30.4 | 2.7 | 1,244 | 40,815 | 4,457 |
| Washington | 32.4 | 7.4 | 2,213 | 68,192 | 2,990 |
| West Virginia | 18.5 | 2.4 | 448 | 24,181 | 1,812 |
| Wisconsin | 32.6 | 4.4 | 1,834 | 56,154 | 4,144 |
| Wyoming | 5.4 | 15.7 | 534 | 97,914 | 340 |
| Outside United States | ..... | .... | 282 | 3,568 | 2,612 |
| Puerto Rico | ..... | .... | 197 | 3,435 | 2,577 |
| Virgin Islands | ..... | .... | 23 | 133 | 35 |
| Other countries | ..... | .... | 62 | | |

* *Statistical Abstract of the United States*, 1965.  U.S. Bureau of the Census and Office of Management Services, FAA.

In formulating the National Airport Plan for 1966–1970, the Federal Aviation Administration has included for development 4,106 landing facilities.  Included in this total are 887 new facilities.  Qualitative improvements to the remaining airports, heliports, and seaplane facilities are included in the plan.

Distribution of airports on record with FAA by states is given in Table 11-1.  Of the total, air-carrier airports numbered 709 at the beginning of 1965 compared with 715 in 1964.  Leveling off in the size of the air-carrier fleet is reflected in the decline in the number of airports catering to the air-carrier segment of the aviation complex.  The growth trend in the number of airports can be attributed primarily to general aviation activity, especially for business purposes.

**11-3  Aircraft**  As of Jan. 1, 1965, the Federal Aviation Administration's records indicated 137,189 registered aircraft, of which about two-thirds, or 90,935, were active.  Aircraft registered with FAA are, like airports, divided into two categories—air carrier and general aviation.  The tran-

**Table 11-3  Eligible aircraft registered with FAA, by type, and by state and other area, Dec. 31, 1965***

| State or other area | Total eligible aircraft | Air carrier | General aviation | | | |
|---|---|---|---|---|---|---|
| | | | Fixed-wing multi-engine | Fixed-wing 1-engine 4-place and over | Rotor-craft | All other |
| Total................. | 97,741 | 2,299 | 11,977 | 49,789 | 1,503 | 32,173 |
| United States.......... | 97,459 | 2,283 | 11,902 | 49,653 | 1,492 | 32,129 |
| Alabama............ | 1,201 | 0 | 167 | 638 | 5 | 391 |
| Alaska............. | 1,702 | 102 | 64 | 725 | 50 | 761 |
| Arizona............ | 1,462 | 1 | 159 | 768 | 27 | 507 |
| Arkansas........... | 1,397 | 0 | 195 | 601 | 7 | 594 |
| California.......... | 13,108 | 172 | 1,352 | 6,961 | 320 | 4,303 |
| Colorado........... | 1,458 | 47 | 147 | 791 | 15 | 458 |
| Connecticut........ | 694 | 2 | 86 | 338 | 16 | 252 |
| Delaware........... | 320 | 6 | 76 | 143 | 1 | 94 |
| District of Columbia.. | 599 | 45 | 192 | 295 | 18 | 49 |
| Florida............. | 3,630 | 137 | 636 | 1,779 | 69 | 1,009 |
| Georgia............ | 1,676 | 122 | 223 | 761 | 6 | 564 |
| Hawaii............. | 169 | 23 | 36 | 44 | 4 | 62 |
| Idaho............. | 929 | 0 | 64 | 497 | 20 | 348 |
| Illinois............. | 4,482 | 282 | 594 | 2,395 | 31 | 1,180 |
| Indiana............ | 2,560 | 25 | 378 | 1,460 | 38 | 659 |
| Iowa.............. | 1,911 | 0 | 158 | 1,130 | 9 | 614 |
| Kansas............ | 2,782 | 0 | 329 | 1,678 | 8 | 767 |
| Kentucky.......... | 704 | 1 | 107 | 391 | 7 | 198 |
| Louisiana.......... | 1,530 | 0 | 195 | 602 | 105 | 628 |
| Maine............. | 438 | 0 | 31 | 179 | 4 | 224 |
| Maryland.......... | 952 | 0 | 109 | 486 | 19 | 338 |
| Massachusetts....... | 1,253 | 23 | 151 | 638 | 25 | 416 |
| Michigan........... | 3,744 | 39 | 445 | 2,028 | 21 | 1,211 |
| Minnesota.......... | 2,540 | 105 | 185 | 1,152 | 17 | 1,081 |
| Mississippi......... | 1,111 | 0 | 125 | 448 | 10 | 528 |
| Missouri........... | 2,472 | 231 | 297 | 1,254 | 15 | 675 |
| Montana........... | 1,199 | 2 | 74 | 613 | 13 | 497 |
| Nebraska.......... | 1,363 | 1 | 118 | 696 | 18 | 530 |
| Nevada............ | 736 | 29 | 121 | 373 | 34 | 179 |
| New Hampshire...... | 263 | 0 | 30 | 131 | 0 | 102 |
| New Jersey......... | 2,074 | 21 | 254 | 1,110 | 29 | 660 |
| New Mexico......... | 1,089 | 0 | 143 | 701 | 3 | 242 |
| New York.......... | 4,368 | 599 | 636 | 1,765 | 65 | 1,303 |
| North Carolina....... | 1,660 | 43 | 188 | 847 | 17 | 565 |
| North Dakota....... | 764 | 0 | 25 | 297 | 3 | 439 |
| Ohio.............. | 4,148 | 1 | 661 | 2,211 | 33 | 1,242 |

Table 11-3   Eligible aircraft registered with FAA, by type, and by state and other area, Dec. 31, 1965* (continued)

| State or other area | Total eligible aircraft | Air carrier | General aviation | | | |
|---|---|---|---|---|---|---|
| | | | Fixed-wing multi-engine | Fixed-wing 1-engine 4-place and over | Rotor-craft | All other |
| Oklahoma | 2,132 | 3 | 287 | 1,084 | 43 | 715 |
| Oregon | 1,925 | 1 | 211 | 1,109 | 46 | 558 |
| Pennsylvania | 3,048 | 0 | 495 | 1,516 | 72 | 965 |
| Rhode Island | 134 | 0 | 16 | 65 | 7 | 46 |
| South Carolina | 692 | 0 | 88 | 369 | 12 | 223 |
| South Dakota | 775 | 0 | 37 | 347 | 1 | 390 |
| Tennessee | 1,266 | 29 | 221 | 638 | 4 | 374 |
| Texas | 7,971 | 152 | 1,137 | 4,023 | 118 | 2,541 |
| Utah | 583 | 1 | 49 | 369 | 8 | 156 |
| Vermont | 172 | 0 | 18 | 88 | 0 | 66 |
| Virginia | 1,244 | 1 | 123 | 665 | 18 | 437 |
| Washington | 2,213 | 36 | 113 | 1,137 | 56 | 871 |
| West Virginia | 448 | 1 | 66 | 216 | 7 | 158 |
| Wisconsin | 1,834 | 0 | 226 | 824 | 13 | 771 |
| Wyoming | 534 | 0 | 64 | 277 | 5 | 188 |
| Outside United States | 282 | 16 | 75 | 136 | 11 | 44 |
| Puerto Rico | 197 | 15 | 48 | 94 | 10 | 30 |
| Virgin Islands | 23 | 0 | 15 | 8 | 0 | 0 |
| Other countries | 62 | 1 | 12 | 34 | 1 | 14 |

* Office of Management Services, FAA.

sition to a predominantly turbine-powered air-carrier fleet has increased its capacity so that future changes may stem more from the fleet's composition than from its numerical size.   An appreciation of aircraft distribution, composition, and growth trends for both air-carrier and general aviation fleets may be gained from a study of Tables 11-2 and 11-3.

According to FAA estimates, if present trends of growth continue, by 1970 the active civil aircraft fleet will total 107,090.   Of this total, it is expected that 2,090 aircraft will constitute the air-carrier fleet.   Since 1959 there has been a steady decline in the number of single-engine aircraft of three places and less, and a steady increase in the numbers of multiengine aircraft and four-place-and-over single-engine aircraft.

Present trends toward larger aircraft for the air-carrier fleet will

Table 11-4  Domestic intercity passenger-miles, by mode of travel and class of service (in millions), 1955–1965*

| Mode and class | 1955 | 1956 | 1957 | 1958 | 1959 | 1960 | 1961 | 1962 | 1963 | 1964 | 1965 |
|---|---|---|---|---|---|---|---|---|---|---|---|
| Total | 648,404 | 680,139 | 681,788 | 690,726 | 722,388 | 745,757 | 757,234 | 788,695 | 815,549 | 843,652 | 873,968 |
| Total common carrier | 62,587 | 64,625 | 65,534 | 61,230 | 62,953 | 65,140 | 65,234 | 68,995 | 74,549 | 80,652 | 87,968 |
| Scheduled air carrier† | 19,740 | 22,276 | 25,246 | 25,256 | 29,152 | 30,375 | 30,879 | 33,436 | 38,252 | 43,903 | 51,608 |
| Regular service | 13,024 | 14,202 | 15,736 | 15,180 | 16,848 | 15,957 | 13,770 | 12,499 | 13,585 | 13,916 | 15,099 |
| Coach service | 6,716 | 8,074 | 9,510 | 10,076 | 12,304 | 14,418 | 17,109 | 20,937 | 24,667 | 29,987 | 36,509 |
| Class I line-haul railways‡ | 23,747 | 23,349 | 20,988 | 18,474 | 17,501 | 17,065 | 16,155 | 15,859 | 14,397 | 14,049 | 13,260 |
| First-class service | 6,440 | 6,275 | 5,185 | 4,249 | 3,798 | 3,643 | 3,262 | 3,102 | 2,611 | 2,416 | 2,191 |
| Coach service | 17,307 | 17,074 | 15,803 | 14,225 | 13,703 | 13,422 | 12,893 | 12,757 | 11,786 | 11,633 | 11,069 |
| Motor carriers§—Class I, II, III. | 19,100 | 19,000 | 19,300 | 17,500 | 16,300 | 17,700 | 18,200 | 19,700 | 21,900 | 22,700¶ | 23,100¶ |
| Private automobile | 585,817 | 615,514 | 616,254 | 629,496 | 659,435 | 680,617 | 692,000 | 719,700 | 741,000 | 763,000¶ | 786,000¶ |
| Per cent air to total | 3.0 | 3.3 | 3.7 | 3.7 | 4.0 | 4.1 | 4.1 | 4.2 | 4.7 | 5.2 | 5.9 |
| Per cent air to total common carrier | 31.5 | 34.5 | 38.5 | 41.2 | 46.3 | 46.6 | 47.3 | 48.5 | 51.3 | 54.4 | 58.7 |
| Per cent air to total rail | 83.1 | 95.4 | 120.3 | 136.7 | 166.6 | 178.0 | 191.1 | 210.8 | 265.7 | 312.5 | 389.2 |
| Per cent air to first-class rail | 306.5 | 355.0 | 486.9 | 594.4 | 767.6 | 833.8 | 946.6 | 1,077.9 | 1,465.0 | 1,817.2 | 2,355.5 |

* Interstate Commerce Commission, Bureau of Transport Economics and Statistics, and National Association of Motor Bus Owners. Compiled by Office of Management Services, FAA.
† Scheduled operations of domestic trunk and local service carriers.
‡ Includes Pullman Co. and excludes commutation.
§ Estimated.  Excludes intra-state and other local movements.
¶ Estimated intercity passenger-miles, revised series.

continue.  It is expected that by 1975 the supersonic transport (SST) will be carrying 300 to 400 passengers on intercontinental flights at a cruising speed of 1,800 mph.  The gross weight of such an aircraft for take-off will be in the order of $\frac{1}{2}$ million lb.

**11-4  Operations**  The place of aviation in domestic intercity passenger travel can be seen in Table 11-4, tabulated in millions of passenger miles. Operations within the conterminous United States, interisland in Hawaii, and intra-Alaska are classified as domestic.  Scheduled air-carrier service provides more than half of the common-carrier portion of this travel service.

The advent of air-carrier aircraft into movement of air cargo (express and freight) is presented in Table 11-5 for the conterminous United States.  Including domestic, international, and territorial traffic, during 1964 nearly 1.1 million tons of air cargo were enplaned and flown 1,379.8 million ton-miles, and 330.4 tons of mail were enplaned and flown 382.3 million ton-miles.

The FAA has developed an air traffic hub structure to measure civil air traffic.  An air traffic hub is coextensive with the metropolitan area of a single city or a standard metropolitan statistical area (SMSA) as defined by the U.S. Bureau of the Census.  This hub structure is used by the FAA as its principal operations control and as a basic control for

**Table 11-5   Domestic airline traffic enplaned at United States stations (excluding Alaska and Hawaii), 1955–1965***

| Year | Air-carrier aircraft departures | Number of enplaned passengers | Tons of enplaned mail | Tons of enplaned cargo |
|---|---|---|---|---|
| 1955.......... | 2,901,758 | 37,226,432 | 124,263.2 | 389,307.9 |
| 1956.......... | 3,094,075 | 40,752,563 | 132,112.7 | 422,517.1 |
| 1957.......... | 3,318,282 | 44,017,548 | 142,052.3 | 434,788.4 |
| 1958.......... | 3,176,102 | 43,568,139 | 150,788.3 | 431,562.3 |
| 1959.......... | 3,420,682 | 49,357,870 | 164,216.2 | 501,713.6 |
| 1960.......... | 3,343,989 | 50,584,135 | 183,663.1 | 510,492.5 |
| 1961.......... | 3,248,467 | 55,011,493 | 201,875.7 | 553,465.2 |
| 1962.......... | 3,209,158 | 58,911,587 | 221,676.7 | 646,663.1 |
| 1963.......... | 3,330,479 | 67,318,615 | 230,878.5 | 701,990.1 |
| 1964.......... | 3,430,382 | 76,657,102 | 206,152.9 | 863,811.4 |
| 1965.......... | 3,825,787 | 89,123,088 | 296,102.8 | 1,080,239.3 |

* 1955–1961, FAA "Air Commerce Traffic Pattern"; 1962–1965, CAB-FAA "Airport Activity Statistics of Certificated Route Air Carriers."

**Table 11-6   FAA air traffic hub classification**

|                       | Per cent of total     |
| Hub classification    | enplaned passengers   |
|-----------------------|-----------------------|
| Large................ | 1.00 or more          |
| Medium............... | 0.25 to 0.99          |
| Small................ | 0.05 to 0.24          |
| Nonhub............... | Less than 0.05        |

**Table 11-7   Comparative accident data, 1948–1965\***

Passenger fatalities per 100,000,000 passenger-miles

| Year | Passenger automobiles and taxis | Buses | Railroad passenger trains | Domestic scheduled air transport planes |
|------|------|------|------|------|
| 1948......... | 2.10 | 0.18 | 0.13 | 1.33 |
| 1949......... | 2.70 | 0.20 | 0.08 | 1.32 |
| 1950......... | 2.90 | 0.18 | 0.58 | 1.15 |
| 1951......... | 3.00 | 0.24 | 0.43 | 1.30 |
| 1952......... | 3.00 | 0.21 | 0.04 | 0.35 |
| 1953......... | 2.90 | 0.18 | 0.16 | 0.56 |
| 1954......... | 2.70 | 0.11 | 0.08 | 0.09 |
| 1955......... | 2.70 | 0.18 | 0.07 | 0.76 |
| 1956......... | 2.70 | 0.16 | 0.20 | 0.62 |
| 1957......... | 2.60 | 0.19 | 0.07 | 0.12 |
| 1958......... | 2.30 | 0.17 | 0.27 | 0.43 |
| 1959......... | 2.30 | 0.21 | 0.05 | 0.69 |
| 1960......... | 2.20 | 0.13 | 0.16 | 0.93 |
| 1961......... | 2.10 | 0.19 | 0.10 | 0.38 |
| 1962......... | 2.20† | 0.11 | 0.14 | 0.34 |
| 1963......... | 2.30 | 0.23 | 0.07 | 0.12 |
| 1964......... | 2.40 | 0.15 | 0.05 | 0.14 |
| 1965......... | 2.40 | 0.16 | 0.06 | 0.38 |

\* Motor-vehicle data (automobiles, taxis, and buses) from the National Safety Council "Accident Facts" based on data from state traffic authorities, Bureau of Public Roads, National Association of Motor Bus Operators, and the American Transit Association.   Railroad data from the National Safety Council "Accident Facts" based on data from the Interstate Commerce Commission. Domestic scheduled air transport data from CAB.   Compiled by Office of Management Services, FAA.
† Revised.

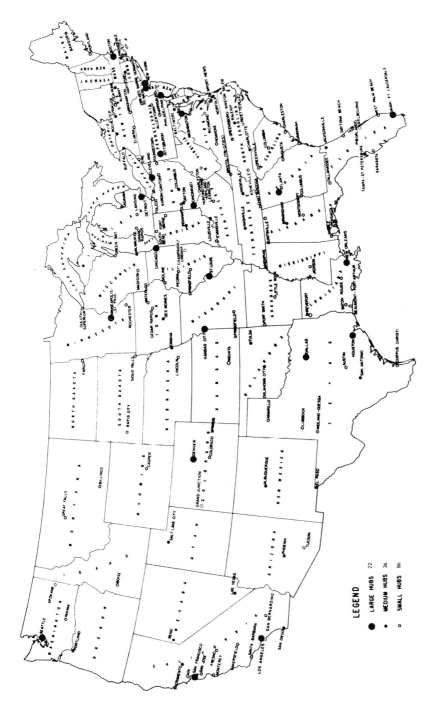

**Fig. 11-1** Air traffic hubs as of Dec. 31, 1965. (*Source: Office of Management Services, F.A.A.*)

most of its economic and operations research procedures.  Hub classification is made on the basis of total enplaned passengers in scheduled
domestic air-carrier service for the preceding calendar year.   This classification is shown in Table 11-6.   For example, as of Jan. 1, 1964, a large
hub had 766,571 or more enplaned passengers during 1964 (see Table
11-5), and a nonhub had less than 38,329 enplaned passengers during
1964.

As of Jan. 1, 1965, there were 142 air traffic hubs, representing 27.6
per cent of the 515 communities in the 48 states and the District of
Columbia receiving scheduled air-carrier service.   Geographic locations
of the air traffic hubs are shown in Fig. 11-1.

A comparative evaluation of safety of domestic scheduled air-carrier
operations is presented in Table 11-7.   It is notable that safety of air
passenger transport has varied over a considerable range from year to
year during the period shown in the table.   Thorough investigation and
analysis of aviation accidents by the Civil Aeronautics Board, followed
by remedial action taken jointly with the Federal Aviation Administration, have led to continued improvement in air traffic safety.

In its "Aviation Forecasts," December 1964, the FAA predicts for
1975: 130 million domestic air-carrier revenue passengers, traveling an
estimated 75 billion passenger-miles, and an increase of about 50 per cent
over 1964 in hours flown in general aviation—business, industrial, instruction and personal flying—to 23 million hr.

Even greater proportionate increases are generally expected for air
cargo transport, particularly with presently planned development of
large convertible airplanes capable of carrying a payload of about 500
passengers, or of being converted to cargo planes carrying payloads of
100 tons of containerized freight.   Gross take-off weight of these airplanes is expected to be upward of 600 kips.

### BIBLIOGRAPHY

"FAA Statistical Handbook of Aviation," Federal Aviation Administration, 1965.
Horonjeff, Robert: "Planning and Design of Airports," McGraw-Hill Book Company,
    New York, 1962.

# 12

# Airport Layout Plan

**12-1 Planning** The airport has grown to be an important part of the total transportation system. No community development plan is complete without adequate provision for present and future airport requirements. Airport needs will increase as the community grows and expands both in population and in economic functions, and the size and number of aircraft will be to a large extent reflected by the size and economic characteristics of the community which the airport is to serve. The student of airport engineering should recognize the importance of careful study of probable future growth of the community for which airport construction is contemplated, so that adequate provision can be made for expansion of airport facilities to meet the growing demands of the community even though only part of the complete development plan will be undertaken initially.

Airport planning must be integrated with highway, rail, and other ground transportation planning for the community that the airport is to serve, considering not only present needs but also transportation needs in the foreseeable future. Local and regional planning agencies can

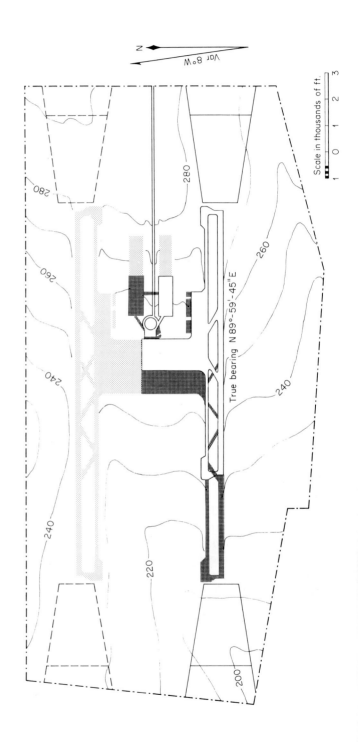

Scale in thousands of ft.

1  0  1  2  3

True bearing N 89°-59'-45"E

N

Var 8°W

280

260

240

220

200

280

260

240

240

**196**

LEGEND

☐ Existing
▦ First phase construction
▓ Second phase construction

NOTES

1. Wind data
   a. Source: National Weather Records Center, Asheville, N.C.          10 yrs. obs.
   b. Total runway wind direction coverage                              98.1%

2. Basic airport data
   a. Site reference temperature
      1. Average temperature of hottest month (°F)                      80
      2. Normal maximum temperature of hottest month (°F)               100
   b. Airport elevation (feet)                                          270
   c. Pavement design strength
   d. Principal subgrade class                                          Ra
   e. Critical aircraft used for runway determination
      1. Present                                                        DC-6B
      2. Future                                                         707-100B
   f. Effective runway gradient
      1. Existing                                                       0.67%
      2. Proposed                                                       0.67%
   g. Designated instrument runway                                      10L-28R

△ = 9.4% Calms, 0-3 M.P.H.

**Fig. 12-1** Typical airport layout plan.   (*Courtesy of F.A.A.*)

197

provide necessary data on present and future population growth, business and industrial development, and general control of land use, all of which are essential to development of an integrated transportation system. In general, smaller airports, catering to general aviation for business, industry, agricultural services, and flight training, and to only the smaller aircraft of the air-carrier fleet, may be well served by highway facilities for collection and distribution of passengers and cargo at each end of the flights. The larger air terminals, engaged in the very near future in handling large volumes of both passengers and freight for domestic and international transport, may require access to freeways and expressways, extension of urban rail transit, and provision of helicopter services to minimize ground time and thus preserve the advantages of rapid flight.

Some of the basic requirements for airport planning are exemplified in Figs. 12-1 and 12-2 (Fig. 12-2 is located on the foldout opposite page 598). It is important, of course, that the airport occupy an easily accessible site located close to the population center or centers that it is to serve—a position favorable to coordination with other modes of transportation—highways, railways, and navigation.

Initial acquisition of land should include sufficient area for future expansion as indicated by forecasts of local or regional planning agencies regarding population growth and economic development for the communities to be served by the airport. Dimensions and orientation of the land area will be governed by required runway length for the anticipated largest aircraft using the airport and by the direction of the prevailing winds. Ownership of land beyond the ends of the runways will give the airport authority direct control of land use in the approach zones. These approach zones should be free—or be made free—of any obstructions to take-off and landing of planes and should not pass directly over residential developments and schools.

**12-2   Selection of site**   Generally, grading quantities will be excessive if ground slopes exceed about $1\frac{1}{2}$ per cent. However, some natural ground slope is desirable to minimize the problem of providing adequate drainage for the airport site.

An elevated site may have some advantages over a valley site. Obstructions in the approach zones will be less likely to exist or develop; drainage of the area will be easier because of more adequate natural drainage, especially if it is a hilltop location; and from the standpoint of aircraft operation, winds are likely to be more uniform and visibility better because of less likelihood of local fog or smoke haze.

A valley site is generally more level, requiring less grading. It may be somewhat more accessible to ground transportation, and installation of public utilities is likely to be less costly.

For guidance in selecting a site and making an airport layout plan, two key phrases stand out—stage construction and room for expansion. Initial stages of construction will be made in accordance with present needs, but careful planning will make provision for expansion of facilities to meet future demands as can reasonably be foreseen and estimated. Furthermore, it will be necessary to estimate the largest aircraft that will serve future demands at this particular airport. Runways may have to be lengthened and taxiways extended to meet them; apron area enlarged to accommodate increasing commercial traffic and larger planes; private-plane parking facilities expanded; buildings enlarged to make room for increased freight and passenger services, maintenance services, lobbies, locker room, and other concessions; and car-parking areas expanded. Many of these expanding requirements are illustrated in Fig. 12-1.

A contour map of the site to a horizontal scale of about 1 in. = 200 ft and contour interval of 3 to 5 ft, depending on general topography, is necessary. Boundaries of soil types within the area and details of soil horizons of each type will be necessary for final adjustment of grade lines so as to make the best use of available materials. Some data regarding these factors and weather information may be assumed, but use should be made of all possible pertinent factual data locally available.

The particular pattern of runways, taxiways, and other facilities is incidental to good coordination. For example, it is likely that two runways will be adequate for proper wind coverage, depending, of course, on actual wind data. These may be placed in the form of an *L*, *X*, or plus (+), or they may not intersect at all; but such factors as topography and soil conditions, the location of the approach road, and the position of building area must be considered so as to provide convenient access from the road and to place runways as near as possible to apron and buildings, thus keeping length of taxiways at a minimum. Some basic elements of the layout plan are illustrated in Fig. 12-3.

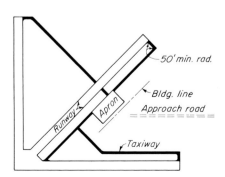

**Fig. 12-3**   Illustration of airport layout.

Requirements for design, construction, and operation of the airport all have some bearing on site selection and layout planning. These requirements are taken up in subsequent chapters.

**12-3   Recommendations for airport development**   In connection with airport planning and layout, the following recommendations[1] are appropriate and may serve to establish a generally acceptable philosophy for airport development.

The Commission feels that definite arrangements should be made and specific governmental agencies designated to develop and to implement the following recommendations:

1. *Support required airport development.*   New airports will be needed and present airports must be improved. State, county, and municipal governments should be prepared to assume their proper share of this expense.
2. *Expand Federal-aid airport program.*   Authorization of matching funds for Federal aid to airports should be implemented by adequate appropriations.   Highest priority in the application of Federal aid should be given to runways and their protective extensions incorporated into the airport, to bring major municipal airports up to standards recommended in this report.
3. *Integrate municipal and airport planning.*   Airports should be made a part of community master plans completely integrated with transportation requirements for passenger, express, freight, and postal services.   Particular attention should be paid to limited-access highways and other transportation facilities to reduce time to the airport from sources of air transport business.
4. *Incorporate cleared runway extension areas into airports.*   The dominant runways of new airport projects should be protected by cleared extensions at each end *at least* one-half mile in length and 1,000 ft wide.   This area should be completely free from housing or any other form of obstruction.   Such extensions should be considered an integral part of the airport.
5. *Establish effective zoning laws.*   A fan-shaped zone, beyond the half-mile cleared extension described in Recommendation 4, at least two miles long and 6,000 ft wide at its outer limits, should be established at new airports by zoning law, air ease-

---

[1] "The Airport and Its Neighbors," Report of the President's Airport Commission, 1952.

ment or land purchase at each end of dominant runways. In this area, the height of buildings and also the use of the land should be controlled to eliminate the erection of places of public assembly, churches, hospitals, schools, etc., and to restrict residences to the more distant locations within the zone.

6. *Improve existing airports.* Existing airports must continue to serve their communities. However, cities should go as far as is practical toward developing the cleared areas and zoned runway approaches recommended for new airports. No further building should be permitted on runway extensions and, wherever possible, objectionable structures should be removed. Operating procedures should be modified in line with Commission recommendations for minimizing hazard and nuisance to persons living in the vicinity of such airports.

7. *Clarify laws and regulations governing use of airspace.* Authority of the Federal, state, or municipal governments with respect to the regulation of the use of airspace should be clarified to avoid conflicting regulation and laws.

8. *Define navigable airspace in approach zones.* The limits of the navigable airspace for glide path or take-off patterns at airports should be defined.

9. *Extend Civil Aeronautics Act to certificate airports.* The Civil Aeronautics Act should be amended to require certification of airports necessary for interstate commerce and to specify the terms and conditions under which airports so certified shall be operated. Certificates should be revoked if minimum standards for safety are not maintained. Closing or abandonment of an airport should be ordered or allowed only if clearly in the public interest.

10. *Maintain positive air traffic control.* Certain air traffic control zones in areas of high air traffic density should be made the subject of special regulations to insure that all aircraft within the zone are under positive air traffic control at all times regardless of weather.

11. *Raise circling and maneuvering minimums.* Present straight-in instrument approach minimums are considered satisfactory but the minimum ceilings and visibilities under which aircraft are permitted to circle or maneuver under the overcast in congested terminal areas should be raised.

12. *Accelerate installation of aids to air navigation.* Research and development programs and installation projects designed to improve aids to navigation and traffic control in the vicinity of airports, especially in congested areas, should be accelerated.

Installation and adequate manning of radar traffic control systems should be given high priority.

13. *Revise present cross-wind component limits.* Existing cross-wind component limitations should be reviewed to establish more liberal cross-wind landing and take-off specifications for each transport-type aircraft.

14. *Develop and use cross-wind equipment.* Although modern transport aircraft can operate successfully in any but very strong cross-winds, the further development and use of special cross-wind landing gears should be accelerated.

15. *Extend use of single runway system.* New airports should adopt a single or parallel runway design. This should be adequate except under strong wind conditions, in which case a shorter runway at 90 deg to the main one may be required. Present airports should plan to develop the dominant runway at the expense of those less used. Airport expansion should be achieved through additional parallel runways.

16. *Meet standard requirements for runway length.* For each category of airport a standard runway length has been established consistent with its future planned use. Airports should bring their runways up to the standard. For intercontinental or transcontinental airports, the length of the dominant runways should be 8,400 ft with possibility of expansion to 10,000 ft if later required and with clear approaches as per Recommendations 4 and 5.

17. *Accelerate ground noise reduction programs.* Engine run-up schedules and run-up locations should be adjusted to minimize noise near airports. Adequate acoustical treatment in run-up areas and at test stands should be provided.

18. *Instruct flight personnel concerning nuisance factors.* A tight discipline with respect to airport approach and departure procedures to minimize noise nuisance to people on the ground (within the limits of safe operating procedures) should be maintained at all times.

19. *Arrange flight patterns to reduce ground noise.* Airways and flight patterns near airports should be arranged to avoid unnecessary flight over thickly settled areas to minimize noise, but only within the limits of safe flight practice.

20. *Minimize training flights at congested airports.* Flight crew training should be conducted, as far as practicable, away from thickly settled areas and with a minimum number of flights into and out of busy airports.

21. *Minimize test flights near metropolitan areas.* Production fly-
    away from aircraft factories under proper conditions is accept-
    able, but all flights of experimental aircraft and test flying of
    production models near built-up areas should be reduced as
    far as possible.
22. *Avoid military training over congested areas.* Although the
    basing of reserve air units at airports near cities has been con-
    sidered generally desirable, and the location of certain combat
    units there is sometimes necessary, training maneuvers, par-
    ticularly with armed military aircraft, should be conducted
    only over open spaces. Rapid shuttle service to an outlying
    military training field offers minimum interference with civil
    air operations and maximum safety and freedom from nuisance
    to people on the ground.
23. *Separate military and civil flying at congested airports.* Military
    aircraft should not be based on congested civil airports except
    when it is not economically or otherwise feasible to provide
    separate facilities for them nor should commercial aircraft
    operate regularly from busy military airports.
24. *Provide more flight crew training.* Every flight crew should be
    required to have frequent drills in instrument and emergency
    procedures. This can be accomplished in part in flight simu-
    lators. These flight simulators should be located at con-
    venient points and should be available to all operators on a
    fair basis.
25. *Develop helicopters for civil use.* Concurrent with military heli-
    copter development, interested government agencies should
    encourage civil helicopter development for interairport shuttle
    services, and for short-haul use, emphasizing safety, reliability,
    and public toleration factors.

**12-4 Development of site** Cost estimates should be made for electric
power, water supply, and sanitary sewers. It will not always be economically feasible to extend these services from the municipality. Comparative costs for independent development of these facilities for the
different sites should be studied.

Comparison of construction costs which may materially differ at one
site as compared with another—grading, drainage, paving, building
foundations, parking areas, and access roads—can best be made by a
careful study of engineering properties of the soils in the area. Much
will depend on the extent to which the soil of the site can be used in these
construction items.

Topographic maps, soil maps, and aerial photographs, available for nearly every section of the country, will prove invaluable in locating boundaries of soil types, available borrow pits, mineral-aggregate pits, and drainage characteristics of the area.   More detailed soil surveys can then be made to determine for the different soil horizons such critical engineering properties as load-carrying capacity, moisture characteristics, and compactive qualities.

**12-5  Preliminary design procedure**   The foregoing preliminary considerations will lead to preparation of preliminary plans, estimate of quantities, and determination of costs.   A small-scale vicinity map showing the general topography of the area will be adequate to show the location of the airport in relation to the community it is to serve and to nearby airports, airlines, highways, railroads, power and telephone lines, and any nearby public utilities.

A larger-scale map will then be required to show more in detail the vertical control with respect to surrounding topography.   This map should include the airport site and enough of the approach zones to locate any possible obstructions.

**QUESTIONS AND PROBLEMS**

**12-1.** Make a selection, from the several possible locations, of an airport site for a community with which you are familiar.   This may be done even though one or more airports exist, taking into consideration such factors as:

    *a.* Availability of sufficient land area
    *b.* Accessibility and nearness to population center
    *c.* Coordination with other transportation facilities
    *d.* Topography and soil conditions
    *e.* Difficulties attending development of public utilities
    *f.* Obstructions, natural and man-made
    *g.* Traffic patterns of nearby airports

**BIBLIOGRAPHY**

Horonjeff, Robert: "Planning and Design of Airports," McGraw-Hill Book Company, New York, 1962.
"Preparation of Airport Layout Plans," Federal Aviation Administration, Advisory Circular 150/5310-1, September, 1965.

# 13
# Aircraft Data Related to
# Airport Classification
# and Design

The most important characteristic of aircraft affecting design of an airport is size. Specifically the designer needs to know the size of the largest aircraft that will, except in emergency, be using the facilities of the particular airport. Size, in turn, is essentially a function of weight. This may vary from less than a thousand pounds for the lightest fixed-wing personal planes to perhaps a million pounds for future supersonic transport (SST) planes. Increasing traffic demands in all categories of this vast array of aircraft imposes a need for consideration of aircraft size in both classification and design of airports.

**13-1 Airport classification** Airports, in general, cater to the aviation needs of two main categories of aircraft: (1) general aviation aircraft and (2) the air-carrier fleet. The latter category, air-carrier fleet, provides two general classes of service: (1) short- to medium-range flights on trunk routes and for continental flights up to 2,000 miles and (2) long-range international flights serving nonstop flights in the transcontinental, trans-

# Table 13-1a Typical aircraft data for general aviation (VFR)*

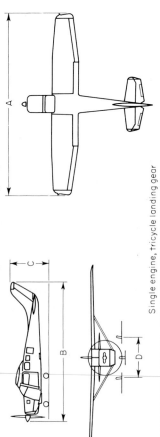

Single engine, tricycle landing gear

| Type certificate holder | Model | Popular name | No. seats | Horse-power | Max gross weight, lb | A wing span | B length | C height | D tread | Min turning radius |
|---|---|---|---|---|---|---|---|---|---|---|
| Air products. | 415 D,E,G | Evcoupe | 2 | 75 | 1,400 | 30 ft 0 in. | 20 ft 9 in. | 5 ft 11 in. | 7 ft 5 in. | 18 ft 9 in. |
| Beech........ | A35,B,R | Bonanza | 4 | 185 | 2,650 | 32 ft 10 in. | 25 ft 2 in. | 6 ft 6 in. | 9 ft 7 in. | 21 ft 3 in. |
| Beech........ | N-35 | Bonanza | 5 | 260 | 3,125 | 33 ft 5 in. | 25 ft 2 in. | 7 ft 7 in. | 9 ft 7 in. | 21 ft 6 in. |
| Beech........ | 35-A33 | Debonair | 4 | 225 | 3,000 | 32 ft 10 in. | 25 ft 6 in. | 8 ft 3 in. | 9 ft 7 in. | 21 ft 3 in. |
| Cessna........ | 150 | 150 | 2 | 100 | 1,500 | 33 ft 4 in. | 21 ft 11 in. | 6 ft 11 in. | 6 ft 5 in. | 19 ft 10 in. |
| Cessna........ | 210 | 210 | 4 | 260 | 2,900 | 36 ft 7 in. | 27 ft 9 in. | 8 ft 8 in. | 8 ft 2 in. | 22 ft 5 in. |
| Mooney...... | M-18C | Mite | 1 | 65 | 850 | 26 ft 10 in. | 17 ft 7 in. | 6 ft 2 in. | 5 ft 9 in. | 16 ft 4 in. |
| Mooney...... | M-20 | Mark 20 | 4 | 150 | 2,450 | 35 ft 0 in. | 23 ft 3 in. | 8 ft 3 in. | 9 ft 2 in. | 22 ft 1 in. |
| Navion...... | A | Navion | 4 | 205 | 2,750 | 33 ft 5 in. | 27 ft 8 in. | 8 ft 8 in. | 8 ft 2 in. | 20 ft 10 in. |
| Piper........ | PA-22-108 | Colt | 2 | 108 | 1,650 | 30 ft 0 in. | 20 ft 0 in. | 6 ft 3 in. | 9 ft 11 in. | 19 ft 11 in. |

* FAA.

**Table 13-1b   Typical aircraft data for general aviation (VFR)***

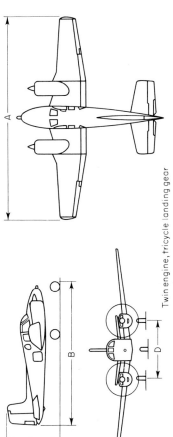

Twin engine, tricycle landing gear

| Type certificate holder | Model | Popular name | No. seats | Horse-power | Max gross weight | A wing span | B length | C height | D tread | Min turning radius |
|---|---|---|---|---|---|---|---|---|---|---|
| Aero........ | 500A | Commander | 7 | 260 | 6,250 | 49 ft 6 in. | 35 ft 1 in. | 14 ft 6 in. | 12 ft 11 in. | 31 ft 2 in. |
| Aero........ | 500B | Commander | 7 | 290 | 6,750 | 49 ft 6 in. | 35 ft 1 in. | 14 ft 6 in. | 12 ft 11 in. | 31 ft 2 in. |
| Aero........ | 560E | Commander | 7 | 295 | 6,500 | 49 ft 6 in. | 35 ft 1 in. | 14 ft 6 in. | 12 ft 11 in. | 31 ft 2 in. |
| Aero........ | 680E | Commander | 7 | 340 | 7,500 | 49 ft 6 in. | 35 ft 1 in. | 14 ft 6 in. | 12 ft 11 in. | 31 ft 2 in. |
| Aero........ | 680F | Commander | 7 | 380 | 8,000 | 49 ft 0 in. | 35 ft 1 in. | 14 ft 6 in. | 12 ft 11 in. | 30 ft 11 in. |
| Beech....... | G50 | Twin-Bonanza | 7 | 340 | 7,150 | 45 ft 3 in. | 31 ft 6 in. | 11 ft 6 in. | 12 ft 9 in. | 29 ft 0 in. |
| Beech....... | A55 | Baron | 4–6 | 260 | 4,880 | 37 ft 10 in. | 26 ft 8 in. | 9 ft 7 in. | 9 ft 7 in. | 23 ft 8 in. |
| Beech....... | 65 | Queen Air | 7–8 | 340 | 7,700 | 45 ft 10 in. | 33 ft 4 in. | 14 ft 2 in. | 12 ft 9 in. | 29 ft 4 in. |
| Piper....... | PA-23-160 | Apache G | 4–5 | 160 | 3,800 | 37 ft 0 in. | 27 ft 1 in. | 9 ft 6 in. | 11 ft 0 in. | 24 ft 0 in. |
| Piper....... | PA-23-250 | Aztec | 5 | 250 | 4,800 | 37 ft 0 in. | 27 ft 7 in. | 10 ft 3 in. | 11 ft 0 in. | 24 ft 0 in. |
| de Havilland | 104 Series 1 | Dove | 10–11<br>13 | 340<br>315 | 8,500 | 57 ft 0 in. | 39 ft 3 in. | 13 ft 4 in. | 13 ft 8 in. | 35 ft 4 in. |

* FAA.

**Table 13-2 Civil aircraft data pertaining to airport design***

| Manufacturer | Designation | Power rating | Cruising speed, mph | Landing speed, mph | Take-off distance, ft† | Landing distance, ft† | Fuel capacity, U.S. gal | Wing area, sq ft | Max take-off gross weight, lb | Wing loading, paf | Power loading |
|---|---|---|---|---|---|---|---|---|---|---|---|
| Cessna | 150 | 1 @ 100 hp | 110 | 48 | 1,385 | 1,075 | 26 | 157 | 1,600 | 10.2 | 16.0 lb/hp |
| Piper | PA-18 | 1 @ 150 hp | 105 | 43 | 500 | 725 | 36 | 178 | 1,750 | 9.8 | 16.7 lb/hp |
| Mooney | M-20D | 1 @ 180 hp | 127 | 57 | 1,800 | 1,550 | 52 | 167 | 2,500 | 15.0 | 13.9 lb/hp |
| Piper | PA-30B | 2 @ 160 hp | 186 | 69 | 1,530 | 1,875 | 90 | 178 | 3,600 | 20.2 | 11.3 lb/hp |
| Ted Smith | AS-400 | 2 @ 200 hp | 223 | 57 | 900 | 1,100 | 100 | 170 | 4,000 | 23.5 | 10.0 lb/hp |
| Cessna | 411 | 2 @ 340 hp | 176 | 84 | 2,010 | 1,815 | 175 | 200 | 6,500 | 32.5 | 9.6 lb/hp |
| Beech | A-65 | 2 @ 170 hp | 215 | 80 | 1,560 | 1,750 | 180 | 277 | 7,700 | 27.8 | 22.6 lb/hp |
| Beech | H-18 | 2 @ 450 hp | 220 | 87 | 2,070 | 1,850 | 318 | 361 | 9,900 | 27.5 | 11.0 lb/hp |
| Douglas | DC-3 | 2 @ 1,475 hp | 216 | 77 | 2,510 | 2,490 | 1,604 | 967 | 36,800 | 38.1 | 12.5 lb/hp |
| Douglas | DC-9-20 | 28,000 lb‡ | 555 | 112 | 6,350 | 5,400 | 2,806 | 925 | 85,700 | 92.7 | 3.06 lb/lb |
| Boeing | 737-1000 | 28,000 lb‡ | 560 | 112 | 6,300 | 5,000 | 2,850 | 922 | 93,500 | 101.4 | 3.34 lb/lb |
| Douglas | DC-6B | 4 @ 2,500 hp | 316 | 93 | 4,100 | 3,010 | 5,380 | 1,463 | 103,000 | 70.4 | 10.3 lb/hp |
| Boeing | 727-200 | 42,000 lb‡ | 533 | 110 | 5,840 | 4,610 | 7,650 | 1,650 | 170,000 | 103.0 | 4.05 lb/lb |
| Convair | 880 | 44,800 lb | 541 | 138 | 8,750 | 6,250 | 10,584 | 2,000 | 184,500 | 92.3 | 4.12 lb/lb |
| Convair | 990-A | 64,200 lb | 554 | 125 | 9,800 | 5,400 | 15,563 | 2,250 | 253,000 | 112.5 | 4.06 lb/lb |
| Boeing | 707-120B | 68,000 lb | 534 | 157 | 7,450 | 6,550 | 17,334 | 2,433 | 257,000 | 105.6 | 3.67 lb/lb |
| Douglas | DC-8-10 | 63,200 lb | 544 | 148 | 9,640 | 6,410 | 17,600 | 2,773 | 276,000 | 99.6 | 4.37 lb/lb |
| Boeing | 707-320 | 70,000 lb | 530 | 161 | 10,650 | 7,280 | 24,855 | 3,010 | 316,000 | 106.6 | 4.59 lb/lb |
| Boeing | 747 | ............ | 625 | 164 | 10,000 | 6,800 | ...... | ...... | 680,000 | | |
| Boeing | SST | ............ | 1,800 | 156 | 7,500 | 6,800 | ...... | ...... | 670,000 | | |

*Aviation Week*, Mar. 6, 1967.

† Landing and take-off distances computed to and from a full stop over a 50-ft barrier, sea level, no wind.

‡ Pounds of thrust.

oceanic, and intercontinental categories. This suggests three main classes of airports that for discussion purposes may be designated as

1. General aviation airports
2. Continental airports
3. International airports

General aviation airports serve on local service routes in the "short haul" category normally not exceeding about 500 miles. They accommodate intercity flights for business and industry, air taxi, or for charter aircraft. They also accommodate local flying operations that include instructional flying, aerial photography, fire patrol, utility patrol, and agricultural pursuits such as crop dusting. Principal dimensions of some of the aircraft normally operating out of these airports are given in Table 13-1a and b. Other pertinent data are given in Table 13-2. In general, they are VFR airports, operating under "visual flight rules" according to FAA standards and regulations. Aircraft operating under these rules are under 12,500-lb maximum weight and include airplanes with a seating capacity up to 12. The minimum turning radius for the VFR aircraft of Table 13-1a and b is a function of wing span and tread dimensions. The aircraft is assumed to turn on one locked wheel with no forward roll. Note that this radius is one-half of the tread dimension plus one-half of the wing span.

Air-carrier aircraft with short- to medium-range capability include, typically, airplanes of the following dimensions and weights:*

| Designation | Number of passengers | Length, ft | Wing span, ft | Normal take-off weight, lb | Range, miles |
|---|---|---|---|---|---|
| DC-3............. | 21–28 | 64.5 | 95 | 25,000 | |
| DC-6B.......... | 64–92 | 105.6 | 117.5 | 103,000 | |
| DC-9-10......... | 90 | 104.4 | 89.4 | 77,700 | 1,470 |
| B-737-200........ | 117 | 100 | 93 | 107,000 | 960 |
| Convair-600....... | 46 | 74.8 | 91.8 | 46,200 | 1,800 |

The continental airport, accommodating trunk line and continental flights up to a 2,000-mile range, would be designed for the largest of these planes from the standpoints of both size and weight, and in accordance with actual performance test flights as described in the next chapter on design standards. In this connection, it is of interest to note that the smallest of the air-carrier fleet is considerably larger than the largest of

* *Aviation Week,* Mar. 6, 1967.

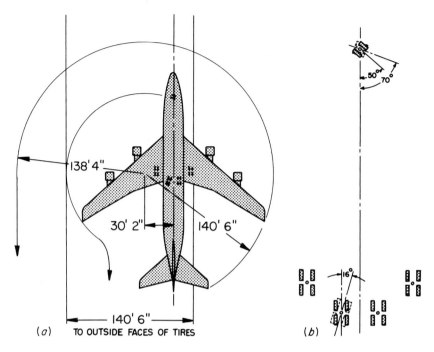

($a$)    TO OUTSIDE FACES OF TIRES                    ($b$)

**Fig. 13-1**   Minimum turning radius (180° turn), Boeing 747.   ($a$) 5-mph powered turn; ($b$) steering diagram.   (*Source: FAA.*)

the general aviation aircraft.   However, if the general aviation airport is to serve small air-carrier aircraft such as the DC-3, then the airport must be designed for that aircraft.

Planes of long-range capability, operating out of international airports, range upward in size and weight from about the Boeing 727 series with a maximum still-air range of about 3,200 miles.   See Table 13-2. The largest of the air-carrier fleet, the Boeing 747, destined to enter the fleet in 1969, has a passenger capacity of 490 and a maximum still-air range of 8,000 miles.   It has a wing span of 195.6 ft, a length of 231.3 ft, and a minimum turning radius of 140.5 ft.   See Table 13.2 and Fig. 13-1. Performance characteristics may indicate no need for longer runways at presently existing international airports.   Landing-gear configuration as illustrated in Fig. 13-1 indicates that local pavement stress concentrations may not exceed those of a plane of about one-half the gross weight, thus requiring little, if any, "beefing up" of pavements at airports presently serving planes with 300 to 350 kips gross take-off weight.

# 14
# Design Standards

Specific design requirements are peculiar to each airport location, and no fixed design criteria can be applied to all locations. Airport design must be in accord with anticipated operating requirements of present and future aircraft and with other airports of the aviation network or system. In addition, airports may vary considerably with regard to such important considerations as adjacent or nearby built-up areas, restricted areas, topography, and obstructions.

**14-1  Orientation of runways**  The FAA specifies that runways should be oriented so that aircraft may take off or land at least 95 per cent of the time with cross-wind components not exceeding 15 mph. A cross-wind component is one that acts at a right angle to the longitudinal axis of the runway or landing strip. The wind data of Table 14-1 are given as an example of the data required to determine the number and orientation of runways.

Percentages of wind direction are given for 16 points of the compass as one might expect to obtain them from a weather station near an air-

**Table 14-1  Example of wind data**

| Wind direction | Percentage of winds | | | |
|---|---|---|---|---|
| | 3–15 mph | 15–30 mph | Over 30 mph | Total |
| N............. | 2.3 | 0.2 | 0.1 | 2.6 |
| NNE........ | 2.2 | 0.3 | ... | 2.5 |
| NE.......... | 4.8 | 0.6 | ... | 5.4 |
| ENE........ | 5.9 | 2.6 | 0.2 | 8.7 |
| E............ | 1.1 | 0.4 | ... | 1.5 |
| ESE......... | 1.5 | 0.5 | ... | 2.0 |
| SE.......... | 9.0 | 4.3 | 0.3 | 13.6 |
| SSE......... | 6.5 | 1.7 | 0.1 | 8.3 |
| S............ | 1.2 | 0.7 | 0.1 | 2.0 |
| SSW........ | 1.3 | 0.5 | ... | 1.8 |
| SW.......... | 7.2 | 0.9 | ... | 8.1 |
| WSW....... | 4.7 | 3.1 | ... | 7.8 |
| W........... | 2.3 | 2.0 | ... | 4.3 |
| WNW...... | 1.4 | 1.6 | 0.2 | 3.2 |
| NW......... | 14.3 | 4.0 | 0.3 | 18.6 |
| NNW....... | 2.0 | 2.8 | 0.2 | 5.0 |
| Calms...... | .... | .... | ... | 4.6 |
| Total..... | 67.7 | 26.2 | 1.5 | 100.0 |

port location being considered for new construction or improvement. The percentage given for any one direction covers an angle of $22\frac{1}{2}$ deg as shown in Fig. 14-1. It is assumed that the wind may come from any point within the $22\frac{1}{2}$-deg sector and that the time distribution of winds for any particular sector is uniform over the arc so covered. Note that the "coverage" for any runway orientation is the sum of two directions

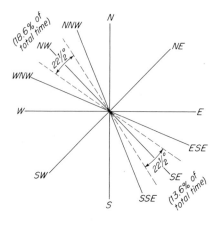

**Fig. 14-1** Illustration of wind coverage. Any one direction of wind from Table 14-1 covers directions on either side as indicated.

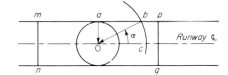

**Fig. 14-2** Preparation of template for
determination of wind coverage.

180 deg apart as indicated in Fig. 14-1 since planes may operate "into
the wind" in either of the two directions.

The cross-wind component of a wind from any direction is equal to
the velocity of the wind multiplied by the sine of the angle that the wind
direction makes with the longitudinal axis of the runway. In Fig. 14-2
the wind may be assumed to be blowing from any angle $\alpha$ toward point
$o$ on the runway center line. If $\alpha = 90$ deg, the cross-wind component
is equal to the wind velocity as represented by the vector $ao$. And if $ao$
represents 15 mph to some convenient scale, then the area enclosed by a
circle of radius $ao$ will represent the coverage, in per cent of time, by this
runway of all winds with velocities of 15 mph or less, including the percentage of calm time (wind velocity below 3 mph).

Similarly, if $\alpha$ is any angle less than 90 deg; the vector $ob$, drawn
to the same scale as $ao$, will represent the velocity $V$ of wind from that
direction with a cross-wind component of 15 mph. And the area of
sector $boc$, less the corresponding sector of the small circle, will represent
the coverage by this runway of all winds from that direction to $\alpha$-zero
and of velocities $V$ to 15 mph. The summation of percentages covered
by a transparent template such as $mnpq$ on a "wind rose" is the wind
coverage for the runway. This procedure and the wind rose are illustrated in Figs. 12-1 and 12-2. Fractional areas are determined by visual
estimate to the nearest tenth of 1 per cent. This is consistent with
normal accuracy of wind data. The most effective orientation of a
runway is thus determined in accordance with the outer—compass—circle
of the wind rose.

Consideration should be given to provision for parallel runways for
this direction in the ultimate development plan for the airport. In this
regard it should be kept in mind that the traffic capacity for a single
runway will probably not exceed 40 to 50 operations (landings and take-
offs) per hour.

It is possible that winds for a particular location have not been
recorded. In this case the data from two or more of the nearest wind-
recording stations must be used to establish wind characteristics for the
site. This can readily be done if the intervening terrain is level or
slightly rolling. However, if the intervening terrain is mountainous,
such estimate of wind data might prove to be entirely unreliable. When

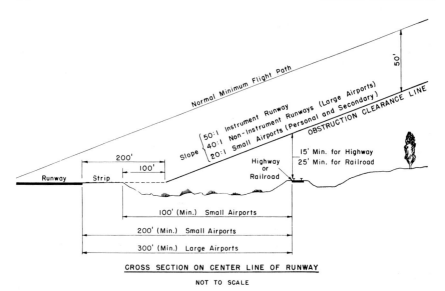

**Fig. 14-3**  Airport standards for highway and railroad clearances.  (*Courtesy of FAA.*)

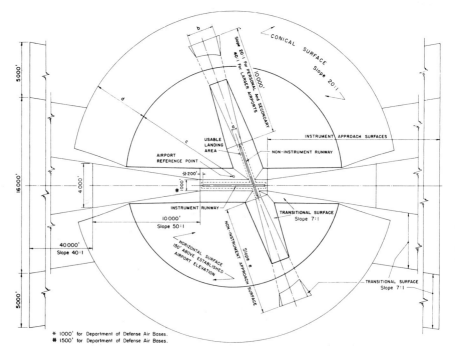

**Fig. 14-4**  Standards for determining obstructions to air navigation.  (*Courtesy of FAA.*)

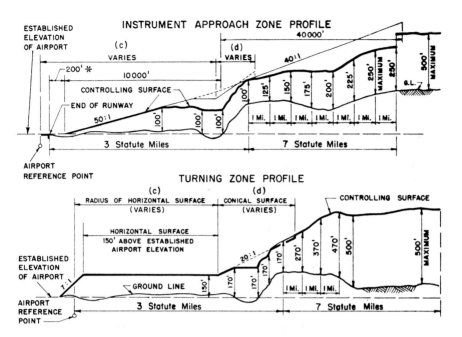

## IMAGINARY SURFACES

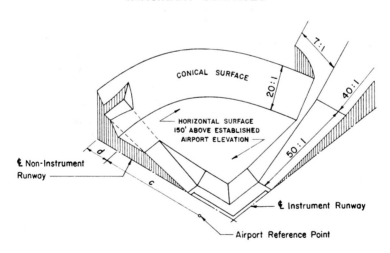

| TYPE OF AIRPORT | Distances in Feet | | | | Slope |
|---|---|---|---|---|---|
| | a | b | c | d | e |
| LOCAL | 250 | 2250 | 5000 | 3000 | 20:1 |
| TRUNK LINE | 400 | 2400 | 7000 | 5000 | 40:1 |
| CONTINENTAL | 500 | 2500 | 10000 | 5000 | 40:1 |
| INTERNATIONAL | 500 | 2500 | 13000 | 7000 | 40:1 |

**Fig. 14-4**   *(Continued)*

there is doubt as to the reliability of distant-weather-station data as applied to the site in question, wind observations should be made at the site for at least 1 year to assist in interpretation of weather-station data as applied to the site.

The directions of runways giving the most wind coverage may have to be shifted somewhat because of obstructions in the approach zones, when such obstructions project above the required clearance line in the approach zones at the ends of the runways.

Some vertical-clearance requirements as recommended by the FAA are shown in Figs. 14-3 and 14-4.

**14-2  Runway lengths**  Since airport design standards are based upon the performance requirements of the aircraft intended to use them, it follows that provision for runway length—a vital element of airport design— must be determined from aircraft performance data under various conditions of airport elevation, airport temperature, take-off weight, and length of flight in time or distance.  The following definitions by FAA are pertinent to determination of runway-length requirements:

1. *Airport elevation.*  The elevation of the airport above mean sea level is to be used for runway-length determination and for this purpose is considered equivalent to the airport pressure altitude.
2. *Temperature.*  The temperature recommended for airport runway-length design is the normal maximum Fahrenheit temperature at the airport for the hottest month.  Normal maximum temperatures may be obtained from U.S. Department of Commerce Weather Bureau publications such as *Weather Bureau Technical Paper* 31.
3. *Take-off weight.*  The take-off weight corresponding to the distance (length of haul) shown on the take-off aircraft performance curves for large airplanes is the lesser of (*a*) and/or (*b*) described below.
   a. The aircraft's zero fuel weight plus the weight of fuel required to fly to the airport of destination plus the weight of fuel reserve required for 1 hr 15 min of flying time.
   b. The aircraft's maximum landing weight plus the weight of fuel required to fly to the airport of destination.
4. *Distance.*  The distance indicated on the aircraft take-off performance curves for large airplanes is the distance that the airplane can fly from one airport to the next with a maximum payload and a minimum fuel load required for the distance.
5. *Runway length required for take-off.*  The runway length determined from the take-off aircraft performance curves for large airplanes is equal to the airplane's accelerate-stop distance or the airplane's take-off distance, whichever is greater.  Condition of no wind is assumed.

6. *Runway length required for landing.* The runway length determined from the landing aircraft performance curves for large airplanes is the airplane's landing roll based on a 5-knot tailwind.

In order that the student may tie these basic airport requirements in with aircraft data shown in Table 13-2, example and actual performance curves for take-off and landing are given in Figs. 14-5 through 14-9. Figure 14-5b shows example performance curves for a large airplane, illustrating the influence or runway-length requirement of airport elevation above mean sea level, normal maximum temperature (average maximum daily temperature) during hottest month, gross take-off weight, and anticipated flight distance.

Runway lengths required for take-off as determined from these charts are increased by 20 per cent for each 1 per cent of effective gradient. The effective runway gradient is determined by dividing the maximum difference in runway centerline elevation by the total runway length.

The procedure for use of the performance charts is illustrated by the example traces of Fig. 14-5 in accordance with the following outline:

1. *Runway length required for landing.* The example trace (a) on the landing aircraft performance curve corresponds to the procedure steps listed below.

   a. Enter the landing aircraft performance curve on the abscissa axis at the aircraft's maximum landing weight (175,000 lb).

   b. Project this point vertically to the intersection with the slanted line corresponding to the airport elevation, interpolating where necessary.

   c. Extend this point of intersection horizontally to the right to the intersection of the runway-length scale and read the runway length required for landing—in this case, 6,700 ft for a maximum landing weight of 175,000 lb.

2. *Runway length required for take-off.* The example trace (a) on the take-off aircraft performance curve corresponds to the procedure steps listed below.

   a. Enter the temperature scale on the abscissa axis at the temperature (85°F).

   b. Project this point vertically to the intersection with the slanted line corresponding to the airport elevation (1,800 ft), interpolating between elevation lines where necessary.

   c. Extend this point of intersection horizontally to the right until coincident with the reference line (RL).

   d. Then, proceed up and to the right or down and to the left, inter-polating between the slanted lines as necessary, until reaching a

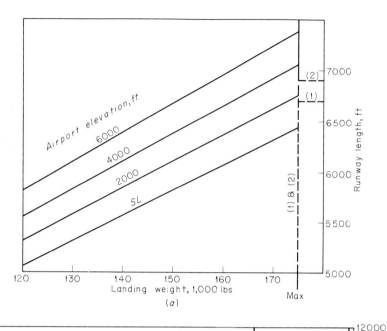

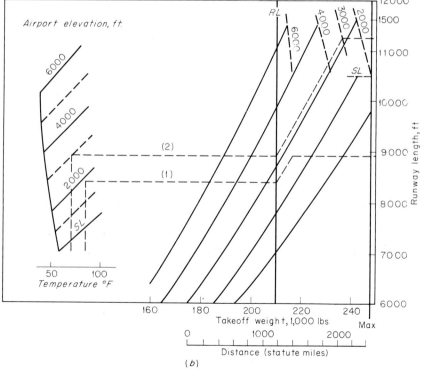

**Fig. 14-5** (*a*) Aircraft performance curve, landing (example, large airplane); (*b*) aircraft performance curve, take-off (example, large airplane). (*Courtesy of FAA.*)

NONTRANSPORT
PRATT & WHITNEY SIC 3G ENGINE
WRIGHT G-202A
(Piston powered)

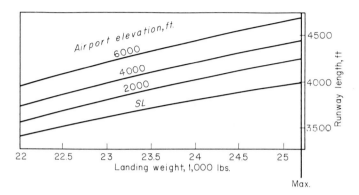

(a)

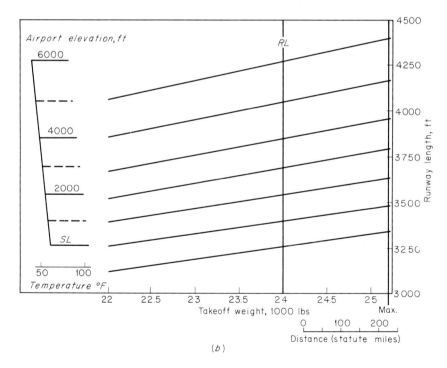

(b)

**Fig. 14-6**   (a) Aircraft performance curve, landing (Douglas DC-3); (b) aircraft performance curve, take-off (Douglas DC-3).   (*Courtesy of FAA.*)

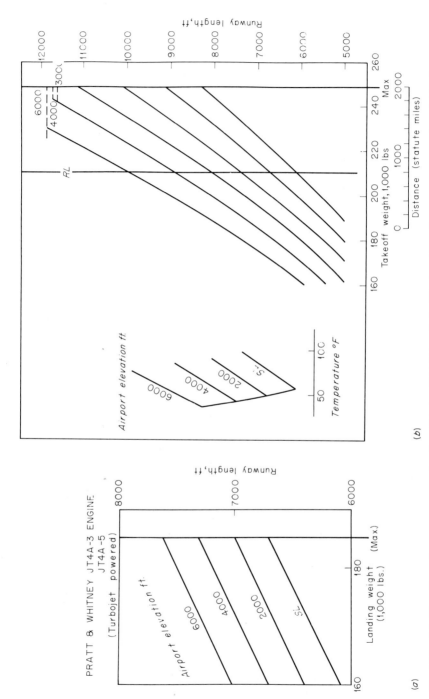

**Fig. 14-7** (a) Aircraft performance curve, landing (Boeing 707-200 series); (b) aircraft performance curve, take-off (Boeing 707-200 series). (Courtesy of F.A.A.)

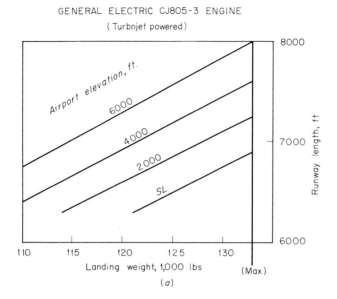

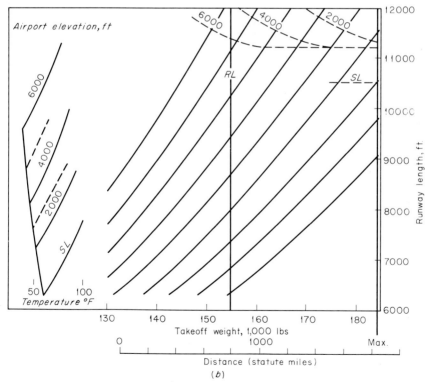

**Fig. 14-8** (*a*) Aircraft performance curve, landing (Convair 880); (*b*) aircraft performance curve, take-off (Convair 880). (*Courtesy of FAA.*)

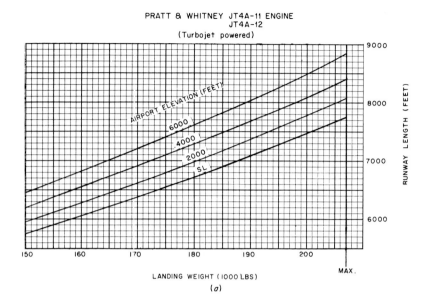

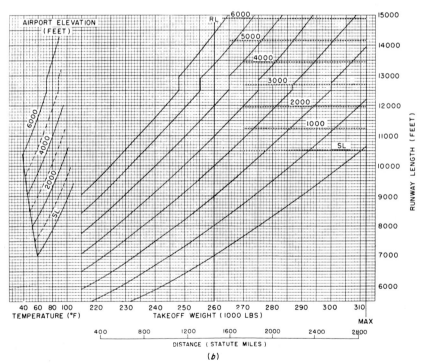

**Fig. 14-9**  (a) Aircraft performance curve, landing (Boeing 707-300 series); (b) aircraft performance curve, take-off (Boeing 707-300 series). (*Courtesy of FAA.*)

point directly above the aircraft's take-off weight or distance (1,400 miles).

e. Project this point horizontally to the right to the intersection with the runway-length scale and read the runway length (8,940 ft).

f. Increase this runway length for effective runway gradient (0.5 per cent): 8,940 + (8,940 × 0.20 × 0.5) = 9,830 ft.

*Example answer.* The runway length required for take-off (9,830 ft) is greater than the runway length required for landing (6,700 ft). Therefore, a runway length of 9,830 ft should be selected to meet the design conditions specified in the example statement.

3. *Runway length required for landing.* From trace (*b*) on the landing aircraft-performance curve, read runway length of 6,900 ft (Fig. 14-5*a*).

4. *Runway length required for take-off.* The example trace (*b*) on the take-off performance curve corresponds to the procedure steps listed below (Fig. 14-5*b*).

a. Enter the temperature scale on the abscissa axis at the temperature (70°F).

b. Project this point vertically to the intersection with slanted line corresponding to the airport elevation (3,000 ft), interpolating between elevation lines where necessary.

c. Extend this point of intersection horizontally to the right until coincident with the reference line (RL).

d. Then, proceed up and to the right or down and to the left, interpolating between the slanted lines as necessary, to the intersection of the 3,000-ft elevation limit line or until reaching a point directly above the aircraft's take-off weight or distance (2,200 miles), whichever occurs first, in this case the 3,000-ft elevation limit line.

e. Project this point horizontally to the right to the intersection with the runway-length scale and read the runway length (11,250 ft).

f. Increase this runway length for effective runway gradient (0.5 per cent): 11,250 + (11,250 × 0.20 × 0.5) = 12,370 ft.

*Example answer.* The runway length required for take-off (12,370 ft) is greater than the runway length required for landing (6,900 ft). Therefore, a runway length of 12,370 ft should be selected to meet the design conditions specified in the example statement.

The effective runway gradient should not exceed the following maximums:

1. One and one-half per cent for a landing strip equal to, or less than, 3,400 ft.

2. One per cent for a landing strip longer than 3,400 ft.

3. For airports that will serve only small aircraft operating under visual flight rules, the effective runway gradient may be increased to 2 per cent.

**14-3 Width and clearance recommendations** FAA recommendations for runway/taxiway width and clearance requirements are given in

**Table 14-2   Runway/taxiway width and lateral-clearance recommendations**

| Minimum standards | VFR air-ports* | General utility airports† | | All other airports† | |
|---|---|---|---|---|---|
| | | A | B | Runway length 3,201 to 4,200 ft | Runway length 4,201 to more ft |
| Width of: | | | | | |
| Landing strip, ft............... | 100 | 150 | 200 | 400 | 500 |
| Runway, ft................... | 50 | 75 | 100 | 100 | 150 |
| Taxiway, ft................... | 20 | 40 | 40 | 50 | 75‡ |
| Taxiway center line to: | | | | | |
| Runway center line, ft........... | 100 | 200 | 225 | 250 | 400 |
| Parallel taxiway center line, ft .... | ... | ... | ... | 200 | 300 |
| Aircraft parking area, ft.......... | ... | 100 | 100 | 175 | 250 |
| Obstacle, ft................... | ... | 75 | 75 | 100 | 200 |
| Runway center line to: | | | | | |
| Aircraft parking area, ft.......... | 140 | 300 | 325 | 425 | 650 |
| Obstacle, ft§................... | 100 | 125 | 125 | 200 | 250 |
| Runway center line to building line: | | | | | |
| Instrument runway, ft........... | ... | ... | ... | 750 | 750 |
| Noninstrument runway, ft........ | 200 | 300 | 320 | 500 | 750 |

\* Dimensions for VFR airports are minimum.
† A = less that 5,000-ft elevation and less than 15-mph (13 knots) cross wind 95% of time.
B = greater than 5,000-ft elevation or greater than 15-mph (13 knots) cross wind 5% of time.
‡ These lengths are for runways serving large aircraft based on sea-level elevation, zero effective runway gradient, and Fahrenheit temperature of 59° standard sea level plus 41°.
§ For an ILS (instrument landing system) runway, this lateral clearance is 500 ft. Beyond these lateral clearances, heights of objects are still subject to limitations of the transitional surfaces. Exceptions are made for certain navigational, meteorological, and visual aids approved by the FAA Administrator, the location and height of which are fixed by their functional purposes.

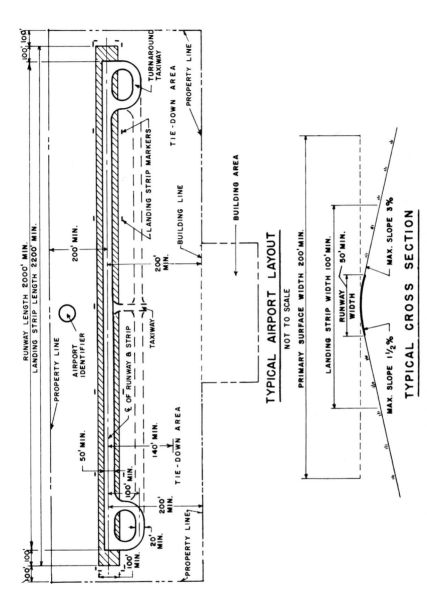

**Fig. 14-10** Minimum standards, VFR airports. *(Courtesy of FAA.)*

225

Table 14-2.  Parallel runway separation is specified in the following outline:

1. For simultaneous ILS (instrument landing system) or precision approaches, the minimum separation between center lines of parallel runways is 5,000 ft.  For actual operations under these conditions, specific electronic navigational aids and monitoring equipment, Air Traffic Service control, and approved procedures are required by the FAA.
2. For simultaneous VFR (visual flight rules) landings or take-offs, the minimum separation between center lines of parallel runways for each airplane category is:
    *a.* 300 ft when the airplanes involved are light-weight single-engine propeller-driven.
    *b.* 500 ft when the airplanes are twin-engine propeller-driven.
    *c.* 700 ft for all others.
    *d.* If airplanes of different categories are involved, use the separation method required for the larger airplane.

**14-4  Design standards for general aviation**  Airport requirements for the smaller aircraft which constitute the general aviation fleet—planes

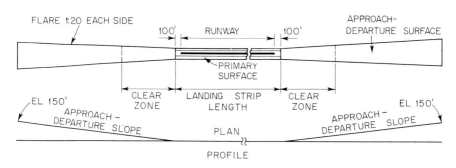

AIRPORT SURFACES

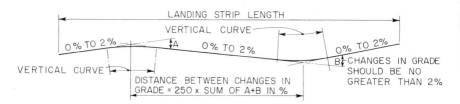

GRADE LIMITATIONS IN LENGTH OF LANDING STRIP

**Fig. 14-11**  Design requirements, VFR airports.  (*Courtesy of FAA.*)

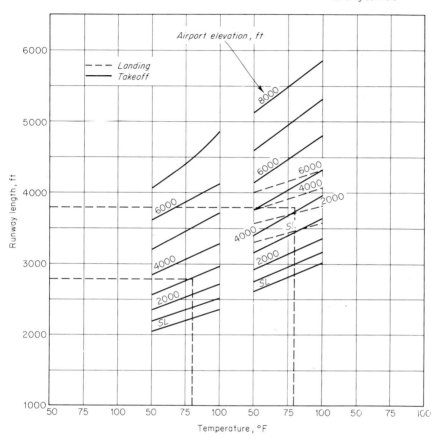

**Fig. 14-12**  Example curves, general aviation aircraft.  (*Courtesy of FAA.*)

with gross weight for take-off up to 12,500 lb—may vary over a considerable range depending on the requirements needed for the larger planes based at a particular airport or regularly using it in scheduled or non-scheduled service.  These airports will generally include VFR and general utility categories as shown in Table 14-2, but may also require runway lengths exceeding 3,200 ft.  Minimum design requirements for VFR airports are illustrated in Fig. 14-10.  Note that minimum runway

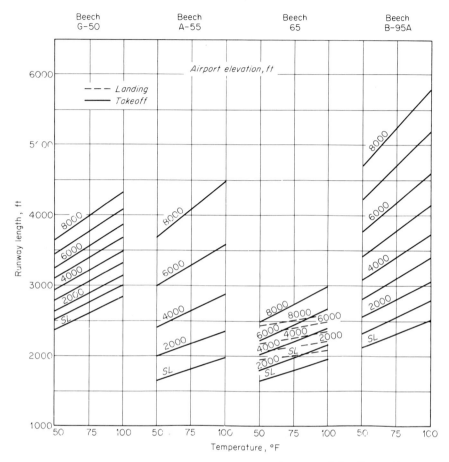

**Fig. 14-13** Typical aircraft performance curves, take-off and landing. (*Courtesy of FAA.*)

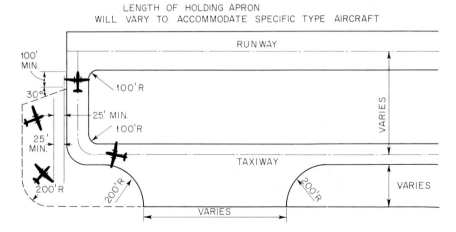

**Fig. 14-14** Typical holding apron. (*Courtesy of FAA.*)

length is 2,000 ft. Further design requirements and limitations are given in Fig. 14-11.

Runway length is established from performance curves by procedure similar to that for larger aircraft. Example curves and example traces are given in Fig. 14-12; actual performance curves are given in Fig. 14-13. The take-off length determined from these curves must be increased at a rate of 20 per cent for each 1 per cent of effective runway gradient.

**14-5 Taxiways** A taxiway is provided from the terminal area to each end of each runway. The connection at the runway end is determined by a holding apron for the larger airports. A typical situation is illustrated in Fig. 14-14. Minimum requirements will be governed by minimum clearance of runway and taxiway center lines and by provision of sufficient length of holding apron to accommodate aircraft traffic demands at a particular airport.

For smaller airports or those of lesser aircraft traffic, the runway may be used as a taxiway by providing a taxiway turnaround as illustrated in Fig. 14-15. Note that minimum requirements for anticipated future clearance for runway and taxiway center lines will determine minimum radii indicated in the figure.

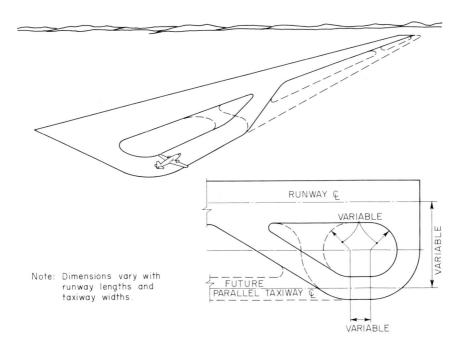

Note: Dimensions vary with runway lengths and taxiway widths.

RUNWAY ℄

VARIABLE

FUTURE PARALLEL TAXIWAY ℄

VARIABLE

VARIABLE

**Fig. 14-15** Typical taxiway turnaround. (*Courtesy of FAA.*)

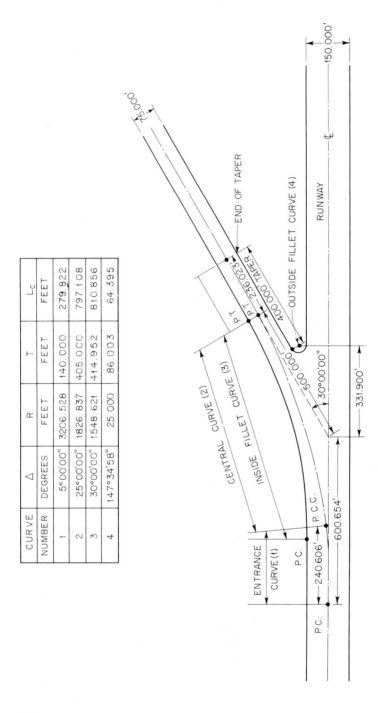

| CURVE | Δ | R | T | L_c |
|---|---|---|---|---|
| NUMBER | DEGREES | FEET | FEET | FEET |
| 1 | 5°00'00" | 3206.528 | 140.000 | 279.822 |
| 2 | 25°00'00" | 1826.837 | 405.000 | 797.108 |
| 3 | 30°00'00" | 1548.621 | 414.952 | 810.856 |
| 4 | 147°34'58" | 25.000 | 86.003 | 64.395 |

**Fig. 14-16** Exit taxiways. (*Courtesy of FAA.*)

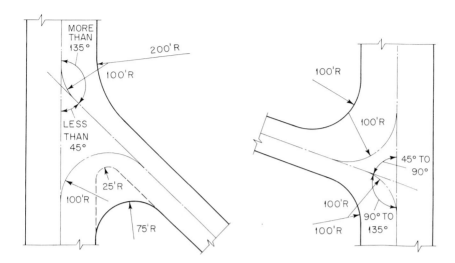

**Fig. 14-17** Typical runway and taxiway fillets. (*Courtesy of FAA.*)

Exit taxiways are provided so that the smaller plane may turn off the runway near the end of its ground run in landing, an accommodation for larger aircraft as well. This provision of exit taxiways and specified dimensions are given in Fig. 14-16. Specifications for other runway and taxiway connections are given in Fig. 14-17.

## QUESTIONS AND PROBLEMS

**14-1.** The centerline grade for a N-S runway is 0.6 per cent where it intersects with an E-W runway for which the grade is 0.4 per cent. The elevation of the point of intersection of the center lines may be taken as 100 at top of pavement. Determine elevations of the 24 remaining points of intersection mentioned on page 252 and draw final contours, assuming the intersection for a "continental" airport.

**14-2.** Runway pavement for the airport of Prob. 14-1 is to be a 10-in. uniform thickness portland-cement concrete slab. Shouldering is select borrow varying in thickness from 10 in. at edge of pavement to 4 in. at edge of landing strip. Determine subgrade elevations to be used by the grading contractor at the points of intersection.

**14-3.** On a contour map for a possible airport site, make a complete layout to scale of facilities for an airport, providing for Boeing 707-200 series turbojet operation. This will be a considerable project, and time spent on the various details will necessarily be geared to time available. The following outline of procedure is suggested:

    *a.* Determine the boundaries of property to be acquired for the airport and locate the runway center lines, with due regard for topography and soil conditions and with proper orientation with respect to locally available or assumed wind data, obstructions, etc.

    *b.* Plot the original ground-line profiles using 1, 3, or 5 profiles for each runway as time permits, plotting the soil profile for each ground-line profile, showing at least the CBR value for each soil, so that attention may be given to selective use of soils in laying the grade lines, in accordance with Sec. 5-3 and Fig. 5-5.

   c. Draw final contours on original contour map, properly adjusting both final contours and grade lines at intersections.

   d. Compute grading quantities by the contour method of horizontal sections. See Chap. 16 and, in particular, Sec. 16-5 and Fig. 16-5.

   e. Place apron, buildings, plane-parking areas, automobile-parking areas, drainage systems, lighting layout, and marking layout on the layout plan. Some of these will probably require separate tracing-paper overlays on the general-plan layout as the topics are taken up in succeeding chapters.


**BIBLIOGRAPHY**

"Airport Design," Civil Aeronautics Administration, 1961.
"Runway Length Requirements for Airport Design," Federal Aviation Administration, 1965.
"Small Airport," Civil Aeronautics Administration, 1945.
"VFR Airports," Federal Aviation Administration, 1965.

# 15
# Airport Terminal Development

**15-1 Planning** Planning for airport buildings will be greatly influenced by the over-all plan for a community's airport requirements. In many small communities traffic will be limited to small aircraft of private flying activities. In others, mixed activities of private flying and commercial operation will prevail for most of the foreseeable future.

Within the scope of the development plan of many other communities it is to be expected that scheduled air traffic of commercial operations will reach such volume that the activities of private flying and flying instruction must be transferred to another airport or airports of the general aviation classification.

Provision for expansion must be made, not only for airport buildings but for all the airport facilities. Experience has shown that careful planning for future needs requires provision of ample space for circulation of automobile traffic and for automobile parking in the building area. As building requirements increase, motor-vehicle traffic facilities may be crowded back to a narrowing fringe between buildings and property lines.

**15-2  Minimum requirements**  It will be noted from Table 14-2 that minimum clearance between center line of runway and any building line varies from 200 to 750 ft for different types of airports, considering only noninstrument operation, and 750 ft for instrument operation on any airport.  These distances are minimum requirements for safe operation on the runways and are not intended to provide apron space for parking and circulation of planes.  Here again full consideration should be given to ultimate expansion of airport facilities before permanent building lines are established.

The minimum depth of apron for any airport having scheduled service should provide for one line of plane-parking berths next to the building line and one-way traffic.  Berths will require about 150 ft of depth and one-way traffic another 150 ft, making the minimum apron depth 300 ft.  At any major air terminal—continental type and larger—where larger planes may be expected to operate, minimum depth should be 600 ft.

Length of apron will be determined by the amount of traffic at the airport.  As an approximate estimate of space requirements, from 6 to 12 parking and loading berths should be provided for at small airports, from 12 to 20 berths at medium-size airports, and from 20 to 30 berths at major terminals.  Berth diameters will vary with wing spans, 150 to 200 ft being the most common diameters used for scheduled commercial operation except at major terminals.

A concourse-to-house pedestrian circulation between building and apron should be provided where any considerable scheduled service is anticipated.  This concourse may be part of the building or it may be a separate space along the apron.  Gates may be necessary to control pedestrian traffic between lobby and planes and to keep the concourse free from congestion due to uncontrolled general public traffic.  Depth of concourse may vary from 10 to 30 ft or more depending on anticipated volumes of passenger traffic.

Thus the building line may be established with reasonable certainty that expansion of buildings and facilities may be accomplished without encroachment on traffic and parking areas, taking into account the type of service to which the airport may ultimately develop.

In general, the dimension from front to back of the building area will be more or less fixed, and room for lateral expansion should be provided in the original plan so that any single function in any building may be expanded without interfering with any other function or any other building.  It is important, then, to plan for the maximum probable development of the airport.

**15-3  Size and arrangement of facilities**  Arrangement of buildings and service facilities will necessarily conform to the available space, but

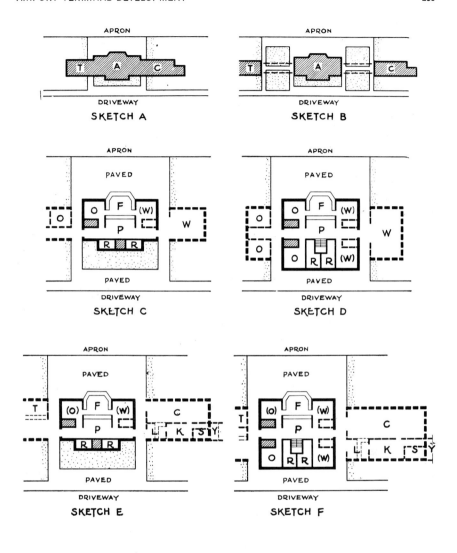

**Fig. 15-1**  Administration building, personal flying.  (*Courtesy of FAA.*)

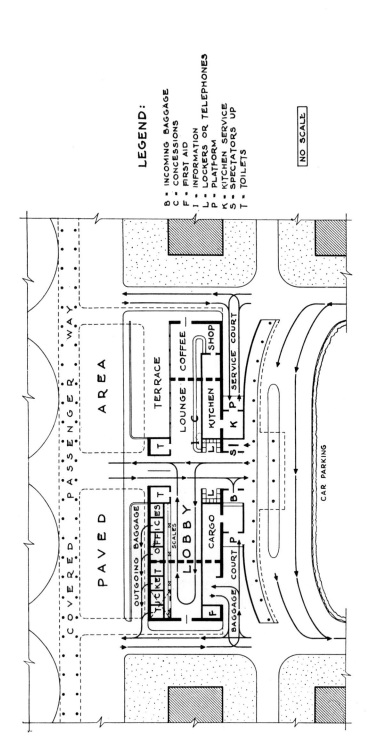

LEGEND:

B = INCOMING BAGGAGE
C = CONCESSIONS
F = FIRST AID
I = INFORMATION
L = LOCKERS OR TELEPHONES
P = PLATFORM
K = KITCHEN SERVICE
S = SPECTATORS UP
T = TOILETS

NO SCALE

▬ ▬ ▬ = ORIGINAL CONSTRUCTION

**Fig. 15-2** Small terminal building. *(Courtesy of F.A.A.)*

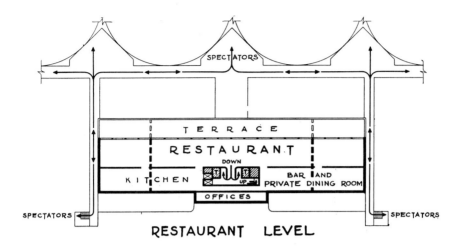

RESTAURANT LEVEL

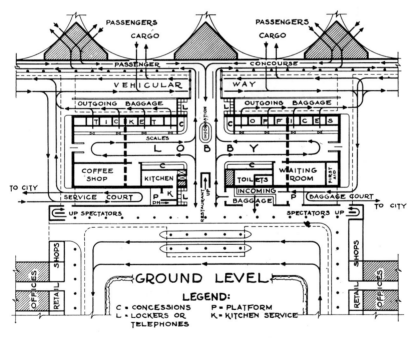

Fig. 15-3  Terminal building, one-level operation.  (*Courtesy of FAA.*)

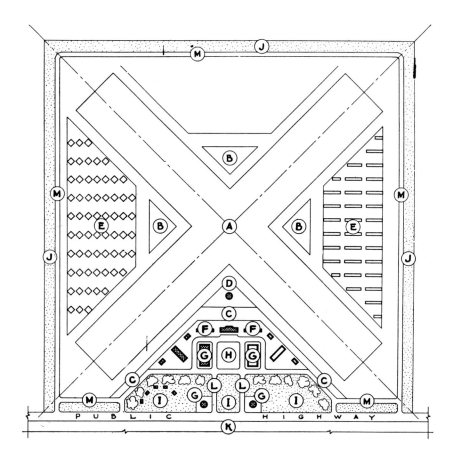

## LEGEND

A = LANDING AREA                          H = CAR PARKING
B = PLANE PARKING                         I = PARK
C = APRON                                 J = PERIMETER PLANTING
D = PLANE SERVICE STATION                 K = ZONED AREA
E = PERSONAL PLANE STORAGE                L = ENTRANCE ROAD
F = ADMINISTRATIVE BUILDING AREA          M = PERIMETER ROAD
G = REVENUE BUILDINGS

▓▓▓ = ORIGINAL CONSTRUCTION

[ NO SCALE ]

**Fig. 15-4**   Personal flying airports.   (*Courtesy of FAA.*)

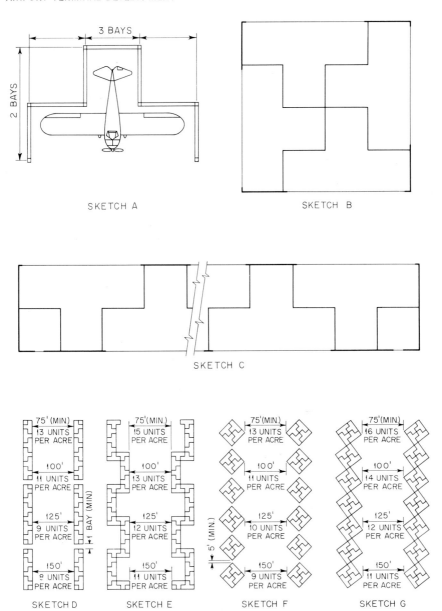

**Fig. 15-5**   Unit hangars.   (*Courtesy of FAA.*)

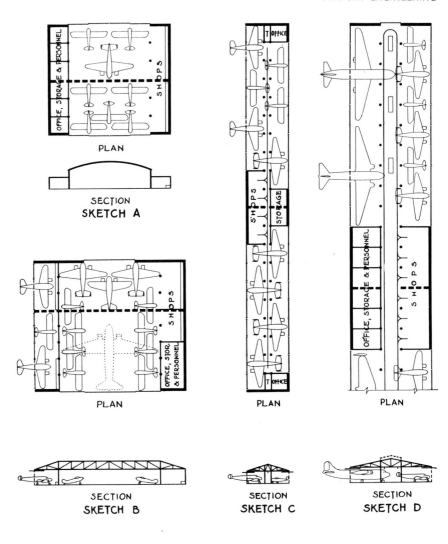

**Fig. 15-6**  Maintenance hangars.  *(Courtesy of FAA.)*

LEGEND

1. OFFICE
2. HEATING ROOM
3. STORAGE
4. TOILET & SHOWER ROOM

5. LOCKERS & MULTI-PURPOSE ROOM
6. VEHICULAR SERVICE
7. EQUIPMENT GARAGE
8. SAND STORAGE
9. FIRE APPARATUS ROOM

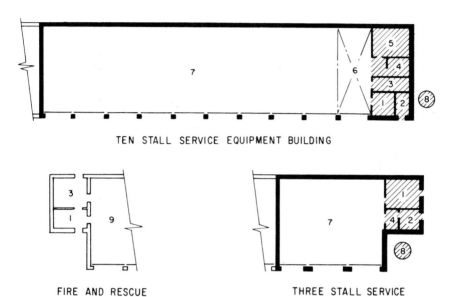

TEN STALL SERVICE EQUIPMENT BUILDING

FIRE AND RESCUE
EQUIPMENT UNIT

THREE STALL SERVICE
EQUIPMENT BUILDING

NO SCALE

**Fig. 15-7** Combination equipment buildings. *(Courtesy of FAA.)*

it should be kept in mind that the administration building will be the main center and it should front on the apron area, with important service and maintenance facilities as closely associated with the administration center as feasible without undue crowding. Figures 15-1 to 15-6 give many helpful suggestions for arrangement and for stage construction during airport growth and expansion. Some estimated requirements for airport-terminal building facilities are given in Table 15-1.

An airport maintenance building should be provided to accommodate offices, personnel, and grounds-maintenance equipment. In addition to offices for the superintendent of grounds and personnel, there should be adequate workshops for carpentry, electrical work, plumbing, metal work, and painting. Ample space should be provided for storage

**Table 15-1  Estimated requirements for airport-terminal buildings
per 100 passengers (typical peak hour\*)**

Ticket counter, lin ft............................................ 40
Ticket-counter work area, sq ft.......................... 350
Ticket lobby, sq ft............................................ 700
Baggage counter, lin ft..................................... 15
Baggage work area, sq ft................................. 220
Baggage lobby area, sq ft................................ 220
Waiting-room area, sq ft.............................. 1,800
Waiting-room seats.......................................... 45
Men's rest-room area, sq ft............................. 350
Women's rest-room and lounge area, sq ft........ 400
Kitchen and storage area, sq ft....................... 650
Eating area, sq ft........................................ 1,400
News, novelties, and gifts area, sq ft............... 200
Telephones...................................................... 7
Airline operations and employee facilities, sq ft........... 3,200

\* Typical peak-hour passengers—the average number of passengers
originating and terminating during the several busiest or peak hours
of the season with the highest passenger activity—may be determined as about 0.04 per cent of the estimated annual passenger
volume at the end of the period of forecast.

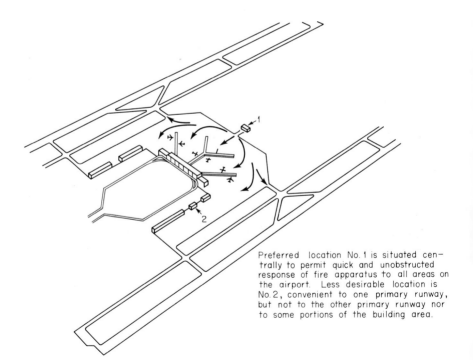

Preferred location No. 1 is situated centrally to permit quick and unobstructed
response of fire apparatus to all areas on
the airport. Less desirable location is
No. 2, convenient to one primary runway,
but not to the other primary runway nor
to some portions of the building area.

**Fig. 15-8**  Typical fire building site location.   (*Courtesy of FAA.*)

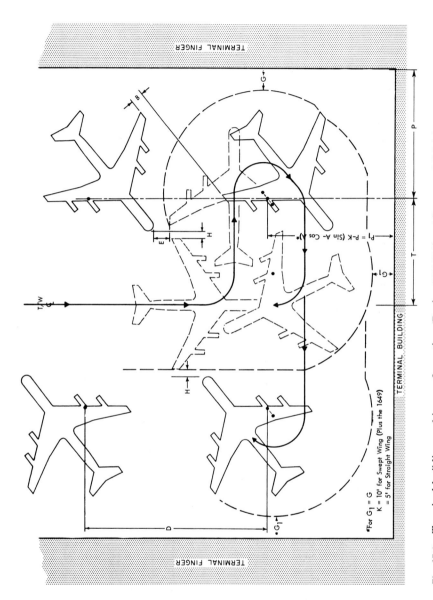

**Fig. 15-9** Terminal building parking configuration (Boeing 707-300). (*Courtesy of F.A.A.*)

243

of fire trucks, ambulances, grounds-maintenance equipment, delivery trucks, and official cars. It should be kept in mind that garage service is likely to be a good revenue producer.

The amount of hangar space required for airline planes should not be overestimated. For profitable operation these planes must be kept in the air most of the time, and during the relatively small amount of time that they are grounded they will be parked in the open. Therefore plane parking area should be generous, but hangar space for storage of large planes is not apt to be an important cost item.

Maintenance hangars for repair and servicing of large planes will usually be built for a specific airline according to its specifications, and most major repairs will be done at a plane's home base. Thus large conventional types of hangars will not be discussed here.

At many airports with mixed flying, a modified nose type of maintenance hangar, as illustrated in Fig. 15-6, will provide space for repair and servicing of both large and small planes, and it is to be expected that such a hangar will be self-sustaining from a financial standpoint.

Storage of privately owned planes will need to be provided for. Space for this will not usually be available in the immediate building area but may be in any available area of the airport where convenient access may be had. Two practical unit hangars are suggested in Fig. 15-5, a square hangar of four units and a rectangular hangar of six units. In order to accommodate some variation in wing span and height of planes, two door sizes will usually need to be provided, namely, 34 by 7 ft and 40 by 8 ft. These sizes and larger should provide a desirable 2-ft clearance at the end of each wing and sufficient height for ease in moving the planes in and out of the hangar.

Details of location and arrangement of buildings and apron for proper coordination of all airport functions are illustrated in Figs. 15-1 to 15-9.

## BIBLIOGRAPHY

"Administration Building," Federal Aviation Administration, 1960.
"Airport Cargo Facilities," Federal Aviation Administration, 1964.
"Airport, Fire and Rescue Equipment Buildings Guide," Federal Aviation Administration, 1961.
"Airport Service Equipment Buildings," Federal Aviation Administration, 1964.
"Airport Terminal Buildings," Federal Aviation Administration, 1960.

# 16
# Grading

**16-1 General requirements** It is to be expected that grading will be a major cost item in the construction of an airport. The general topography of the airport site will largely determine the grading quantities; that is, the allowable maximum grades on the runways and taxiways are all relatively low. Allowable maximum grades for the various types of airports are generally 1.5 per cent, with 2 per cent allowed for small airports. Requirements for vertical curves and grade changes are illustrated in Fig. 16-1. From this it can be seen that any large variation in ground elevations on the site will mean large grading quantities.

As a further requirement in order to promote adequate sight distance on the runways, landing-strip grade changes for a landing strip equal to or less than 3,400 ft should be such that there will be an unobstructed line of sight from any point 5 ft above the landing-strip center line to any other point 5 ft above the landing-strip center line within a distance of one-half the length of the landing strip or 2,000 ft, whichever is less. When the landing strip is longer than 3,400 ft, change 5 ft to 10 ft and 2,000 ft to 4,000 ft in the preceding statement.

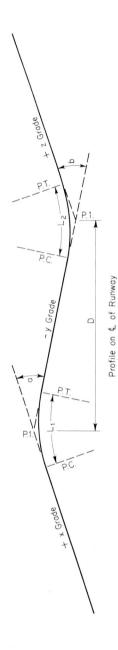

Profile on ℄ of Runway

VERTICAL CURVES REQUIRED WHEN GRADE CHANGES SUCH AS a OR b EXCEED 0.40%

| | Airports with the longest landing strip equal to or less than 3400 ft | Airports with the longest landing strip longer than 3400 ft |
|---|---|---|
| Maximum grade change such as a or b not to exceed | 2% | $1\frac{1}{2}$% |
| Length of Vertical Curve ($L_1$ or $L_2$) | 300 ft for each 1% grade change | 1000 ft for each 1% grade change |
| Distance between points of intersections for Vertical Curves, or D | 25,000 (a+b) | 100,000 (a+b) |

EXAMPLE: Assume $x = +1.0\%$, $y = -0.5\%$, and $z = +0.4\%$
Then
   a (algebraic difference between x and y) = 1.5% and b = 0.9%

| Airports with the longest landing strip equal to or less than 3400 ft | Airports with the longest landing strip longer than 3400 ft |
|---|---|
| $L_1$ (length of Vertical Curve) = 300 x 1.5 = 450 ft and $L_2$ = 300 x 0.9 = 270 ft | $L_1$ = 1000 x 1.5 = 1500 ft, $L_2$ = 1000 x 0.9 = 900 ft |
| D (distance between points of intersections) = 25,000 (.015 + .009) = 600 ft | D = 100,000 (.015 + .009) = 2400 ft |

**Fig. 16-1** Maximum grade change, minimum length of vertical curve, and distances between vertical curves on runways for various types of airports. (*Courtesy of F.A.A.*)

Thus the determination of final grade lines becomes an important factor in controlling grading quantities, and a reasonable attempt should be made toward balancing cut and fill quantities. This in itself, however, is not enough for good engineering design. The best possible use should be made of the soil as it is found to exist on the site. To this end a comprehensive study of soil types to be found on the site is in order.

Engineering properties of each horizon of the various soil profiles should be determined so that an evaluation may be made for each soil as to its load-carrying capacity, moisture characteristics, and compactive qualities. Then from this evaluation the performance of each soil as subgrade and foundation material may be predicted and the extent of its use may be determined.

**16-2  Selective use of soils**  In general the topsoil or A horizon of the soil profile, if it contains considerable organic matter and supports grass and other vegetation, should be salvaged and stockpiled for use in turfing of unpaved areas of the airport. Turfing of unpaved areas is important for control of dust, for control of erosion of slopes, and for improving appearance of grounds around buildings and parking areas. Turf may also serve as a wearing surface for traffic areas on smaller airports.

The poorer soils, as determined from engineering properties, deemed unsuitable for subgrade construction, may be used to fill low places outside of traffic areas which would otherwise collect water or distract from the general appearance of the airport.

Selective use should then be made of the better soils in accordance with evaluations previously made for each soil. The less stable soils are used where they will be lightly loaded—well below the pavement. Procedure for this selective use of material is outlined in Chap. 2.

This procedure not only provides for the most efficient use of local material but also makes for greater uniformity of subgrade, which in turn simplifies base-course and pavement design and construction. In general it is considered better engineering practice to obtain uniform subgrade-strength characteristics by selective grading with a single design of runway pavement, rather than provide a variety of pavement thicknesses to fit the needs of widely varying subgrade support. The standard runway thickness can then be increased appropriately for taxiways and aprons.

Having made the best use of the material available on the site, it will often be necessary to establish a borrow pit from which additional fill material may be imported to bring the subgrade to the required grade line and to a fair degree of uniformity in strength characteristics. The cost of overhaul of such borrow should be considered. In determining the economical distance that select material may be hauled, it should be kept in mind that the use of a good grade of imported material in building up the subgrade will to some extent reduce the more expensive pavement-structure requirements. This will of course depend on the quality of the borrow-pit material, and a careful evaluation of the material will be fully justified.

**16-3   FAA soil classification**   Soil classification combined with a rating or evaluation of each group is a valuable guide in determining pavement requirements. Such a classification, based on size gradation of soil particles, liquid limit, and plasticity index has been prepared by the FAA. From these properties each soil may be identified and placed in its proper group classification in accordance with limiting values, as shown in Table 16-1. Pavement requirements are then determined for each soil in accordance with drainage and frost characteristics of the site with reference to the last four columns of the table. Use of these columns for pavement design is explained in Chap. 18, Pavements.

In general, groups E-1 to E-5 comprise the granular or coarse-grained soils and groups E-6 to E-12 comprise the fine-grained soils. The division between granular and fine-grained soils is made on the basis that granular soils have less than 45 per cent of silt and clay combined. The portion of soil passing the No. 10 sieve is considered to be the critical portion with respect to climatic influences; consequently sand, silt, and clay fractions for any soil group are determined on the basis of that portion of soil passing the No. 10 sieve.

The presence of material coarser than the No. 10 sieve will serve to improve the over-all stability of the soil. Consequently, a soil may be upgraded one to two groups in classification when the percentage of soil retained on the No. 10 sieve exceeds 45 per cent for groups E-1 to E-5 and 55 per cent for the fine-grained soils, provided that this coarse fraction is composed of reasonably well graded and sound material. Stones or rock fragments scattered through the soil will not be of sufficient benefit to justify upgrading.

A brief description of each soil group in the classification is given in the following paragraphs.

Group E-1 includes well-graded coarse granular soils that are stable even under poor drainage conditions and are not subject to detrimental frost heave. Soils of this group may conform to requirements for soil-type base courses, such as well-graded sand clays with excellent binder.

Group E-2 is similar to group E-1 but has less coarse sand and may contain greater percentages of silt and clay. Consequently, soils of this group may become unstable when poorly drained as well as being subject to frost heave to a limited extent.

Groups E-3 and E-4 include the fine sandy soils of inferior grading.

**Table 16-1 Classification of soils for airport construction***

| Soil group | Mechanical analysis — Material finer than No. 10 sieve | | | | Liquid limit | Plasticity index | Subgrade class — Good drainage | | Subgrade class — Poor drainage | |
| --- | --- | --- | --- | --- | --- | --- | --- | --- | --- | --- |
| | Retained on No. 10 sieve,† per cent | Coarse sands, pass No. 10, ret. No. 60, per cent | Fine sand, pass No. 60, ret. No. 270, per cent | Combined silt and clay pass No. 270, per cent | | | No frost | Severe frost | No frost | Severe frost |
| E-1 | 0–45 | 40+ | 60– | 15– | 25– | 6– | Fa Ra | Fa Ra | Fa Ra | Fa Ra |
| E-2 | 0–45 | 15+ | 85– | 25– | 25– | 6– | Fa Ra | Fa Ra | F1 Ra | F2 Ra |
| E-3 | 0–45 | .... | .... | 25– | 25– | 6– | F1 Ra | F1 Ra | F2 Ra | F2 Ra |
| E-4 | 0–45 | .... | .... | 35– | 35– | 10– | F1 Ra | F1 Ra | F2 Rb | F3 Rb |
| E-5 | 0–55 | .... | .... | 45– | 40– | 15– | F1 Ra | F2 Rb | F3 Rb | F4 Rb |
| E-6 | 0–55 | .... | .... | 45+ | 40– | 10– | F2 Rb | F3 Rb | F4 Rb | F5 Rc |
| E-7 | 0–55 | .... | .... | 45+ | 50– | 10–30 | F3 Rb | F4 Rb | F5 Rb | F6 Rc |
| E-8 | 0–55 | .... | .... | 45+ | 60– | 15–40 | F3 Rb | F5 Rc | F6 Rc | F7 Rd |
| E-9 | 0–55 | .... | .... | 45+ | 40+ | 30– | F4 Rb | F6 Rc | F7 Rc | F8 Rd |
| E-10 | 0–55 | .... | .... | 45+ | 70– | 20–50 | F5 Rc | F6 Rc | F7 Rc | F8 Rd |
| E-11 | 0–55 | .... | .... | 45+ | 80– | 30+ | F5 Rc | F7 Rd | F8 Rd | F9 Re |
| E-12 | 0–55 | .... | .... | 45+ | 80+ | ..... | F6 Rc | F8 Re | F9 Re | F10 Re |
| E-13 | Muck and peat—field examination | | | | | | Not suitable for subgrade | | | |

\* United States Federal Aviation Administration system.

† Classification is based on sieve analysis of the portion of the sample passing the No. 10 sieve. When a sample contains material coarser than the No. 10 sieve in amounts equal to or greater than the maximum limit shown in the table, a raise in classification may be allowed provided the coarse material is reasonably sound and fairly well graded.

249

They may consist of fine cohesionless sand or sand-clay types with a fair to good quality of binder. They are less stable than group E-2 soils under adverse conditions of drainage and frost action.

Group E-5 comprises all poorly graded granular soils having more than 35 per cent and less than 45 per cent of silt and clay combined. This group includes, also, all soils with less than 45 per cent of silt and clay and plasticity indices greater than 10. A plasticity index greater than 15, even though the soil may have more than 55 per cent of sand, would cause it to be classified with the fine-grained soils.

The E-6 group consists of the silts and silty loam soils having zero to low plasticity. These soils are friable and quite stable when dry or at low moisture contents. They lose stability and become very spongy when wet and for this reason are difficult to compact unless the moisture content is carefully controlled. Capillary rise in the soils of this group is very rapid and, more than soils of any of the other groups, they are subject to detrimental frost heave.

Group E-7 includes the clay loams, silty clays, clays, and some sandy clays. They range from friable to hard consistency when dry and are plastic when wet. These soils are stiff and dense when compacted at the proper moisture content. Variations in moisture are apt to produce detrimental volume change. Capillary forces acting in the soil are strong, but the rate of capillary rise is relatively slow, and frost heave, while detrimental, is not as severe as in the E-6 soils.

Group E-8 soils are similar to the E-7 soils but the higher liquid limits indicate a greater degree of compressibility, expansion, and shrinkage and lower stability under adverse moisture conditions.

Group E-9 comprises the silts and clays containing micaceous and diatomaceous materials. They are highly elastic and very difficult to compact. They have low stability in both the wet and dry state and are subject to frost heave.

Group E-10 includes the silty clay and clay soils that form hard clods when dry and are very plastic when wet. They are very compressible, possess the properties of expansion, shrinkage, and elasticity to a high degree, and are subject to frost heave. Soils of this group are more difficult to compact than those of the E-7 and E-8 groups and require careful control of moisture to produce a dense, stable fill.

Group E-11 soils are similar to those of the E-10 group but have higher liquid limits. This group includes all soils with liquid limits between 70 and 80 and plasticity indices over 30.

Group E-12 comprises all soils having liquid limits over 80 regardless of their plasticity indices. They may be highly plastic clays that are extremely unstable in the presence of moisture or they may be very elastic soils containing mica, diatoms, or organic matter in excessive

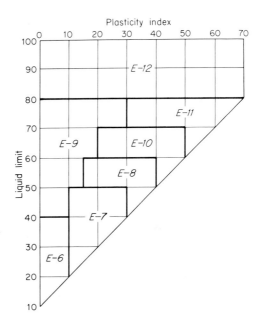

**Fig. 16-2** Classification chart for fine-grained soils, FAA system.

amounts. Whatever the cause of their instability, they will require the maximum in corrective measures.

Group E-13 takes in organic swamp soils such as muck and peat, which are recognized by examination in the field. Their range in test values is too great to be of any value in a system of identification and classification. They are characterized by very low stability, very low density in their natural state, and very high moisture contents.

A separate classification chart based on liquid limit and plasticity index is used to assist further in selecting the proper group classification for fine-grained soils where some overlapping of groups might occur. This chart is given in Fig. 16-2.

**16-4   Design control**   In the grading of traffic areas, transverse grades are provided so that surface runoff will drain away from pavements and into the drainage system.   Maximum transverse grades are illustrated in Fig. 16-3.   Grades near the maximum allowable will most efficiently remove surface runoff, but it will not always be possible to maintain a constant section.   Some flexibility in transverse grades is necessary for proper handling of runway intersections.   It becomes necessary, then, to establish desirable minimum grades.   Best practice indicates that slopes should not be less than $\frac{1}{2}$ per cent on paved surfaces and, if possible, not less than 1 per cent on turfed areas to ensure proper drainage.

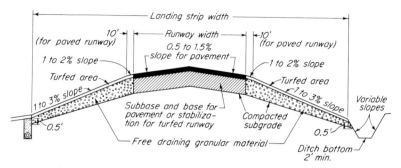

**Fig. 16-3**  Typical cross section of landing strip.

In general, five profiles will be plotted for each runway, namely, center line, pavement edges, and landing-strip edges.  Thus there will be 25 points of intersection in the crossing of two runways.  The stationing of these points may be readily obtained from the plan drawing.  The elevation of the centerline intersection will have been determined when the centerline grades for each runway were established.  For each of the other 24 points, however, elevations must be determined so that drainage of the intersection is effectively provided for and so that a smooth transition of grades can be made.

This warping of grades at the intersection is best done by constructing a contour map of the intersection.  A contour interval of 0.2 ft will generally prove most effective.  An estimated elevation for each intersection point, determined as the average of those taken from the separate profiles, is indicated on the map.  Contour lines are then drawn in and elevations adjusted to give the most satisfactory arrangement for proper drainage and uniformity of contour spacing.

These adjusted elevations are then placed on the separate profiles, and grades are adjusted so that a smooth transition is made in the approach to the intersection from any direction.  An example of typical intersection details is given in Fig. 16-4.

Centerline profiles will usually be sufficient for taxiways.  Grading of building and apron areas will generally be established directly from accurate contour maps with typical cross sections to indicate grading and drainage construction details.

Engineering properties of soils and basic principles governing construction procedures in grading operations are treated in Chap. 2.  Compaction of subgrade soils for maximum density at optimum moisture content to develop the best strength characteristics of available materials is of special significance in airport construction where wheel loads may be many times greater than for highways.

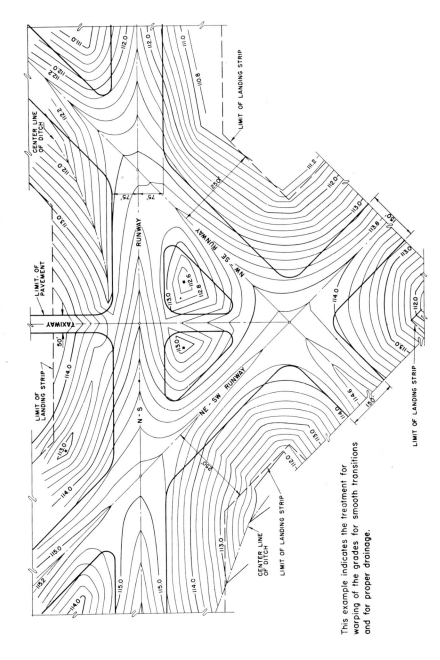

This example indicates the treatment for warping of the grades for smooth transitions and for proper drainage.

**Fig. 16-4** Example of typical intersection details. (*Courtesy of FAA.*)

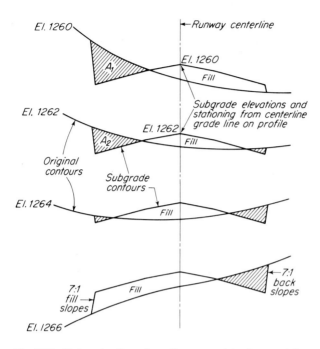

**Fig. 16-5** Determination of grading quantities from original and final contours. Note that areas $A_1$ and $A_2$ are horizontal, not vertical, sections.

**16-5 Grading quantities** Computation of grading quantities will make possible a comparative study of variations in layout and provide an estimate for bidding purposes. The thoroughness and accuracy with which this is done will necessarily be influenced by available time for the project.

A convenient method for estimating earthwork quantities makes use of original and final contours. The method is illustrated in Fig. 16-5. Original and final contour lines for any particular elevation will enclose horizontal areas such as $A_1$ and $A_2$, which in the figure are in a cut section. The volume of excavation represented by these areas is the average of the two areas multiplied by the contour interval. Fill sections may be handled in the same manner. Where cut or fill sections "zero out" between two contours, the vertical distance must be estimated. A summation of excavation and embankment quantities can then be made, and with the use of a proper shrinkage factor the amount of borrow or waste determined.

Quantities for base-course and pavement construction may then be readily computed in accordance with provisions for adequate subgrade protection as outlined in Chap. 18, Pavements. In this regard it should be kept in mind that base-course construction will extend across the entire landing strip so that emergency operation of planes may safely be made off the runway proper. The wearing course or pavement will be placed on only the designated runway area.

## QUESTIONS AND PROBLEMS

**16-1.** The station of the point of intersection of two grade lines of a runway is 87 + 56.2 at elevation 103.6. The grade changes from minus 0.4 per cent to plus 0.7 per cent. If the runway is for an international airport, determine:

    *a.* Length of vertical curve
    *b.* Stationing of P.C. and P.T.
    *c.* Elevation of P.C. and P.T.
    *d.* Elevation of 100-ft stations along vertical curve

**16-2.** *a.* What minimum distance should separate the point of intersection of Prob. 16-1 and a succeeding point of intersection where the grade changes from plus 0.7 per cent to minus 0.1 per cent?

    *b.* What limitations, if any, should be placed on the stationing of either end of the runway, if the provision of the second paragraph of Sec. 16-1 is complied with?

    *c.* Can a runway of sufficient length for an international airport be provided, using the above grades?

**16-3.** Classify the soils of Prob. 2-6 according to the FAA classification.

## BIBLIOGRAPHY

"Airport Turfing," Civil Aeronautics Administration, 1945.
"Manual on Airfields," The Asphalt Institute, 1947.

# 17
# Drainage

**17-1 Intercepting drains**  An adequate drainage system for the airport site will provide for efficient removal of surface water from the area and for effective control of ground water.   It is likely to be a large item of cost in the construction of the airport.   The feasibility of providing an adequate drainage system should receive serious consideration in selecting the site for the airport, and throughout the design drainage figures as an ever-present and significant factor.

Stability of the soil supporting the loads on runways, taxiways, and aprons might be seriously reduced by high moisture content.  The placing of relatively impervious pavements over these large surfaces cuts off most of the evaporation which normally took place over the area, and even though the surface water is effectively removed, it is possible for the soil underlying the pavement structure to accumulate high moisture content by capillarity from the ground-water table below and by lateral flow of ground water from surrounding areas.

These facts serve further to emphasize the need for selective use of materials in the grading operations and for providing adequate protective cover for those materials which are inherently unstable under adverse moisture conditions.

Consideration should be given to the possibility of diversion of "foreign" water from surrounding area by the construction of intercepting drains. Riparian rights of adjoining property holders should not be lost sight of in this regard. Diversion of water from one watershed to another and unduly concentrating outlets upon adjoining property should be avoided. However, the size and extent of drainage structures and facilities on the airport site may often be greatly reduced by proper use of diversion ditches for surface-water and underground drains for intercepting ground-water flow into the area.

**17-2   The rational formula**   The natural topography of the site, modified by grading operations, will divide the area into separate watersheds for which drainage structures must be provided to carry *surface runoff* across runways and taxiways. A rational approach to determination of size of waterway openings for these structures may be made by the use of the formula

$$Q = CIA$$

where $Q$ = maximum runoff from the given watershed, cfs, which determines the discharge capacity of the drainage structure

$C$ = a coefficient which represents ratio of runoff to rainfall

$I$ = average rainfall intensity, in./hr, during time of concentration of runoff from the watershed

$A$ = drainage area, acres

It will be seen that the formula is for all practical purposes dimensionally correct:

$$Q \text{ (cfs)} = C \text{ (a dimensionless ratio)} \times I \ \{[\text{in.}/(12 \text{ in./ft})]$$
$$[1/(\text{hr} \times 3{,}600 \text{ sec/hr})]\} \times (A \times 43{,}560 \text{ sq ft/acre})$$
$$= CIA \text{ (cfs)}$$

The product of 12 and 3,600 practically cancels 43,560 to simplify the formula. This apparent simplicity of formula is, however, grossly misleading. Of the three variables on the right-hand side of the equation, $A$ is the only one that may be determined with any degree of certainty. The complexity of the problem becomes apparent by a study of the other variables.

**17-3   Runoff coefficient**   The ratio of runoff to rainfall $C$ is dependent on surface slope and roughness, permeability of the soil, and degree of saturation, vegetative cover, and to some extent on rainfall intensity. Surface roughness and vegetative cover tend to create surface detention,

**Table 17-1   Runoff by surface type**

| *Type of surface* | *Runoff coefficient C* |
|---|---|
| Paved surfaces and roof areas................ | 0.85–1.00 |
| Gravel and macadam surfaces............... | 0.35–0.70 |
| Slightly pervious bare earth................. | 0.50–0.85 |
| Slightly pervious earth, turfed............... | 0.30–0.70 |
| Moderately pervious bare earth.............. | 0.25–0.50 |
| Moderately pervious earth, turfed............ | 0.00–0.20 |

which materially reduces the rate of runoff.   Pervious soils absorb water at a significant rate if they are not already saturated and if absorption is not impeded or prevented by the presence of frost.

These factors are further modified by rainfall intensity.   Rate of runoff from the watershed is dependent on depth of surface accumulation of water much in the same way that rate of discharge of a conduit is affected by depth of flow as well as by slope and surface roughness of the conduit.

Values for runoff coefficient are given in Table 17-1.   These values are particularly suited to airport sites, where average slopes in the watersheds are likely to be within the limits of $\frac{1}{2}$ to 2 per cent and the watershed areas are not particularly large.   Drainage structures for stream flow in large watersheds are taken up in Part 1, Roads and Pavements.   It will be noted that a range of values is given for each surface type.   The maximum rainfall intensity to be provided for at the site might be any figure between $\frac{1}{2}$ and 3 in./hr.   It is recommended that the low value for any surface type be used for low rainfall intensity and the high value for high rainfall intensity, giving some weight to average slope of the area.

**17-4   Storm frequency**   Rainfall intensity to be provided for at a particular site should be determined from records of local weather stations with particular attention to rate of rainfall, duration of storm, and frequency of occurrence.   In general, high rainfall rate will occur over a relatively short period of time of a few minutes to an hour, whereas lower rainfall rates may extend over a period of several hours.   Also, the highest rate of rainfall for a particular locality may occur only once in 40 or 50 years, whereas somewhat lower intensities would in general occur at more regular intervals.

A decision must be made, then, as to which storm frequency to use so that a proper balance can be made between cost of providing protection against the more severe storms and cost of damage and inconvenience caused by these storms.   The FAA recommends a design storm frequency of 5 years.   Thus a study of the maximum rainfall intensities occurring once in 5 years for the locality in question is in order.

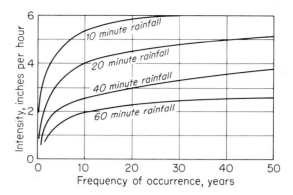

**Fig. 17-1**  Example of storm-frequency data.

A plot of rainfall intensity in inches per hour as ordinates against time duration of storm in minutes as abscissas for storms of 5-year frequency is made from the best available weather data for the site.   Figures 17-1 to 17-3 depict rainfall-intensity characteristics involving the three parameters: (1) rate of rainfall, in inches per hour; (2) duration of the rate, in minutes; and (3) the frequency interval, in years.   The curves are adapted from Mead[1] and represent rainfall at St. Paul, Minn., for 159 storms over a period of 43 years.

Rainfall intensity for design purposes may then be determined from the 5-year frequency curve of Fig. 17-2 in accordance with the time duration which will produce maximum flow at the inlet of the drainage struc-

[1] D. W. Mead, "Hydrology," McGraw-Hill Book Company, New York, 1919.

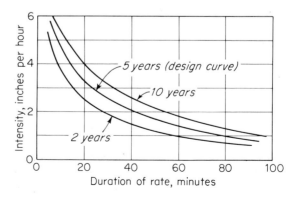

**Fig. 17-2**  Example of storm-duration data.

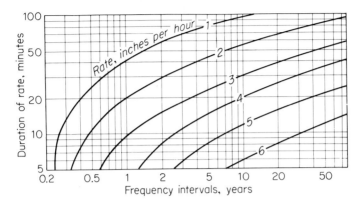

**Fig. 17-3**  Example of duration of rate of rainfall as related to frequency in years.

ture.  This maximum flow will depend on the time of concentration of runoff at the inlet.

**17-5  Concentration time**  Time of concentration is defined as the time required for a particle of water from the remotest part of the drainage area to reach the inlet of the drainage structure.  The maximum flow to be provided for by the structure will occur when the rainfall has continued for a period of time equal to the time of concentration, because then every part of the watershed will be contributing simultaneously to flow at the structure.

Time of concentration will be determined, then, by the distance from the remotest part of the drainage area to the structure and by the average velocity of flow of runoff in the area.  According to Mead the velocity of overland flow may vary from 5 to 15 fpm for turf and from 20 to 50 fpm for paved surfaces.  Experience of the FAA substantiates these values for average slopes varying from 0.5 to 2 per cent.

**17-6  Ponding**  The size of drainage structure as determined by consideration of the foregoing principles may be materially reduced by allowing some ponding time at the inlet.  Ponding is not to be recommended as general practice because of the adverse effect that the water so impounded may have on moisture conditions in the subgrade and the risk of flooding adjacent areas.  There will often be instances, however, when a substantial ponding basin exists wherein runoff water may be temporarily stored to accommodate slower rate of outflow through a smaller structure.  Thus substantial economies in design may be achieved if even a small amount of temporary storage can be safely provided. Best practice indicates that ponding time should not exceed 2 hr.

**17-7   Selection of drain capacity**   It can be seen that a foolproof procedure cannot be established for the complex and difficult problem of drainage design.   In the matter of deciding on the 5-year rainfall rate alone, for instance, it must be recognized that the storm of 50-year frequency may occur the day after installations are made.   Use of the so-called rational formula $Q = CIA$ has, however, provided a satisfactory solution to problems of drainage design, and the best that can be said of any formula or procedure is that it works.

In order to summarize the foregoing principles of design, let it be required to determine the capacity of a drainage structure for the following conditions: The drainage area of 22 acres has an average slope of $1\frac{1}{2}$ per cent and length—to remotest part—of 1,200 ft.   The area is 15 per cent or 3.3 acres of pavement and the remainder is fairly good turf on slightly pervious soil.   The design curve of Fig. 17-2 will be used to estimate rainfall intensity.

Velocity of overland flow in the area is estimated as 12 fpm for the turfed portion and 40 fpm for the pavement.   The weighted average of velocity will then be about 15 fpm, giving a time of concentration of 80 min.   From the design curve of Fig. 17-2 the maximum rate of rainfall $I$ to be provided for is 1 in./hr (using 80-min duration).

**17-8   Design of drainage structure**   Runoff coefficients for this slope and rainfall intensity are taken from Table 17-1 as 0.90 for the paved area and 0.50 for the turfed area.   From these factors the capacity of the drainage structure is determined as

$$Q = (0.90 \times 1 \times 3.3) + (0.50 \times 1 \times 18.7) = 12.3 \text{ cfs}$$

Charts for determining size of pipe culvert for given values of $Q$ are presented in Figs. 17-4 and 17-5.   The charts are based on Manning's formula

$$Q = A(1.486/n)R^{\frac{2}{3}}S^{\frac{1}{2}}$$

where $Q$ = discharge, cfs
  $A$ = cross-section area of flow, sq ft
  $R$ = hydraulic radius, ft = area of section/wetted perimeter
  $S$ = slope of gradient, ft/ft
  $n$ = coefficient of roughness of pipe

Values of $n$ are taken as 0.015 for concrete pipe and vitrified clay tile and 0.021 for corrugated-metal pipe.

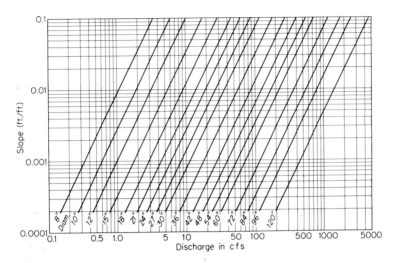

**Fig. 17-4**  Discharge of pipes, based on Manning's formula $n = 0.015$.

Assuming a concrete pipe to be laid on a slope of, say, 0.003 ft/ft, it will be seen from Fig. 17-4 that a 24-in.-diameter pipe will be needed.

If ponding area and depth are available at or near the pipe inlet, a reduction in size of pipe culvert may be made on the following basis. It has been assumed that rainfall has continued uniformly over the area for 80 min at the rate of 1 in./hr and then stopped. Thus flow through the

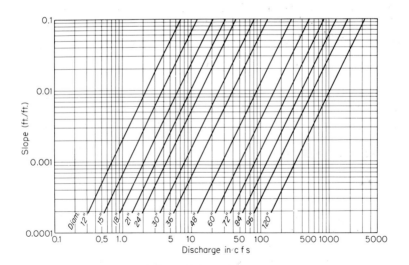

**Fig. 17-5**  Discharge of pipes, based on Manning's formula $n = 0.021$.

pipe would increase from zero to capacity flow in one time of concentration, then decrease from capacity flow to zero in one time of concentration. In effect, then, the total discharge would be determined by considering the pipe flowing at full capacity for one time of concentration. Using the time of concentration, then, as the basic time for discharge of the runoff water, calculated additional time for discharge of the same amount of water will be considered as ponding time.

Thus if $1\frac{1}{2}$ hr (ponding time) are to be considered and $Q^1$ is the capacity of pipe required, then, since the same amount of water will be discharged,

$$Q^1(80 + 90) \times 60 = 12.3 \times 80 \times 60$$

from which $Q^1 = 5.8$ cfs, and from Fig. 17-4 it will be seen that an 18-in.-diameter pipe will be adequate.

The actual net effect of including ponding time is not readily determined. It is significant, for instance, that the smaller pipe will be discharging at full capacity in less than one time of concentration and that because of the ponding it will almost certainly be operating with a considerable head at the inlet, thus decreasing the calculated time of discharge.

On this basis, then, storage capacity of the "pond" should be adequately provided for by considering the amount of water that can be discharged by the smaller pipe during the allowed ponding time at normal capacity. This amount in the example given would be

$$5.8 \times 1.5 \times 3,600 = 31,320 \text{ cu ft}$$

or 0.723 acre-ft, which is equivalent to 6 in. of water over an area 250 ft square. In any case the elevation of the surface of the temporarily impounded water should be at least 3 ft below the elevation of the shoulder of the runway pavement.

**17-9 Inlets** In the relatively flat topography of an airport site, entrance of runoff water into the culvert will often need to be provided for by a drop inlet covered with a grate. This will be especially true for the smaller drainage areas in and around traffic areas. The grate should be designed to support a uniform load of 85 to 100 psi of gross area so that it will be strong enough to support moving loads to which it may be subjected.

Sufficient grate opening should be provided to admit water at a rate equal to the capacity of the pipe. It may be calculated by the use of the formula for flow through orifices as follows:

$$Q = CA \sqrt{2gh}$$

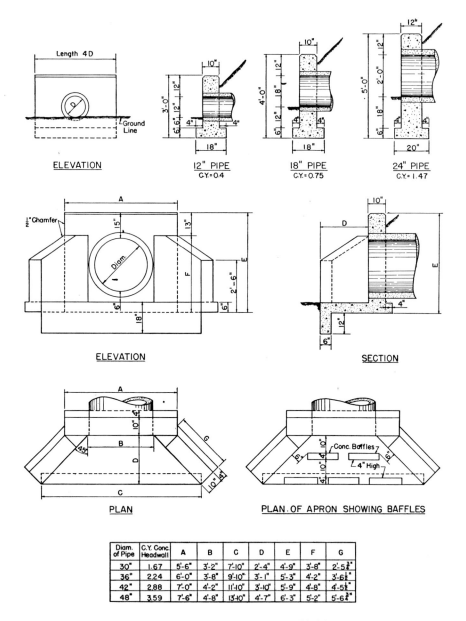

| Diam. of Pipe | C.Y. Conc. Headwall | A | B | C | D | E | F | G |
|---|---|---|---|---|---|---|---|---|
| 30" | 1.67 | 5'-6" | 3'-2" | 7'-10" | 2'-4" | 4'-9" | 3'-8" | 2'-5¾" |
| 36" | 2.24 | 6'-0" | 3'-8" | 9'-10" | 3'-1" | 5'-3" | 4'-2" | 3'-6½" |
| 42" | 2.88 | 7'-0" | 4'-2" | 11'-10" | 3'-10" | 5'-9" | 4'-8" | 4'-5½" |
| 48" | 3.59 | 7'-6" | 4'-8" | 13'-10" | 4'-7" | 6'-3" | 5'-2" | 5'-6¾" |

**Fig. 17-6**  Suggested types of headwalls.  (*Courtesy of FAA.*)

where $Q$ = discharge, cfs

$C$ = coefficient of discharge (approximately 0.7)

$A$ = area of openings, sq ft

$g$ = acceleration of gravity

$h$ = head on grate, ft

As an example let it be required to determine the area of inlet openings for a pipe whose discharge capacity is 4.5 cfs. Assume that the area immediately surrounding the inlet may be sloped toward the inlet so that a head of 0.3 ft will be provided on the grate. Then

$$4.5 = 0.7A \sqrt{2 \times 32.2 \times 0.3}$$

from which $A = 1.46$ sq ft, or 210 sq in. This area should be increased by 50 per cent to provide for partial clogging of the grate by drifting debris.

In general, inlets will be placed at all low points in the surface created by grading. In the case of long pipe runs for surface drainage, with the slope in one general direction inlets should be provided at intervals of about 300 ft so that runoff may enter the drainage system without excessive accumulation or long surface runs.

**17-10 Catch basins** The use of catch basins in airport drainage systems may be largely eliminated by laying pipelines on self-cleaning grades. Experience has shown that depositing of suspended material in the pipe is prevented if the mean velocity of the water in the pipe is at least $\frac{1}{2}$ ft/sec. This requirement will be easily fulfilled in all but a few special cases where slopes of pipes must be kept very low. Thus velocity of flow in all pipes should be included in the tabulation of pipe data so that any need for catch basins may be determined.

Manholes or combination manholes and inlets should be provided at all pipeline grade changes, direction changes, pipe-size changes, junctures of two or more lines, and at reasonable intervals for cleaning out and inspection.

Suggested designs as recommended by FAA for headwalls, inlet grates, inlets, and manholes are shown in Figs. 17-6 to 17-9. Recommended pavement-gutter section is shown in Fig. 17-10.

**17-11 The underground drainage system** Much of *subsurface drainage* for the airport area consists of interception and diversion of underground flows. As mentioned earlier in the chapter, much can be done to lower and control ground-water tables by intercepting underground flow of water into the area from higher adjacent ground with suitable intercepting drains to carry the flow *around* critical traffic areas.

It seems appropriate to repeat here again that the placement of rela-

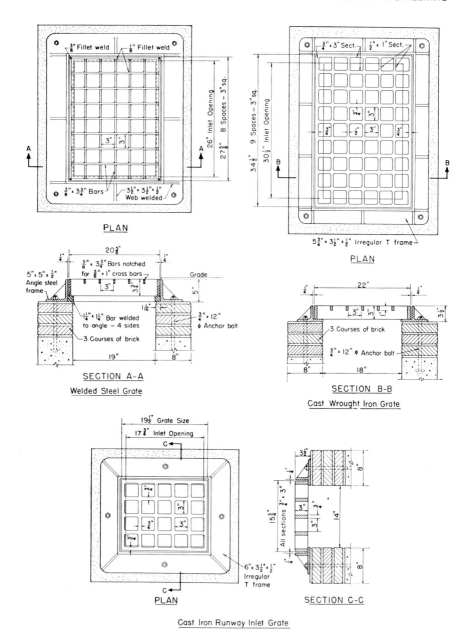

**Fig. 17-7**  Examples of typical inlet grates.  (*Courtesy of FAA.*)

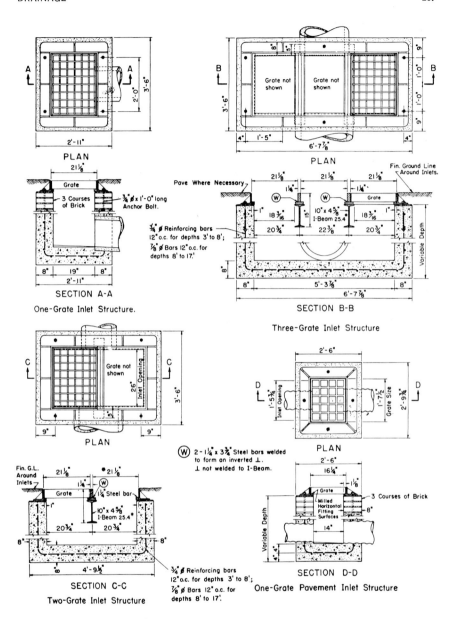

**Fig. 17-8** Examples of inlet design. *(Courtesy of FAA.)*

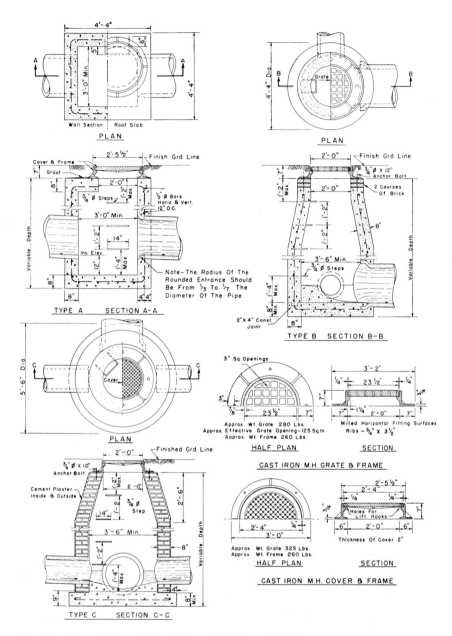

**Fig. 17-9**   Suggested manhole design.   (*Courtesy of FAA.*)

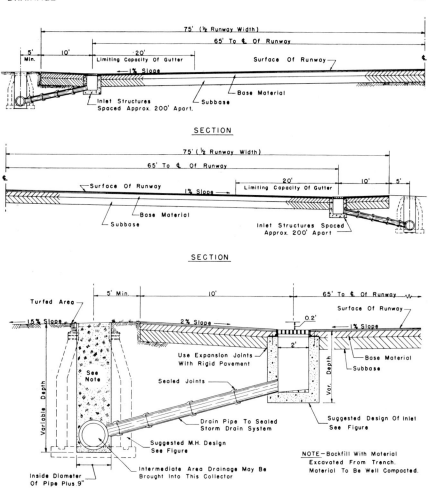

**Fig. 17-10** Recommended pavement-gutter section. (*Courtesy of FAA.*)

tively impervious pavements over wide expanses of runways, taxiways, and aprons prohibits much evaporation of ground water which would otherwise take place. This tends to increase the moisture content of subgrade soils by capillarity and infiltration of ground water and may have an adverse effect on the stability of subgrade, sub-base, and base-course materials.

The use of porous granular material in sub-base and base courses under pavements will greatly facilitate removal of accumulated free water in these courses. Suitable underground drains should be provided along

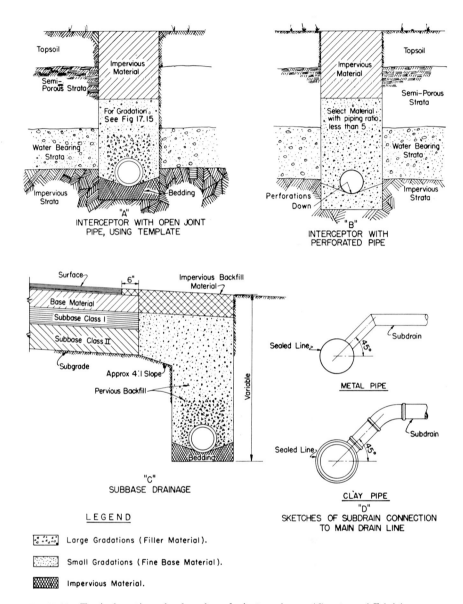

**Fig. 17-11**  Typical section of subsurface drain trenches.  (*Courtesy of FAA.*)

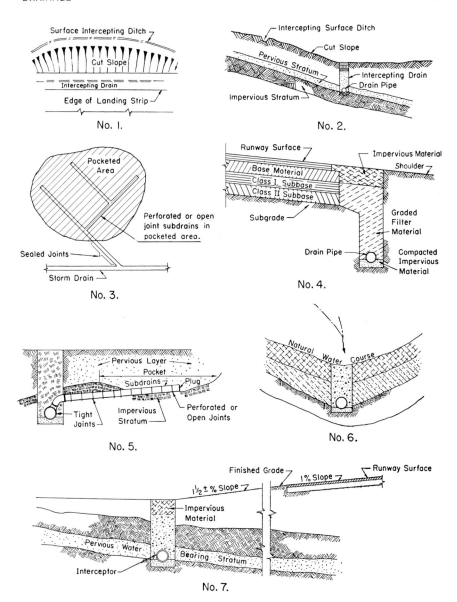

**Fig. 17-12** Types of subsurface drainage. *(Courtesy of FAA.)*

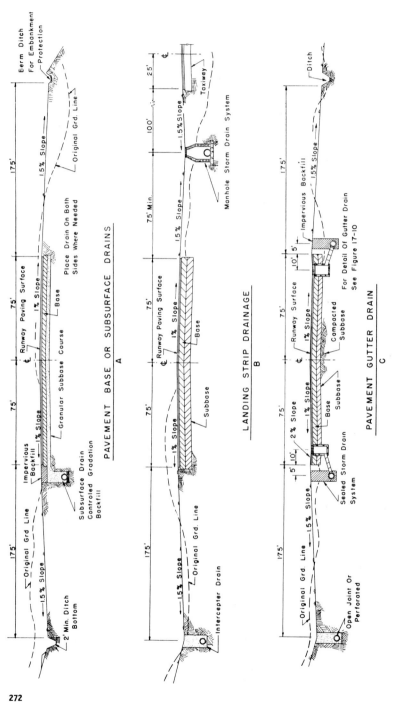

**Fig. 17-13** Standard typical runway cross section. (*Courtesy of FAA.*)

272

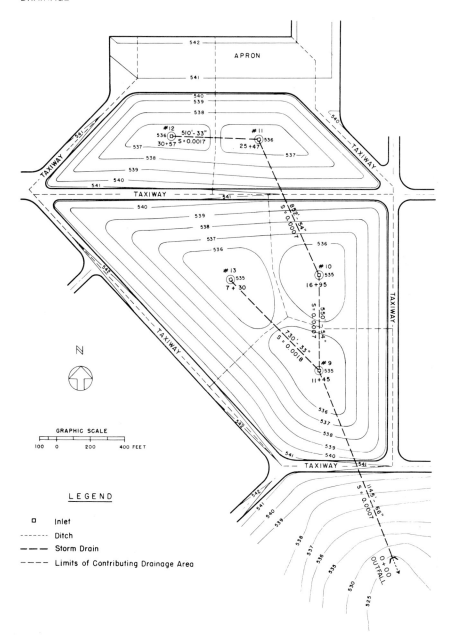

**Fig. 17-14**  Portion of airport showing drainage design.  (*Courtesy of FAA.*)

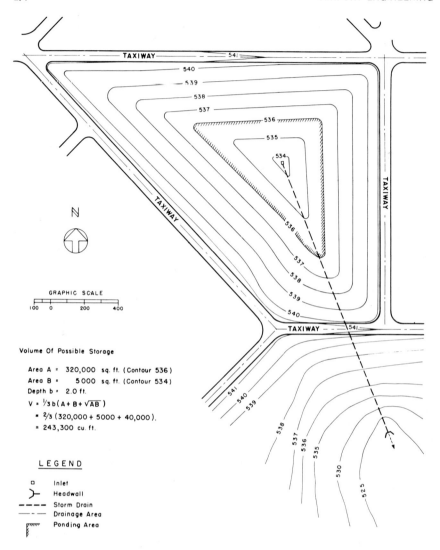

**Fig. 17-15**   Example of providing for ponding area.   (*Courtesy of FAA.*)

pavement edges to carry off this water, and the subgrade should be built to the transverse grade of the pavement section to facilitate flow of water to the drains.   Open joint drains or perforated pipe will generally be used for this purpose.   Examples of subsurface drainage design are shown in Figs. 17-11 to 17-15.

Coupled with intercepting and diverting ground-water flow and accumulation is the problem of draining wet areas where the water table is above the frost line or where wet-weather springs may develop. A grid system of drains under traffic areas with outlets to storm drains or open ditches may have to be provided. Open-joint and perforated drains will generally be used for this purpose.

It is important to recognize that some types of soil are self-draining, others are drainable, and still others are difficult or impossible to drain. Soils in FAA classification groups E-1 to E-5 are in general self-draining to the extent that they are coarse-grained soils with high permeability and low capillarity. The possibility of successfully draining the fine-grained soils of E-6 to E-12 groups will largely depend on the percentages of sand present in these soils. Those with relatively high percentages of sand will be drainable, while those with high percentages of clay and/or silt and organic matter will in general be difficult or impossible to drain because of inherently high capillarity and low permeability even under considerable hydrostatic head in the case of deep ditches or deeply laid drain pipe.

**Table 17-2  Recommended depth and spacing of subdrains for various soil classes***

| Soil classes | Percentages of soil separates | | | Depth to bottom of drain, ft | Distance between subdrains, ft |
|---|---|---|---|---|---|
| | Sand | Silt | Clay | | |
| Sand............... | 80–100 | 0–20 | 0–20 | 3–4 | 150–300 |
| | | | | 2–3 | 100–150 |
| Sandy loam.......... | 50–80 | 0–50 | 0–20 | 3–4 | 100–150 |
| | | | | 2–3 | 85–100 |
| Loam............... | 30–50 | 30–50 | 0–20 | 3–4 | 85–100 |
| | | | | 2–3 | 75–85 |
| Silt loam............ | 0–50 | 50–100 | 0–20 | 3–4 | 75–85 |
| | | | | 2–3 | 65–75 |
| Sandy clay loam...... | 50–80 | 0–30 | 20–30 | 3–4 | 65–75 |
| | | | | 2–3 | 55–65 |
| Clay loam........... | 20–50 | 20–50 | 20–30 | 3–4 | 55–65 |
| | | | | 2–3 | 45–55 |
| Silty clay loam....... | 0–30 | 50–80 | 20–30 | 3–4 | 45–55 |
| | | | | 2–3 | 40–45 |
| Sandy clay.......... | 50–70 | 0–20 | 30–50 | 3–4 | 40–45 |
| | | | | 2–3 | 35–40 |
| Silty clay........... | 0–20 | 50–70 | 30–50 | 3–4 | 35–40 |
| | | | | 2–3 | 30–35 |
| Clay............... | 0–50 | 0–50 | 30–100 | 3–4 | 30–35 |
| | | | | 2–3 | 25–30 |

* "Handbook of Drainage and Construction Products," Armco Drainage and Metal Products, Inc., 1955.

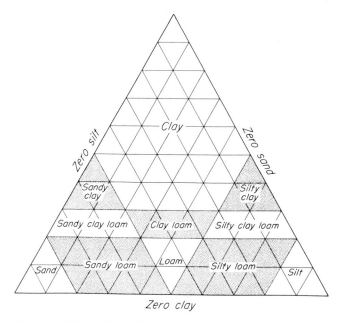

**Fig. 17-16**   Textural classification of soil.

A better appreciation of the problem and its possible solution is obtained from a study of depth and spacing of subdrains for various classes of soil as given in Table 17-2, based on textural classification of soils shown in Fig. 17-16.

The advantage of selective use of material in construction of the subgrade is further emphasized in the problems of subsurface drainage. Bringing the better available soils (E-1 to E-5) to upper layers of the subgrade and thus providing adequate *protective cover* on the poorer fine-grained soils may mean the difference between a single line of drains on either side of the runways and other paved areas and a closely spaced network of drains under the pavement.   The effectiveness of even closely spaced drains will be questionable in cases where the soil is predominantly silt or clay or a combination of these.

**17-12   Drainage filters**   Filter material to be used for backfill over drain pipes should prevent infiltration of the surrounding natural soil into the interstices of the porous backfill material and into the drain pipes through perforations or open joints.   The usual control is that the 15 per cent size of the filter material should not exceed five times the 85 per cent size of the soil being drained.   Stated another way, the *piping ratio*, 15 per cent size of filter material/85 per cent size of soil, ≤5. If the soil is a predominantly one-size material like a fine sand or a silty soil, the piping ratio should not exceed 4.

In addition to the above requirement for piping ratio, the grain-size curve for filter material should be approximately parallel to that for the soil so that all the finer particles of soil will be effectively prevented from washing into the filter material.

It will not be practical to try to obtain a gradation of filter material exactly as indicated by the gradation curve so drawn. This theoretical curve should, however, represent the coarse side of the gradation range of particle size. The fine side of the gradation range will then be limited by the physical and economic feasibility of modifying the grading of available filter material balanced against the fact that material with a coarser grading will, in general, perform more effectively in admitting water to the drain pipe.

A typical example of determination of a satisfactory range of gradation for filter material for a given soil gradation is given in Fig. 17-17. The grain-size curve for the soil is drawn and the 85 per cent size located. If the soil is reasonably well graded, this size may be multiplied by 5 to obtain the 15 per cent size of the filter material. Then a parallel curve is drawn through this point to determine the coarse side of the gradation range. The fine side of the gradation range shown in Fig. 17-17 is the fine side of specification limits for commercial-size concrete sand.

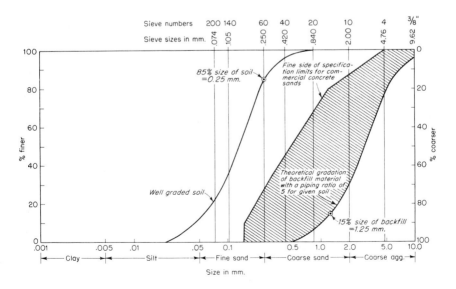

**Fig. 17-17**  Example of protective-filter grading.

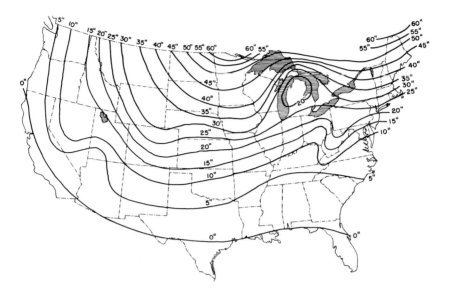

**Fig. 17-18**   Average annual frost penetration.

The principle of piping ratio may also be effectively used in providing a filter sand layer to prevent fine subgrade soils from "pumping" into the interstices of sub-base and base-course aggregates in case of adverse moisture conditions in the subgrade. This will permit water to get through to the porous base material and thence to drain pipes and at the same time prevent the pumping of silt and clay into the pavement structure. A sand layer 4 to 6 in. in thickness has proved adequate for this purpose provided piping ratios are taken into consideration.

**17-13   Depth of drain pipe**   Drainage pipelines should be placed below the depth of frost penetration whenever possible. Average annual frost penetration for various parts of the country is given in Fig. 17-18. It may be used as a guide for determining minimum depth for pipeline installation. Local data should be obtained, however, as to annual frost penetration if such records are available.

Further minimum depth requirements must be provided to protect adequately the drainage pipelines against wheel loads in traffic areas. A very helpful guide for determination of minimum cover over the top of conduits for various kinds of pipe and for a variety of wheel loads is given in Table 17-3.

**17-14   Backfill**   Backfilling the trench over the pipelines is one of the most important items in construction of the drainage system. After a

# Table 17-3 Recommended minimum depth of cover for conduits

• Pipe is not generally available or is not included in Specifications.
− Pipe is not considered safe for wheel load indicated.

Width of trench:- 1.5D + 9"
Dead load:- 120#/cu. ft.

| Kind of Pipe | Number of AASHO Specifications | For Wheel Load of 15,000 Pounds — Nominal Diameter of Pipes | | | | | | | | | | | | | | | | For Wheel Load of 37,000 Pounds — Nominal Diameter of Pipes | | | | | | | | | | | | | | | | For Wheel Load of 60,000 Pounds — Nominal Diameter of Pipes | | | | | | | | | | | | | | | |
|---|---|---|---|---|---|---|---|---|---|---|---|---|---|---|---|---|---|---|---|---|---|---|---|---|---|---|---|---|---|---|---|---|---|---|---|---|---|---|---|---|---|---|---|---|---|---|---|---|---|
| | | 6 | 8 | 10 | 12 | 15 | 18 | 21 | 24 | 27 | 30 | 33 | 36 | 42 | 48 | 54 | 60 | 6 | 8 | 10 | 12 | 15 | 18 | 21 | 24 | 27 | 30 | 33 | 36 | 42 | 48 | 54 | 60 | 6 | 8 | 10 | 12 | 15 | 18 | 21 | 24 | 27 | 30 | 33 | 36 | 42 | 48 | 54 | 60 |
| Farm Drain Pipe | Extra Quality M 66-42 | $\frac{1}{2}$ | 2 | $2\frac{1}{2}$ | 3 | 4 | – | – | not recommended | | | | | | | | | $3\frac{1}{2}$ | $3\frac{1}{2}$ | $4\frac{1}{2}$ | $6\frac{1}{2}$ | – | – | – | not recommended | | | | | | | | | – | – | • | • | – | – | – | not recommended | | | | | | | | |
| Vitrified Clay Culvert Pipe Extra Strength | M 65-42 | $\frac{1}{2}$ | 2 | $2\frac{1}{2}$ | $2\frac{1}{2}$ | $2\frac{1}{2}$ | $2\frac{1}{2}$ | $2\frac{1}{2}$ | 2 | 2 | 2 | 2 | 2 | • | • | • | – | 3 | 4 | $4\frac{1}{2}$ | $4\frac{1}{2}$ | $5\frac{1}{2}$ | $5\frac{1}{2}$ | 5 | 5 | • | – | – | – | • | • | • | – | 4 | 5 | $6\frac{1}{2}$ | 8 | – | – | – | – | – | – | – | – | • | • | 5 | – |
| Plain Concrete Culvert Pipe | M 86-42 | 2 | $2\frac{1}{2}$ | 3 | 3 | $3\frac{1}{2}$ | $3\frac{1}{2}$ | 4 | 4 | • | • | • | • | • | • | • | • | • | $4\frac{1}{2}$ | 5 | • | • | • | • | • | • | • | • | • | • | • | • | • | • | 6 | $7\frac{1}{2}$ | • | • | • | • | • | • | • | • | • | • | • | • |
| Reinforced Concrete Culvert Pipe Standard | 3500 lb Concrete M 41-42 | • | • | • | 2 | 2 | 2 | $2\frac{1}{2}$ | $2\frac{1}{2}$ | 3 | 3 | $2\frac{1}{2}$ | $2\frac{1}{2}$ | $2\frac{1}{2}$ | • | – | – | • | • | • | 4 | 4 | • | • | • | • | 4 | • | • | • | • | • | – | – | • | • | 5 | $5\frac{1}{2}$ | 6 | • | • | • | • | • | • | • | • | 5 | – |
| Reinforced Concrete Culvert Pipe Extra Strength | 4500 lb Concrete M 41-42 | • | • | • | • | • | • | • | $1\frac{1}{2}$ | • | • | • | • | • | • | 2 | – | • | • | • | • | 4 | • | • | • | 4 | • | $3\frac{1}{2}$ | $3\frac{1}{2}$ | $3\frac{1}{2}$ | 3 | 3 | 3 | • | • | • | $5\frac{1}{2}$ | • | • | $5\frac{1}{2}$ | $5\frac{1}{2}$ | $5\frac{1}{2}$ | 6 | 6 | 5 | 5 | – |
| | 18 Gage | – | – | • | • | – | – | – | – | • | – | • | • | • | • | • | • | $1\frac{1}{2}$ | – | • | • | – | – | • | • | • | • | • | • | • | • | • | • | $1\frac{1}{2}$ | – | $1\frac{1}{2}$ | $1\frac{1}{2}$ | – | $2\frac{1}{2}$ | $2\frac{2}{3}$ | $3\frac{1}{2}$ | 4 | $4\frac{1}{2}$ | $4\frac{1}{2}$ | 5 | 6 | 6 | 6 | – |
| | 16 Gage | • | – | – | – | – | – | – | – | • | • | • | $\frac{1}{2}$ | • | • | • | – | – | $\frac{1}{2}$ | $\frac{1}{2}$ | $\frac{1}{2}$ | – | • | – | 2 | • | 2 | 2 | 2 | • | • | • | • | – | – | – | • | – | 2 | 2 | 3 | $3\frac{1}{2}$ | $4\frac{1}{2}$ | 5 | 5 | 6 | • |
| Corrugated Metal Culvert Pipe | 14 Gage | • | – | – | – | – | – | – | – | – | – | $\frac{1}{2}$ | • | • | • | • | • | – | – | • | – | • | – | • | • | • | $\frac{1}{2}$ | $\frac{1}{2}$ | $\frac{1}{2}$ | • | • | • | • | • | • | – | – | • | • | • | 2 | $2\frac{1}{2}$ | 3 | $3\frac{1}{2}$ | 4 | 5 | 6 | 5 |
| | M 36-42 12 Gage | • | • | • | – | – | – | • | • | • | • | • | • | • | 2 | 2 | 2 | • | • | • | • | • | • | • | • | • | • | • | • | $2\frac{1}{2}$ | $2\frac{1}{2}$ | 3 | 3 | • | • | • | – | • | • | 2 | 2 | $2\frac{1}{2}$ | 3 | 4 | $4\frac{1}{2}$ | $4\frac{1}{2}$ | $4\frac{1}{2}$ |
| | 10 Gage | • | • | • | • | • | • | • | • | • | • | • | • | • | – | – | – | • | • | • | • | • | • | • | • | • | • | • | – | 2 | 2 | 2 | 2 | • | • | • | • | • | • | • | • | • | $2\frac{1}{2}$ | 3 | $3\frac{1}{2}$ | $3\frac{1}{2}$ | $3\frac{1}{2}$ |
| | 8 Gage | • | • | • | • | • | • | • | • | • | • | • | • | • | – | – | – | • | • | • | • | • | • | • | • | • | • | • | – | $\frac{1}{2}$ | $1\frac{1}{2}$ | $2\frac{1}{2}$ | $2\frac{1}{2}$ | • | • | • | • | • | • | • | • | • | • | • | $2\frac{1}{2}$ | 3 | 3 | 3 |

279

pipeline has been inspected and approved as to proper bedding, alignment, and joint construction, selected material from excavation or from borrow pits is tamped in place around and over the pipe in layers not exceeding 6 in.  Care should be exercised so as to ensure thorough compaction of fill under the haunches of the pipe.  This procedure of tamping, with either hand tampers or mechanically operated tampers, is usually carried to at least 6 in. above the top of the pipe.  The remaining depth of trench is then backfilled and compacted to a density equal to or greater than that of surrounding soil, and in graded areas compaction should be consistent with density requirements for the subgrade construction.

In the case of subdrains filter material should be compacted uniformly around and over the drain pipes in layers by tamping procedure similar to those mentioned above.

## QUESTIONS AND PROBLEMS

**17-1.** A 24-in. ID concrete pipe culvert is laid on a slope of 0.01 ft/ft.  Find the capacity of the pipe (a) running full, (b) running one-half full.

**17-2.** A small drainage area of 15 acres on an airport site has an average slope of about 2 per cent.  The watershed is 1,800 ft long and is to be turfed.  The runoff from this area is to be carried under a landing strip with a concrete pipe culvert, laid on a grade of 0.33 per cent.  Determine the size of pipe to be used (a) if no ponding is to be allowed, (b) if 1 hr of ponding time is allowed.

**17-3.** Determine the net area of openings for a drop inlet to be used at an intersection of runways where the runoff may be 3.6 cfs.  Assume that a head of 4 in. may develop at the inlet.

**17-4.** A network of subdrains is to be placed in a subgrade soil to improve its load-carrying capacity.  The soils have the following grading:

| Particle size | Percentage finer |
|---|---|
| No. 10 sieve | 96 |
| No. 20 sieve | 88 |
| No. 40 sieve | 79 |
| No. 60 sieve | 64 |
| No. 140 sieve | 33 |
| No. 200 sieve | 26 |
| 0.05 mm | 20 |
| 0.005 mm | 7 |

a. Plot the theoretical gradation curve for suitable filter material which will give a piping ratio of 5.

b. Determine a grading range of suitable filter material, assuming that a fairly fine commercial concrete sand will be the finest material to be considered.

**17-5.** a. What is the textural classification of the subgrade soil of Prob. 17-4?

b. What is the textural classification of the coarsest filter material that should be used?

## BIBLIOGRAPHY

"Airport Drainage," Federal Aviation Administration, 1965.
"Concrete Pipe Handbook," American Concrete Pipe Association, 1958.
"Handbook of Culvert and Drainage Practice," Armco Drainage Products Association, 1958.
Use of the Rational Formula in Airport Drainage, *Civil Aeronautics Administration Tech. Develop. Rept.* 131, 1950.

# 18
# Pavements

**18-1 Pavement types** The theory of pavement design has been discussed in some detail in the treatment of roads and pavements and will not be repeated here. However, it should be remembered that a pavement is primarily a device for spreading the wheel load so that subgrade pressures are reduced to a tolerable intensity. Thus pavement design is dependent upon two factors: wheel loads and subgrade characteristics.

There are many procedures for expressing the influence of the subgrade on the selection of pavement thickness. The present discussion will be limited to those in accepted practice for airport construction. For design purposes the gross plane weight is used in the FAA procedure.

The CBR method of design is presented in Part 1, Roads and Pavements. The FAA method of design based on its soil classifications E-1 to E-13 will be taken up here.

Of the two general types of pavement, rigid and flexible, portland-cement concrete pavement is the principal rigid type. Bituminous-surface courses on granular-base courses comprise the flexible type. Bituminous material may also be used in the base course.

Standard specifications cover a wide range of base courses consisting of crushed rock, gravel, slag, shell, sand clay, and soil cement, and surface courses including portland-cement concrete, bituminous concrete, sand-bituminous mixtures, and bituminous-surface treatments.

**18-2  Use of design charts**  A key to pavement-thickness requirements for each type of subgrade soil is given in the last four columns of Table 16-1. Anticipated effects of moisture and frost on the subgrade soil are taken into account.

Good drainage is said to exist if topography and internal drainage characteristics of the soil are such that there will be no accumulation of water which would contribute to instability of the subgrade. Antici-pated effectiveness of the drainage system to handle surface runoff and to control ground-water flow, water-table elevation, and capillarity should be considered in deciding between good drainage and poor drainage.

Severe frost is considered to exist if the depth of frost penetration is greater than the total thickness of pavement structure required for "no frost" condition. Otherwise the condition of no frost prevails. Average annual frost-penetration depths for the United States are given in the chapter on drainage.

Small airports which accommodate personal aircraft and small air-craft engaged in nonscheduled business and industrial activities may not require paved operational areas. Conditions at the site may be adapta-ble for development of a turf surface adequate for limited operations of these light aircraft.

In general, soil-stabilization and bituminous-surface treatment may prove adequate for aircraft up to 12,500 lb gross weight. For aircraft up to 30,000 lb gross weight, a low-cost flexible pavement structure com-posed of a surface course, base course, and sub-base will be required in accordance with the subgrade soil classification and the flexible $F$ desig-nation in the last four columns of Table 16-1. Design curves for these pavements are given in Fig. 18-1.

The dynamic effect (impact) of moving wheel loads is overcome by the fact that when a plane is moving—when taking off or landing or when taxiing at any appreciable speed—it is at least partially air-borne. More-over, the soil is strained less by a transient loading through a pavement structure.

It is to be expected, too, that the effect of load repetition will not be critical on the runway proper because of the wide distribution of traffic on this area. Considerable repetition of slow-moving and static loads will occur, however, on taxiways, aprons, turnarounds, and runway ends, and experience has shown that these are the most critical areas from the standpoint of pavement design.

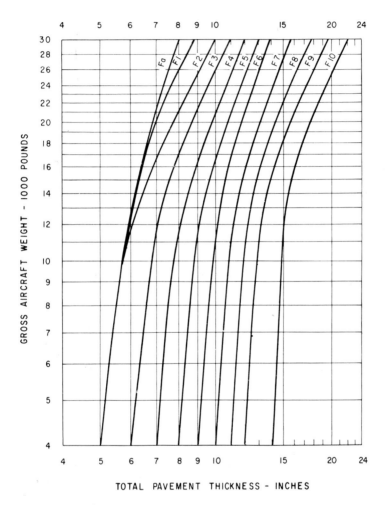

**Fig. 18-1** Design curves for flexible pavements—light aircraft. (*Courtesy of FAA.*)

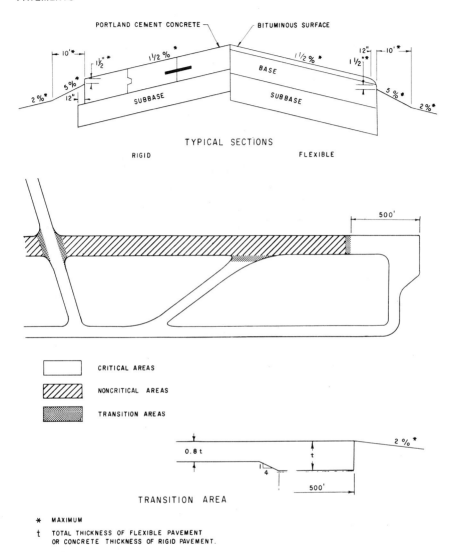

**Fig. 18-2** Typical sections and critical areas. (*Courtesy of FAA.*)

For this reason two sets of design curves are provided for determination of pavement thicknesses for various subgrade soil ratings and various wheel loads, one set for critical areas and one for noncritical areas. These are illustrated in Fig. 18-2.

Thickness design curves for flexible pavements are given in Figs. 18-3 to 18-5. Thickness design curves for rigid pavements are shown in

NONCRITICAL AREAS - TOTAL PAVEMENT THICKNESS - INCHES

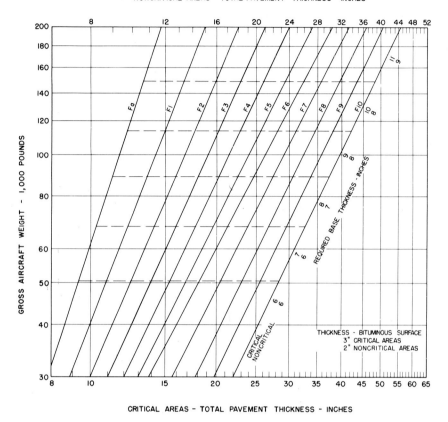

CRITICAL AREAS - TOTAL PAVEMENT THICKNESS - INCHES

**Fig. 18-3**   Design curves—flexible pavement—single gear.   (*Courtesy of FAA.*)

Fig. 18-6.   For rigid pavements the required thickness for noncritical areas is obtained by taking 80 per cent of the required critical area pavement thickness.   No reduction is made in the sub-base thickness.   All design curves are determined on the basis that 5 per cent of the gross weight of the aircraft is supported by the nose gear and that the remaining 95 per cent is distributed equally between the two main undercarriage assemblies.

**18-3   Pavement joint details**   Rigid pavement joint details are illustrated in Fig. 18-7, and description and use are given in Table 18-1.   Arrangement of joints is illustrated in Fig. 18-8.

**18-4   Pavement overlays**   As larger and heavier aircraft are built and become part of the aviation fleet, an airport is often required to serve heavier aircraft than those for which its pavements were originally designed and built. Continued serviceability may be achieved by the construction of overlays—pavements constructed on top of existing pavements. Various combinations of existing pavement and overlay construction are illustrated in Fig. 18-9. A rigid or concrete overlay is one consisting of portland-cement concrete. A flexible overlay consists of a combination of high-quality base course and a bituminous-surface course. A bituminous overlay consists entirely of bituminous concrete.

Overlay design procedure requires preliminary determination of the soil group and subgrade class of the soil underlying the existing pavement on the basis of soil tests, drainage, and frost conditions. Total pavement thickness required for this underlying soil condition for the heavier aircraft is then determined from the preceding design curves. The actual

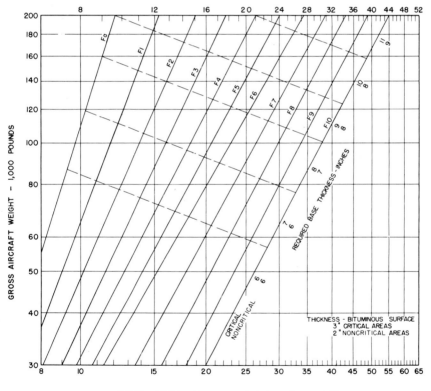

NONCRITICAL AREAS – TOTAL PAVEMENT THICKNESS – INCHES

CRITICAL AREAS – TOTAL PAVEMENT THICKNESS – INCHES

**Fig. 18-4**   Design curves—flexible pavement—dual gear.   (*Courtesy of FAA.*)

thickness and condition of each layer of existing pavement is then subtracted from the total thickness requirements in accordance with the following criteria to determine overlay thicknesses:

1. Sub-base courses should not be used in pavement overlays.
2. Bituminous overlays should have a minimum thickness of 3 in.
3. An existing dense-graded, plant-mix bituminous surface in sound condition may be evaluated for base-course purposes, on the basis that each inch of surface is equivalent to $1\frac{1}{2}$ in. of base course, provided the entire overlay will consist of bituminous concrete.
4. Under all other conditions, the existing surface course will be considered, inch for inch, as base course.

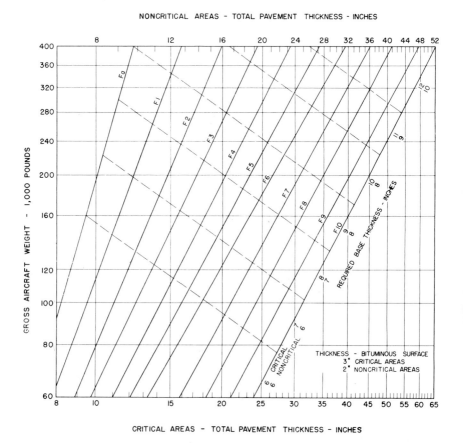

**Fig. 18-5**  Design curves—flexible pavement—dual-tandem gear.  (*Courtesy of FAA.*)

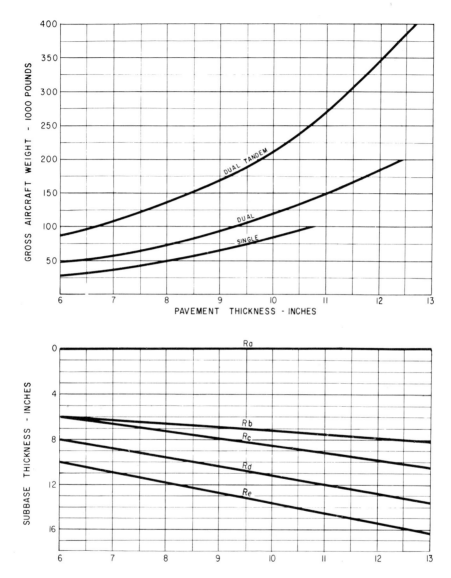

**Fig. 18-6** Design curves—rigid pavement—critical area. (*Courtesy of FAA.*)

5. If a bituminous base is to be utilized, a thickness adjustment may be made on the basis of 1 in. of such base being equivalent to $1\frac{1}{2}$ in. of nonbituminous base.

For flexible and bituminous overlays on rigid pavements, FAA specifies the following formulas:

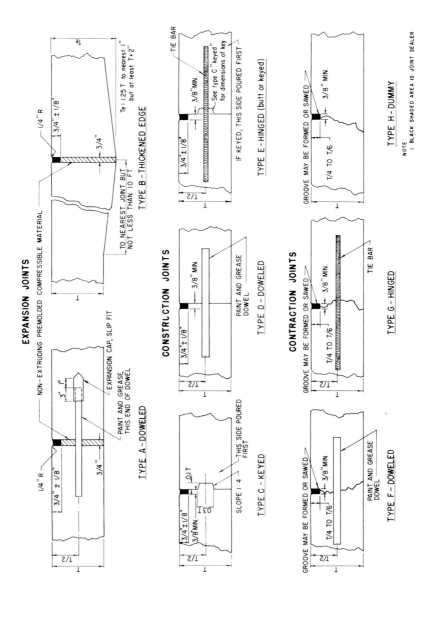

**Fig. 18-7** Details of joints in rigid pavements.

1. *For flexible overlays:*

$$t_f = 2.5(Fh - h_e)$$

where $t_f$ = required thickness of flexible overlay
$F$ = factor which varies with subgrade class
$h$ = required thickness of an equivalent single-slab place directly on the subgrade or sub-base
$h_e$ = thickness of existing slab

2. *For bituminous overlays:*

$$t_b = \frac{t_f + 0.5t_s}{1.5}$$

**Table 18-1   Joint types***
Description and use

| Type | Description | Longitudinal | Transverse |
|---|---|---|---|
| A | Doweled expansion joint | | Use near intersections to isolate them |
| B | Thickened edge expansion joint | Use at intersections where dowels are not suitable and where pavements abut structures | Provide thickened edge (or keyway) where pavement enlargement is likely |
| C or D | Keyed or doweled construction joint | Use for all construction joints except where type E is used | Use type D where paving operations are delayed or stopped |
| E | Hinged construction joint | Use for all construction joints of the taxiways and for all other construction joints that are 25 ft or less from the pavement edge | |
| F | Doweled contraction joint | | Use for all contraction joints in critical areas, for all reinforced pavement areas, and for the first two joints on each side of expansion joints |
| G | Hinged contraction joint | Use for all contraction joints of the taxiway and for all other contraction joints placed 25 ft or less from the pavement edge | |
| H | Dummy contraction joint | Use for all other contraction joints in pavement | Use for all remaining contraction joints in nonreinforced pavements |

* FAA.

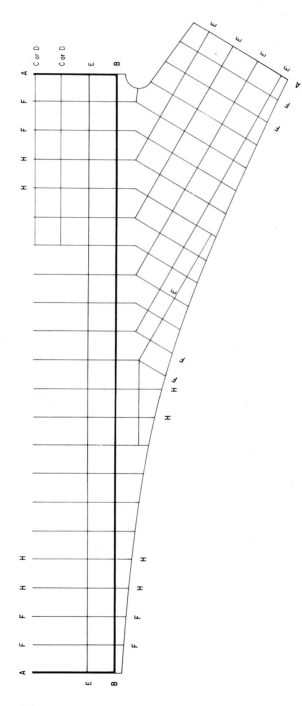

**Fig. 18-8** Arrangement of joints at intersection of runway and taxiway (nonreinforced). *(Courtesy of FAA.)*

292

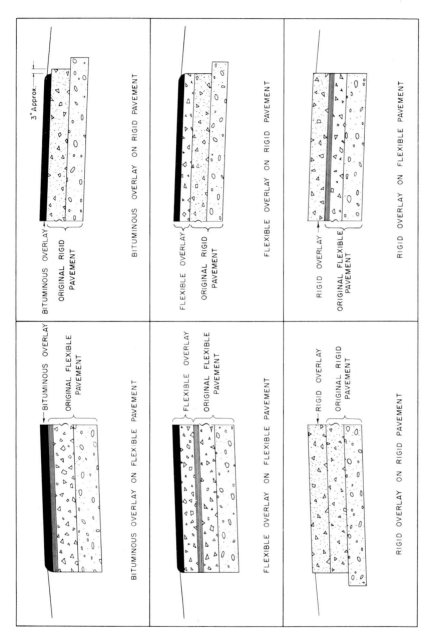

3" Approx.

BITUMINOUS OVERLAY

ORIGINAL RIGID
PAVEMENT

BITUMINOUS OVERLAY ON RIGID PAVEMENT

FLEXIBLE OVERLAY

ORIGINAL RIGID
PAVEMENT

FLEXIBLE OVERLAY ON RIGID PAVEMENT

RIGID OVERLAY

ORIGINAL FLEXIBLE
PAVEMENT

RIGID OVERLAY ON FLEXIBLE PAVEMENT

BITUMINOUS OVERLAY

ORIGINAL FLEXIBLE
PAVEMENT

BITUMINOUS OVERLAY ON FLEXIBLE PAVEMENT

FLEXIBLE OVERLAY

ORIGINAL FLEXIBLE
PAVEMENT

FLEXIBLE OVERLAY ON FLEXIBLE PAVEMENT

RIGID OVERLAY

ORIGINAL RIGID
PAVEMENT

RIGID OVERLAY ON RIGID PAVEMENT

**Fig. 18-9** Typical sections of overlay pavements. *(Courtesy of F.A.A.)*

293

where $t_b$ = required thickness of bituminous overlay

$t_f$ = required thickness of flexible overlay

$t_s$ = required thickness of surface course

The following minimums apply to flexible and bituminous overlays on rigid pavements:

4 in. for nonbituminous base course

3 in. for a bituminous overlay

| Existing subgrade class | Value of $F$ when sub-base under existing pavement conforms to requirements for class of subgrade indicated below | | | | |
|---|---|---|---|---|---|
| | Ra* | Rb | Rc | Rd | Re |
| Ra | 0.80 | | | | |
| Rb | 0.90 | 0.80 | | | |
| Rc | 0.94 | 0.90 | 0.80 | | |
| Rd | 0.98 | 0.94 | 0.90 | 0.80 | |
| Re | 1.00 | 0.98 | 0.94 | 0.90 | 0.80 |

* Figures in this column apply when no sub-base has been provided.

Thickness design of concrete overlays on existing flexible pavements is taken from the curves in Fig. 18-6. The existing flexible pavement is considered as sub-base for the overlay slab.

*Concrete overlay on rigid pavement* The design of concrete overlays on the existing rigid pavements is also based on the curves in Fig. 18-6. The rigid pavement design curves will disclose the thickness of concrete slab and sub-base required to satisfy the design conditions for a pavement constructed directly on the existing subgrade. Since the concrete slab thickness, so determined, is predicated on the provision of a sub-base varying in thickness with the subgrade class and aircraft loading, the thickness requirement for an equivalent single slab $h$ must be adjusted for an existing subgrade class other than Ra. A satisfactory adjustment for the basic design curves can be made as follows:

1. If less than 6 in. of sub-base has been provided for a pavement supported by a subgrade other than Ra, add 1 in. of slab thickness to the required single-slab thickness $h$.
2. No adjustment to the basic design curve thickness is required if the subgrade is classed as Ra or if a minimum of 6 in. of sub-base has been provided for on any other subgrade.

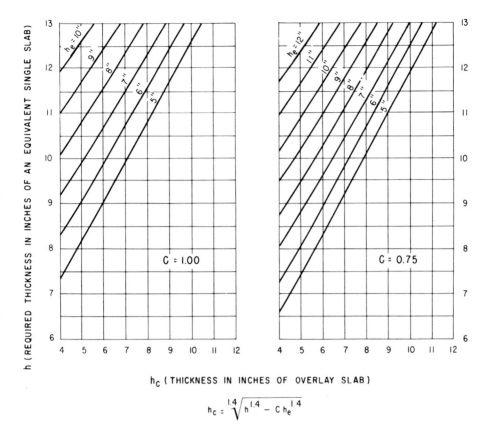

$$h_c = \sqrt[1.4]{h^{1.4} - C\,h_e^{1.4}}$$

**Fig. 18-10**  Concrete overlay on rigid pavement.  (*Courtesy of FAA.*)

3. Although these adjustments seem incompatible with the requirements of an original design based on Fig. 18-6, with respect to sub-base thickness, the condition of the existing pavement and the correction coefficient as discussed below should compensate for these differences.
4. Based on the above, and preliminary data obtained, the thickness of the concrete overlay slab to be applied to the existing rigid pavement is determined by the curves of Fig. 18-10.   Values of the coefficient $C$ are based on the condition of the existing pavement as determined from the pavement condition survey.   Recommended values are

$C = 1.00$ for existing pavement in good condition
$C = 0.75$ for existing pavement with initial corner cracks due to loading, but no progressive cracking
$C = 0.35$ for existing pavement badly cracked or crushed

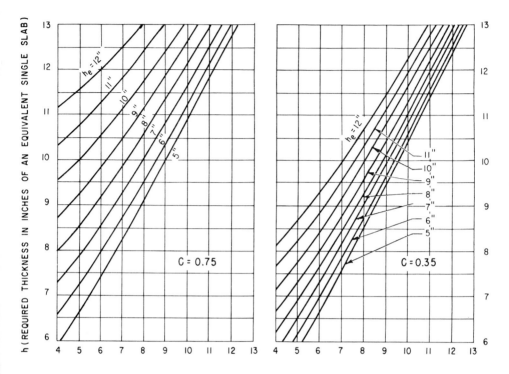

$$h_c = \sqrt{h^2 - C\,h_e^2}$$

**Fig. 18-11**   Concrete overlay on rigid pavement with separating course.  (*Courtesy of FAA.*)

Conditions at a particular location may indicate the desirability of adopting intermediate values for $C$ within the recommended range.

5. Under some circumstances it may be desirable to apply a separating course or layer of bituminous concrete or granular material to the surface prior to the application of the rigid overlay.  If such is the case, an increase in the overlay thickness is warranted and the curves in Fig. 18-11 may be employed to establish the thickness of the overlay slab.

**QUESTIONS AND PROBLEMS**

**18-1.** The subgrade soil at an airport site in central Kansas has the following characteristics: 33 per cent retained on the No. 10 sieve; 28 per cent combined silt and clay; liquid limit of 32; and a plasticity index of 8.  Drainage may be considered as good.  Determine required thickness of flexible-pavement structure for runways and taxiways using a design aircraft of 125,000 lb gross weight on dual-tandem gear.

**18-2.** Determine required thickness of rigid-pavement structure for the conditions of Prob. 18-1.

**18-3.** Determine overlay-thickness design for the pavements of Probs. 18-1 and 18-2 to provide for aircraft of 200,000 lb gross weight, using the procedures outlines in Sec. 18-4.

## BIBLIOGRAPHY

"Airport Paving," Federal Aviation Administration Advisory Circular, 1964.

A Mathematical Study of Shearing Stress Produced in a Pavement by the Locked Wheels of an Airplane during the Warm-up of its Engines, *Civil Aeron. Admin. Tech. Develop. Note* 47, 1951.

"Thickness Design for Concrete Pavements," HB 35, Portland Cement Association, 1966.

# 19

# Marking and Lighting

**19-1  Aids to traffic**  In the interest of safety to air navigation the FAA has formulated certain requirements for marking and lighting of airport facilities.  The airport should be marked so that pilots may easily "spot" it and so that they will be able to identify distinctly the landing area, determine its extent, avoid hazards, and ascertain the direction of the wind.

Airport boundary markers should clearly indicate the landing and take-off area.  Placement and suggested construction of boundary markers are shown in Fig. 19-1.

To aid the pilot further in locating the airport and in spotting indicators and signal devices, the segmented-circle marker system has been standardized.  The segmented circle with a conventional wind cone constitutes the minimum installation requirement for an operating airport. Other signals and traffic-control features are included as needed for a particular airport.

The segments may be constructed and painted in accordance with the suggested construction of boundary markers shown in Fig. 19-1. Other details of this marking system are shown in Fig. 19-2.

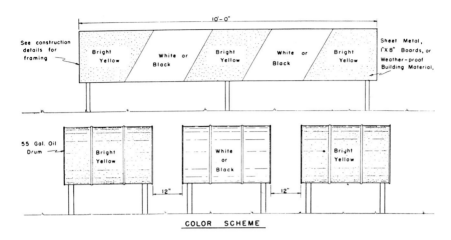

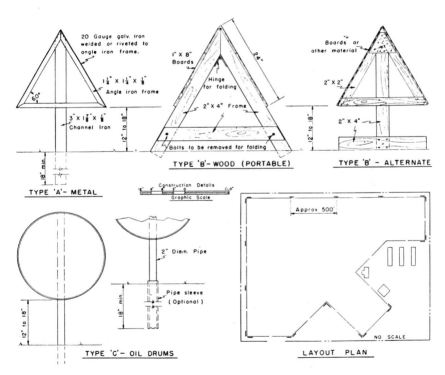

**Fig. 19-1** Placement and suggested construction of boundary markers. (*Courtesy of FAA.*)

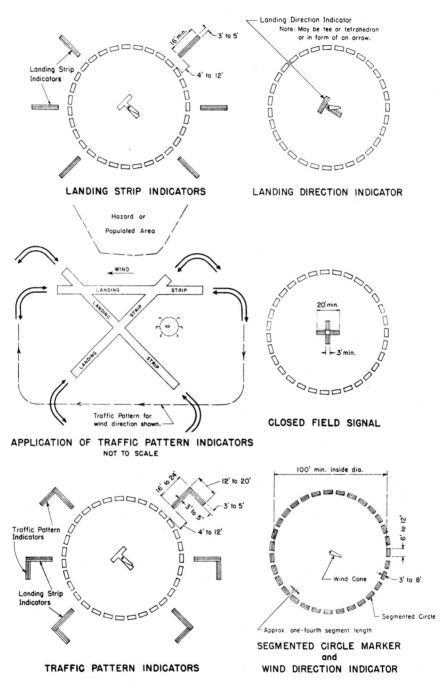

**Fig. 19-2**  Segmented-circle airport marker system.  *(Courtesy of FAA.)*

**19-2  Runway identification**  Standard numbering and marking of runways and taxiways are indicated in Fig. 19-3.  Detailed requirements for this numbering and marking are:

1. The number at each runway end is the whole number nearest one-tenth of the magnetic azimuth of the runway center line measured clockwise from magnetic north.
2. The letters $L$, $C$, and $R$ are used to differentiate between parallel runways.
3. Length symbols are used to indicate the runway length in feet.
4. Distance markers are used to indicate points 1,500 ft from the ends of runways.
5. Details of longitudinal center stripes for runways and taxiways are shown in Fig. 19-3.

Runway markings are white.  A background of black paint may be used if necessary.  Taxiway markings are yellow.

**19-3  Lighting**  For safe landings and take-offs at night or during restricted daytime visibility, a system of signal lights is used to convey information to pilots by color and configuration of these lights.

Only a brief outline of lighting requirements will be attempted here. Figures 19-4 and 19-5 depict some of the main features which will normally be shown on the master-plan layout of the airport project.

In the early stages of development of the airport, floodlighting may constitute most of the airport illumination.  Hangar and building areas including the apron may be effectively lighted with floodlights spaced about 50 ft apart.  Floodlights of higher intensity may also be used for landing-field lighting.  Two or more floodlights are placed at the end of each runway and directed so as to illuminate effectively the landing area to be used.  Only the floodlights at the end of the runway from which landing or take-off is to be made are turned on.

As further developments are made on the airport and as traffic increases, the landing-field floodlights are replaced by runway lights, as shown in Fig. 19-5.  These runway lights are placed in pairs—one on each side of the runway 10 ft outward from the edge of the pavement. The longitudinal spacing of these pairs is made as uniform as possible and at about 200-ft intervals.

Threshold lights, marking the ends of the runway, are green.  They are placed at right angles to the center line of the runway and grouped as shown in Fig. 19-5.

The next step in developing a comprehensive lighting system is the installation of taxiway lights.  These are blue lights placed in pairs on either side of the taxiway.  A pair of lights spaced 5 ft apart is placed

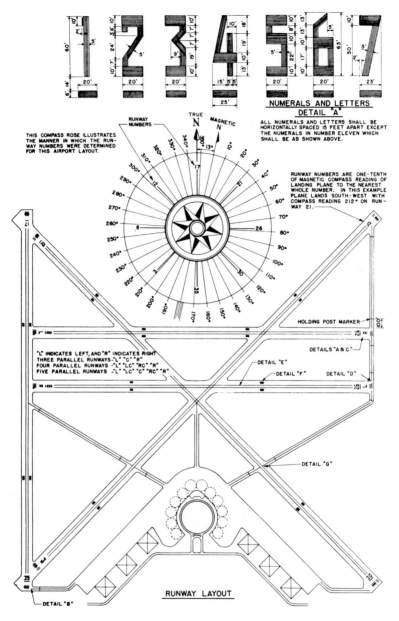

**Fig. 19-3** Airport runway and taxiway standard numbering and marking. (*Courtesy of FAA.*)

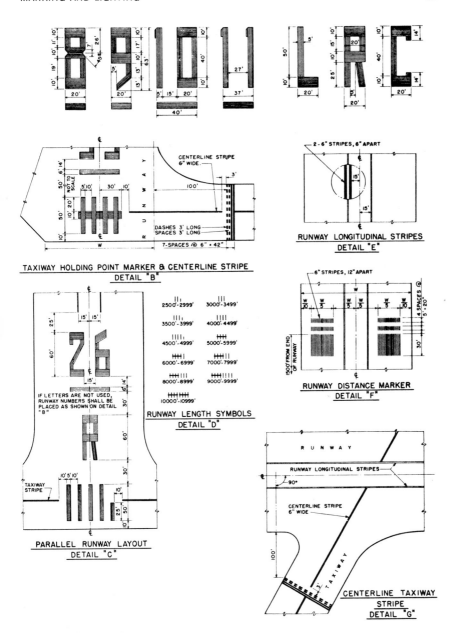

TAXIWAY HOLDING POINT MARKER & CENTERLINE STRIPE
DETAIL "B"

RUNWAY LENGTH SYMBOLS
DETAIL "D"

PARALLEL RUNWAY LAYOUT
DETAIL "C"

RUNWAY LONGITUDINAL STRIPES
DETAIL "E"

RUNWAY DISTANCE MARKER
DETAIL "F"

CENTERLINE TAXIWAY
STRIPE
DETAIL "G"

**Fig. 19-3**  (*Continued*)

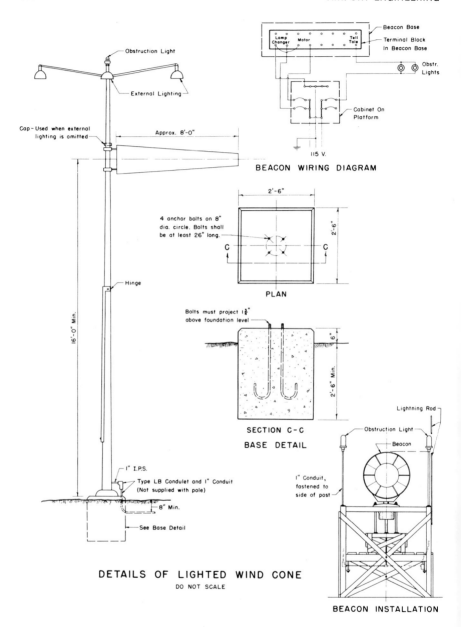

BEACON WIRING DIAGRAM

PLAN

SECTION C-C
BASE DETAIL

DETAILS OF LIGHTED WIND CONE
DO NOT SCALE

BEACON INSTALLATION

**Fig. 19-4**   Lighting details.   *(Courtesy of FAA.)*

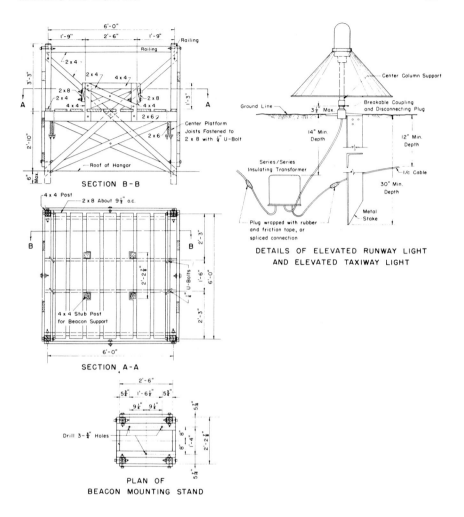

SECTION B-B

SECTION A-A

PLAN OF
BEACON MOUNTING STAND

DETAILS OF ELEVATED RUNWAY LIGHT
AND ELEVATED TAXIWAY LIGHT

BEACON PLATFORM DETAILS

**Fig. 19-4** (*Continued*)

at the end of any line of taxiway lights to mark the beginning or end of the taxiway. The larger air-carrier airports are provided with touch-down and centerline lighting with the use of lights placed in the pavement flush with the surface. The layout plan for this installation is shown in Fig. 19-6. A typical layout for a medium-intensity approach-light system is shown in Fig. 19-7.

An illuminated wind tee or tetrahedron replaces the wind cone at

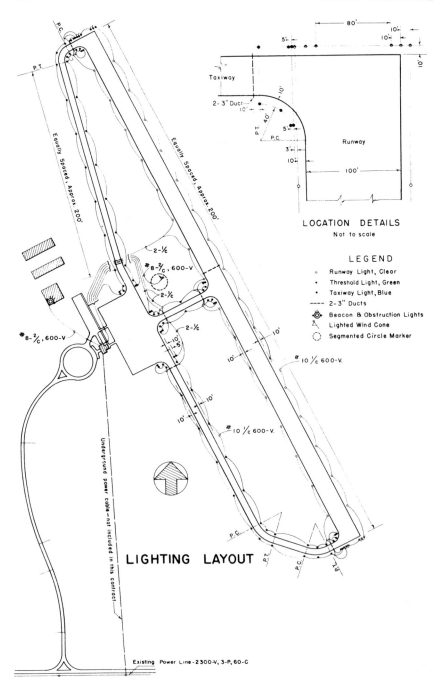

LOCATION DETAILS
Not to scale

LEGEND

○   Runway Light, Clear
·   Threshold Light, Green
•   Taxiway Light, Blue
----   2-3" Ducts
◉   Beacon & Obstruction Lights
⋀   Lighted Wind Cone
◌   Segmented Circle Marker

LIGHTING LAYOUT

**Fig. 19-5**  Lighting layout.  (*Courtesy of FAA.*)

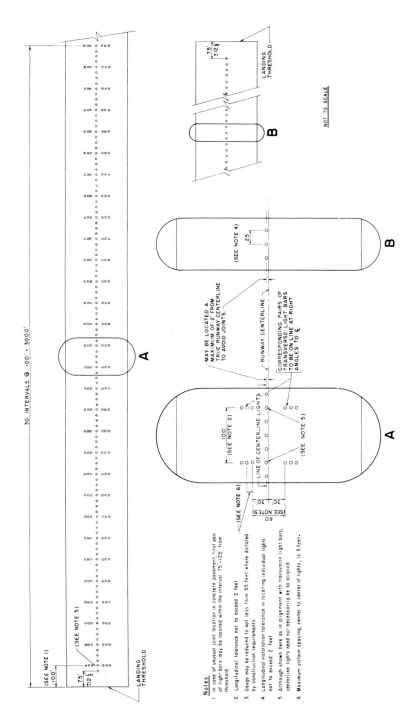

**Fig. 19-6** Runway touchdown zone and centerline lighting layout. *(Courtesy of F.A.A.)*

30 INTERVALS @ 100' = 3000'

(SEE NOTE 1)
(SEE NOTE 5)
100'
7.5'
±12½'
LANDING THRESHOLD

A

B

75'
±12½'
LANDING THRESHOLD

NOT TO SCALE

25'
(SEE NOTE 4)

B

MAY BE LOCATED A MAXIMUM OF 2' FROM TRUE RUNWAY CENTERLINE TO AVOID JOINTS.

RUNWAY CENTERLINE

CORRESPONDING PAIRS OF TRANSVERSE LIGHT BARS TO BE ON LINE AT RIGHT ANGLES TO ℄

100'
(SEE NOTE 2)

LINE OF CENTERLINE LIGHTS

(SEE NOTE 5)

(SEE NOTE 6)

30', 30'
(SEE NOTE 3)
60'

A

**Notes:**

1. In case of unusual joint location in concrete pavement, first pair of light bars may be located within the interval 75'–125' from threshold.

2. Longitudinal tolerance not to exceed 2 feet

3. Gauge may be reduced to not less than 55 feet where dictated by construction requirements

4. Longitudinal installation tolerance in locating individual lights not to exceed 2 feet

5. Although shown here as in alignment with transverse light bars, centerline lights need not necessarily be so aligned.

6. Maximum uniform spacing, center to center of lights, is 5 feet.

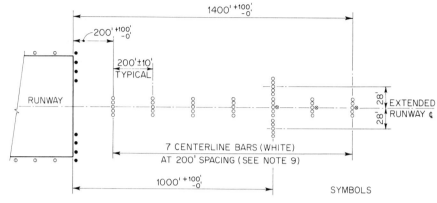

○ Existing runway edge lights
● Existing runway threshold lights
○ Steady burning approach lights
⊗ Sequenced flashing lights

1. The optimum location of the approach lights is in a horizontal plane at runway end elevation.

2. A maximum 2 percent upward longitudinal slope tolerance may be used to raise the light pattern above objects within its area.

3. A maximum 1 percent downward longitudinal slope tolerance may be used to reduce the height of supporting structures.

4. All steady burning and flashing lights are aimed with their beam axes parallel to the runway centerline and intercepting the established approach slope at a horizontal distance of 1600 feet in advance of the light.

5. All obstructions, as determined by applicable criteria for determining obstructions to air navigation, are lighted and marked as required.

6. All steady burning and flashing lights in the system emit white light.

7. Intensity control is provided for the steady burning lights.

8. The three flashing lights flash in sequence.

9. The MALS light bar closest to the runway threshold is located at a distance of 200 feet $^{+100'}_{-0'}$. All other light bars should be installed at 200-foot intervals with a ±10-foot tolerance at each light bar station. The above tolerances can be used where it is impractical to install the light bars at the optimum locations.

**Fig. 19-7**  Typical layout for medium-intensity approach-light system with sequenced flashing lights.  (*Courtesy of FAA.*)

larger airports.  This wind-direction indicator may be free-swinging or it may be operated by remote control.

### QUESTIONS AND PROBLEMS

**19-1.** A general aviation airport is to consist of a single runway.  Make a plan layout of lighting and marking facilities adequate for night operation.

### BIBLIOGRAPHY

"Airport Design," Civil Aeronautics Administration, 1961.
Configuration Details of In-Runway Lighting, *Federal Aviation Agency Advisory Circ.* 150/5340-3, 1963.
A Study of the Visibility and Glare Ranges of Slope-line Approach Lights, *Civil Aeronautics Administration Tech. Develop. Rept.* 150, 1951.

# part three

# Railroad Engineering

# 20

# Historical Development and Present Trends

**20-1  Growth of railroads**  The railway first began to serve the public as a common carrier in the year 1830.  The Baltimore & Ohio began early in that year to handle regular freight and passenger traffic by rail with horses used for motive power.  Near the end of the year the South Carolina Railroad was the first company to begin operating with steam power.

Growth of railway mileage in the United States since these humble beginnings of 1830 set the pace for growth and development of the country.  Settlement and economic development of frontier regions were pushed forward by the railroads, bringing the settlers and their building materials and machinery and transporting their produce to Eastern markets.

The phenomenal growth of railways in the decades 1870 to 1890 extended the network of rail transportation over the entire nation. Some appreciation of the rate of growth of railway mileage in the United States may be had by referring to Table 20-1.  It should be understood that a mile of railroad may consist of a single track or two or more parallel tracks. "Miles of track" is a comprehensive term which includes main-line tracks, passing sidings, way-switching tracks, spur tracks, and yard tracks.

It is understandable that in the rapid development of this vast network of rail transportation over the country some duplication of service would occur, some lines would prove to be nonprofitable, and some entirely unnecessary. In general, two requirements were involved in the economic adjustment of railway mileage: (1) there should be enough railway lines to serve adequately all economically important areas of the country; and (2) there should be sufficient number of companies represented so that any general area or region would have the benefit of competition between railway companies for transportation services.

No mastermind worked out the niceties of the economic adjustments of the railway network, but neither did these develop without the guidance of engineering skill. The adequacy of our present rail transportation network in its service to the entire nation is a tribute to the professional competence of early railroad engineers, not only in overcoming the physical obstacles of alignment and gradient but also in estimating the economic potentials of areas through which routes were established.

A study of the railway network of any considerable area of the country will show main lines of two or more companies, more or less parallel in general direction for long-distance transportation, and an interlacing of branch lines competing for service to important local areas.

Table 20-1  United States railway mileage

| Year | Miles of road | Miles of track |
|------|--------------:|---------------:|
| 1830 | 23      |         |
| 1840 | 2,818   |         |
| 1850 | 9,021   |         |
| 1860 | 30,626  |         |
| 1870 | 52,922  |         |
| 1880 | 93,296  | 115,647 |
| 1890 | 163,597 | 199,876 |
| 1900 | 193,346 | 258,784 |
| 1910 | 240,439 | 315,767 |
| 1920 | 252,845 | 406,580 |
| 1930 | 249,052 | 429,883 |
| 1940 | 233,670 | 405,975 |
| 1950 | 226,519 | 397,305 |
| 1965 | 212,130 | 372,300 |

Location and development of population centers have been largely influenced by the location of railroads.

It can be seen, then, that the network of railways is substantially complete in extent, and reference to Table 20-1 will show that railway mileage has decreased during the past few decades while adjusting to the economic need of the transportation industry.   Thus the engineer is not primarily concerned with railway location from the standpoint of establishment of new lines, but rather with *relocation* of existing lines both as to alignment and gradient to improve operating characteristics and reduce operating costs.   Efficient operation of the railroads is necessary for movement of great volumes of low-class freight at low rates and for maintaining effective competition with other forms of transportation. The adequacy of present railroad facilities for handling long-distance freight has been demonstrated in the critical periods of the past two world wars.

**20-2   Freight and passenger data**   Publications of the Bureau of Transport Economics and Statistics of the Interstate Commerce Commission provide statistics on operating costs, revenues, volumes of traffic, and other data concerning the operation of railways and other forms of transportation.   Some of these data taken from 1966 publications are shown in Table 20-2.

Movement of freight constitutes the main source of revenue for the railway.   Statistics of the Interstate Commerce Commission (ICC) show that in terms of total operating revenue the percentage of passenger revenue has steadily decreased since before the turn of the century, except during the abnormal years of 1942 to 1944 when heavy troop movements were handled.   Much has been done in recent years to make passenger service more attractive.   Improvements to railway-passenger

Table 20-2   Volume of intercity freight traffic in ton-miles by kinds of transportation, 1965*

| Transport agency | Ton-miles, billions | Per cent of annual total |
|---|---|---|
| Railways, including mail and express................. | 708.7 | 43.02 |
| Highways, for hire and private trucks................. | 370.8 | 22.50 |
| Inland waterways, including Great Lakes.............. | 256.0 | 15.54 |
| Pipelines, oil........................................ | 310.1 | 18.82 |
| Airways, including mail and express................. | 1.9 | 0.12 |
| Total........................................... | 1,647.5 | 100.0 |

* Interstate Commerce Commission, Bureau of Transport Economics and Statistics.

equipment have included air conditioning, roominess and flexibility of movement in larger passenger cars with luggage-storage rooms at the ends of the cars, large attractive windows, attractive and convenient furnishings and streamlined trains, and motive power for fast transcontinental and commuting services. It is notable that railway facilities offer one of the safest means of passenger travel. See Table 11-6 in Airport Engineering section.

Operating railroads are classified by the Interstate Commerce Commission according to the amount of their annual operating revenues. Effective Jan. 1, 1965: Class I railroads are those with annual operating revenues of 5 million dollars or more. Class I railroads operate approximately 94 per cent of the total railway mileage of the United States and earn about 98 per cent of the operating revenues of all the line-haul railroads—excluding switching and terminal companies. There were 76 Class I railroads at the close of 1965.

Operating statistics for freight traffic for calendar year 1963 on Class I railroads are shown in Table 20-3. The figures indicate that the average ton of freight was hauled a distance of 460 miles. While the average cost to the shipper is given as 1.31 cents per ton-mile, there is a considerable range of rates for different classes of freight and for different items within such classes as "manufactures" and "merchandise."

Forwarder traffic includes piggyback services—carrying truck

**Table 20-3   Freight traffic, calendar year 1963***

Railroads of Class I in the United States

| Freight traffic | Gross freight revenue, millions | Revenue tons originated, millions | Revenue per ton originated | Average load per car originated, tons |
|---|---|---|---|---|
| Products of agriculture............. | $1,192.6 | $    160.6 | $  7.43 | 42.6 |
| Animals and products.............. | 205.2 | 9.4 | 21.88 | 19.0 |
| Products of mines................. | 1,966.3 | 662.4 | 2.97 | 64.3 |
| Products of forests................ | 661.0 | 78.2 | 8.45 | 41.9 |
| Manufactures and miscellaneous.... | 4,295.6 | 368.0 | 11.67 | 34.4 |
| Forwarder traffic†................. | 164.0 | 4.6 | 35.91 | 12.8 |
| Merchandise‡..................... | 74.8 | 1.7 | 44.59 | |
| Total........................ | $8,559.5 | $1,284.9 | $  6.66 | 46.7 |

Total revenue ton-miles: 621,737 million
Average receipts per ton-mile: 1.310 cents
* Statistics of Railroads of Class I, *Association of American Railroads*, Sept., 1966.
† Carload lots.
‡ Less-than-carload lots.

trailers or containers on flatcars. This type of service is playing an important role in modernization improvements of the railroad transportation services and may continue even more importantly in the integration of air-freight transport into the national transportation system.

Since about 1960, the unit train has come into general use. Basically, the unit train is a "through" freight train loaded with a single shipment at the point of origin and sent straight to the shipment's destination without handling at intermediate freight classification yards. The train is then routed back to its point of origin and made ready for another trip. The unit train is being used by the nation's largest railroads to haul coal, ore, grain, sand, gravel, cement, lumber, steel, and other bulk materials. This concept has greatly reduced rates to the shipper and has given the railroad a better competitive position in the transportation of freight.

Statistics of the Bureau of Economics of the Association of American Railroads, September, 1966, show that for the calendar year 1965, Class I railroads carried revenue passengers a total of 17.5 billion passenger-miles. Of this total, commutation accounted for about 4.1 billion passenger-miles. Average receipts for the total service amounted to 3.182 cents per passenger-mile with an average journey per passenger of 58.18 miles.

While total passenger service decreased steadily from 1960 to 1965 from 21.6 to 17.5 billion passenger-miles, the commutation service held steady at about 4.1 billion passenger-miles, with an average journey per passenger of about 21 miles. Continued improvements in this commuter service—in speed, comfort, convenience, and safety—are likely to place the railroads in an increasingly important role in handling commuter traffic for increasingly greater distances of commutation.

A 15-year record of intercity transportation by modes of transport is presented in Table 20-4, indicating recent national trends in both freight and passenger traffic.

**20-3  Motive power**  An important trend in improvements to railroad facilities has been the streamlining of motive power, chiefly through replacement of steam locomotives by diesel-electric locomotives. The first diesel-electric locomotive in railroad service was a switch engine installed in 1925. In 1934 diesel-electric locomotives were introduced into regular freight service. The number of locomotives in service at the end of years 1955 to 1965 is shown in Table 20-5.

The diesel-electric locomotive may be composed of one (or as many as five or six) diesel-electric unit. This flexibility is one of the factors that make it an efficient power plant. Further description and performance characteristics of diesel-electric and electric locomotives are given

Table 20-4   Intercity transportation by modes, 1950–1966, in United States

## Intercity Passenger Miles by Mode of Travel

| | Auto-mobiles ① | Motor Coaches ① | Total Motor Vehicles ① | Railways, Revenue Passengers | Inland Waterways | Airways, Domestic Revenue Services ② | Total |
|---|---|---|---|---|---|---|---|
| **Passenger-Miles, in Billions** | | | | | | | |
| 1950............ | 438.3 | 26.4 | 464.7 | 32.5 | 1.2 | 10.1 | 508.5 |
| 1955............ | 637.4 | 25.5 | 662.9 | 28.7 | 1.7 | 22.7 | 716.0 |
| 1958............ | 684.9 | 20.8 | 705.7 | 23.6 | 2.1 | 28.5 | 759.9 |
| 1959............ | 687.4 | 20.4 | 707.8 | 22.4 | 2.0 | 32.6 | 764.8 |
| 1960③.......... | 706.1 | 19.9 | 726.0 | 21.6 | 2.7 | 34.0 ④ | 784.3 |
| 1961............ | 713.6 | 19.7 | 733.3 | 20.5 | 2.3 | 34.6 | 790.7 |
| 1962............ | 735.9 | 21.3 | 757.2 | 20.2 | 2.7 | 37.6 | 817.7 |
| 1963............ | 765.9 | 21.9 | 787.8 | 18.6 | 2.8 | 42.8 | 852.0 |
| 1964............ | 801.8 | 22.7 | 824.5 | 18.4 | 2.8 | 49.2 | 894.9 |
| 1965............ | 838.1 | 23.3 | 861.4 | 17.5 | 3.1 | 58.1 | 940.1 |
| 1966*........... | 870.0 | 24.0 | 894.0 | 16.7 | 3.4 | 68.0 | 982.0 |
| **Passenger-Miles, percent by Mode of Travel** | | | | | | | |
| 1950............ | 86.19 | 5.20 | 91.39 | 6.39 | 0.23 | 1.99 | 100% |
| 1955............ | 89.02 | 3.56 | 92.58 | 4.01 | 0.24 | 3.17 | 100% |
| 1958............ | 90.13 | 2.74 | 92.87 | 3.10 | 0.28 | 3.75 | 100% |
| 1959............ | 89.88 | 2.67 | 92.55 | 2.93 | 0.26 | 4.26 | 100% |
| 1960③.......... | 90.03 | 2.54 | 92.57 | 2.75 | 0.34 | 4.34 ④ | 100% |
| 1961............ | 90.25 | 2.49 | 92.74 | 2.59 | 0.29 | 4.38 | 100% |
| 1962............ | 90.00 | 2.60 | 92.60 | 2.47 | 0.33 | 4.60 | 100% |
| 1963............ | 89.89 | 2.57 | 92.46 | 2.18 | 0.33 | 5.03 | 100% |
| 1964............ | 89.60 | 2.53 | 92.13 | 2.06 | 0.31 | 5.50 | 100% |
| 1965............ | 89.15 | 2.48 | 91.63 | 1.86 | 0.33 | 6.18 | 100% |
| 1966*........... | 88.59 | 2.44 | 90.03 | 1.70 | 0.35 | 6.92 | 100% |

* Estimated

① Includes intra-city portions of intercity trips. Omits rural to rural trips; strictly intra-city trips with both origin and destination confined to same city; local bus or transit movements; non-revenue school and government bus operations. ② Includes (a) fixed-base operators, (b) non-revenue planes owned by business concerns, (c) pleasure flying by individuals or clubs. Omits miles flown in "aerial application" in dusting, spraying, etc.; rural survey and other industrial activities; instructional flying, civil air patrol, experimental and test flying; all Government-owned aircraft; territorial, overseas and foreign. ③ 1960 and later years include Alaska and Hawaii. ④ Based on Civil Aeronautics Board of Statistics, Federal Aviation Agency surveys and other data. Covers domestic except movements over international waters or foreign countries. These figures, as they include Alaska and Hawaii, are not comparable with data in previous years. The addition in 1960 of Alaskan and Hawaiian passenger-miles in terms of increases were less than half of one percent.

## Ton-Miles of Intercity Freight

| | Motor Trucks ① | Railways ② | Inland Waterways ③ | Pipe Lines | Domestic Airways | Total |
|---|---|---|---|---|---|---|
| **Ton-Miles, in Billions** | | | | | | |
| 1950................. | 172.9 | 596.9 | 163.3 | 129.2 | .318 | 1,062.6 |
| 1955................. | 223.3 | 631.4 | 216.5 | 203.2 | .481 | 1,274.9 |
| 1958................. | 255.5 | 558.7 | 189.0 | 211.3 | .579 | 1,215.2 |
| 1959................. | 278.9 | 582.5 | 196.6 | 227.0 | .739 | 1,285.7 |
| 1960④............... | 285.5 | 579.1 | 220.3 | 228.6 | .778⑤ | 1,314.3 |
| 1961................. | 296.5 | 567.0 | 209.7 | 233.2 | .895 | 1,310.3 |
| 1962................. | 309.4 | 600.0 | 223.1 | 237.7 | 1.289 | 1,371.5 |
| 1963................. | 331.8 | 629.3 | 234.2 | 253.4 | 1.296 | 1,450.0 |
| 1964................. | 349.8 | 666.2 | 250.2 | 268.7 | 1.504 | 1,536.4 |
| 1965................. | 370.8 | 708.7 | 256.0 | 310.1 | 1.910 | 1,647.5 |
| 1966*................ | 400.0 | 750.0 | 260.0 | 350.0 | 2.300 | 1,760.0 |
| **Ton-Miles, percent by Type of Transport** | | | | | | |
| 1950................. | 16.27 | 56.17 | 15.37 | 12.16 | .03 | 100% |
| 1955................. | 17.51 | 49.53 | 16.98 | 15.94 | .04 | 100% |
| 1958................. | 21.03 | 45.98 | 15.55 | 17.39 | .05 | 100% |
| 1959................. | 21.70 | 45.31 | 15.29 | 17.66 | .06 | 100% |
| 1960④............... | 21.72 | 44.06 | 16.76 | 17.40 | .06⑤ | 100% |
| 1961................. | 22.63 | 43.50 | 16.00 | 17.80 | .07 | 100% |
| 1962................. | 22.56 | 43.75 | 16.27 | 17.53 | .09 | 100% |
| 1963................. | 22.88 | 43.40 | 16.15 | 17.48 | .09 | 100% |
| 1964................. | 22.77 | 43.36 | 16.28 | 17.49 | .10 | 100% |
| 1965................. | 22.50 | 43.02 | 15.54 | 18.82 | .12 | 100% |
| 1966*................ | 22.70 | 42.56 | 14.75 | 19.86 | .13 | 100% |

* Estimated

① Ton-miles between cities and between rural and urban areas included, whether private or for hire. Rural-to-rural movements and city deliveries are omitted. ② Revenue ton-miles. ③ Does not include coastwise and inter-coastal ton-miles. ④ 1960 and later years include Alaska and Hawaii. ⑤ Based on Civil Aeronautics Board of Statistics, Federal Aviation Agency surveys and other data. Covers domestic except movements over international waters or foreign countries. These figures, as they include Alaska and Hawaii, are not comparable with data in previous years. The addition in 1960 of Alaskan and Hawaiian ton-miles in terms of increases were less than half of one percent.
SOURCE: Interstate Commerce Commission, American Trucking Associations, and Automobile Manufacturers Association.

**Table 20-5   Locomotives in service***

Class I railroads of the United States

| Year | Diesel-electric units | Electric units | Steam locomotives | Other | Total |
|------|------|------|------|------|------|
| 1955 | 24,786 | 627 | 5,982 | 34 | 31,429 |
| 1956 | 26,081 | 606 | 3,714 | 32 | 30,433 |
| 1957 | 27,186 | 585 | 2,447 | 30 | 30,248 |
| 1958 | 27,575 | 556 | 1,350 | 32 | 29,513 |
| 1959 | 28,163 | 539 | 754 | 37 | 29,493 |
| 1960 | 28,278 | 492 | 261 | 49 | 29,080 |
| 1961 | 28,169 | 478 | 112 | 56 | 28,815 |
| 1962 | 28,104 | 434 | 51 | 50 | 28,639 |
| 1963 | 27,945 | 429 | 36 | 39 | 28,449 |
| 1964 | 27,837 | 393 | 34 | 36 | 28,300 |
| 1965 | 27,389 | 362 | 29 | 36 | 27,816 |

* Statistics of Railroads of Class I, September, *Association of American Railroads*, 1966.

in Chap. 23.   Development of locomotives of higher power has served to decrease the number of locomotives in service, as indicated in Table 20-5.

Main-line electrification was first introduced on American railroads in 1895.   It is not to be expected that general adoption of electric motive power will take place.   This is true mainly because of the high initial cost of installation of power plants, transmission lines, and generally more costly locomotives.   Thus general adoption of electrification is not economically feasible, and will be restricted, as in the past, to specific locations, such as mountainous terrain with long relatively steep grades and large metropolitan areas with high-density traffic, where electrification can be economically justified.   A better appreciation of this fact can be gained by consideration of some of the advantages and disadvantages of railway electrification.

Some of the advantages are:

1. Clean operation, free from smoke and gas.   This makes it particularly applicable to tunnel (there are about 320 miles of railroad tunnels in the United States) and subway operation and to metropolitan areas where smoke and fumes may be a nuisance or may be legally prohibited.
2. High starting torque and steady uniform tractive effort with transmission of power directly to the driving wheels.
3. The braking effect of the motors on a train going downgrade produces a considerable saving on mechanical brakes.

4. While the motors are being driven by the wheels of the locomotive, they function as generators, and much of the energy of the down-grade train is regenerated back into the transmission line for operation of upgrade trains. It can be seen that this is of particular advantage on long approach grades to mountain passes.

Some disadvantages of an electrified system are:

1. High cost of initial installation. Interest on a heavy investment will limit installations to heavy traffic areas or to specific areas where conditions such as control of smoke and fumes and regeneration of power are significant.
2. The stationary power plant requires a source of cheap energy such as waterpower or coal deposits. This ties in with operation in mountainous areas where regeneration of power may be effectively accomplished and where tunnels may prove economically effective in the control of alignment and gradient.
3. Lack of flexibility of the electrified system becomes apparent when one considers that route and terminals are fixed by the existence of transmission lines, and limitations of the power plant and attendant substations largely determine length of haul.

Most of the advantages of electric motive power and the advantages of an efficient built-in power plant combine to make the diesel-electric unit a very efficient locomotive.

**20-4  Recent developments**  Among the most important developments in railroading is the piggyback service, which has expanded in recent years. This service is essentially a trailer-on-flatcar service, transferring highway truck trailers to specially equipped railway flatcars for intercity transport. The trailer is then unloaded for local delivery of the load, using the regular truck-tractor unit. The service has expanded to include transfer of any large containers of freight from one transportation mode to another.

Increasing demands on the country's transportation system for intercity movement of freight (see Table 20-4) has brought about a need for containerization—placing freight in containers of standard sizes for convenience of transfer from one mode of transport to another. And with the advent of aviation into the freight-moving business, it is necessary to include that mode of transport in the standardization of containerization. A container 8 ft square in cross section and of various standard lengths of, say, 8, 10, 12, 16, and 20 ft may serve to meet the several requirements and limitations of the vehicles involved, viz., the highway stake trucks, the beds of railway flatcars or boxcars, the fuselages of large aircraft, the decks of boats and barges, and the holds of oceangoing vessels.

Developments in railroad freight cars include piggyback flatcars and container cars; large-size covered hoppers, tankers, and box-cars; two- and three-level auto-rack cars for transporting automobiles; heavy-duty flatcars; and refrigerator cars with self-contained refrigerator system providing zero temperatures when needed without servicing en route. Freight-car capacity has increased to an average cargo loading of about 50 tons, with maximum cargo loadings possibly exceeding 150 tons, with

**Table 20-6  Property investment and operating income account, calendar year 1965***

Class I railroads in the United States

| Item | Amount |
|---|---|
| Property investment, end of year: | |
| Investment in railroad property used in transportation service.. | $34,874,658,111 |
| Material and supplies | $    454,806,226 |
| Cash | $    361,319,955 |
| Total | $35,690,784,292 |
| Net investment: | |
| After accrued depreciation | $26,318,531,861 |
| Average beginning and end of year | $26,040,645,372 |
| | |
| Operating income account: | |
| Average miles represented by income account | 213,413 |
| Total operating revenues | $10,207,849,727 |
| Freight | $ 8,835,958,014 |
| Passenger | $    553,056,215 |
| Mail | $    311,341,450 |
| Express | $     76,453,778 |
| All other revenues | $    431,040,270 |
| Total operating expenses | $ 7,849,840,601 |
| Maintenance of way and structures | $ 1,235,801,374 |
| Maintenance of equipment | $ 1,774,877,670 |
| Traffic | $    261,787,241 |
| Transportation | $ 4,020,161,430 |
| General | $    490,173,558 |
| All other expenses | $     67,039,328 |
| Operating ratio, per cent | 76.90 |
| Net operating revenue | $ 2,358,009,126 |
| Railway tax accruals | $    916,493,750 |
| Railway operating income | $ 1,441,515,376 |
| Hire of equipment, debit balance | $    443,482,401 |
| Joint facility rents, debit balance | $     36,517,277 |
| Net railway operating income | $    961,515,698 |
| Rate of return on average net investment, per cent | 3.69 |

* Statistics of Railroads of Class I in the United States, *Association of American Railroads*, 1966.

the use of high-strength durable steels in both wheels and rails; and the use of two three-axle trucks on each carload and stress concentrations produced by such loadings has made necessary more general use of the continuously welded rail for main-line operation on many heavy-traffic rail routes. No really successful replacement has been found for the treated-wood tie and its 140-year-old rail fastening, the lowly railroad spike.

**20-5  Property investment and operating income**  A summary of investment in railroad property, revenues, operating income, and rate of return is given in Table 20-6.

### QUESTIONS AND PROBLEMS

**20-1.** *a.* What was the rate of intercity freight movement in the United States in 1965 expressed in ton-miles per capita?

*b.* How do you account for this annual rate of freight transport demand?

*c.* In what other countries of the world would you expect to find a similar per capita demand? Explain the reasons for your answer.

**20-2.** *a.* Which of the five principal modes of intercity freight transport experienced increases in annual ton-mile volume of such movement in the 10-year period 1955 to 1965?

*b.* By what transport mode is freight generally collected and distributed within and around a terminal city, and what effect does this have on average length of haul: by truck? by railway? by airplane?

### BIBLIOGRAPHY

Hay, William W.: "An Introduction to Transportation Engineering," John Wiley & Sons, Inc., New York, 1961.

Kuhn, Tillo E.: "Public Enterprise Economics and Transport Problems," University of California Press, Berkeley, 1962.

Raymond, W. G., H. E. Riggs, and W. C. Sadler: "Elements of Railroad Engineering," 6th ed., John Wiley & Sons, Inc., New York, 1947.

"Statistics of United States Railways," Interstate Commerce Commission, 1965.

Wellington, A. M.: "The Economic Theory of the Location of Railways," 6th ed., chaps. 1–5, John Wiley & Sons, Inc., New York, 1946.

# 21

# The Permanent Way

**21-1  Alignment and grades**  Of primary concern to the civil engineer are problems of railway location and construction, particularly with respect to alignment and grades, as affecting operating characteristics and operating costs.  His major interest, then, so far as railways engineering is concerned, will be in the application of his knowledge of surveying, structural design, and constructional methods and materials to problems of the railroad way and its attendant structures and facilities, with due regard for the nature of the loads which are to be hauled over it and for the limitations of the motive power.

Thus the engineer's skill in laying out and building railway curves with proper easement and superelevation will enable him to provide curvature in alignment; but his appreciation of attendant increased construction costs and reduced operating efficiency due to greater train resistance on curves will induce him to keep curvature at a minimum.

Maximum permissible degree of curvature on main lines varies with the nature of the terrain, ranging from 1- or 2-deg curves in relatively flat or gently rolling country to 10 deg or more in mountain districts.

On light traffic lines and some branch lines sharper curves are common, and in railroad yards 40-deg curves may be found.

Grades from 0.0 to 1.0 per cent predominate on main lines throughout the country, and 2.5 per cent is a common maximum even in mountain districts. Certain logging and mining railroads can operate on grades upward of 5 per cent, mainly because in such cases the movement of the commodity is downhill.

Length of vertical curve at any grade break may be determined by the American Railway Engineering Association (AREA) specification that on main lines rates of grade change of 0.1 per cent per station on summits and 0.05 per cent per station in sags should not be exceeded. On minor railroads 0.2 per cent per station on summits and 0.1 per cent per station in sags may be used.

Rate of grade change is more critical in sags—that is, where the grade changes from a descending grade to a lesser descending or to an ascending grade—because there is a tendency for slack to develop in the couplings as the forward end of the train is slowed on the grade change. Thus considerable jerking will take place as the slack is taken up if the grade change is too abrupt.

**21-2  Grading**  Grading of the roadbed should adhere to the principles of selective use of available materials and effective compaction to 90 or 95 per cent of maximum density at near optimum moisture content, as discussed in earlier chapters on highway construction. A typical cross section of roadbed showing intercepting ditches and pipe drains is shown in Fig. 21-1. Roadbed grading provides the surface upon which the ballast is placed. Turfing of slopes will ordinarily prove to be

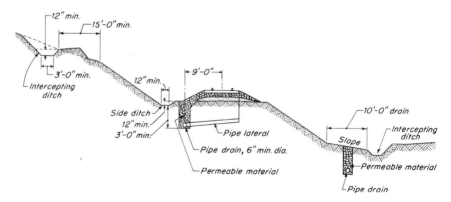

**Fig. 21-1**  Section of roadbed, showing intercepting ditches and pipe drains. (*Source: AREA Manual.*)

economically justified by lower maintenance costs resulting from minimized erosion and, to some extent, control of weed and brush growth.

**21-3 Ballast** Ballast is provided to distribute loads uniformly over the
subgrade, to hold the track structure to line and grade, to reduce dust,
to prevent growth of brush and weeds, and to provide good drainage
for the track structure. In order to accomplish these ends, the ballast
should be composed of clean, sound, durable fragments and particles,
reasonably well graded from coarse to fine. Depth of ballast and sub-
ballast courses will vary with the supporting characteristics of the subgrade, relatively unstable subgrade soils requiring greater thickness of
protective cover. In general the thickness of ballast will be about 24 in.
below the bottom of the ties.

In order of desirable properties, materials generally used for ballast
are crushed rock, washed and screened gravel, pit-run gravel, slag, and
cinders. The interlocking effect of crushed-rock aggregate particles
provides a high degree of stability under the wave action of successive
wheel loads of a moving train. This action tends to produce considerable
displacement of rounded aggregate particles, particularly at the ends of
the ties, causing the ties to become "center-bound" or suspended at the
center and necessitating occasional tamping of ballast by hand or with
mechanical tampers in roadbed maintenance.

Initial cost of crushed rock for ballast may prove excessive and
availability of usable local materials may often dictate what is to be used.
Pit-run gravel may often be improved both in desirable grading and in
stability by crushing of oversize instead of removing it by screening.
American Railway Engineering Association specifications require that
prepared ballast shall be crushed stone, crushed air-cooled blast-furnace
slag (see Table 21-1), or gravel (crushed or uncrushed, as specified) composed of hard, strong durable particles, free from injurious amounts of
deleterious substances; types and sizes to be designated by the engineer.

Quality requirements for prepared ballast according to AREA
specifications require that deleterious substances shall not be present
in prepared ballast in excess of the following amounts:

|  | *Per cent* |
| --- | --- |
| Soft and friable pieces...................... | 5 |
| Material finer than No. 200 sieve.............. | 1 |
| Clay lumps................................. | 0.5 |

The percentage of wear as determined by the Los Angeles Rattler
Test should not exceed 40 per cent.

**Table 21-1   Recommended grading requirements for crushed stone and crushed air-cooled blast-furnace slag**

Percentages by weight passing each sieve size

| 3 in. | $2\frac{1}{2}$ in. | 2 in. | $1\frac{1}{2}$ in. | 1 in. | $\frac{3}{4}$ in. | $\frac{1}{2}$ in. | $\frac{3}{8}$ in. | No. 4 | No. 8 |
|---|---|---|---|---|---|---|---|---|---|
| 100 | 90–100 | . . . . . . | 25–60 | . . . . . . | 0–10 | 0–5 | | | |
| | 100 | 90–100 | 35–70 | 0–15 | . . . . . | 0–5 | | | |
| | | 100 | 90–100 | 20–55 | 0–15 | . . . . . | 0–5 | | |
| | | | 100 | 90–100 | 40–75 | 15–35 | 0–15 | 0–5 | |
| | | | 100 | 90–100 | . . . . . | 25–60 | . . . . | 0–10 | 0–5 |

Gravel for prepared ballast should conform to one of the sets of grading requirements, depending on the percentage of crushed particles specified, as listed in Table 21-2.   Particles having one or more faces resulting from fracture are considered as crushed particles.   The percentage by weight of crushed particles in the material coarser than the No. 4 sieve is considered as the percentage of crushed particles in the sample.

Typical ballast sections are shown in Figs. 21-2 and 21-3.

**21-4   Crossties**   Wheel loads are distributed to the ballast by crossties. These ties also serve to hold the rails to line and gauge.   Size of ties most widely used under heavy traffic on main lines is 7 in. by 9 in. by 8 ft 6 in. The present trend is toward the use of 9-ft ties.   A space of 10 in. between tops of ties allows sufficient room for tamping of ballast.   Thus it can be seen that the maximum bearing area on the ballast may be secured by the use of wider and longer ties laid with this spacing.   On the average this requires about 3,200 ties per mile of track.

Ties are usually made of wood.   Attempts to use other materials such as reinforced concrete and steel have not uncovered any particularly successful substitutes for wood.   Twenty-seven species of wood are designated by AREA as suitable for ties.   Oak and southern pine are

**Table 21-2   Recommended grading of gravel for ballast**

| Per cent crushed particles | Percentages by weight finer than each sieve size | | | | | | | |
|---|---|---|---|---|---|---|---|---|
| | $-\frac{1}{2}$ in. | 1 in. | $\frac{1}{2}$ in. | No. 4 | No. 8 | No. 16 | No. 50 | No. 100 |
| 0–20 | 100 | 80–100 | 50–85 | 20–40 | 15–35 | 5–25 | 0–10 | 0–2 |
| 21–40 | 100 | 65–100 | 35–75 | 10–35 | 0–10 | 0–5 | | |
| 41–100 | 100 | 60–95 | 25–50 | 0–15 | 0–5 | | | |

by far the most common species used, comprising about 60 per cent of the total.   About 10 per cent of the total is Douglas fir.

Tie renewal constitutes a major item of railway maintenance cost. For this reason considerable research has been carried on in the past three decades, particularly by the U.S. Forest Products Laboratory at Madison, Wis., to determine life characteristics of ties and the effects of treatment with wood preservatives.   Nearly all tie installations during recent years have been made with treated ties, the most common preservatives being creosote and zinc chloride.

Experiments have shown that proper treatment of wood which is to be in contact with the ground will increase its life by several times that of the untreated wood.   For instance, in a comparison of wood preservatives in the Mississippi Post Study made by the U.S. Forest Products Laboratory it was found that untreated southern yellow pine posts had an average life of 3.3 years, while posts treated with creosote-petroleum mixtures have indicated an average life of 11 to 15 years.   The experiment is still in progress, and many groups of posts which were pressure-treated with various chemical preservatives showed no failures after 15 years of service.   The research department of the Association of American Railroads has established that as of 1965, a treated tie has an average service life of about 37 years.

The Association of American Railroads has estimated that Class I railroads of the United States install more than 27 million crossties a year on the average, and it is understandable that the cost of such replacements, coupled with the cost of periodical inspection of ties in place, constitutes a substantial item of maintenance expense.   A determination of annual cost of tie replacement requires the application of some basic principles of engineering economy, which will be very briefly discussed here.

The total cost of a capital investment for an item which will have to be replaced without salvage value at the end of a definite number of years may be determined by the formula

$$A = P(1 + i)^n \qquad\qquad\qquad (21\text{-}1)$$

where $A$ is the total cost or compound amount of an initial investment $P$ at an interest rate $i$ convertible annually over a period of $n$ years representing the estimated life of the item of expense.

The annual cost may then be considered to be the annual deposit which would have to be made at the end of each year, at the same rate of compound interest, to equal the total cost in $n$ years.   This annual cost is then given by $D$ in the following formula:

$$A = D[1 + (1 + i) + (1 + i)^2 + \cdots + (1 + i)^{n-1}]$$

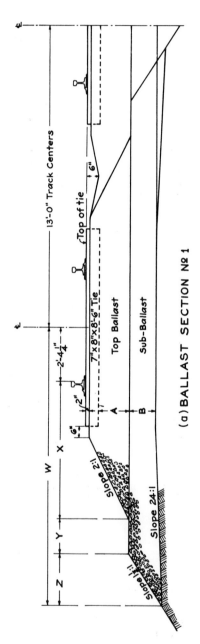

## (a) BALLAST SECTION Nº 1

## TABLE OF VARIABLES FOR SECTION Nº 1

| Section | W | A | B | X | Y | Z | Top Ballast Quantities per Mile** | | |
|---|---|---|---|---|---|---|---|---|---|
| | | | | | | | Per Single Track Cu. Yds. | Per Double Track Cu. Yds. | Per Add'n. Track Cu. Yds. |
| a | 12'-0" | 16" | 14" | 5'-11" | 1'-6" | 2'-3" | 4803 | 9467 | 4664 |
| b | | 14" | 13" | 5'-7" | 1'-11" | 2'-2" | 4195 | 8371 | 4176 |
| c | | 14" | 12" | 5'-7" | 2'-1" | 2'-0" | 4195 | 8371 | 4176 |
| d | | 12" | 12" | 5'-3" | 2'-5" | 2'-0" | 3612 | 7302 | 3690 |
| e | | 12" | 12" | 5'-3" | 1'-6" | 1'-11" | 3612 | 7302 | 3690 |
| f | 11'-0" | 12" | 10" | 5'-3" | 1'-9" | 1'-8" | 3612 | 7302 | 3690 |
| g | | 10" | 10" | 4'-11" | 2'-1" | 1'-8" | 3058 | 6260 | 3202 |
| h | | 10" | 8" | 4'-11" | 2'-4" | 1'-5" | 3058 | 6260 | 3202 |
| i | 10'-0" | 10" | 8" | 4'-11" | 1'-6" | 1'-3" | 3058 | 6260 | 3202 |
| j | | 8" | 8" | 4'-7" | 1'-9" | 1'-4" | 2525 | 5239 | 2714 |
| k | | 8" | 6" | 4'-7" | 2'-0" | 1'-1" | 2525 | 5239 | 2714 |
| l | | 6" | 8" | 4'-3" | 2'-1" | 1'-4" | 2019 | 4247 | 2228 |
| m | | 6" | 6" | 4'-3" | 2'-4" | 1'-1" | 2019 | 4247 | 2228 |

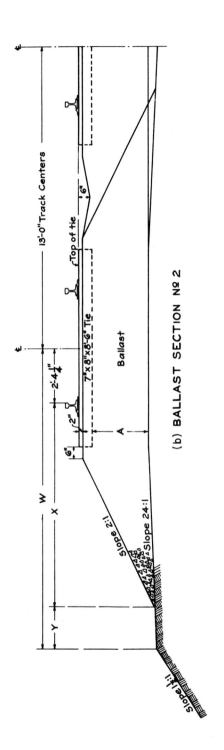

(b) BALLAST SECTION № 2

**TABLE OF VARIABLES FOR SECTION № 2**

| Section | W | A | X | Y | Quantities per Mile** Per Single Track Cu.Yds. | Per Double Track Cu.Yds. | Per Add'n. Track Cu.Yds. |
|---|---|---|---|---|---|---|---|
| a | 13'-0" | 30" | 8'-10" | 1'-10" | 10165 | 18239 | 8074 |
| b | | 28" | 8'-5" | 2'-3" | 9308 | 16896 | 7588 |
| c | | 26" | 8'-1" | 2'-7" | 8530 | 15630 | 7100 |
| d | | 26" | 8'-1" | 1'-7" | 8530 | 15630 | 7100 |
| e | | 24" | 7'-9" | 1'-11" | 7779 | 14393 | 6614 |
| f | 12'-0" | 22" | 7'-4" | 2'-4" | 7010 | 13136 | 6126 |
| g | | 20" | 7'-0" | 2'-8" | 6310 | 11948 | 5638 |

| | W | A | X | Y | Per Single | Per Double | Per Add'n |
|---|---|---|---|---|---|---|---|
| h | 11'-0" | 20" | 7'-0" | 1'-8" | 6310 | 11948 | 5638 |
| i | | 18" | 6'-8" | 2'-0" | 5636 | 10788 | 5152 |
| j | | 16" | 6'-3" | 2'-5" | 4954 | 9618 | 4664 |
| k | | 14" | 5'-11" | 2'-9" | 4333 | 8509 | 4176 |
| l | | 14" | 5'-11" | 1'-9" | 4333 | 8509 | 4176 |
| m | 10'-0" | 12" | 5'-7" | 2'-1" | 3738 | 7428 | 3690 |
| n | | 10" | 5'-2" | 2'-6" | 3143 | 6345 | 3202 |
| o | | 8" | 4'-10" | 2'-10" | 2600 | 5314 | 2714 |
| p | | 6" | 4'-5" | 3'-3" | 2065 | 4293 | 2228 |

** Quantities computed on basis of 3200 ties per mile and 15% allowance for shrinkage.

**Fig. 21-2** Typical ballast section (AREA). Single- and multiple-tangent track.

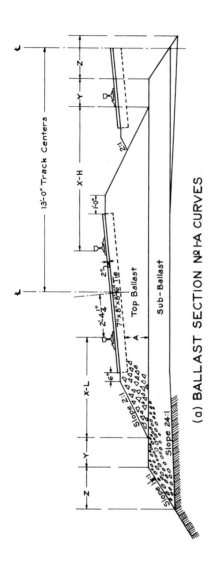

(a) BALLAST SECTION Nº I-A CURVES

TABLE OF VARIABLES AND QUANTITIES PER 100 FEET SINGLE TRACK-FOR SECTION Nº I-A CURVES

| Section | A | SUPERELEVATION | | | | | | | | | | | | | | | | | |
|---|---|---|---|---|---|---|---|---|---|---|---|---|---|---|---|---|---|---|
| | | 1" | | | 2" | | | 3" | | | 4" | | | 5" | | | 6" | | |
| | | X-L | X-H | Cu.Yds. | X-L | X-H | Cu.Yds. | X-L | X-H | Cu.Yds. | X-L | X-H | Cu.Yds. | X-L | X-H | Cu.Yds. | X-L | X-H | Cu.Yds. |
| I-A a | 16" | 5'-11" | 6'-7" | 99 | 5'-10" | 6'-10" | 101 | 5'-9" | 7'-1" | 104 | 5'-8" | 7'-4" | 109 | 5'-7" | 7'-6" | 113 | 5'-6" | 7'-8" | 116 |
| I-A b-c | 14" | 5'-7" | 6'-3" | 86 | 5'-6" | 6'-6" | 89 | 5'-5" | 6'-9" | 92 | 5'-4" | 7'-0" | 95 | 5'-3" | 7'-2" | 100 | 5'-2" | 7'-4" | 103 |
| I-A d-e-f | 12" | 5'-3" | 5'-11" | 74 | 5'-2" | 6'-2" | 78 | 5'-1" | 6'-5" | 81 | 5'-0" | 6'-8" | 84 | 4'-11" | 6'-10" | 87 | 4'-10" | 7'-0" | 91 |
| I-A g-h-i | 10" | 4'-11" | 5'-7" | 64 | 4'-10" | 5'-10" | 66 | 4'-9" | 6'-1" | 69 | 4'-8" | 6'-4" | 72 | 4'-7" | 6'-6" | 76 | 4'-6" | 6'-8" | 79 |
| I-A j-k | 8" | 4'-7" | 5'-3" | 53 | 4'-6" | 5'-6" | 56 | 4'-5" | 5'-9" | 59 | 4'-4" | 6'-0" | 62 | 4'-3" | 6'-2" | 65 | 4'-2" | 6'-4" | 68 |
| I-A l-m | 6" | 4'-3" | 4'-11" | 43 | 4'-2" | 5'-2" | 45 | 4'-1" | 5'-5" | 49 | 4'-0" | 5'-8" | 51 | 3'-11" | 5'-10" | 54 | 3'-10" | 6'-0" | 57 |

TABLE OF VARIABLES AND QUANTITIES PER 100 FEET SINGLE TRACK-FOR SECTION-Nº 2-A CURVES

| Section | A | SUPERELEVATION | | | | | | | | | | | | | | | | | |
|---|---|---|---|---|---|---|---|---|---|---|---|---|---|---|---|---|---|---|
| | | 1" | | | 2" | | | 3" | | | 4" | | | 5" | | | 6" | | |
| | | X-L | X-H | Cu.Yds. | X-L | X-H | Cu.Yds. | X-L | X-H | Cu.Yds. | X-L | X-H | Cu.Yds. | X-L | X-H | Cu.Yds. | X-L | X-H | Cu.Yds. |
| 2-A a | 30" | 9'-0" | 9'-10" | 207 | 8'-11" | 10'-1" | 211 | 8'-10" | 10'-4" | 216 | 8'-9" | 10'-7" | 220 | 8'-8" | 10'-10" | 225 | 8'-7" | 11'-1" | 230 |
| 2-A b | 28" | 8'-7" | 9'-6" | 191 | 8'-6" | 9'-9" | 195 | 8'-5" | 10'-0" | 199 | 8'-4" | 10'-3" | 203 | 8'-3" | 10'-6" | 208 | 8'-2" | 10'-9" | 213 |
| 2-A c-d | 26" | 8'-3" | 9'-1" | 174 | 8'-2" | 9'-4" | 178 | 8'-1" | 9'-7" | 183 | 8'-0" | 9'-10" | 187 | 7'-11" | 10'-1" | 191 | 7'-10" | 10'-4" | 197 |
| 2-A e | 24" | 7'-10" | 8'-8" | 160 | 7'-9" | 8'-11" | 162 | 7'-8" | 9'-2" | 166 | 7'-7" | 9'-5" | 171 | 7'-6" | 9'-8" | 174 | 7'-5" | 9'-11" | 179 |
| 2-A f | 22" | 7'-6" | 8'-4" | 145 | 7'-5" | 8'-7" | 148 | 7'-4" | 8'-10" | 152 | 7'-3" | 9'-1" | 156 | 7'-2" | 9'-4" | 161 | 7'-1" | 9'-7" | 165 |
| 2-A g-h | 20" | 7'-1" | 7'-11" | 130 | 7'-0" | 8'-2" | 134 | 6'-11" | 8'-5" | 138 | 6'-10" | 8'-8" | 141 | 6'-9" | 8'-11" | 145 | 6'-8" | 9'-2" | 150 |
| 2-A i | 18" | 6'-10" | 7'-7" | 116 | 6'-9" | 7'-10" | 120 | 6'-8" | 8'-1" | 124 | 6'-7" | 8'-4" | 128 | 6'-6" | 8'-7" | 131 | 6'-5" | 8'-10" | 135 |
| 2-A j | 16" | 6'-5" | 7'-3" | 103 | 6'-4" | 7'-6" | 107 | 6'-3" | 7'-9" | 111 | 6'-2" | 8'-0" | 114 | 6'-1" | 8'-3" | 118 | 6'-0" | 8'-6" | 122 |
| 2-A k-l | 14" | 6'-0" | 6'-10" | 90 | 5'-11" | 7'-1" | 93 | 5'-10" | 7'-4" | 98 | 5'-9" | 7'-7" | 102 | 5'-8" | 7'-10" | 105 | 5'-7" | 8'-1" | 108 |
| 2-A m | 12" | 5'-9" | 6'-6" | 79 | 5'-8" | 6'-9" | 82 | 5'-7" | 7'-0" | 85 | 5'-6" | 7'-3" | 88 | 5'-5" | 7'-6" | 93 | 5'-4" | 7'-9" | 96 |
| 2-A n | 10" | 5'-4" | 6'-2" | 67 | 5'-3" | 6'-5" | 70 | 5'-2" | 6'-8" | 73 | 5'-1" | 6'-11" | 77 | 5'-0" | 7'-2" | 80 | 4'-11" | 7'-5" | 84 |
| 2-A o | 8" | 4'-11" | 5'-9" | 56 | 4'-10" | 6'-0" | 58 | 4'-9" | 6'-3" | 61 | 4'-8" | 6'-6" | 65 | 4'-7" | 6'-9" | 68 | 4'-6" | 7'-0" | 71 |
| 2-A p | 6" | 4'-8" | 5'-5" | 45 | 4'-7" | 5'-8" | 48 | 4'-6" | 5'-11" | 51 | 4'-5" | 6'-2" | 54 | 4'-4" | 6'-5" | 58 | 4'-3" | 6'-8" | 61 |

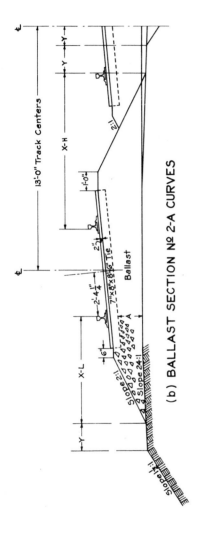

(b) BALLAST SECTION Nº 2-A CURVES

Fig. 21-3 Typical ballast section (AREA). Single and multiple track on curves (for superelevation in excess of 3 in.).

329

or

$$A = \frac{(1 + i)^n - 1}{i} D \tag{21-2}$$

and, combining Eqs. (21-1) and (21-2),

$$D = P \frac{i(1 + i)^n}{(1 + i)^n - 1} \tag{21-3}$$

The expression $[i(1 + i)^n]/[(1 + i)^n - 1]$ is a capital recovery factor, and values for this factor are given in Table 21-3 for various interest rates and lifespans, in years.

As an example of the use of this table, let it be required to determine how much can economically be invested in the treatment of ties if the average life expectancy of untreated ties is 7 years and the average life expectancy of treated ties is 20 years, and the initial cost in place for the untreated tie is $2.50, including freight, handling, and installation.

From Table 21-3 it will be seen that the *annual* cost of the untreated tie (assuming an interest rate of 4 per cent per annum) is

2.50 × 0.1666 = $0.4165

Now, if annual cost may be accepted as a sound basis for comparison,

**Table 21-3   Values of $[i(1 + i)^n]/[(1 + i)^n - 1]$**

| n years | i = 4% | i = 5% | i = 6% |
|---------|--------|--------|--------|
| 1  | 1.0400 | 1.0500 | 1.0600 |
| 2  | 0.5304 | 0.5380 | 0.5456 |
| 3  | 0.3604 | 0.3673 | 0.3741 |
| 4  | 0.2755 | 0.2821 | 0.2885 |
| 5  | 0.2247 | 0.2309 | 0.2374 |
| 6  | 0.1907 | 0.1970 | 0.2034 |
| 7  | 0.1666 | 0.1728 | 0.1792 |
| 8  | 0.1486 | 0.1547 | 0.1611 |
| 9  | 0.1345 | 0.1407 | 0.1470 |
| 10 | 0.1233 | 0.1297 | 0.1359 |
| 15 | 0.0899 | 0.0963 | 0.1030 |
| 20 | 0.0736 | 0.0802 | 0.0872 |
| 25 | 0.0640 | 0.0709 | 0.0782 |
| 30 | 0.0578 | 0.0650 | 0.0726 |
| 35 | 0.0536 | 0.0611 | 0.0689 |
| 40 | 0.0505 | 0.0583 | 0.0664 |
| 45 | 0.0482 | 0.0562 | 0.0647 |
| 50 | 0.0466 | 0.0548 | 0.0634 |

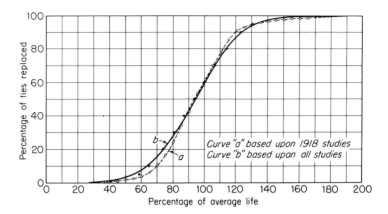

**Fig. 21-4** Total replacement of railroad ties. (*Courtesy of U.S. Forest Products Laboratory, Madison, Wis.*)

the initial cost of the treated tie can be

$$\frac{0.4165}{0.0736} = \$5.66$$

and assuming that cost of original tie, freight, handling, and installation would be the same for treated or untreated ties, then $5.66 − $2.50 or $3.16 might economically be spent per tie for preservative treatment. Pressure treatment of ties with an effective preservative would normally be only a fraction of this amount, and it is not unreasonable to expect that the life of the tie would be increased to three times that of the untreated tie.

Tie-replacement data based on the probable average life of ties may be obtained from Figs. 21-4 and 21-5. It should be noted that the accuracy of prediction from these figures increases as the number and homogeneity of ties in the group being considered increase. The curve of Fig. 21-4 shows that 60 per cent of the ties in a group will be replaced at the end of the average life of the ties. Detailed tie records have been kept by railroad companies, and the average life expectancy for any particular species of wood, type of treatment, and climatic conditions may be readily obtained.

The chart of Fig. 21-5 is a summary of a long-time study of probable life of ties made by the U.S. Forest Products Laboratory at Madison, Wis., and provides a convenient means for predicting tie replacements for a reasonably large and homogeneous installation of ties. For example, a tie-replacement schedule by years may be made for, say, 10,000 ties installed this year, assuming an average life expectancy of, say, 30 years.

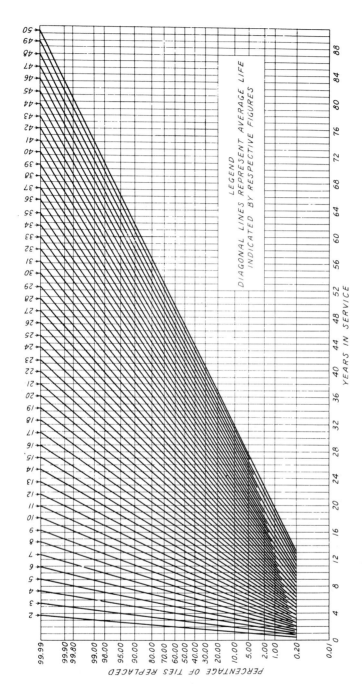

**Fig. 21-5** Chart for determining probable life of ties. (*Courtesy of U.S. Forest Products Laboratory, Madison, Wis.*)

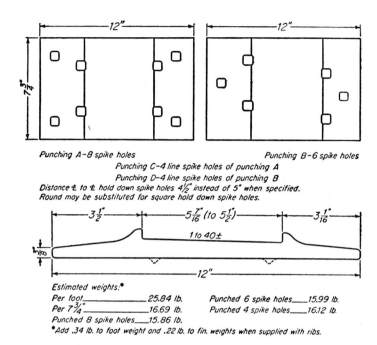

Punching A-8 spike holes                    Punching B-6 spike holes
Punching C-4 line spike holes of punching A
Punching D-4 line spike holes of punching B
Distance ₵ to ₵ hold down spike holes 4½" instead of 5" when specified.
Round may be substituted for square hold down spike holes.

Estimated weights:*

| | | |
|---|---|---|
| Per foot_____25.84 lb. | Punched 6 spike holes____15.99 lb. |
| Per 7¾"_____16.69 lb. | Punched 4 spike holes____16.12 lb. |
| Punched 8 spike holes___15.86 lb. | |

*Add .34 lb. to foot weight and .22 lb. to fin. weights when supplied with ribs.

**Fig. 21-6**  AREA 12-in. tie plate for use with rails with $5\frac{3}{8}$-in. base width.


**21-5  Tie plates**  Mechanical wear on ties due to the pounding of heavy wheel loads and the side sway of trains is largely overcome by the use of tie plates, as illustrated in Fig. 21-6.  In addition, rail braces are also used on curves and at switches to give further support against the lateral thrust of locomotive and cars on the rails.

**21-6  Rails**  The present standard length of rail is 39 ft.  Rails are designated by weight per running yard.  Weights ranging from 90 to 155 lb/yd are in use, and the present average on Class I railroads is about 110 lb/yd.  A typical rail section is shown in Fig. 21-7 and the joint bar and assembly in Fig. 21-8.

**21-7  Railroad gauge**  Standard railroad gauge, measured between the inside of rail heads, is 4 ft $8\frac{1}{2}$ in.  Many different gauges varying from 3 to 6 ft have been used in the United States, but free interchange of freight and passenger cars made standardization necessary so that at the present time, apart from the standard gauge, only a few hundred miles of 3-ft or narrow-gauge railroad are in existence.

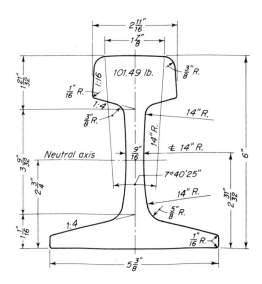

|       | Area, sq in. | %     |
|-------|--------------|-------|
| Head  | 3.80         | 38.2  |
| Web   | 2.25         | 22.6  |
| Base  | 3.90         | 39.2  |
| Total | 9.95         | 100.0 |

| | |
|---|---|
| Moment of inertia | 49.0 |
| Section modulus, head | 15.1 |
| Section modulus, base | 17.8 |
| Ratio, moment of inertia to area | 4.92 |
| Ratio, section-modulus head to area | 1.52 |
| Ratio, height to base | 1.12 |

**Fig. 21-7** A 100-lb AREA rail section. (*Courtesy of AREA.*)

**21-8  Rail failure**  Many rail failures develop from tiny defects which may be detected in early stages by a low-voltage electric current, setting up a magnetic field around the railhead with instrumentation to pick up variations in this magnetic field as contact brushes move along the rail. The Sperry detector car operating on this basis is widely used for "in-place" inspection and marking of defective rails.

About one-half of the railhead is considered to be available for wear. Normally, less than half of this amount of wear is permitted in main-line operation, and then the rails which are otherwise sound are replaced and moved to yard track and less critical areas. The Association of American Railroads estimates that somewhat more than 2 million gross tons of steel rails are normally laid annually by Class I railroads in the United States. Here again the student's attention is directed toward the economic justification for intensive research to discover ways and means of lengthening the service life of an expendable installation.

**21-9  Spikes**  The hook-headed spike (Fig. 21-9), which is used by railroads throughout the world to fasten steel rails to crossties, was designed in 1831.  While it is admittedly a poor fastening device, no better one has been found to replace it.  Ordinarily four spikes are driven into each tie.  On curves and at switches as many as eight per tie will be required.

**21-10  Superelevation**  Banking of curves is usually accomplished by elevating the outer rail.  The student will recall from applied dynamics that the ideal angle of elevation at which to bank a curve is determined

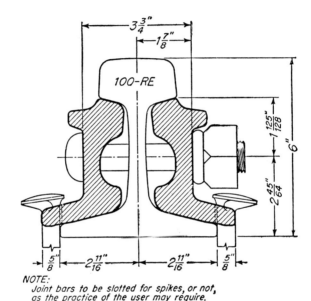

NOTE:
*Joint bars to be slotted for spikes, or not,
as the practice of the user may require.*

| Physical properties | One bar | Two bars |
|---|---|---|
| Moment of inertia, in.⁴ ................ | 8.94 | 17.88 |
| Section modulus: | | |
|     Above n.a., in.³ ................. | 4.01 | 8.02 |
|     Below n.a., in.³ ................. | 4.36 | 8.72 |
| Area, sq in ........................ | 5.04 | 10.08 |
| Net weight, 24-in. length, lb .......... | 33.47* | 66.94 |
| Net weight, 36-in. length, lb .......... | 50.21* | 100.42 |

\* If slotted for spikes, deduct 0.09 lb per slot.

**Fig. 21-8**  Joint bar and assembly for 100-lb RE rail.  (*Courtesy of AREA.*)

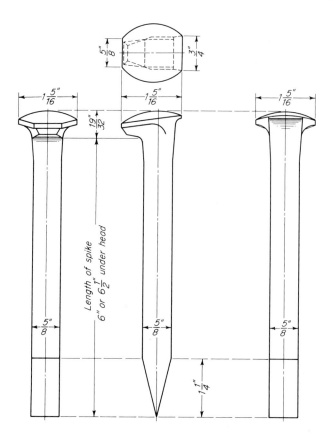

**Fig. 21-9**   Design of cut track spike.   (*From AREA Manual.*)

when the resultant of the weight of the vehicle $W$ and the centrifugal force $(W/g)a$ is perpendicular to the roadway so that there is no tendency to slide up or down nor to overturn.   It may be seen from Fig. 21-10 that these conditions prevail when $\tan \theta = (Wa/g)/W = a/g$, where $a$ is the normal acceleration due to curvature and is given by the equation $a = v^2/R$, in which $v$ is velocity in feet per second and $R$ is the radius of curvature in feet.

Then $\tan \theta = v^2/gR$, and if velocity $V$ is expressed in miles per hour and $g$ is taken as 32.2 ft/sec², $\tan \theta = 0.0667V^2/R$.   Since the angle $\theta$ will generally be small, the tangent and sine may be taken as approximately equal and $\tan \theta = e/d$, where $d$ is the distance between center lines of rails (taken as 4.92 ft) and $e$ is the amount of elevation of the outer rail in feet.   From these two equations for $\tan \theta$, it follows that

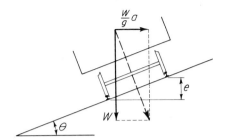

**Fig. 21-10** Body forces acting on car at a curve.

$e = 0.328V^2/R$. Figure 21-11, based on this formula, gives elevation of outer rail in inches for various degrees of curvature and speeds.

In order to provide a smooth approach to superelevated curves, the superelevation is usually built up at the rate of about 1 in. to 50 ft on the approach tangent and then reduced at the same rate on the tangent after leaving the curve. When spiral easement is provided in the curve, the superelevation is, at each point on the spiral, made consistent with the radius of curvature at that point.

Each curve, then, will be banked for a particular speed, and any variation from this speed will cause an outward or inward component of force. A compromise will usually have to be made to provide for both high-speed passenger trains and slow-moving freight trains. Generally the theoretical $e$ is reduced by 3 in. It should be recognized that long freight trains operating upgrade on a heavily banked, relatively sharp curve will cause a considerable thrust against the inner rail, and it is not improbable that a top-heavy load might tip inward on the curve because of the inward component of the pull on the car couplings in addition to the banking.

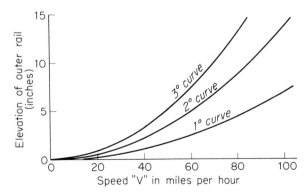

**Fig. 21-11** Rail elevation on curves.

For this reason and because of excessive wear on the inside rails and attendant increased curve resistance, the amount of superelevation of the outer rail above the inside rail is usually limited to about 8 in. where both passenger and freight trains are to operate, the exact limitation depending somewhat on the gradients and on anticipated freight-train loadings. The need for low-degree curves in horizontal alignment can be seen in Fig. 21-11.

## QUESTIONS AND PROBLEMS

**21-1.** About 10 miles of railroad is to be reconstructed, using 30,000 new treated ties for which the estimated average life expectancy is 40 years. How many of these original ties will have been replaced after 30 years?

**21-2.** In the preceding problem, assume that the first tie replacements will be made after 10 years and every 5 years thereafter.

    *a.* Make a table showing the total number of replacements for each of such years up to 25 years.

    *b.* After how many years will all of the original ties be replaced?

**21-3.** The main-line track between two towns, used for both freight and passenger service, is to be relocated with some reduction in curvature so that the passenger trains may operate on the entire length at 50 mph. What maximum degree of curvature should be specified?

**21-4.** A proposed improvement involving reduction of grades and curvature on a certain railroad will have no direct effect on revenue but will result in an estimated net saving in operating expense of $30 per day based on 365 days a year. What expenditure would be justified for this improvement if it is desired to realize 5 per cent return on the capital investment? Assume an estimated service life of 40 years for the proposed improvement.

**21-5.** The estimated life of an untreated tie is 6 years. If the cost of the tie is $1.25, freight and handling $0.60, insertion in track $0.65, how many years should the tie last to justify an expenditure of $1.30 for treatment with interest rate at 5 per cent?

## BIBLIOGRAPHY

*Am. Ry. Eng. Assoc. Proc.*, **34** (1933), **49** (1948), **50** (1949), **51** (1950).

"AREA Manual," vol. 1, chaps. 1, 3–5; vol. 2, chap. 15, American Railway Engineering Association, 1953.

Magee, G. M.: Locomotive Design and Rail Stresses, *Railway Age*, **102**(20):825–828 (1937).

Rail Damage and the Relation of Locomotive Thereto, *Railway Age*, **104**(15):653–657 (1938).

# 22
# Turnouts, Sidetracks, and Yards

**22-1  Definition**  A turnout is a curved track which is used to connect one track with another.  Switch rails and frogs constitute the main parts of a turnout.  A switch mechanism, operating on the movable ends of the switch rails, makes it possible for a train to turn off from one track and onto another.  Two types of switches are used for this purpose, the *stub switch* and the *split switch*.

**22-2  The stub switch**  Main parts and operation of the stub switch are illustrated in Fig. 22-1.  In this type of switch, both switch rails are main-line rails.  The free ends of the switch rails at $A$ constitute the *toe of switch*.  The *heel of switch* at point $B$ is the point of tangency of the curved portion of the turnout with the main-line rails.  The portion of switch rails to the right of $B$ are securely spiked to the ties, while the portion between $A$ and $B$—the length of switch—is left free so that the rails may be sprung by the switch mechanism to conform to the curvature of the turnout.  The distance through which the toe of switch moves to accomplish the turnout is called the *throw of switch*.  It will generally be

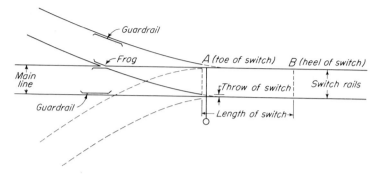

**Fig. 22-1** Stub switch.

about 5 to 6 in., depending somewhat on the curvature of the turnout. Considerable flexibility is inherent in the stub switch in that the same switch may be used for one turnout to the right and one to the left, or two turnouts to the right or left. These possibilities may be seen in Fig. 22-1.

The stub switch is not a foolproof device, nor is it considered safe for fast train operation. It is used to some extent in switch yards and for temporary work tracks, but it has been largely replaced by the split switch for general use.

**22-3 The split switch** The transition from main track to turnout with the split switch is accomplished by using one main-line rail and one turnout rail as switch rails. Parts and operation of the split switch are shown in Fig. 22-2. The switch rails are connected to the track rails at $B$—the heel of switch—by joint bars but are free to swing about these connections as pivots. The free ends at $A$, the *switch point*, are beveled to a wedge point about $\frac{1}{4}$ in. in thickness so that a smooth transition is made when a switch rail is placed against a main rail or a turnout rail. The switch rails $AB$ are straight. A complete diagram illustrating preferred names of parts for split switches is given in Fig. 22-3.

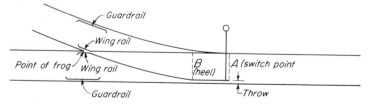

**Fig. 22-2** Split switch.

**22-4  Frogs**  Two general types of frogs are commonly used, the *rigid* frog and the *spring* frog.   In the rigid frog both wing rails (Fig. 22-2) are rigidly fastened to the main part of the frog with proper clearance for wheel flanges.

In the spring frog the wing rail on the main track is securely fastened to the main part of the frog so that the main flangeway is always open; but the wing rail on the turnout track is held against the frog point by a spring to afford continuous bearing for the wheels on the main track. This movable wing is pushed aside by the flange of each wheel as it passes the frog on a turnout.

A diagram illustrating preferred names of parts for bolted rigid frogs is given in Fig. 22-4.   Frogs may be designated by their angles, but they are more commonly designated by numbers.   A frog number $n$ is defined as

$$n = \tfrac{1}{2} \cot \tfrac{1}{2}\phi$$

where $\phi$ is the frog angle.   It is the ratio of heel length, measured along bisector of $\phi$ to heel spread (Fig. 22-4).

**22-5  Turnouts and crossovers**  A simple turnout is illustrated by Figs. 22-1 and 22-2.   They are used for turning out from main-line track to branch lines or to crossover track connecting parallel or nonparallel main tracks or siding and yard track.

In order to avoid cutting of rails AREA has adopted certain combinations of switches and frogs for turnouts and crossovers.   In practice, railroad companies choose as standard certain frog numbers and types of frogs having these numbers.   In general, the larger numbers—16 to 20—are used for high-speed movements on main-line tracks, 10 or 12 for main-line slow-speed movements, and 8 for yards and sidings.   The spring-type frog is usually designated for important track.

A tabulation of turnout and crossover data for straight split switches and various frog numbers is given in Fig. 22-5.   From this tabulation, the entire turnout and crossover track can be laid out by simple transit and tape measurements from a mark which designates the position of the frog.   This position is usually marked by setting a stake on the center line of the main track or sidetrack opposite the position of the actual frog point or, if track is in place, by marking this point on the rail.   The actual point of frog is made $\tfrac{1}{2}$ in. thick and is usually designated as the $\tfrac{1}{2}$-in. point of frog.   From this point, when a standard frog is chosen, the length of actual lead locates the actual point of switch from which tangent distances and offsets locate the turnout rails and frog position, in accordance with the data of Fig. 22-5.

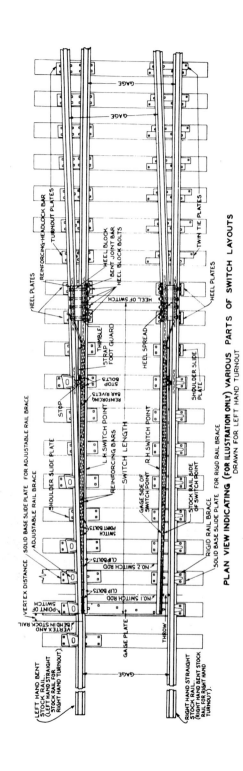

PLAN VIEW INDICATING (FOR ILLUSTRATION ONLY) VARIOUS PARTS OF SWITCH LAYOUTS

DRAWN FOR LEFT HAND TURNOUT

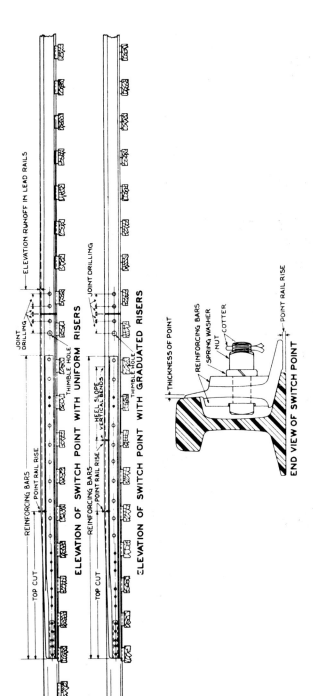

**Fig. 22-3**  Component parts of split switches.  (*From AREA Manual.*)

ELEVATION OF SWITCH POINT WITH UNIFORM RISERS

ELEVATION OF SWITCH POINT WITH GRADUATED RISERS

END VIEW OF SWITCH POINT

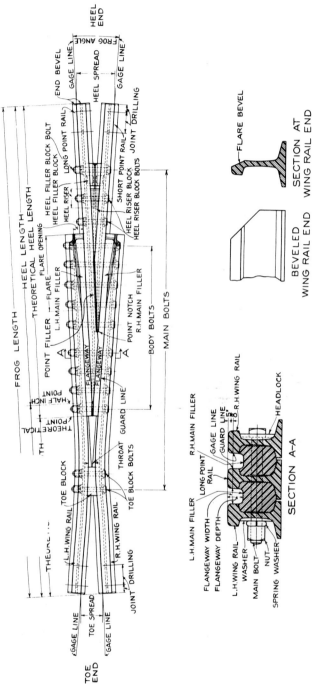

**Fig. 22-4** Component parts of frogs. (*From AREA Manual.*)

344

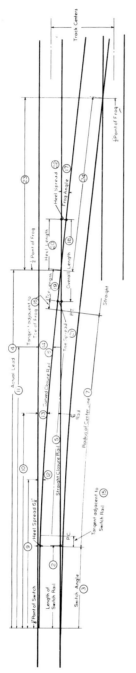

## TURNOUT AND CROSSOVER DATA

**Fig. 22-5** Turnout and crossover data for straight split switches. (*From AREA Manual.*)

**22-6  Crossings**  When two tracks cross, guardrails and wing rails, which constitute essentially four frogs, are necessary for safe operation on the crossing.  Usually it is only by chance that a standard frog can be used.  Consequently, special frogs are made to accommodate a particular angle of crossing, curved tracks, etc.  Details of layout are illustrated in Fig. 22-6.

**22-7  Sidetracks and yards**  Sidetracks or sidings are used to provide for meeting or passing trains and to permit cars to be left alongside depot loading platforms, warehouses, and factories.  Convenience to passenger traffic and efficient loading and switching of freight cars constitute the main problems to be worked out for any particular situation.  Properly placed turnouts and crossovers will provide the general solution to these problems.  For the more complicated switching operations, yards are necessary.

Yards provide shops and facilities for maintenance and repair of locomotives and cars, housing for idle locomotives, storage tracks for temporarily idle cars, and facilities for receiving and dispatching of passenger and freight trains.  Efficient and economical movement of cars for receiving, breaking up, making up, and dispatching of trains constitutes the principal consideration to be given to design of a system of yard tracks at freight and passenger division points.  Yard tracks are not primarily storage tracks, but should rather be considered as sorting tracks from which trains may be dispatched in the shortest possible time with a minimum of expense.

For a single-track road, yard tracks should be placed on one side of the main tracks so that it will not be necessary to cross the main track for yard work.  Similarly, for a double-track road it is best to spread the main tracks and place buildings and yard tracks between the two main tracks.  Length and width of yard will be governed by such factors as length and number of incoming and outgoing trains; number of classifications to be made for through and local freight and the number of cars for each class per day; building area for roundhouse, shops, substations, control towers, and other facilities; and storage tracks for temporarily idle cars.

It can be seen that for effective and economical operation the yard should be compact and well integrated so that sorting operations may be carried on to as large an extent as possible from the control tower.  Excessively long yard tracks should be avoided because of delay in switching operations and high speeds necessary to send cars the full length of such tracks.

**22-8  Switching**  A well-designed yard will make use of gravity for switching operations so as to avoid high-speed starts or long pushing of

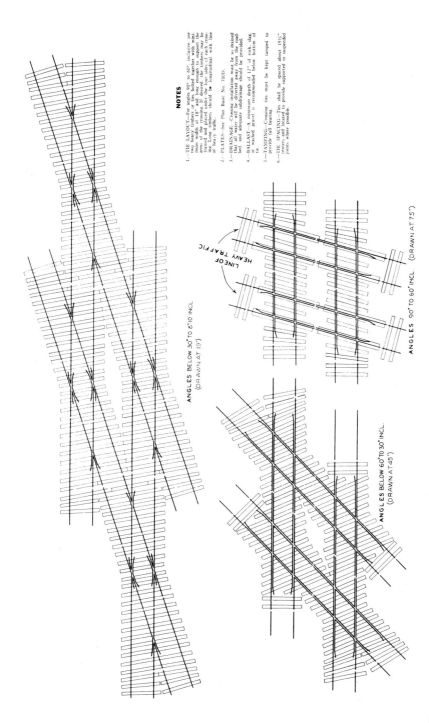

**NOTES**

1—THE LAYOUT—For angles 90° to 60° inclusive, use two heavy timbers or ties bolted together with minimum width of 18" and long enough to support the arms of the crossing. If desired, the timbers may be framed and placed under the four sides of each crossing. Long timbers should be longitudinal with line of heavy traffic.

2—PLATES—See Plan Basic No. 7001).

3—DRAINAGE—Crossing installation must be so drained that all water will be diverted away from the roadbed, and adequate subdrainage should be provided.

4—BALLAST—A minimum depth of 12" of rock, slag, or washed gravel is recommended below bottom of tie.

5—TAMPING—Crossing ties must be kept tamped to provide full bearing.

6—TIE SPACING—Ties shall be spaced about 19½" centers, and located to provide supported or suspended joints where possible.

**ANGLES BELOW 30° TO 8°10 INCL.**
(DRAWN AT 19°)

LINE OF
HEAVY TRAFFIC

**ANGLES 90° TO 60° INCL.** (DRAWN AT 75°)

**ANGLES BELOW 60° TO 30° INCL.**
(DRAWN AT 45°)

**Fig. 22-6** Tie layout for railroad crossing. *(From AREA Manual.)*

347

cars with the locomotive. An effective gravity type of switching used in most large terminals is called *hump switching*. A hump is built up so that the yard slopes upward from each end toward the center. The approach track from the receiving tracks up onto the hump is built on a 1 per cent grade up which cars are pushed to the top of the hump. They then roll under the influence of gravity downgrade to the classification tracks, being guided to the proper track by remote-control switching operated from the central control tower. The downgrade is made about 4 per cent at the start for a distance of 30 to 50 ft and is gradually reduced to about 0.2 to 0.5 per cent for the classification tracks. The student will recognize from his study of dynamics that the car is subject to a net accelerating force of about 72 lb/ton on the 4 per cent downgrade, allowing 8 lb/ton for total rolling friction; and in the classification yard it will be subjected to a net decelerating force of friction. Nevertheless it will be necessary for a "rider" to apply brakes to the car, or for the car to be otherwise retarded as it approaches the cars which are to make up the train.

Car retarders, operated from the control tower, have gradually replaced car riders at larger terminals. The retarders operate electrically to press a pair of blocks against the insides of the wheels a few inches above the rails, the intensity of pressure controlling the speed of the moving car.

Weighing of cars is provided for by automatic scales on which each

**Fig. 22-7** Proviso yard, Chicago. (*Chicago & North Western Railroad.*)

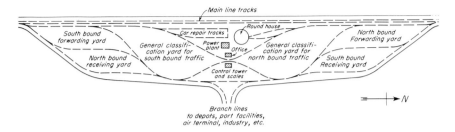

**Fig. 22-8**   General scheme for yard layout.

car is weighed as it passes slowly over the scale track.   A separate track
is laid for this purpose so that only the cars to be weighed are run over
the scale track.

**22-9   Yard capacity**   Some details of yard layout and operation are
shown by the picture of Fig. 22-7 showing the proviso yard of the Chicago
& North Western Railroad in Chicago.

A general scheme of yard layout is shown in Fig. 22-8.   General
track arrangement should be such that main-line tracks do not pass
through the yard.   Connections to the main tracks from forwarding and
receiving tracks should be as simple and direct as practicable.   Curvature
of tracks should not, in general, exceed about 12°30′.   Tracks in the
yards should be spaced not less than 18 ft center to center.   In computing
yard capacity, 45 ft of track is usually allowed per car.

Redesigning and enlargement of yards, improved communication
facilities and inspection procedures, and consolidation of small yard
units are doing much to improve facilities for expediting freight traffic
through yards and terminals.

Included in the over-all cost of operation of yard and terminal
facilities are (1) delay costs—costs of actual waiting time in receiving
yards—and (2) processing costs—cost of receiving, classification, and
dispatching of trains.   In general, as yard capacity is increased, delay
costs are reduced, but processing costs are increased, while a low yard
capacity will result in high delay costs and low processing costs.   According to AREA, the total standing capacity of the receiving, classification,
and departure tracks may be roughly assumed to be about equal to the
working capacity of the yard.   Recent studies indicate that the optimal
yard capacity will be about 10 per cent greater than the average arrival
rate.   Thus if the average daily arrival or "input" of cars into a terminal
is 1,000 cars per day, the optimal yard capacity is about 1,100 cars.

Where the number of switching operations is relatively small, the
yard is usually level, or nearly so, and it may be only a single yard—one-

half of the layout shown in Fig. 22-8. Separate departure or forwarding tracks may not be necessary if there are sufficient classification tracks to take care of the required number of trains. Where hump switching is used, the grades of receiving and forwarding yards should be kept at a minimum—below 0.2 per cent—so that trains will stand in these yards without braking.

**22-10  Coordination of transportation facilities**  The engineer is often confronted with problems related to community and regional planning for orderly development of growing communities and metropolitan areas. An important phase of such planning is the coordination of transportation facilities.

Harbor development for waterway traffic and airports, once established, will occupy a fairly permanent or fixed position in the traffic pattern; that is, the high initial cost of such installations will generally prohibit their abandonment in favor of more strategic location once they have been established in the transportation scheme. Also, site selection is critical with respect to natural physical characteristics. This is also true for terminal facilities for highway and railway traffic, but to a lesser degree. As parking facilities for automobiles in a business district become inadequate, for instance, there is the possibility of developing a fringe parking area at a considerable radius from the congested district, with cheap and efficient mass rapid transit between this belt or fringe and the business district.

An outer circumferential route or belt including both highway and railroad may also be established to route traffic around a city or municipal area instead of directly through it. Freight and passenger terminals for both highway and railway traffic are established on this belt line, and rapid transit by bus and truck provides the connecting link with business and industry. This outer circumferential route is so placed as to make the best use of existing roadways and tracks and also to make a direct connection with water and air terminals. Where only one through railroad exists, it may be tangent to the belt line on only a small portion of the periphery, but a complete belt of tracks, paralleling the highway belt, will often be most effective where two or more through railroads radiate from the population center in several directions.

It should be recognized that coordination of transportation facilities is a continuing process requiring comprehensive long-range planning. Immediate over-all transfer and rearrangement of routes and terminal facilities will not ordinarily be possible, but as existing facilities become obsolete or inadequate, their improvement and reconstruction should fit into the long-range community and regional plans. The establishment of union stations, for instance, has greatly facilitated handling of passenger and express traffic. In many places the railroad yards consist of several scattered units, some of which are individually inadequate. Consolidation of such yards on the periphery of a belt line would expedite freight traffic through yards and terminals and make possible an integrated system of loading and transfer of freight as between all modes of transportation (see Prob. 22-4).

## QUESTIONS AND PROBLEMS

**22-1.** Make a pencil mechanical drawing, complete with dimensions and names of parts, of a crossover from the data of Fig. 22-5.

**22-2.** Make a field layout for construction of the crossover of Prob. 22-1, connecting two parallel tracks.

**22-3.** *a.* Draw to scale a plan layout similar to Fig. 22-8 for a yard to handle an input of 1,200 cars per day, consisting of 20 trains of 60 cars each.

*b.* Draw to scale a lengthwise profile of the yard through the center, assuming this yard to provide for hump switching.

**22-4.** From a map which shows the existing transportation routes and facilities for a medium-size city, say, 50,000 to 500,000 population, work out a comprehensive system of coordinated facilities taking into account such factors as (*a*) relief of traffic congestion in the business district, (*b*) establishment of an efficient intracity rapid-transit system, and (*c*) integrating terminal facilities for all modes of transportation.

## BIBLIOGRAPHY

*Am. Ry. Eng. Assoc. Proc.,* **48** (1947).
"Track Work Plans," American Railway Engineering Association, 1948.

# 23
# Diesel-electric Locomotives

**23-1 Locomotive capacity** Assuming that a proposed relocation will not affect the total revenue to be realized from the transportation of freight and passengers, it is apparent that the cost of construction must be met by resulting savings in the cost of operation. Line and grade revision may reduce the amount of fuel consumed by a train between two points on the line, or by reducing the time required for the run between two points there may be a saving in the cost of the train crew; but in freight service, at least, the greatest savings are apt to result from a reduction in the number of trains required to transport a specific quantity of freight. The effect of relocation on fuel consumption, schedules, or train tonnage cannot be estimated without a knowledge of locomotive performance and train resistances.

The horsepower output of a locomotive is limited by the traction that can be developed between the driving wheels and the rails and by the output of the engine. In either case

$$\text{Power} = \text{force} \times \text{velocity} \tag{23-1}$$

Where the force $T$ is expressed in pounds, and velocity $V$ is given in miles per hour, the above expression becomes

$$P = T \frac{V}{375} \quad \text{hp} \tag{23-2}$$

*Tractive capacity* is the maximum horsepower that can be developed through rail friction. At low speeds the engine capacity or motor output should equal the tractive capacity; that is, the full power of the loco-motive is sufficient to spin its wheels. The *basic speed* $V'$ is the maximum speed at which rail friction $T'$ is just equal to the tractive effort that the engine can exert at the driving wheels, ignoring engine losses. The maximum value of rail friction is

$$T' = 2,000fW' \tag{23-3}$$

where $f$ is the coefficient of rail friction and $W'$ is total weight in tons, supported by the driving wheels. The actual value of $f$ varies widely with temperature, moisture, and rail conditions, but by dropping sand on the rails it is almost always possible to attain an effective friction above 0.25 for starting and for low speeds. A value of 0.30 is commonly accepted for use in routine computations for diesel-electric locomotives, whereas 0.25 is generally accepted for electric locomotives for starting and moderate speeds up to about 30 mph. At higher speeds for electric locomotives and motorized cars engaged in rapid passenger service, values of friction coefficient for determining available traction for braking and high-speed tractive effort are generally taken as about 0.15. An approximate relationship of friction coefficient to speed is shown in Fig. 23-1.

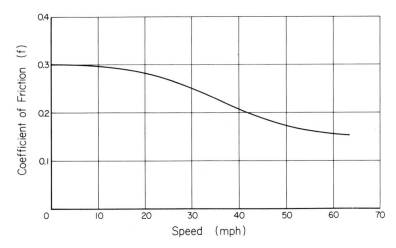

**Fig. 23-1** Approximate relationship of coefficient of friction to speed for steel wheels on steel rails.

With a value of $f = 0.25$, and combining Eqs. (23-2) and (23-3),

$$P' = 4W' \frac{V}{3} \qquad\qquad (23\text{-}4)$$

where $P'$ is the tractive capacity.

**23-2   General characteristics of diesel locomotives**   In recent years there has been a strong trend toward the use of diesel-electric locomotives. For the first time in 125 years, during the year 1953 no American railroad ordered a steam locomotive.   Among the several advantages which have led to this preference are economy of operation and high tractive effort at low speed.   Diesels are not lighter in weight than comparable steam locomotives, but the electric drive permits placing a larger proportion of the total weight on the driving axles.   As a concession to this better performance in starting, it is customary in speed-output computations to estimate the tractive effort of adhesion at 30 per cent instead of 25 per cent of the weight on the drivers.   The amount of power available is limited by the capacity of the diesel engine.   The normal horsepower output of the engine is set by the manufacturer's rating, which is well below the maximum capacity.   Generators transform the engine output into electrical energy, which is delivered to motors geared to the driving axles.   The normal nominal horsepower output is generally taken to be 80 per cent of the normal capacity, with the margin covering motor and generator efficiencies and an allowance for auxiliaries.

There are three primary parts to the diesel-electric locomotive: (1) the diesel engine; (2) the generator, directly connected to the diesel engine, designed to produce simultaneously sufficient direct current for the traction motors and alternating current for lighting, electrical controls, air compressor, blowers, and other auxiliaries; and (3) the traction motors, which operate the driving wheels through a chain of gears.   The locomotive units are so assembled, thus greatly facilitating repair and maintenance.

**Fig. 23-2**   Model S-8 800-hp switcher.   (*Baldwin-Lima-Hamilton Corp.*)

## LOCATION OF PRINCIPAL PARTS

1. Switchmen's Steps
2. Headlight
3. Engine Air Intake Filter
4. Cooling System, Radiator Cores, Fan & Drive
5. Lube Oil Filter
6. Traction Motor Blower

7. Lube Oil Filler
8. Engine Compartment
9. Diesel Engine
10. Main Generator
11. Air Compressor
12. Storage Batteries
13. Electrical Equipment Cabinet

14. Fireman's Seat
15. Fire Extinguisher
16. Handbrake
17. Cab Heater
18. Operator's Controls
19. Operator's Seat
20. Air Brake Equipment

21. Motor Truck
22. Fuel Oil Tank
23. Fuel Oil Tank Filler
24. Air Reservoir
25. Water Filler, Diesel Engine
26. Sand Box Filler

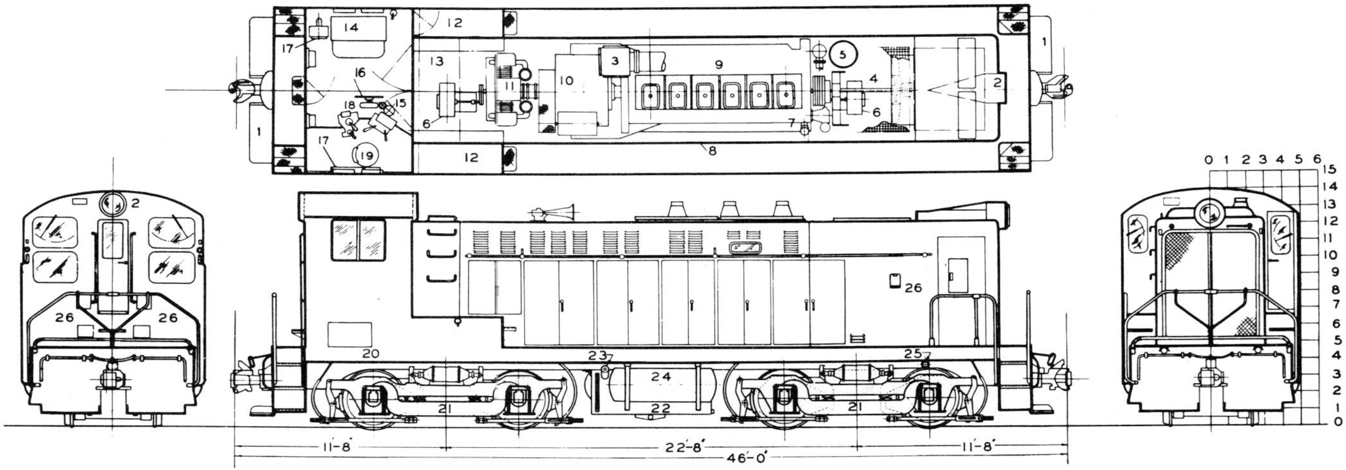

**Fig. 23-3**  Diesel-locomotive design.  S-8 switching locomotive, 800 hp available for traction, four motors, four axles.

The switching locomotive is generally a single locomotive unit. One such unit, the 800-hp, 100-ton model S-8 switcher built by the Baldwin-Lima-Hamilton Corporation, is illustrated in Figs. 23-2 and 23-3. Some further design features of this locomotive are as follows:

Weight loaded (working order)...................... 198,500 lb
Weight on drivers................................. 198,500 lb
Wheel diameter....................................     40 in.
Fuel-oil capacity.................................    650 gal
Sand capacity.....................................     30 cu ft
Gear ratio........................................      14:68
Tractive effort for starting, 25% adhesion........  49,625 lb
Rated continuous tractive effort, at 6.8 mph......  34,000 lb
Minimum-radius curvature, with train..............     44 deg

Approximate tractive effort at various speeds for this locomotive is given in Fig. 23-4. Power is generated with the use of a six-cylinder diesel engine and supplied to each of the four axles by four traction motors, one for each axle.

A great variety of diesel-electric locomotives are built to serve various needs. Among these are switchers, road switchers, freight, passenger, combination freight and passenger, and all-purpose units. A locomotive for road service may be made up of one or several units, usually from one to four. When two or more units are operated as a single locomotive, the leading unit has a cab with the control panel while the trailing units usually do not. An illustration and general description of such units—the Baldwin-Westinghouse road freight locomotive RF-16, comprising 1,600-hp units—are given in Figs. 23-5 and 23-6. Some

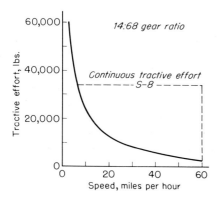

**Fig. 23-4** Approximate tractive effort, Baldwin-Westinghouse model S-8 switching locomotive.

**Fig. 23-5** Baldwin-Westinghouse 6,400-hp diesel-electric road freight locomotive (four 1,600-hp units). (*Baldwin-Lima-Hamilton Corp.*)

additional design features are as follows:

|  | *A* units (leading) | *B* units (trailing) |
|---|---|---|
| Weight loaded, working order...... | 248,000 lb | 244,000 lb |
| Weight on drivers............... | 248,000 lb | 244,000 lb |
| Wheel diameter................. | 42 in. | 42 in. |
| Fuel-oil capacity............... | 1,200 gal | 1,200 gal |
| Sand capacity.................. | 16 cu ft | 16 cu ft |
| Gear ratios.................... | 15:68, 15:63, or 17:62 | 15:68, 15:63, or 17:62 |
| Starting tractive effort, 25% adhesion......................... | 62,000 lb | 61,000 lb |
| Minimum-radius curvature, with train........................ | 21 deg | 21 deg |

Approximate tractive effort at various speeds for the RF-16 road freight unit at various speeds with indicated rated tractive effort and maximum speeds for the three possible gear ratios are given in Fig. 23-7.

Not all of the diesel-electric locomotive units have 100 per cent of the weight on the driving wheels. The 1,600-hp all-service locomotive AS-416 of the Baldwin-Lima-Hamilton Corporation, for instance, weighs, loaded (working order), 255,000 lb and has 176,000 lb on the drivers. This is because each of the two trucks has three axles, two of which are driven with traction motors. Figure 23-8 shows some typical locomotive trucks.

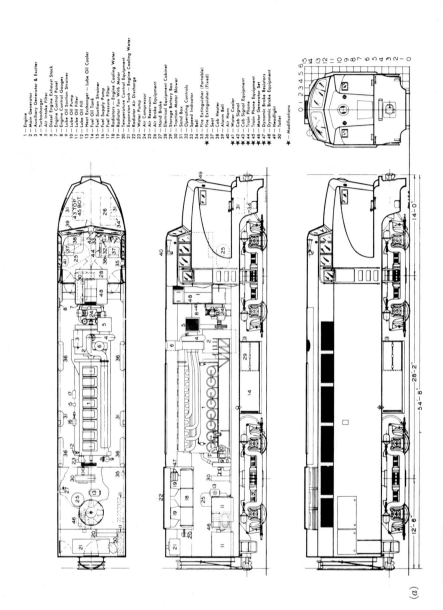

1 — Engine
2 — Main Generator
3 — Auxiliary Generator & Exciter
4 — Turbocharger
5 — Air Intake Filter
6 — Diesel Engine Exhaust Stack
7 — Engine Control Panel
8 — Engine Control Gauges
9 — Lube Oil Suction Strainer
10 — Lube Oil Pump
11 — Lube Oil Filter
12 — Lube Oil Fill
13 — Heat Exchanger — Lube Oil Cooler
14 — Fuel Oil Tank
15 — Fuel Suction Strainer
16 — Fuel Supply Pump
17 — Fuel Pressure Filter
18 — Radiators — Engine Cooling Water
19 — Radiator Fan With Motor
20 — Temperature Control Equipment
21 — Expansion Tank — Engine Cooling Water
22 — Radiator Air Discharge
23 — Water Pump
24 — Air Compressor
25 — Air Reservoir
26 — Air Brake Equipment
27 — Hand Brake
28 — Electrical Equipment Cabinet
29 — Storage Battery Box
30 — Traction Motor Blower
31 — Sand Box
32 — Operating Controls
33 — Speed Indicator
34 — Bell
★ 35 — Fire Extinguisher (Portable)
★ 36 — Fire Extinguisher (Fixed)
37 — Seat
38 — Cab Heater
39 — Alarm Bell
40 — Air Horn
41 — Water Cooler
★ 42 — Cab Signal
★ 43 — Cab Signal Equipment
★ 44 — Train Phone
★ 45 — Train Phone Equipment
★ 46 — Motor Generator Set
★ 47 — Dynamic Brake Resistors
★ 48 — Dynamic Brake Equipment
49 — Headlight
50 — Toilet

★ — Modifications

(a)

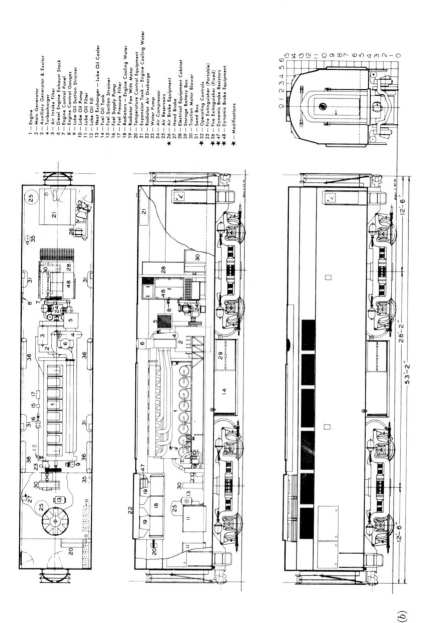

1 — Engine
2 — Main Generator
3 — Auxiliary Generator & Exciter
4 — Turbocharger
5 — Air Intake Filter
6 — Diesel Engine Exhaust Stack
7 — Engine Control Panel
8 — Engine Control Gauges
9 — Lube Oil Suction Strainer
10 — Lube Oil Pump
11 — Lube Oil Filter
12 — Lube Oil Fill
13 — Heat Exchanger — Lube Oil Cooler
14 — Fuel Oil Tank
15 — Fuel Suction Strainer
16 — Fuel Supply Pump
17 — Fuel Pressure Filter
18 — Radiators — Engine Cooling Water
19 — Radiator Fan With Motor
20 — Temperature Control Equipment
21 — Expansion Tank — Engine Cooling Water
22 — Radiator Air Discharge
23 — Water Pump
24 — Air Compressor
25 — Air Reservoirs
★ 26 — Air Brake Equipment
27 — Hand Brake
28 — Electrical Equipment Cabinet
29 — Storage Battery Box
30 — Traction Motor Blower
31 — Sand Box
★ 32 — Operating Controls
35 — Fire Extinguisher (Portable)
36 — Fire Extinguisher (Fixed)
★★ 47 — Dynamic Brake Resistors
★★ 48 — Dynamic Brake Equipment
★ — Modifications

(b)

**Fig. 23-6**  Layout of a Baldwin RF-16 road freight locomotive.  (a) 1,600-hp A unit; (b) 1,600-hp B unit.

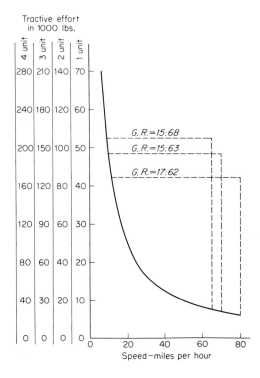

**Fig. 23-7** Approximate tractive. effort, Baldwin RF-16 road freight unit (with optional gear ratios).

**23-3 Locomotive performance** The driver horsepower output per ton on drivers is given by empirical formula by the AREA for diesel-electric locomotives as follows:

$$P = K(1 - e^{-y}) \qquad (23\text{-}5)$$

where $K$ is taken as 80 per cent of rated horsepower per ton on drivers, $e$ is the base of natural logarithms, and

$$y = \frac{V}{V'} \qquad (23\text{-}6)$$

At basic speed $V = V'$, and

$$P = 0.632K \qquad (23\text{-}7)$$

Also, by Eq. (23-2),

$$P = 600 \frac{V'}{375} \qquad (23\text{-}8)$$

whence

$$V' = 0.395K \qquad (23\text{-}9)$$

(a)

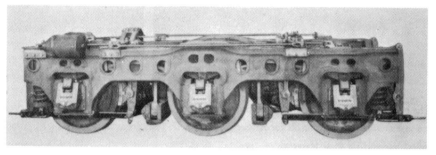

(b)

(c)

**Fig. 23-8**  Locomotive trucks.   (a) AS-16, two axles, two motors, with roller bearings;
(b) AS-416, three axles, two motors; (c) AS-616, three axles, three motors.   (*Baldwin-Lima-Hamilton Corp.*)

(1) Weight on drivers_____tons     (2) Weight of locomotive=_____tons
(3) Rated horsepower_____hp     (4) Nominal output=80% of (3)=_____hp
(5) $\frac{(4)}{(1)}$=K=_____hp per ton on drivers     (6) V'=0.395 (5) =_____mph

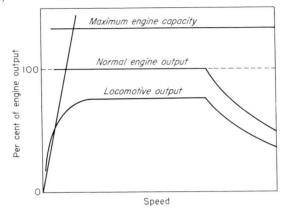

| Normal gearing | | | | Special gearing | | Horsepower output | |
|---|---|---|---|---|---|---|---|
| $\frac{V}{V'}$ | V (mph) | Tractive effort (#) | | V (mph) | Tractive effort total | Per ton on drivers | Total |
| | | Per ton on drivers | Total | | | | |
| 0 | 0 | 600 | | 0 | | 0 | 0 |
| 1.00 | | 600 | | | | 0.632 K | |
| 1.25 | | 546 | | | | 0.713 K | |
| 1.50 | | 492 | | | | 0.777 K | |
| 1.75 | | 448 | | | | 0.826 K | |
| 2.00 | | 411 | | | | 0.865 K | |
| 2.50 | | 348 | | | | 0.918 K | |
| 3.00 | | 300 | | | | 0.950 K | |
| 3.50 | | 263 | | | | 0.970 K | |
| 4.00 | | 233 | | | | 0.982 K | |
| 4.50 | | 209 | | | | 0.989 K | |
| 5.00 | | 189 | | | | 0.993 K | |
| 6.00 | | 158 | | | | 0.998 K | |
| 7.00 | | 135 | | | | 0.999 K | |
| 8.00 | | 119 | | | | 1.000 K | |
| 10.00 | | 95 | | | | 1.000 K | |
| 12.00 | | 79 | | | | 1.000 K | |
| 14.00 | | 67 | | | | 1.000 K | |
| Overspeed | | | | | | | |
| 10% | | | | | | 0.911 K | |
| 20% | | | | | | 0.833 K | |
| 30% | | | | | | 0.761 K | |
| 40% | | | | | | 0.714 K | |

**Fig. 23-9** Computation form, diesel-electric locomotive. (*From AREA Manual.*)

The relation between speed and power expressed in Eq. (23-5) holds for normal gearing of the motor.  For high-speed locomotives it becomes desirable to change the gear ratio.  For double-speed gearing, for example, the speed corresponding to a specified horsepower would be twice that obtained from the formula, while the corresponding tractive effort would be half the normal value.

With normal gearing the maximum speed at which the full capacity of the engine is available lies somewhere between 8 and $14V'$, depending upon design details.  At speeds 10, 20, 30, and 40 per cent greater than this maximum, the driver horsepower output becomes 0.911, 0.833, 0.761, and $0.714K$, respectively.

A summary of basic performance information is shown in Fig. 23-9, based on the formulas and data of preceding paragraphs, with tractive effort of adhesion taken as 30 per cent.

**23-4   New dimensions in motive power**   Larger units of main-line diesel-electric locomotives are presently being built.  The power ratings of these larger units range from 2,000 to 3,600 hp.  An example is the six-axle, six-motor, 3,600-hp unit shown in Fig. 23-10.  Performance of this

**Fig. 23-10**  Electro-Motive's most powerful single-engine diesel-electric locomotive, the 3,600-hp General Motors SK-45, offers a higher plateau of power for high-speed, heavy-hauling locomotive service on the nation's railroads.  This mighty prime mover uses the GM 645 series, 20-cylinder diesel engine.  (*Electro-Motive Division of General Motors.*)

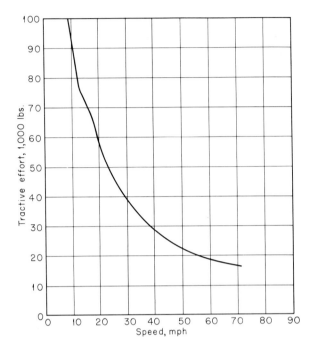

**Fig. 23-11**  Tractive-effort curve, 3,600-hp model SD-45 locomotive.  Equipment: One 20-645 engine, one AR10 generator, and six D77 traction motors with a 62:15 gear ratio and 40-in.-diameter wheels.  (*Electro-Motive Division of General Motors.*)

locomotive unit is given by the tractive-effort curve in Fig. 23-11. Specifications and components are as follows:

*Specifications*

Dimensions:
- Length over coupler pulling faces.................................. 65 ft 8 in.
- Distance between bolster centers................................. 40 ft 0 in.
- Truck rigid-wheel base.......................................... 13 ft 7 in.
- Width over grab irons........................................... 10 ft $3\frac{1}{8}$ in.
- Height above rails.............................................. 15 ft $5\frac{1}{4}$ in.
- Wheel diameter................................................. 40 in.
- Minimum curve radius........................................... 274 ft

Weight:
- Total loaded weight on rails, approx............................. 368,000 lb

Supplies:
- Fuel.......................................................... 3,200 gal
- Sand.......................................................... 56 cu ft
- Cooling water................................................. 295 gal
- Lubricating oil................................................ 294 gal

*Components*

Engine:

Twenty-cylinder GM 645 diesel, two-cycle 45°V, $9\frac{1}{16}$-in. bore, 10-in. stroke, unit injection, turbocharger scavenging through cylinder-wall intake and multivalve exhaust.

Main generator:

Electro-Motive AR-10 alternator with rectified output for delivery to traction motors; nominal 600-volt direct-current rating, ventilated by blower.

Traction motors:

Six Electro-Motive D-77 direct-current series wound, force ventilated, axle-hung motors with roller-type armature bearings.

Truck assemblies:

Two fully flexible three-motor six-wheel truck assemblies with center bearing load distributed by an H-shaped bolster and transferred to the truck frame through four double-coil vertical spring packs located at the bolster corners.

## QUESTIONS AND PROBLEMS

**23-1.** Determine the basic speed of the switching locomotive S-8 described in this chapter and plot tractive effort versus speed and horsepower output versus speed from a tabulation as illustrated in Fig. 23-9.

**23-2.** A diesel-electric locomotive unit with manufacturer's rating of 2,400 hp weighs 256,000 lb and has 256,000 lb on the driving wheels. Normal gearing is 15:68. Tabulate tractive effort and horsepower output versus $V/V'$ for normal gearing and for a gear ratio of 15:63.

**23-3.** Plot curves of tractive effort versus speed and horsepower output versus speed for a locomotive composed of four units described in Prob. 23-2. Use normal gearing.

## BIBLIOGRAPHY

*Am. Ry. Eng. Assoc. Bull.* 455, February, 1945.

*Am. Ry. Eng. Assoc. Proc.*, **41** (1940) and **44** (1943).

"AREA Manual," vol. 2, chap. 16, American Railway Engineering Association, Chicago, 1953.

Diesel-electric Locomotive Units in Railway Service, *Railway Age*, **128**(21) (1950).

# 24
# Electric Locomotives

**24-1 Characteristics and performance of electric locomotives** The limiting factor for electric locomotives is the capacity of the power line, a relatively unlimited outside source of energy. This great advantage of the electric locomotive is also a serious disadvantage in the sense that the construction of transmission lines and power plants is an important item of additional expense in areas remote from existing installations.

Electric locomotives are capable of exerting full tractive effort at relatively high speeds. They can be fitted with regenerative braking, which enables the motors to act as generators on descents or when slowing down, feeding power back into the line to be used at some other point. They are especially desirable on mountain divisions where one train can be kept at a safe speed and at the same time help some other train on the ascent. The Chicago, Milwaukee, St. Paul and Pacific is electrified through the Rocky and Cascade Mountains, while the Great Northern is electrified through the Cascades. The heavy passenger locomotives of the former have 12 drive wheels with 1,000-volt d-c motors of about 3,000 hp. The total weight is 521,000 lb, of which 457,600 lb rests on the drivers. The Great Northern locomotives have 16 drive wheels, a total weight of 724,000 lb, and 556,000 lb on the drivers. They develop a tractive effort of about 166,900 lb at operating speeds.

The capacity of electric locomotives is based upon an allowable heat rise in the motors over a given period of time. The tractive capacity in starting, as in the case of other locomotives, depends upon rail friction.

The tractive effort exerted at varying speeds has been expressed by empirical AREA formulas for the several types of electric drive. The power supply may either be direct current or alternating current. Alternating-current motors may be the single-phase, the split-phase induction type, or the single-phase motor-generator sets.

The tractive efforts of d-c locomotives are approximately inversely proportional to the cube of the speeds. That is,

$$TV^3 = \text{a constant} \tag{24-1}$$

and

$$PV^2 = \text{a constant} \tag{24-2}$$

The tractive effort per ton on drivers at basic speed is assumed to be 500 lb, as in the case of steam locomotives. If the tractive effort for some other specific speed is known, the basic speed can be computed by Eq. (24-1), and the tractive effort for various multiples of $V'$ can be tabulated. In these computations it is customary to assume that the trolley voltage is 90 per cent of the substation voltage. This is equivalent to using 90 per cent of the speed corresponding to the rated tractive effort in finding $V'$. Direct-current locomotives generally have a second tractive-effort curve, which is obtained by the use of a shunted field.

In calculations and tabulations, driver tractive effort and locomotive horsepower (per ton on drivers and total) are generally computed for the following multiples of $V'$: 0.94, 1.00, 1.10, 1.20, 1.30, 1.40, 1.50, 1.60, 1.80, 2.20, 2.30, 2.40, 2.60.

For single-phase a-c locomotives, the approximate relation between tractive effort and speed is given by the expression

$$T^3V^5 = \text{a constant} \tag{24-3}$$

Then

$$P^3V^2 = \text{a constant} \tag{24-4}$$

Basic speed is computed as before, except that because of variable-voltage taps on the transformer it is not necessary to adjust the speed for average trolley voltage. Horsepower rating is usually based upon 60 per cent of maximum speed.

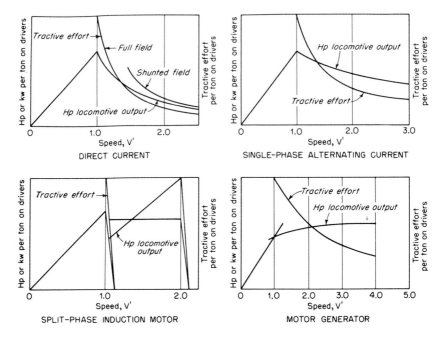

**Fig. 24-1**  Electric locomotive performance.  (*From AREA Manual.*)

The split-phase induction-motor locomotive is inherently a constant-speed locomotive.  The speed is doubled by cutting out half the magnetic poles.

The motor-generator locomotive uses a d-c motor drive with an a-c power supply.

Performance characteristics, horsepower, and tractive effort for various speeds in terms of basic speed are given for four types of electric locomotives in Fig. 24-1.

## QUESTIONS AND PROBLEMS

**24-1.** The rated tractive effort of a single-phase a-c electric locomotive is 160 lb/ton on drivers at 30 mph.
    *a.* What is its basic speed?
    *b.* What is its tractive effort at 20 mph?

**24-2.** The rated tractive effort of a d-c electric locomotive is 280 lb/ton on drivers at 20 mph.  Determine its basic speed and tabulate values of tractive effort and horsepower output for multiples of $V'$, as explained in Sec. 23-3.  Plot curves of tractive effort versus speed, and horsepower output versus speed.

## BIBLIOGRAPHY

*Am. Ry. Eng. Assoc. Proc.*, **41** (1940).

"AREA Manual," vol. 2, chap. 16, American Railway Engineering Association, 1953.

Healy, K. T.: "Electrification of Steam Railroads," McGraw-Hill Book Company, New York, 1929.

Johnson, R. P.: "The Steam Locomotive," Simmons-Boardman Publishing Corporation, New York, 1942.

Kerr, Charles, Jr.: Full Power at All Speeds, *Railway Age*, **128**(13):51–55 (1950).

"A Practical Evaluation of Railroad Motive Power," Simmons-Boardman Publishing Corporation, New York, 1947.

*Trans. ASME*, **54** (1932).

Yellott, John I.: The Experimental Coal-burning Gas Turbine, *Railway Age*, **128**(15): 59–62 (1950).

# 25
# Train Resistances

**25-1  Frictional resistances**  The value of tractive resistance in pounds per ton of car weight on tangent level track under mild weather conditions is given by the following formula:[1]

$$R = \frac{9.4}{w^{\frac{1}{2}}} + \frac{12.5}{w} + JV + \frac{KAV^2}{wn} \tag{25-1}$$

where $w$ = average weight per axle, tons
$J,K$ = coefficients depending upon the type of car
$V$ = speed, mph
$A$ = cross-sectional area of car, including trucks, sq ft
$n$ = number of axles per car

For locomotives, $J = 0.03$, $K = 0.0024$, $A = 120$ sq ft, if $wn$ exceeds 100 tons.

For freight cars, $J = 0.045$, $K = 0.0005$, $A = 85$–$90$ sq ft.

For passenger cars, $J = 0.03$, $K = 0.00034$, $A = 120$ sq ft.

---

[1] W. J. Davis, Jr., *Am. Ry. Eng. Assoc. Proc.*, **44**:190, 670, 679 (1943).

For multiple-unit trains, $J = 0.045$, $A = 100$–$110$ sq ft, $K = 0.0024$ for the leading car and $0.00034$ for trailing cars.

Where less accuracy is needed, the AREA suggests that the frictional resistance of freight trains under normal conditions in warm weather with modern equipment running at speeds between 7 and 35 mph may be determined for the purpose of comparing different gradients and locations by the formula

$$R' = 2.2T + 121.6C \qquad (25\text{-}2)$$

where $R'$ = total resistance on level tangent, lb
  $T$ = total weight of cars and contents, tons
  $C$ = total number of cars in train

The values thus obtained will usually be from 3 to 8 lb/ton, and a fair average for mixed traffic may be taken at 6 lb/ton. At extremely low temperatures tractive resistance may reach the value of 30 lb/ton for empty freight cars. Loaded cars have less resistance per ton than empty cars.

**Table 25-1  Tractive resistance of diesel-electric locomotives***

| Model | No. of units | Tons | Resistance in pounds per ton at speeds in miles per hour† (by Davis formula) | | | | | | | | | | | | |
|---|---|---|---|---|---|---|---|---|---|---|---|---|---|---|---|
| | | | 5 | 10 | 15 | 20 | 25 | 30 | 35 | 40 | 45 | 50 | 60 | 70 | 80 |
| S-8 | 1 | 99 | 2.7 | 3.0 | 3.6 | 4.2 | 5.0 | 6.0 | | | | | | | |
| | 2 | 198 | 2.7 | 2.9 | 3.3 | 3.6 | 4.1 | 4.7 | | | | | | | |
| S-12 | 1 | 117.5 | 2.5 | 2.9 | 3.4 | 4.0 | 4.7 | 5.5 | | | | | | | |
| | 2 | 235 | 2.5 | 2.7 | 3.0 | 3.4 | 3.8 | 4.3 | | | | | | | |
| RS-12 | 1 | 112 | 2.6 | 2.9 | 3.3 | 4.0 | 4.7 | 5.6 | 6.6 | 7.7 | 9.0 | 10.3 | 11.8 | 13.5 | |
| | 2 | 224 | 2.5 | 2.8 | 3.1 | 3.5 | 4.0 | 4.4 | 5.0 | 5.6 | 6.3 | 7.1 | 7.9 | 8.8 | |
| AS-16 | 1 | 120 | 2.4 | 2.8 | 3.3 | 3.8 | 4.5 | 5.3 | 6.3 | 7.3 | 8.5 | 9.8 | 12.7 | 16.1 | 20.0 |
| | 2 | 240 | 2.4 | 2.7 | 3.0 | 3.4 | 3.8 | 4.3 | 4.8 | 5.4 | 6.0 | 6.8 | 8.4 | 10.3 | 12.4 |
| AS-416 | 1 | 128 | 2.8 | 3.2 | 3.6 | 4.2 | 4.8 | 5.6 | 6.5 | 7.5 | 8.6 | 9.8 | 12.6 | | |
| | 2 | 256 | 2.8 | 3.1 | 3.4 | 3.7 | 4.1 | 4.6 | 5.1 | 5.7 | 6.3 | 7.0 | 8.5 | | |
| AS-616 | 1 | 162.5 | 2.7 | 2.8 | 3.2 | 3.7 | 4.2 | 4.9 | 5.5 | 6.4 | 7.3 | | | | |
| | 2 | 275 | 2.7 | 2.8 | 3.0 | 3.4 | 3.8 | 4.2 | 4.7 | 5.2 | 5.8 | | | | |
| RT-624 | 1 | 180 | 2.5 | 2.7 | 3.1 | 3.5 | 4.0 | 4.6 | 5.3 | 6.0 | 6.8 | | | | |
| | 2 | 360 | 2.5 | 2.6 | 2.9 | 3.2 | 3.5 | 3.9 | 4.3 | 4.7 | 5.2 | | | | |
| RF-16 | 1 | 125 | 2.2 | 2.5 | 2.8 | 3.3 | 3.8 | 4.4 | 5.1 | 6.1 | 7.0 | 7.9 | 10.0 | 12.5 | 15.2 |
| | 2 | 250 | 2.2 | 2.4 | 2.6 | 3.0 | 3.3 | 3.6 | 4.1 | 4.8 | 5.3 | 5.8 | 7.0 | 8.4 | 9.9 |
| | 3 | 375 | 2.2 | 2.4 | 2.6 | 2.9 | 3.1 | 3.4 | 3.7 | 4.3 | 4.8 | 5.1 | 6.0 | 7.1 | 8.1 |
| | 4 | 500 | 2.2 | 2.4 | 2.5 | 2.8 | 3.0 | 3.3 | 3.6 | 4.1 | 4.5 | 4.8 | 5.5 | 6.4 | 7.2 |
| | 5 | 625 | 2.2 | 2.4 | 2.5 | 2.8 | 2.9 | 3.2 | 3.5 | 4.0 | 4.3 | 4.6 | 5.2 | 6.0 | 6.7 |

* Baldwin Locomotive Works.
† Reduce journal resistance 10 per cent from 5 to 35 mph when roller-bearing-equipped. (Road locomotives equipped with roller bearings.)

**Table 25-2   Tractive resistance of freight cars (by Davis formula)***

| Weight of cars, tons | Resistance in pounds per ton at speeds in miles per hour† | | | | | | | | | | | | | | |
|---|---|---|---|---|---|---|---|---|---|---|---|---|---|---|---|
| | 5 | 10 | 15 | 20 | 25 | 30 | 35 | 40 | 45 | 50 | 60 | 70 | 80 | 90 | 100 |
| 20 | 7.3 | 7.7 | 8.2 | 8.8 | 9.5 | 10.3 | 11.3 | 12.3 | 13.5 | 14.7 | 17.6 | 20.9 | 24.62 | 29.0 | 33.4 |
| 25 | 6.2 | 6.6 | 7.0 | 7.6 | 8.2 | 8.9 | 9.7 | 10.6 | 11.5 | 12.6 | 15.0 | 17.6 | 20.7 | 24.3 | 27.8 |
| 30 | 5.4 | 5.8 | 6.2 | 6.7 | 7.2 | 7.8 | 8.5 | 9.3 | 10.2 | 11.0 | 13.2 | 15.3 | 17.9 | 21.2 | 24.2 |
| 35 | 4.8 | 5.2 | 5.6 | 6.0 | 6.5 | 7.1 | 7.7 | 8.4 | 9.2 | 10.0 | 11.9 | 13.8 | 16.1 | 19.0 | 21.5 |
| 40 | 4.4 | 4.8 | 5.1 | 5.5 | 6.0 | 6.5 | 7.1 | 7.8 | 8.4 | 9.2 | 10.6 | 12.6 | 14.7 | 17.3 | 19.6 |
| 45 | 4.1 | 4.4 | 4.8 | 5.2 | 5.8 | 6.1 | 6.6 | 7.2 | 7.9 | 8.6 | 10.1 | 11.7 | 13.6 | 16.0 | 18.1 |
| 50 | 3.8 | 4.2 | 4.5 | 4.9 | 5.3 | 5.7 | 6.2 | 6.8 | 7.4 | 8.1 | 9.5 | 10.9 | 12.7 | 15.0 | 16.8 |
| 60 | 3.4 | 3.7 | 4.0 | 4.4 | 4.5 | 5.2 | 5.7 | 6.2 | 6.7 | 7.3 | 8.5 | 9.9 | 11.4 | 13.4 | 15.0 |
| 70 | 3.2 | 3.4 | 3.7 | 4.1 | 4.4 | 4.8 | 5.3 | 5.7 | 6.2 | 6.7 | 7.9 | 9.1 | 10.5 | 12.3 | 13.7 |
| 80 | 3.0 | 3.3 | 3.5 | 3.9 | 4.2 | 4.6 | 5.0 | 5.4 | 5.9 | 6.4 | 7.4 | 8.5 | 9.8 | 11.4 | 12.7 |
| 90 | 2.8 | 3.1 | 3.4 | 3.7 | 4.0 | 4.4 | 4.7 | 5.1 | 5.6 | 6.0 | 7.0 | 8.1 | 9.2 | 10.8 | 11.9 |
| 100 | 2.7 | 3.0 | 3.2 | 3.5 | 3.8 | 4.2 | 4.6 | 4.9 | 5.4 | 5.8 | 6.7 | 7.7 | 8.8 | 10.3 | 11.3 |
| 120 | 2.5 | 2.7 | 3.0 | 3.3 | 3.6 | 3.9 | 4.3 | 4.6 | 5.0 | 5.4 | 6.3 | 7.2 | 8.2 | 9.5 | 10.4 |
| 140 | 2.4 | 2.6 | 2.9 | 3.2 | 3.5 | 3.8 | 4.1 | 4.5 | 4.8 | 5.2 | 5.9 | 6.8 | 7.7 | 8.9 | 9.7 |

Three Axles per Truck—Double-truck Cars ($A$ = 87 Sq Ft)

| | 5 | 10 | 15 | 20 | 25 | 30 | 35 | 40 | 45 | 50 | 60 | 70 | 80 | 90 | 100 |
|---|---|---|---|---|---|---|---|---|---|---|---|---|---|---|---|
| 140 | 2.8 | 3.0 | 3.3 | 3.6 | 3.9 | 4.2 | 4.5 | 4.8 | 5.2 | 5.6 | 6.4 | 7.2 | 8.1 | 9.1 | 10.1 |
| 160 | 2.6 | 2.9 | 3.1 | 3.4 | 3.7 | 4.0 | 4.3 | 4.6 | 5.0 | 5.3 | 6.1 | 6.9 | 7.7 | 8.6 | 9.6 |
| 180 | 2.5 | 2.7 | 3.0 | 3.3 | 3.5 | 3.8 | 4.1 | 4.5 | 4.8 | 5.1 | 5.8 | 6.6 | 7.4 | 8.3 | 9.2 |
| 200 | 2.4 | 2.6 | 2.9 | 3.2 | 3.4 | 3.7 | 4.0 | 4.3 | 4.6 | 5.0 | 5.7 | 6.4 | 7.2 | 8.0 | 8.8 |

* Baldwin Locomotive Works.
† Reduce journal resistances 10 per cent from 5 to 35 mph when roller-bearing-equipped.

A tabulation of resistance in pounds per ton at various speeds in miles per hour for diesel-electric locomotives operating on tangent level track is given in Table 25-1.   A similar tabulation is given in Table 25-2 for freight cars of various weights in tons.   Note that journal resistance may be reduced 10 per cent for speeds from 5 to 35 mph when rolling stock is equipped with roller bearings.   Minimum train resistances occur at a speed of about 6 mph.   Resistances at higher speeds will be greater, as indicated by Tables 25-1 and 25-2.   Journal friction for starting, however, may be as high as 15 or 20 lb/ton; then it drops off rapidly as the train gets under way (Fig. 25-1).   Much of this high journal friction for starting from rest is offset by the fact that in starting a train each car is "started" individually as the slack in the couplings is taken up progressively from one end of the train to the other.   It is desirable to

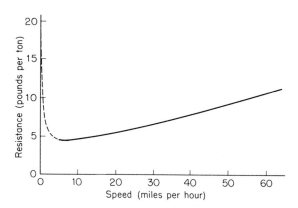

**Fig. 25-1**  Resistance at various speeds for 40-ton freight car.

place passing sidings, freight stations, etc., on summits or artificial humps, for advantage in both stopping and starting trains.

**25-2  Grade resistance**  Grade resistance amounts to 20 lb/ton times the per cent of grade.  This is equivalent to saying that the tangential force required to pull a body up a slope is equal to the product of its weight multiplied by the tangent of the angle of slope.

**25-3  Curve resistance**  An additional source of resistance is encountered in pulling a train around a curve.  Car wheels are rigidly attached to their axles, and since the outer wheel traverses a greater distance than the inner wheel in rounding a curve, while still experiencing the same number of revolutions, one or the other wheel must slip.  Undoubtedly the magnitude of the resistance depends to some extent upon the speed and the length of the curve, but still other factors complicate the situation, and it is customary to assume an average value per degree of curvature.  For many purposes it is satisfactory to consider each degree of central angle equivalent to a rise of 0.04 ft.  In terms of degree of curvature, this means that a 1-deg curve is equivalent to a 0.04 per cent grade, and the corresponding resistance is 0.8 lb/ton.

It is customary to "compensate" or offset curve resistance by equivalent reduction of grades on curves.  This is particularly significant on heavy grades so that trains will not be unduly slowed on the curves while operating upgrade.  The usual allowance for compensating grades on curves is 0.04 to 0.05 per cent of grade reduction per degree of curvature.

**25-4 Accelerating force**  If the train is accelerating, its inertia constitutes another source of resistance.  The accelerating force (or the opposing inertia reaction) is

$$F = \frac{Wa}{g} \tag{25-3}$$

The force tending to accelerate a car has its source in the pull of the
coupling; its magnitude is equal to the difference between this pull and
the sum of the resistances.  It must be noted that as the car acquires a
linear acceleration, the car wheels must experience a corresponding
angular acceleration.  If the moment of inertia of the rotating parts is
known, it becomes a simple problem in mechanics to compute the increase
in tractive resistance which will produce any specific value of angular
acceleration.  However, this item is not of great importance and is
usually cared for by an arbitrary increase in the $F$ of Eq. (25-3).  This is
done by using the value of 95.5 lb instead of the theoretical figure of
91.1 lb for the force required to impart an acceleration of 1 mph per
second to a body weighing 1 ton.  As it stands, Eq. (25-3) demands use
of consistent units.  With the above modification, and with $F$ in pounds,
$W$ the weight of the train in tons, and $a$ in miles per hour per second, the
equation becomes

$$F = 95.5Wa \tag{25-4}$$

Like curve resistance, inertia can be expressed in terms of an equivalent increase in grade $G'$:

$$G' = \frac{95.5a}{20} = 4.775a \tag{25-5}$$

### QUESTIONS AND PROBLEMS

**25-1.** Make tabulation of locomotive resistances at various speeds for the four-unit
diesel-electric locomotive of Prob. 23-3; assume 2 three-axle trucks per unit.

**25-2.** *a.* Tabulate resistances for various speeds up to 70 mph for a train of 60 freight
cars with a total loaded weight of 5,000 tons, using the data of Table 25-2.  Assume
40 loaded cars at 100 tons each, 10 cars at 60 tons each, and 10 cars at 40 tons each.
   *b.* For what speed does the tabulation check the resistance computed by use
of Eq. (25-2)?

**25-3.** Tabulate resistances for various speeds up to 90 mph for a 600-ton train of 10
passenger cars having four axles each.

**25-4.** *a.* Determine the grade resistance for the locomotive of Prob. 25-1 operating
on a 2 per cent grade.
   *b.* Determine the grade resistance for the train of cars of Prob. 25-2 on a 2 per
cent grade.

**25-5.** The average grade between two stations 10 miles apart is +0.3 per cent.  Determine an equivalent grade if total curvature on the line amounts to 240 deg of central angle.

**25-6.** What equivalent increase in grade corresponds to an acceleration of 0.1 mph per second?

## BIBLIOGRAPHY

*Am. Ry. Eng. Assoc. Proc.*, **50** (1949).

"AREA Manual," vol. 2, chap. 16, American Railway Engineering Association, Chicago, 1953.

Davis, W. J., Jr.: Tractive Resistance of Electric Locomotive and Cars, *General Electric Review*, October, 1926.

Tuthill, John K.: High Speed Freight Train Resistance: Its Relation to Average Car Weight, *Univ. Illinois Eng. Expt. Sta. Bull.* 376, 1948.

Wellington, A. M.: "Economic Theory of Railway Location," 6th ed., John Wiley & Sons, Inc., New York, 1898.

# 26

# The Velocity Profile
# and Performance Calculations

**26-1 Energy heads** In order to study the performance characteristics of locomotive and train of cars operating on a particular stretch of track, there is available a method which makes use of the fact that the total energy of a moving train consists of elevation head and velocity head. The method has been proved by actual train performance and by dynamometer tests, and it is applicable to both freight and passenger trains, using steam, electric, or diesel-electric locomotives.

The elevation head that a train acquires is determined by the rise and fall of the track gradient above or below a datum elevation. Thus a train operating uphill on a 2 per cent grade acquires elevation head at the rate of 2 ft per 100 ft of horizontal travel distance. The net tractive effort—over and above that required for frictional resistances—required for this rate of accumulation of elevation head is 40 lb/ton of total train weight.

The velocity head $H$ of a moving train is given by the formula $H = v^2/2g$. The AREA recommends that this be increased by about 6 per cent for the rotative energy of the wheels, and where $v$ in feet per second is changed to $V$ in miles per hour, $H = 0.03555V^2$. A tabulation of velocity heads for various speeds is given in Table 26-1.

**Table 26-1  Velocity heads, in feet, for various speeds, in miles per hour**

Based on the Formula $H = 0.0355V^2$

| $V$ | Velocity head | $V$ | Velocity head | $V$ | Velocity head |
|------|------|------|------|------|------|
| 2.5 | 0.22 | 25 | 22.2 | 60 | 127.8 |
| 5.0 | 0.89 | 30 | 32.0 | 65 | 150.0 |
| 7.5 | 2.00 | 35 | 43.5 | 70 | 174.0 |
| 10.0 | 3.55 | 40 | 56.8 | 75 | 199.7 |
| 12.5 | 5.55 | 45 | 71.9 | 80 | 227.2 |
| 15 | 7.99 | 50 | 88.8 | 90 | 287.5 |
| 20 | 14.20 | 55 | 107.4 | 100 | 355.0 |

A plot of the ground line or track gradients will constitute the elevation profile above a datum elevation. A plot of velocity heads above the ground line or elevation profile constitutes the velocity profile.

In preparation for plotting the velocity profile it will be convenient to tabulate certain data in accordance with the following outline:

| (1) | (2) | (3) | (4) | (5) | (6) (7) Net tractive effort available for energy increase | | (8) |
|------|------|------|------|------|------|------|------|
| $V$, mph | $T$, lb | Locomotive resistance, lb | Drawbar pull, lb | Train resistance, lb | Lb | Lb/ton | Acceleration grade, per cent |
| 0 | | | | | | | |
| 5 | | | | | | | |

(1) Tabulate speeds from 0 to maximum speed in increments of 5 mph.

(2) Record total tractive effort at drivers as determined in accordance with discussion in Chaps. 23 and 24.

(3) Locomotive resistance is determined by use of the Davis formula, as outlined in Chap. 25.

(4) Column 2 minus column 1 gives the drawbar pull of the locomotive.

(5) Train resistance, like locomotive resistance recorded in column 3, is determined from data of Chap. 25 for operation on tangent level track.

(6) Column 4 minus column 5 gives the net tractive effort available for increasing energy head of the total train.

(7) The net tractive effort available for increasing energy head is expressed in pounds per ton, obtained by dividing the figures of column 6 by the total weight of train (including locomotive) in tons. This is the amount of tractive effort, in pounds per ton, which is available for climbing grades or accelerating the train. The maximum grade in per cent which the particular locomotive and train of cars can negotiate at any given speed—also the maximum rate at which the particular locomotive and train of cars can accumulate velocity head, in feet per 100 ft of horizontal travel distance—is called the acceleration grade.

(8) This acceleration grade is obtained by dividing the figures of column 7 by 20, in accordance with Sec. 25-2.

**26-2   Ruling grades**   In general, a ruling grade is the maximum grade up which a particular locomotive and train of cars can operate at a given speed.   It is determined as the acceleration grade of column 8 of the previous section.   More specifically, a ruling grade is usually established for freight-train operation, as the maximum grade up which a specified locomotive can haul a specified maximum train at a constant speed of 10 mph.   This will generally govern in establishing desirable maximum grades to be used in location and construction of the railroad.

**26-3   Momentum grades**   For any particular locomotive and train of cars, any grade greater than its acceleration grade for a given speed is a momentum grade; that is, part of the momentum that the train has by virtue of its velocity at the bottom of such a grade is used in ascending the grade.   In other words, the velocity head of the train is reduced on a momentum grade as the train ascends.

**26-4   Plotting the velocity profile**   Figure 26-1 will be used to illustrate the method of plotting the velocity profile.   Line $ABCD$ represents the ground line or track profile.   The train is assumed to start from zero speed at $A$.   The velocity-profile line $ab$ is drawn as the acceleration grade, column 8, for a speed of 5 mph; that is, in plotting the velocity-profile grade from any point, the acceleration grade for the next higher speed is used, and speed changes are made in increments of 5 mph.   This velocity grade is extended to the point where the velocity head is that for a speed of 5 mph, or 0.89 ft, measured vertically from the track profile as $bb'$.   From this point $b$ the line $bc$ is drawn at the acceleration grade for 10 mph, to $d$ where the velocity head is 3.55 ft ($cc'$), and so on until maximum speed is reached or until a ruling of momentum grade is reached.   On a ruling grade the velocity profile is parallel to the track profile at full power, as line $de$ of Fig. 26-1, which is parallel to $BC$.

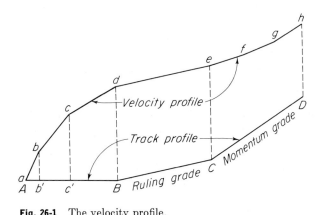

**Fig. 26-1**   The velocity profile.

As speed is reduced on a momentum grade the net available tractive effort, column 7, increases.  Therefore speed is reduced in decrements of 5 mph, again plotting the profile grade as the acceleration grade for the next higher speed.  Such a plot results in a line similar to *efgh* on which the grade is gradually increased—operating at full power—until ultimately it would become parallel to the grade of *CD* at the speed for which the acceleration grade of the locomotive and train is equal to the track grade, operating at full power.  The speed of the train at *D* is determined from the velocity head *hD*.

It is important to note that the rate of increase of velocity head per unit of horizontal travel distance is determined by the algebraic difference between the velocity-profile grade and the track grade.  This is in accord with the discussion of energy heads in Sec. 26-1.

**26-5  Time consumed by train operation**  A fairly complete analysis of such operating characteristics as time of travel over the route and work done can be made by tabulating certain data derived from the velocity profile.  It will be convenient to tabulate such data as follows:

| (1) | (2) | (3) | (4) | (5) | (6) |
|---|---|---|---|---|---|
| Station to station, 1,000 ft | Average $v$, fps | Time, sec | Distance, ft | Tractive effort, lb | Work done, million ft-lb |
|  |  |  |  |  |  |

(1) In column 1 are tabulated the station-to-station numbers at each break or change of grade of the velocity profile.

(2) Column 2 gives the average speed between the station breaks.  It will be convenient to convert the units of speed to feet per second for determination of the time lapse between stations.

(3) and (4) Column 3 is obtained by dividing the figures of column 4 by those of column 2.  A summation of column 3 then gives the time consumed by operation of the train over a given run.

**26-6  Energy of train operation**  The actual work done by the locomotive in moving itself and train of cars over a length of track is the summation of the products of driver tractive effort and distance traveled.  Driver tractive effort is recorded in column 5.  When the locomotive is operating at full capacity, while accelerating and while operating on acceleration or momentum grades, the full driver tractive effort of the locomotive is used.  When the locomotive is not being used to full capacity, as in the case of operation on a grade less than the acceleration grade for a particular speed, the driver tractive effort will be the sum of all locomotive and train resistances for that speed.  Note that in the case of a minus track grade, the grade resistance will be helping to move the train.

The product of columns 5 and 4 is tabulated as work done in column
6. The total of column 6 divided by 1.98 will give the driver output in
horsepower-hours. One horsepower-hour equals 1,980,000 ft-lb of
energy.

For diesel-electric locomotives it is customary to compute fuel consumption on the basis of about 30 per cent efficiency. This is equivalent
to an allowance of approximately 0.6 lb of fuel oil per driver horsepower-
hour.

## QUESTIONS AND PROBLEMS

**26-1.** Tabulate data in preparation for plotting the velocity profile for the diesel-
electric locomotive of Prob. 25-1 and the train of cars of Prob. 25-2.

**26-2.** On a sheet of profile paper, using a scale of 1 in. = 40 ft (vertical) and 1 in. =
4,000 ft (horizontal), plot the velocity profile for the locomotive and train of Prob.
26-1 on the following track profile, starting with zero speed at Sta. 0, and using
1,000-ft stations.

| Station to station | Track grade and curvature |
|---|---|
| 0–10. . . . . . . . . . . . . . | Level, no curvature. |
| 10–25. . . . . . . . . . . . . | Ruling grade for speed at Sta. 10. |
| 25–40. . . . . . . . . . . . . | 1.75 per cent grade. Determine speed of train at Sta. 40. |
| 40–80. . . . . . . . . . . . . | Limited speed zone (30 mph). One per cent track grade. 840 deg of curvature. Note that the equivalent grade will be 1.084 per cent. |
| 80–100. . . . . . . . . . . . | Adjust track (and velocity) profile to ground line at Sta. 80. Establish a minus grade, *grade of repose*, for constant speed of 20 mph with zero tractive effort. Operate on this grade at maximum speed of 40 mph. What tractive effort is required? |
| 100–116. . . . . . . . . . . | Level, no curvature. Operate at 40 mph. |
| 116–120. . . . . . . . . . . | Bring train to a stop at Sta. 120. Determine average braking force for approximately constant deceleration. Level, no curvature. |

**26-3.** *a.* Determine the time and energy required for the freight train run of Prob. 26-2.
*b.* How many gallons of fuel oil are required for the run, assuming $6\frac{1}{2}$ lb per
gal of fuel oil.

**26-4.** Assume that relocation of this line will eliminate curvature and rise and fall.
Also assume that the relocation will save 10,000 ft of distance, thus moving the end
of the run back to Sta. 110. Establish a uniform grade from Sta. 10 to 100. Operate
at full power from Sta. 0 to 106 and bring the train to a stop at Sta. 110. Determine
the saving in time and in fuel.

## BIBLIOGRAPHY

"AREA Manual," vol. 2, chap. 16, American Railway Engineering Association,
Chicago, 1953.
Smart, V. I.: The Use of Velocity Profiles, *Railway Review*, **74**(5):209–215 (1924).

# 27
# Size and Number of Trains

**27-1** **Most economical speed** The maximum train will cover a given run in a definite period of time. The same locomotive will haul somewhat less load in a considerably shorter time. The most efficient tonnage rating, or the economic speed, results in the transportation of the largest possible amount of freight per day per locomotive over the specified run. It corresponds to the maximum value of the product of average speed multiplied by the train weight. In the case of a single-track line the determination of economic speed is complicated by the time lost on sidings to allow passing of trains headed in the opposite direction. For either single-track or double-track a thorough investigation of economic speed requires study of actual speeds over the existing or proposed profile, as well as a knowledge of any special local-schedule requirements.

When the total tonnage to be moved in one direction between two division points is known, an approximate solution to the problem of determination of size and number of trains, assuming no schedule interference, may be made by determining an average grade—the differ-

ence in elevations of the beginning and end of the run divided by the
length of the run plus an adjustment for curvature—and calculating
the maximum train weight in tons that can be hauled on this grade at
various speeds in miles per hour by a specified locomotive. The product
of speed and train weight then gives the rate of haul in ton-miles per
hour, and a plot of this rate against speed will indicate the most economical speed—giving the maximum rate of haul—from which the most
economical size of train may be determined for hauling by a particular
locomotive.

A tabulation such as the following will be effective in handling the
problem:

| (1) | (2) | (3) | (4) | (5) | (6) |
|---|---|---|---|---|---|
| Speed, mph | Net drawbar pull, lb | Train resistance, lb/ton | Train weight, total tons | Cargo weight, net tons | Net rate, ton-mile/hr |
| 10 | | | | | |
| 20 | | | | | |
| 30 | | | | | |
| 40 | | | | | |

(1) At least four points should be plotted so that a smooth curve can be established
on the plot of rate versus speed. It is likely that in most cases the economical speed
will be between 10 and 40 mph, so that this range of speed in increments of 10 mph
may normally be used.

(2) The net drawbar pull of the locomotive is tabulated in column 2 for the particular locomotive to be used on the run. The net drawbar pull will be the driver tractive
effort minus the locomotive resistance (from Table 25-1 or from tabulation for previous
problems) and minus the grade resistance for the equivalent average grade which has
been established for the run.

(3) The total train resistance in pounds per ton, including frictional resistance—
from Table 25-2—and grade resistance on established equivalent average grade is
tabulated in column 3.

(4) The total weight of train in tons that can be pulled at various speeds is then
determined as column 2 divided by column 3 and tabulated in column 4.

(5) The net cargo weight is then determined by the size and weight of freight cars
used. For example, if 60-ton freight cars, weighing 20 tons empty, are used, then
the net cargo weight for the train is two-thirds of the total train weight. This is
tabulated in column 5.

(6) The rate of haul, column 6, in ton-miles per hour, is the product of columns 1
and 5. A plot of rate of haul versus speed will then indicate the most economical
speed to be used.

A study of economic speed in connection with the problems at the
end of this chapter will indicate the need for freight classification in the
yards, particularly as between "fast-freight" and "slow-freight" oper-

ation.   The factor of time is important, for instance, in the handling of perishable or semiperishable goods and certain "packaged" merchandise shipped in less than carload lots.   In competition with other forms of transportation for this service, the economic speed for handling such fast freight will be the *fastest* safe speed, as in the case of passenger service.   Empty freight cars may also be handled as special trains where the volume of empty-car movement warrants it.

It is significant, then, that the bulk of freight volume is most economically hauled at a comparatively low speed.   A reduction in average grade (say, by reduction in curvature on upgrade operation) would also tend to reduce the economic speed.

## QUESTIONS AND PROBLEMS

**27-1.** The difference in elevation between two division points 100 miles apart is 1,520 ft. Curvature amounts to 1,600 deg of central angle.   It is assumed that the rise and fall, or irregularity of the grade line over the route, is not excessive.   Establish an equivalent average grade for the run, in the uphill direction.

**27-2.** The freight volume to be hauled in the uphill direction on the above route is 30,000 tons per day.   Determine the optimum size and number of trains to be operated daily in that direction, using 120-ton freight cars which weigh 30 tons empty and the locomotive of Prob. 26-2.   Assume no schedule interference.

## BIBLIOGRAPHY

"AREA Manual," vol. 2, chap. 16, American Railway Engineering Association, Chicago, 1953.
Heskett, J. Z.: Tonnage Ratings for Diesel Power, *Railway Age*, **129**(14):21–26 (1950).
Raymond, W. G., H. E. Riggs, and W. C. Sadler: "Elements of Railroad Engineering," John Wiley & Sons, Inc., New York, 1947.

# 28

# Economics of Relocation

**28-1  Problems of relocation**  In preceding chapters some basic characteristics of train performance have been discussed, with a view toward
establishing principles and methods by means of which problems of railroad location and relocation may be analyzed.  In this chapter a summary review will be made of factors affecting train performance and
the influence these factors may have on choice of alternate routes and
operating procedures or on decisions regarding the economic feasibility
of grade and alignment improvements.  The main factors—curvature
reduction, grade reduction, and distance reduction—as affecting operating
costs will be considered separately for purposes of clarity, but it should be
understood that problems of train operation are all interrelated and
must often be considered together.

**28-2  Curvature reduction**  The advent of high-speed operation—upward
of 70 mph for both passenger and fast-freight services—has made the
problem of curvature reduction one of importance.  Fast train operation
provides the type of service essential to meet the competition of other

forms of transportation. Such service is most nearly attained when the given high speed can be maintained throughout the length of train run. This means that maximum curvature must be reduced to the point where the desired speed can be continuously maintained without excessive superelevation.

The question that arises then is, What constitutes excessive superelevation? If only fast passenger and freight trains need to be considered, then superelevation—usually constructed as elevation of the outer rail—may be made as much as about 9 in. Reference to Fig. 21-11 shows that, theoretically, for continuous operation of 65 mph the maximum degree of curvature to be specified for the line should be 3-deg curves, if 9 in. of superelevation may be allowed. Curves built to this specification would, however, be safe for maximum speeds of about 70 mph; that is, maximum safe operating speeds may be taken as 5 to 8 mph faster than that indicated by the curves of Fig. 21-11, because some considerable safety against overturning or outward derailment is provided by the inward moment of a train by virtue of its own weight on a superelevated track, by rail friction, and by the low center of gravity of modern equipment.

Ordinarily it will not be feasible to build separate tracks for fast trains and slow trains; instead, slow heavy-freight trains will generally operate on the same track with passenger trains. For this reason, as explained in Sec. 21-10, it will be necessary to keep the maximum elevation of outer rail at about 8 in., thus limiting curvature to very flat curves where high-speed operation is to be continuously maintained.

Two basic differences in analysis of various relocation problems should be recognized. First, problems concerned primarily with speeding up of schedules for passenger, mail, express, and fast-freight services are problems arising mainly from competition for these services with other forms of transportation. Full and detailed economic justification for each item of improvement is not sought; but rather the over-all cost of establishing desired schedules must be weighed against anticipated gains from high-speed service or against the possible loss of services already established. It is primarily in this first group that problems of curve reduction for speeding up schedules fall. The velocity profile and speed-time-distance tabulations may be used to analyze the effects of reduced curvature on speed of train operation.

Secondly, problems concerned with cost of train operation are of greatest significance in heavy-freight movements. Reference to Probs. 26-3 and 26-4, considering the economical speed for heavy-freight movement, will show that high-speed operation is not effective in increasing the rate of haul. Nor is competition with other means of transportation so vital a factor as in the case of passenger, mail, and express services.

Here economic justification for various items of improvement may well be sought and is usually readily obtained on the basis of saving in operating costs. Chief among the problems of this second category are those of grade and distance reduction. It will readily be seen that there is no real dividing line between problems of relocation with respect to high-speed services and with respect to slow-freight movement. That is, high-speed train operation is also affected by grade and distance reduction, and slow-freight service is affected by curvature reduction to the extent that total train resistance is reduced. A separation is made in this discussion only to bring out the inherent difference in analysis of problems concerned with heavy-freight movement as compared with those concerned with passenger and other high-speed services. It should be remembered that for the most part both are operating on the same track.

**28-3   Grade reduction**   The most economical speed for a given passenger locomotive and train to operate up a 2 per cent grade is the speed for which the acceleration grade for this locomotive and train is 2 per cent. In other words, it is the maximum speed of operation up the 2 per cent grade. True, of course, only if the train may operate safely at this speed, and such safety of operation is in turn dependent primarily on existing curvature. Thus acceleration grades are ruling grades for passenger and other high-speed service, so far as maximum speed—not necessarily safe speed—of operation is concerned (Prob. 26-3).

Strictly, ruling grades are established for slow-freight operation—usually 10 mph—as discussed in Sec. 26-2. The economic savings resulting from reduced gradient or lowering the rise and fall of an existing grade line is capable of analysis by use of speed-time-distance calculations and the velocity profile. This saving in operating cost may then be capitalized at current interest rates. Thus justification of expenditure for improvement may be determined.

Where rise and fall of a grade line are concerned, the energy used in lifting the train through a vertical rise is, to some extent, utilized in operating the train in the corresponding fall. In this connection careful study must be made of any downgrade operation with respect to cost analysis or speed-time-distance calculations. The average resistance of an average freight train operating at 15 mph may, for the sake of example, be taken at 5.5 lb/ton. On a minus 0.275 per cent grade such a train would just hold its own as far as speed is concerned. On a lesser minus grade it would lose speed or require power to maintain the speed, while on a steeper grade it would accelerate, as pointed out previously in Chap. 26. The significant fact is that this grade of repose of 0.275 per cent is very small. On a minus 1.275 per cent grade, for instance, the braking force at the rims of the wheels to hold the speed of the train at 15 mph would average 20 lb/ton of train weight.  All such braking forces continued over considerable distances contribute heavily to the cost of maintaining and replacing brake shoes.

A railroad company's cost records will show how much is spent for brake maintenance and replacements.  This cost may be a considerable item of train-operating expense.  Therefore, careful attention must be given to minimizing rise and fall as well as controlling of maximum grades. It will be recognized that problems of relocation which involve grade reduction and minimizing rise and fall between established stations are best studied, each on its own merits, by use of the velocity profile and speed-time-distance calculations, to determine comparative operating costs and possible savings.  Further generalization will not be made here.

**28-4  Distance reduction**  The most common method used in mountainous terrain for overcoming elevation with minimum gradients in early location of both railways and highways was to lengthen the grade line.  A mountain range may be deeply penetrated on easy grades by a railroad, but ultimately the crest must be crossed by overcoming the intervening elevation.  Distance may be saved by establishing "pusher" grades—excessive grades on which an extra locomotive is used to help trains surmount the barrier.  Such excessive grades make for expensive operation, both up- and downgrade.  In lieu of heavy grades the line may be doubled back and forth upon itself so as to include sufficient distance for permissible grades.

The problem is not solved, however, by the inclusion of extra distance, because extra curvature must be introduced and curvature on heavy grades must be compensated.  Thus it is not difficult to see why tunnels are frequently used to pass a mountain barrier at lower elevation. A tunnel may serve at once to reduce grades, to reduce rise and fall, to reduce curvature, and to reduce distance.  They are, however, expensive to build and maintain, and such cost must be balanced against capitalized net savings in operating costs resulting from their construction and use.

Savings resulting from reduction of distance may be determined by consideration of those items of cost which are affected by distance reduction.  In the case of a new line, both construction costs and operating cost are affected by distance.  It is quite apparent, for instance, that such construction-cost items as right of way, grading, ballast, ties, rails and fastenings, fences, and pole lines vary directly with distance.  In the case of relocation, the salvage value of the abandoned portion of line may offset to a considerable extent the cost of the relocation construction, particularly if several miles of distance are eliminated.

Railroad company records will show ton-mile costs for moving freight.  These records will be available for any study to be made of contemplated relocation projects. Published reports of ICC statistics of railroads in the United States indicate that for estimating purposes this unit cost may be taken as about $\frac{1}{2}$ cent per gross ton-mile, based on total train weight. Of this total unit cost, many items are *fixed costs*, such as general administrative costs and those required for operation of terminal facilities. These fixed costs are not materially affected by distance reduction. Selection of operating-cost items which are wholly or in part affected by distance reduction is an infinitely complicated and intricate matter, admittedly beyond the scope of this book.

A carefully selected percentage spread of train-operating expenses on Class I railroads prepared by AREA shows that changes in distance of less than 10 miles may affect about one-third of operating costs, exclusive of train wages. Thus if a mile of distance is eliminated, one-third of the operating costs, exclusive of train wages, for that mile—or $\frac{1}{3} \times \frac{1}{2} = \frac{1}{6}$ cent—is saved for each gross ton of haul.

Similarly, changes in distance of more than 10 miles may affect about one-half of the operating costs on that distance, or the saving resulting from such distance reduction would be $\frac{1}{2} \times \frac{1}{2}$ or $\frac{1}{4}$ cent per gross ton-mile for the distance in miles eliminated.

The foregoing estimates of the effects of distance reduction are based on average figures for Class I railroads where train-operating wages are not affected. Careful and detailed analysis of cost of operation must necessarily be made for a particular railroad or a given location where economic justification for relocation is sought.

## QUESTIONS AND PROBLEMS

**28-1.** A relocation resulting in a reduction of 6 miles of distance is proposed for a railroad route over mountainous terrain. What annual saving in operating cost may be expected, on the average, to result from this distance reduction if 25,000 gross tons move over the route daily, 365 days in the year?

**28-2.** In order to ensure a net cash profit to the company as a result of the relocation of Prob. 28-1, the annual saving is to be capitalized at 10 per cent. What net expenditure, over and above the net salvage value of the old road, can be made for the relocation on the basis of the saving in distance alone?

**28-3.** To the extent that time is available, work out the details of analysis concerning the economic justification for constructing a tunnel through a mountain ridge. All conditions of distance, alignment, grades, volumes of traffic, and general-cost items may be assumed; but use should be made of available contour maps of the area and pertinent traffic and cost information available from railroad company records. Emphasis should be placed on procedure of analysis with respect to use of velocity profile and speed-time-distance calculations for determination of possible economies resulting from (a) curvature and grade reductions, (b) reduction of rise and fall, (c) distance reduction, (d) speeding up of schedules, and (e) possibilities of handling increased traffic.

## BIBLIOGRAPHY

*Am. Ry. Eng. Assoc. Proc.*, **37** (1936), **42** (1941), **44** (1943), **47** (1946).

Grodinsky, Julius: "Railroad Consolidation," Appleton-Century-Crofts, Inc., New York.

Raymond, W. G., H. E. Riggs, and W. C. Sadler: "Elements of Railroad Engineering," 6th ed., John Wiley & Sons, Inc., New York, 1947.

"Statistics of United States Railways," Interstate Commerce Commission, 1949 and 1950.

Wellington, A. M.: "The Economic Theory of Location of Railways," 6th ed., John Wiley & Sons, Inc., New York, 1898.

White, Joseph L.: "Analysis of Railroad Operations," Simmons-Boardman Publishing Corporation, New York, 1950.

# 29
# Urban Rail Transit

**29-1  Demand for rapid transit**  Interregional high-speed ground transportation connecting major urban centers has become a subject of national and international interest.  Railroads, by making use of available technology for roadway improvements and for modernization of rolling stock, are playing an increasingly important role in providing rapid transit service between major cities that lie within 500 miles of each other.  The trains of Japan's new Tokaido railroad cruise at 130 mph to provide fast, comfortable passenger service between Tokyo and Osaka, a distance of about 320 miles.  In Canada, the readiness of public response to improved rail service is being demonstrated by the new Canadian National train service operating between Toronto and Montreal, a distance of about 335 miles.

In the United States, interest in fast trains is spreading to state governments.  The new trains, providing 125-mph passenger service for medium-distance journeys, would be competitive with airline schedules on a downtown-to-downtown basis.  A simple calculation may serve to illustrate.  Let $D$ represent the distance downtown-to-downtown and assume three-quarters of an hour of ground time at each end of the flight—between downtown and the airport. Then for the same travel time for either plane or train, with the plane cruising at 500 mph

$$\frac{D}{125} = 1.5 + \frac{D}{500}$$

From this, $D = 250$ miles—the minimum distance for which flying would be attractive, from the standpoint of travel time alone. Convenient terminal facilities in the downtown areas for the train service may extend considerably the distance for which train travel would have popular appeal, depending of course on the level of service the trains provide, with regard to such factors as convenience, comfort, cost, and safety, as well as travel time.

Creation by Congress in 1966 of the Department of Transportation has encouraged similar creation of transportation departments by the states, to approach transit and transportation problems on a regional basis. In the larger metropolitan areas, there is increasing need for rail transit and commuter systems to upgrade the facilities for moving people at higher average speeds and in greater comfort. New rail transit lines will need to mesh with existing commuter facilities, and they must also fit into the over-all planning for the region they are to serve, so as to coordinate with highway, waterway, and airport development.

Research and development in high-speed ground transportation indicate potential for speeds of 300 to over 500 mph with various combinations of vehicle, fluid suspension, propulsion, guided way, and control systems. The flanged wheel on steel rail has limitations of high speed at around 150 mph, but this is well above the present speed limit of 70 mph for highway vehicles. Thus the conventional train seems likely to succeed for some time in providing rapid transit service, for commuting traffic between a central city and the outlying towns and communities of surrounding metropolitan area and for intercity travel between the larger urban centers within a few hundred miles of each other.

Coordination and integration of intercity passenger service with local collection and distribution systems must be made so as to minimize the over-all door-to-door time of the commuter or traveler. Otherwise much of the advantage of fast train service is lost in getting to and from stations and terminals. Small buses, known as "minibuses," are proving to be popular in solving local distribution and circulation problems in downtown sections of some large cities and at the larger airports and industrial plants. This type of collection and distribution service could also prove to be effective in built-up residential sections surrounding outlying transit terminals and intermediate stations.

**29-2  Motive power**  Electrification of rail transit lines is necessary for operation in urban transportation corridors because of its advantages of cleaner, quieter, and generally more efficient operation.  Direct current with line voltage in the order of 600 to 800 volts is generally favored. The direct-current traction motors have the particular advantage of high starting torque.  Direct-current equipment is normally available at 750 volts.  With this system voltage, dynamic braking is feasible without complications.  The system is supplied with direct-current power at frequent intervals throughout its length.  Three-phase alternating current is distributed to trackside substations where it is converted to direct current and supplied to the traction motors via a third—or power—rail.

Because of high power requirements for high-speed operation of transit trains, individually motorized cars are used in preference to the locomotive and train-of-cars arrangement.  Generally, a direct-current electric traction motor is geared to each axle of the car.  Thus, it may operate as an individual unit or in a train.  The leading unit of a train houses the controls and operator, and it also breaks the air resistance for the trailing units.  This latter function becomes an important factor at high speeds—when air resistance consumes a large portion of available tractive effort.

**29-3  Performance**  A discussion of capabilities and limitations of a train of electrically motorized cars with flanged steel wheels on conventional steel rails will be taken up in connection with some assumed requirements of service demand.  Suppose, for instance, that a rail transit line is to extend from the central business district—CBD—of a city to an outlying terminal 8 miles distant with one intermediate station approximately midway.  Demand for commuter service in this corridor has been determined to be 30,000 passengers in the peak periods of 7 to 9 A.M. and 4 to 6 P.M.  This establishes a rate of 15,000 passengers per hour for the peak traffic.

In order to make the service convenient, it is decided that the headway—time between trains—will be 2 min.  This, then, requires that each train will accommodate 500 passengers, and in the interest of comfort, all passengers will be seated.  The cars are 66 ft long and each will seat 72 passengers.  Thus each train will be composed of seven cars: one leading unit and six trailing units.  The length of the loading platform is in the order of 460 ft.

In order to maintain a high average speed over the line, acceleration and deceleration will be made as fast as possible.  For seated passengers, the maximum values may be taken at $\frac{1}{4}$ g, or about 8 ft per second per second.  For standees, maximum acceleration and deceleration is generally limited to about 5 ft per second per second or about 3.4 mph per second.  The maximum of 8 ft per second per second—5.45 mph per second—also utilizes the full rail friction or traction available, assuming the coefficient of friction for steel on steel at 0.25.  That is, the available tractive effort is one-fourth of the weight on the driving wheels.  Thus in order to develop maximum acceleration, all the weight should be on driving wheels with no idler wheels.

For this example it will be assumed that each loaded car weighs 30 tons, for a total train weight of 210 tons.  Each car is mounted on 2 two-axle trucks and each of the four axles is driven by a direct-current electric motor.  Available tractive effort for the train, at 500 lb per ton, amounts to 105,000 lb and in order to accelerate to high speed in a relatively short time, the traction motors will need to have the power capability to develop this available tractive effort at a fairly high speed, say 30 mph.  Then from Eq. (23-2)

$$P = \frac{30 \times 105,000}{375} = 8,400 \text{ hp}$$

for the train, or 1,200 hp per car.  Each axle, then, is provided with a traction motor capable of developing 300 hp at the driving wheels at this speed.

From the basic performance equations for direct-current electric locomotives, Eqs. (24-1) and (24-2), the tractive-effort and power curves in relation to speed may be drawn.  The curve of the tractive effort versus speed for this example train is shown in Fig. 29-1.

Shown also in the figure is the curve for total train resistance versus speed.  Train resistances are computed from the Davis formula Eq. (25-1), with $J = 0.03$, $A = 120$ sq ft, and $K = 0.0024$ for the leading unit and $K = 0.00034$ for the trailing units.  A tabulation of these resistances is given in Table 29-1.  Note the influence of high air resistance on the leading unit at high speeds.

Figure 29-1 shows the limit of speed capability of the train to be a little over 80 mph on a level tangent track.  Since continuous downgrade operation is impractical on a round-trip basis, it will be necessary to increase power for higher speed, by increasing the power rating of the traction motors and by using three-axle trucks instead of two-axle trucks.

The power rating of the traction motor is generally established for a given speed of rotation—say 1,000 rpm for the 300-hp motors previously discussed in this section.  The torque is transmitted to the driving axles through a system of gears to establish a desired speed of the train—30 mph for the example train.  The diameter of the driving wheels is generally in the range of 36 to 42 in.  For an assumed wheel diameter of 40 in., the speed of rotation of the driving wheels—and axles—is a little more than 250 rpm.

Thus a gear ratio of 4 to 1 is established for the example, since it is required that each motor be capable of developing 300 hp at the driving wheels at a train speed of 30 mph. The efficiency of the transmission system is likely to be in the order of 90 per cent, consequently a motor of somewhat higher horsepower rating—say 330 hp—may actually be specified for this train.

In order to increase performance capability of the train, as illustrated in Fig. 29-1, to a maximum speed of 150 mph on a level tangent track, it is necessary to increase the tractive effort at that speed by about 14,000 lb to overcome train resistances. A booster engine could supply this

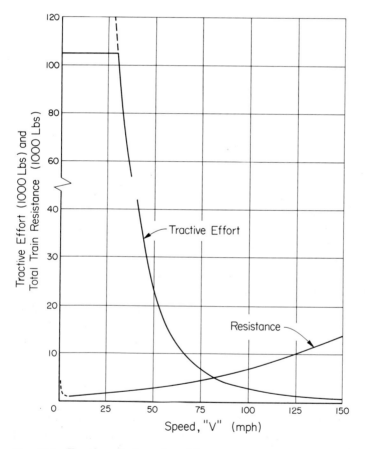

**Fig. 29-1** Tractive-effort and resistance curves, seven-car passenger train, 8,400 hp; tangent, level track.

Table 29-1 Train resistances in pounds per ton for motorized passenger cars (level tangent track)

| $V$, mph | Leading unit | Trailing unit |
|---|---|---|
| 5 | 5.5 | 5.3 |
| 10 | 6.4 | 5.5 |
| 20 | 9.5 | 6.2 |
| 30 | 14.6 | 7.2 |
| 40 | 21.6 | 8.4 |
| 50 | 30.6 | 10.0 |
| 60 | 41.5 | 11.7 |
| 70 | 54.2 | 13.8 |
| 80 | 69.0 | 16.1 |
| 90 | 85.4 | 18.7 |
| 100 | 104.1 | 21.7 |
| 110 | 124.4 | 24.6 |
| 120 | 146.8 | 28.2 |
| 130 | 171.3 | 32.0 |
| 140 | 197.4 | 35.9 |
| 150 | 225.6 | 40.2 |

additional effort.   Present development in rail transit technology includes the use of aircraft turbojet engines to supply the thrust needed for high-speed operation.   In this case, one such engine with a maximum thrust rating of 17,500 lb mounted on top of the leading unit would provide the necessary additional force for 150-mph operation.   These engines are generally operated at a sustained thrust of about 80 per cent of maximum.

Turbojet engines with maximum thrust capability of 60,000 lb are presently in development stage.   Two such engines could adequately supply the thrust needed for maximum acceleration and operation for the train under consideration here.   Electric traction motors would be needed only for departure and approach to stations and terminals, and very likely, in the tunnel sections of the transit line.   Use of these engines reintroduces the problems of noise, blast effects, and air pollution, but even so they may play an important participating role in power develop-ment for rail rapid transit.

**29-4  Station spacing**  Ideally, from an operational standpoint, the rapid transit train should accelerate at maximum rate to its top speed, operate at this speed for as great a distance as practicable, then decelerate at maximum rate.   In order to take advantage of high operating-speed capability, thus establishing higher average speeds for commuting service, station spacing becomes more critical.   Suppose, for instance, that a train is sufficiently powered with booster engines to accelerate to 150 mph at $\frac{1}{4}$ g, or 5.5 mph per second. This would require 27.3 sec of time and 3,000 ft of distance. Allowing the same time and distance for deceleration, about $\frac{1}{2}$ min for stop time at the station, the total time for the stopping maneuver would amount to about $1\frac{1}{2}$ min; whereas, a "through" train could negotiate the 6,000 ft of distance at top speed in 27.3 sec, or about $\frac{1}{2}$ min. Thus, the cost of the stop in time is about 1 min, or a cost in travel distance at top speed of about $2\frac{1}{2}$ miles. From this standpoint then, it seems reasonable to establish a desirable minimum station spacing for this operating speed at, say, 3 to 4 miles.

As rapid transit technology develops to provide top vehicle speeds of 300 mph on fluid suspension (air cushion) in guided ways, or at 1,000 mph with jet propulsion in transit tubes, desirable minimum station spacing becomes proportionately larger. Thus rapid transit, operating at higher average speeds, becomes increasingly a regional service and plays an increasingly important role in regional planning, serving transit needs of a central city from outlying communities and towns and from distant larger cities and urban areas, to a radius of several hundred miles.

Specific location of terminals and intermediate stations for a rail transit line must provide convenient access to the people of the communities they are to serve. Collection and distribution of passengers at both outlying and downtown stations require transport facilities that will effectively minimize door-to-door travel time for all passengers whose place of residence or place of work or business is not within convenient walking distance. At outlying stations, adequate car-parking facilities will be needed for park-and-ride traffic. Effective—and frequent—circulation of minibuses is growing in popularity and demand for collection and distribution of transit passengers at both central and outlying transit stations as well as for circulation of shopping and business traffic to relieve congestion in central business and industrial areas.

**29-5  Control of grades**  High-speed train operation will generally require that grades—and curvature—be kept at a minimum. Determination of grade and curvature resistances are discussed in Chap. 25. Reference to Fig. 29-1 will show that, with electric power, the tractive effort available for overcoming these resistances reduces rapidly as speed increases. Control of grades at minimum values is generally accomplished by a combination of elevated, at-grade, and tunnel sections as dictated by the topography of a particular route.

For aesthetic reasons and because of noise and vibration  generally associated with high-speed rail operation, the tunnel section, or subway, may be desirable—or required—through central business districts and built-up residential areas. In such cases, terminals and intermediate stations may also be placed underground.   However, placing the stations somewhat above the elevation of the main-line rails offers a distinct advantage of rise and fall in the stopping and starting of trains.   For example, the elevation of the rails may be 24 ft below street level in the tunnel and 16 ft (one story) above street level at the station.   The vertical rise and fall of 40 ft will afford a considerable assist for stopping and starting trains.   To control accelerations, the ramp grades for approach and departure should not exceed 20 to 25 per cent.

The elevated station will also leave much of the street area open for circulation of street vehicular and pedestrian traffic and for automobile parking.   Access to the loading and unloading platforms is generally provided by some combination of stairways, escalators, moving belts, and elevators.   In the case of a downtown station or terminal, the loading platforms may tie in directly with the second floor of a centrally located transportation terminal building where coordination of all transport facilities may be effectively carried out.

**29-6   The velocity profile**   Train performance on a particular route involving grades and curvature is best represented and analyzed by use of the velocity profile.   The procedure for this is given in Chap. 26, from data tabulated as illustrated in the following outline:

| (1) | (2) | (3) | (4) (5) Net tractive effort | | (6) |
| :---: | :---: | :---: | :---: | :---: | :---: |
| $V$, mph | $T$, lb | Locomotive resistance, lb | lb | lb/ton | Acceleration grade, % |
| 0 | | | | | |
| 5 | | | | | |
| 10 | | | | | |

(1) Tabulate speeds from 0 to maximum speed in increments of 5 mph.
(2) Record tractive effort determined in accordance with Sec. 29-3.
(3) Record train resistances for operation on tangent, level track.
(4) Column (2) minus column (3) gives net tractive effort available for increasing energy head of the train.   Record this in total pounds in column (4).
(5) Record net tractive effort in pounds per ton of total train weight.
(6) The acceleration grade—the maximum grade in per cent up which the train can operate at the given speed—is determined by dividing the figures of column 5 by 20 in accordance with Sec. 25-2.

Plotting of ground-line and velocity profiles proceeds in accordance with procedure discussed in Chap. 26.

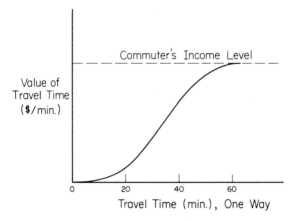

**Fig. 29-2** Approximate relationship of commuter's evaluation of work trip (one way) to travel time.

**29-7  Time and energy calculations**  A tabulation of performance data from the velocity profile may then be made, as outlined in Chap. 26. Summations of time and energy for all or any part of the route make possible a comparative analysis of the effects of line changes such as elimination of curvature, grade reduction, distance reduction, and elimination of rise and fall.

The value of time is of paramount importance to the prospective customer of a rail rapid transit service. The service must in general compete with the automobile in door-to-door travel time for the commuter. It is useful to consider an approximate relationship between the average commuter's evaluation of time and door-to-door travel time. Questionnaires and personal interviews in many studies of commuter traffic throughout the country have indicated that the average commuter is satisfied with a travel time of 15 to 20 min in getting to or from his place of work. After about 20 min he loses patience rapidly, so that somewhere between $\frac{1}{2}$ and 1 hr of travel time (one way) he begins to think of this as part of his workday, consequently approaching in value his rate of income. This general relationship is depicted in Fig. 29-2.

The velocity profile procedure presents data requirements for systematic analysis and design of the rail transit facility. Average speeds and travel times for various segments of the system may be determined from performance characteristics of a particular train operating on selected routes with a variety of conditions of alignment, grades, and distances. For instance, the distance of the remote terminal of a rail line essentially serving commuter traffic, from the central employment area so served, may be determined as the distance traveled in 20 min, in accordance with Fig. 29-2.  It will be important to minimize the time for collection and distribution of passengers at both ends of the trip.  At best, this may average, say, 4 min at each end of the trip, leaving 12 min of train travel time.  Then, if an average operating speed of 60 mph can be maintained for the trains, the remote terminal may be 12 miles from the central employment area.

If the trains have a maximum speed capability of around 150 mph, and the trains from the remote terminal are operated express—without stops at intermediate stations—at an average speed of 120 mph, then the radius of such transit service could extend to 24 miles from the central area, for a 20-min commuting service.  Average operating speeds above 40 mph will generally compete favorably with automobile travel in metropolitan areas.  The importance of providing effective collection and distribution service for passengers becomes apparent, if door-to-door travel time is to be kept at a minimum.

**29-8  Number of trains**  Two-way operation of trains on a rail transit facility will generally consist of a loop extending from an outlying terminal to the central terminal, with provision at the outlying terminal for turning around or reversing and switching of trains.  Similar provision is required at the central terminal unless the facility is extended from there in other directions.  The length $L$ of a loop in miles will be twice the distance in miles between the terminals.  The average distance between trains operating on the loop is equal to the product of average train speed $S$ in miles per hour and the headway $H$ in minutes divided by 60.  The number of trains required for continuous operation on the loop is then determined as

$$N = \frac{60L}{HS} \tag{29-1}$$

Since continuous operation at full capacity will generally be limited to peak periods $T$ in hours (usually $1\frac{1}{2}$ to 2 hr) of heavy commuter-traffic demand, the maximum number of trains required is

$$N \text{ (max)} = \frac{60T}{H} \tag{29-2}$$

This maximum number of trains would be required only for very long runs where the length of the loop is equal to, or greater than, the product $ST$.  It is of interest to note that, from the standpoint of investment in trains, the required number of trains varies inversely with average operating speed.

**29-9  Evaluation of transit service**  In general, transit service competes with the automobile in meeting transport needs of commuters, shoppers, and of trips for business, governmental, and other services located in central districts of business and industry.  Consequently, the potential user of a transit facility is likely to evaluate the facility in terms of the level of service that he enjoys from the use of his private automobile. Usually, his need for a rapid transit service arises from limitations of parking space in the central areas and from limited street capacities in these areas, resulting in vehicular traffic congestion.

Since they must compete in traffic with automobiles, buses may generally provide an even slower service because of the need to make intermediate stops at outlying points along the routes as well as within the central areas.  Rail transit, operating on separate right of way, can offer a rapid service that may offset some of the disadvantages of mass rapid transit sufficiently to make it attractive to more people in outlying communities.

Principally, evaluation of a rail transit service becomes an evaluation of the work trip by the commuter.  In determining the cost of his work trip he should consider costs associated with such factors as inconveniences, discomfort, and safety.  Although it is difficult to place a dollar-and-cents value on some of these factors, it is a necessary step in determining not only the need for transit service but also the level of service to be provided if it is to prove effective and acceptable within a community considering adoption of such service.

Chief among the factors to be compared in making a choice of mode of travel by commuter are:

1. Travel cost
2. Convenience
3. Comfort
4. Dependability
5. Travel time
6. Safety

The commuter may normally determine travel cost as the amount he pays in fares for transit services, including those for getting to and from trains.  Or, if he drives his automobile, the "out-of-pocket" costs for gasoline, oil, tires, minor repairs, and parking fees are considered to be directly chargeable to commuting travel.  A charge for safety considerations may be conveniently included in this travel cost for the commuter using his automobile by prorating his insurance cost on a mileage basis.

The value of travel time to the commuter may vary considerably among individuals. His income level will normally have an important bearing on this, as discussed in Sec. 29-7. The factors of convenience, comfort, and dependability all affect his evaluation because he will need to consider door-to-door travel time for his work trip. Convenience is closely associated with getting to and from trains, as compared with getting into his automobile at his door or to a parking lot at his place of work. Dependability of service is related to the time between trains. It is important that this headway be kept at $1\frac{1}{2}$ to 3 min at the terminals.

The factor of comfort is more intangible, but it has an important bearing on the commuter's evaluation of a transit service. Here again, his basis for comparison and evaluation is his private automobile, and this comparison may greatly affect his determination of the value of time in estimating the cost of his work trip. Travel time may have real utility for the commuter. Generally, this time provides an interval of transition from the cares of the commuter's domestic life to those of his employment or occupation. In his automobile, he can do this in comfortable seating, listening to his radio or conversing with a fellow traveler, or contemplating the many facets of his environment. At what point he begins to consider travel time as a part of the cost of his work trip may vary considerably among commuters. But the comfort in which he travels may greatly influence the value to him of travel time beyond, say, 15 to 20 min—or $\frac{1}{2}$ hr under comfortable conditions.

Rail transit service, then, should provide comfortable seating for all passengers. The need for this was also established earlier in this chapter in the discussion of train performance. Beyond this, the level of service—and cost of service—may be determined by finding answers to such questions as:

1. Should radio be available for individual listening by the passengers?
2. Should the morning paper be available?
3. Should coffee service be provided for the early traveler?
4. Should the evening paper and perhaps a cocktail be available for the homeward journey?
5. Should a steward or stewardess administer these amenities?

Roughness and noise must of course be kept at low levels—smooth enough for coffee service and quiet enough for normal conversation within the cars. The continuously welded rail has done much to affect both of these requirements. Also, in order to improve roadbed tolerance for high-speed operation, investigations are being made into the use of plastic foam to replace the rock ballast directly under the ties.

## QUESTIONS AND PROBLEMS

**29-1.** Tabulate data in preparation for plotting the velocity profile for the seven-car, 8,400-hp rapid transit train discussed in the early part of this chapter.

**29-2.** On a sheet of profile paper, using a scale of 1 in. = 40 ft (vertical) and 1 in. = 2,000 ft (horizontal), plot the velocity profile for the train of Prob. 29-1 on the following track profile, starting with zero speed at Sta. 0, and using 1,000-ft stations.

*Station to station*                            *Track grade and curvature*

0 –10......... Level, no curvature.  Determine speed at Sta. 10.

10–25......... 2.00 per cent grade.  Determine speed at Sta. 25.

25–40......... 1.00 per cent track grade.  450 deg of curvature.  Note that equivalent grade is 1.12 per cent.  Operate on this grade at a maximum speed of 60 mph.  Make 1-min stop at Sta. 36.  (Assume accelerations at 5 mph per second.) Determine tractive effort for operation at 60 mph on 1.12 per cent grade.  Adjust track (and velocity) profile to ground line at Sta. 40.

40–60......... Minus grade of 1.00 per cent.  Operate on this grade at maximum speed of 90 mph.

60–70......... Level.  No curvature.  Operate at full power.  Bring train to stop at Sta. 70.  Decelerate at 5 mph per second.

**29-3.** Determine time and energy requirements for the one-way operation of Prob. 29-2. Calculate kilowatt-hours of electricity required, assuming 10 per cent line and substation loss.  Follow procedure outlined in Chap. 26.

**29-4.** Assume that by use of a combination of elevated, at-grade, and tunnel construction of the rail line of Prob. 29-2, a uniform track grade may be established between Sta. 10 and Sta. 60.  Also assume that curvature as well as rise and fall will be eliminated and that distance will be reduced by 8,000 ft, thus placing the terminal at Sta. 62 (at same elevation) instead of Sta. 70.

Operate at full power on this relocated grade and alignment, include the 1-min stop at Sta. 36, and determine the savings in time and energy resulting from this relocation.

**29-5.** Suppose that you are to estimate the commuter demand for rail transit service in an outlying community of known population density located, say, 12 to 15 miles from a central employment area.

*a.* Prepare a questionnaire for distribution in the community to determine the need for mass rapid transit and also to establish the level of service that would be competitive with existing modes of commuting.

*b.* Select a site within the community for the terminal, providing convenient access to the greatest number of potential passengers, with sufficient area to accommodate park-and-ride commuter traffic.  Note that convenient access to freeways and major arterials may attract park-and-ride passengers from beyond the particular community.

*c.* Determine the routing and approximate scheduling of minibuses to provide 5-min door-to-train service within the community.  This will largely determine the extent of area (and population) served by the terminal.

**29-6.** Prepare programs for computer application of data for:

*a.* Analysis of questionnaire data to establish cost of work trip in accordance with discussion in Sec. 29-8.

*b.* Analysis of velocity profile data for performance of trains under various restraints as indicated in Prob. 29-2.

## BIBLIOGRAPHY

Adams, Warren T.: Factors Influencing Transit and Automobile Use in Urban Areas, *Highway Research Board Bull.* 230, Washington, 1959.
"Chicago Area Transportation Study," Final Report, Chicago, 1959.
Daniel, Mann, Johnson, and Mendenhall: "A Comparative Analysis of Rapid Transit System Equipment and Routes," Report to Los Angeles Metropolitan Transit Authority, Los Angeles, 1960.
Lang, A. Scheffer, and Richard M. Soberman: "Urban Rail Transit, Its Economics and Technology," The M.I.T. Press, Cambridge, Mass., 1964.
Meyer, John R., John J. Kain, and Martin Wohl: "Technology and Urban Transportation," A Report to White House Panel on Civilian Technology, July, 1962.
Schwartz, Arthur: "Forecasting Transit Use," *Highway Research Board Bull.* 297, Washington, 1961.

# part four

# River and Coastal Engineering

# 30
# Water Transportation

**30-1 Introduction** Every civil engineer should be aware of the historic importance of waterways in the development of civilizations and in the growth of nations. Rivers and seas provided primitive man with his first facilities for the mass transportation of goods, and even to this day they integrate the most complex economic and political associations of people. These considerations are important to the engineer because they indicate that the problems of waterway engineering represent the most enduring of markets for engineering services. They are important as well because a moment's reflection on the fundamental nature of this form of transportation may reveal a basic principle that must underlie all successful waterway engineering.

A clue to the inherent importance of water transportation lies in the paradox that a river is at the same time both a transporting agency and the material being transported. Only water channels its own roadbed and strives constantly to improve it for more efficient transportation. In the natural scheme of things only water routes are ready-made—a fact which locating engineers for both railroads and highways have utilized fully. River and harbor engineering is concerned with efforts to make the best possible use of the natural waterways: to maintain channels of adequate depth, to provide terminal facilities and harbors, and to extend or interconnect the natural channels by means of canals. In much of this work there are two distinct methods of attacking the problems involved. The obvious method is the direct one, for example, to dredge a ship channel in the desired location and to exercise continual maintenance. The other method stems from a realization that the existing channel always represents a temporary equilibrium under the action of the forces of nature; rather than ignoring or opposing these forces, it seeks to direct them so as to accomplish the desired end. Thus, instead of dredging, a dike might be constructed at less cost to divert the current in such a fashion that it will scour out the required channel. Work of this sort offers opportunity for the exercise of the highest type of professional ingenuity, and when successful, it provides the fullest measure of professional satisfaction.

In the United States, river and harbor engineering represents a major field of engineering activity. The total investment of the Federal government in channel improvement and harbor works amounts to around 3 billion dollars. Public and private terminal facilities have cost another $3\frac{1}{2}$ billion.[1] Nevertheless, professional training in these fields has received little attention in our educational institutions, because only in recent years have scientific advances in river hydraulics and coastal engineering provided a substitute for practical experience in the exercise of professional judgment.

Federal expenditures for this work are administered by the Defense Department through the U.S. District Engineer Offices, with many hundred civilian engineering employees. Local political subdivisions handle port-development work through various kinds of commissions and bureaus. Most private construction probably is designed and supervised by consultants, but the larger firms will have their own engineering staffs. The interest of the railroads in this field is evidenced by the fact that one of the 28 standing committees of the Association of American Railroads is their Committee on Waterways and Harbors.

**30-2  Inland waterways** River boats and canal barges were very important carriers in the early part of the nineteenth century. Before America's inland waterways had time to reach the stage of development achieved in the more mature economy of Europe, their growth was choked off by the onset of the railroad age. Nevertheless, this form of transportation persists to the present day, providing employment for about

[1] "Transportation and National Policy," p. 47, National Resources Planning Board, 1942.

50,000 wage earners.  The investment in boats and barges amounts to a quarter billion dollars, and the Federal investment in river and canal works is well over a billion.  A controversial aspect of river traffic is the subsidy which exists in the fact of governmental construction and maintenance of waterways.  The wisdom of national policy on river works is not properly an issue in these pages; the fact that this policy creates demand for many thousand man-years of professional engineering employment is sufficient justification for reviewing some of the technical problems which are characteristic of river engineering.  One should not assume that river traffic is wholly dependent upon governmental concessions.  For some products, such as oil, coal, and gravel, and under favorable conditions, the cost of river transportation is so much below the competition that it seems likely that river carriers have a permanent if modest place in transportation, at home as well as abroad.  This view is strengthened by the fact that in 1951 the number of towboats and barges operating on the Mississippi River system and the Gulf intracoastal waterway experienced a 6 per cent increase over the 9,139 vessels operating at the close of 1950.

The inland waterways of the United States total about 24,000 miles, half of which comprise the Mississippi River system.  Eighty per cent of the remaining half is made up by Atlantic and Gulf Coast rivers.  In addition, there are 3,000 miles of intracoastal waterway, stretching along and parallel to the Atlantic and Gulf Coasts.  About 40 per cent of the total length of these inland and intracoastal waterways has an established channel depth of less than 6 ft; 40 per cent range from 6 to 12 ft in depth; and the balance are deeper than 12 ft.

The barges used on American rivers are from 100 to 300 ft long, 20 to 50 ft wide, and carry from 285 to 3,000 tons of cargo.[1]

The civil engineer plays a central role in the design, construction, and maintenance of navigable inland waterways.  River currents must be trained, by walls and weirs and other devices, to keep to an established path and to scour a continuous channel of appropriate size.  Riverbanks must be protected by revetments, groins, and bulkheads.  Canal construction requires stable earthwork and complex locks and dams.  In all these tasks any new construction may cause the most complicated and far-reaching consequences in the behavior of the river.  Many of these activities involve rather simple examples of structural design, but they demand engineering judgment and foresight of the highest order, reinforced by an ability to use the tools of hydraulics and the theory of models.

The system of commercial waterways in the United States is depicted

[1] "Public Aids to Transportation," vol. 3, p. 57, Federal Coordinator of Transportation, 1939.

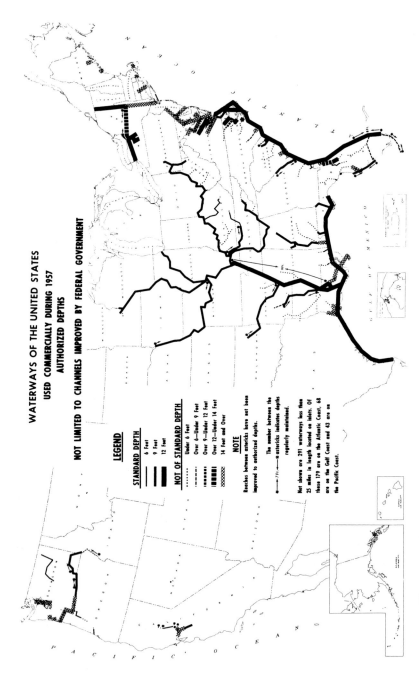

**Fig. 30-1** Waterways of the United States. (*Source: U.S. Department of Commerce.*)

Table 30-1  Commercial waterways of the United States*†

| Region | Authorized depths, feet‡§ | | | | | Total length, miles |
|---|---|---|---|---|---|---|
| | Under 6 | 6 to 9 | 9 to 12 | 12 to 14 | 14 and over | |
| | Total commercial system, miles | | | | | |
| Atlantic Coast .......... | 772 | 1,356 | 633 | 2,038 | 1,191 | 5,990 |
| Gulf Coast .............. | 849 | 688 | 1,119 | 1,214 | 464 | 4,334 |
| Mississippi Basin........ | 1,866 | 538 | 4,404 | 740 | 268 | 7,816 |
| Pacific Coast ........... | 694 | 354 | 241 | 26 | 698 | 2,013 |
| Total................ | 4,181 | 2,936 | 6,397 | 4,018 | 2,621 | 20,153 |
| Per cent.............. | 21 | 14 | 32 | 20 | 13 | 100 |

* U.S. Department of Commerce.
† Waterways were included in this table if they were used for commercial purposes at any time during the period 1956 through 1958.
‡ Depths indicated are authorized depths.  Some waterways have not been improved to the depth authorized.
§ Depths of 6, 9, and 12 feet are considered standard.  In general, the waterways included in the 6- to 9-foot class are 6 feet deep, those in the 9- to 12-foot class are 9 feet deep, and those in the 12- to 14-foot class are 12 feet deep.  There are no standard depths for waterways included in the under-6-foot class and the 14-foot-and-over class.  525 miles were improved by non-Federal action and 668 are used without improvements.

in Fig. 30-1, with status indicated as of Dec. 1, 1960.  Included in the system are all waterways used for commercial purposes at any time during the period 1956 through 1958.  Classifications of the system are given (1) according to authorized depths in Table 30-1 and (2) according to traffic densities in Table 30-2.

**30-3  The Great Lakes**  The heavy ore movement from the Lake Superior iron mines to the steel industry on the shores of Lakes Michigan and Erie make these waterways of unique importance in the industrial development of this country.  From the engineering standpoint the lake problems are quite different from those found in river engineering.  The ships themselves are highly specialized: broad-beamed, over 600 ft long, carrying 10,000 to 15,000 tons of ore.  The total tonnage of the vessels operating on the Great Lakes is a little less than that of U.S. ships engaged in coastwise and intercoastal trade.  The job of the civil engineer is primarily that of providing terminal facilities.  Harbors must be protected from wave action sometimes comparable to ocean storms.  Breakwaters differ from those built in salt water, because the greater perma-

Table 30-2  **Waterway classifications by channel depths and traffic densities***

| Depth, ft | Density class, miles of waterway[†] | | | Total length, miles |
|---|---|---|---|---|
| | A | B | C | |
| Under 6 | ..... | 474 | 3,707 | 4,181 |
| 6 to 9[‡] | ..... | 482 | 2,454 | 2,936 |
| 9 to 12[‡] | 3,064 | 1,061 | 2,272 | 6,397 |
| 12 to 14[‡] | 1,846 | 1,332 | 840 | 4,018 |
| Over 14 | 1,577 | 673 | 371 | 2,621 |
| Total | 6,487 | 4,022 | 9,644 | 20,153 |
| Per cent | 32 | 20 | 48 | 100 |

* U.S. Department of Commerce.
† Annual traffic in ton-miles per mile of waterway:
    Class A: Over 3,000,000
    Class B: 300,000 to 3,000,000
    Class C: 0 to 300,000
‡ Depths indicated are authorized depths. Some waterways have not been improved to the depth authorized.

nence of timber in fresh water gives advantage to wood-crib construction. Channel depths at Great Lakes ports range from 19 ft at Buffalo to 32 ft at Duluth-Superior.

The outstanding technological contribution in transportation engineering on the Lakes is the cargo-handling equipment. A cargo of 10,000 tons of ore is loaded in 3 or 4 hr and discharged in 9 or 10 hr,[1] perhaps a

[1] Walter Havinghurst, "The Long Ships Passing," p. 220, The Macmillan Company, New York, 1945.

Table 30-3  **Development of bulk-ore carriers***

| Year built | Length, ft | Beam, ft | Depth of hold, ft | Power, hp | Capacity, tons | Speed, mph |
|---|---|---|---|---|---|---|
| 1900 | 498 | 52 | 29 | 1,500 | 7,700 | 11 |
| 1905 | 578 | 65 | 31 | 1,800 | 9,890 | $11\frac{1}{2}$ |
| 1910 | 607 | 58 | 33 | 1,800 | 10,800 | $11\frac{1}{2}$ |
| 1920 | 620 | 64 | 33 | 2,400 | 12,000 | $12\frac{1}{2}$ |
| 1930 | 610 | 60 | $32\frac{1}{2}$ | 2,000 | 14,215 | 12 |
| 1942 | 639 | 67 | 35 | 4,000 | 18,300 | 14 |
| 1950 | 678 | 70 | 37 | 7,000 | 21,500 | 16 |
| 1953 | 690 | 70 | 37 | 7,000 | 21,900 | 16 |

* From Captain Gilbert L. Shelley, Recent Developments in Bulk Ore Carriers, *Proc. Am. Merchant Marine Conf.*, **17**:62 (1952).

fifth of the time required in ocean ports. This is a highly specialized operation and results from the great importance of turn-around time in relatively short-haul freight movements.

**30-4   Seaports**   In terms of cargo tons, water-borne traffic of the United States is about equally divided between coastwise, Great Lakes, and foreign commerce. The bulk cargo dominating Lakes traffic affords much less employment than the mixed cargo which makes shipping so important to the prosperity of seaports along the coasts. The more extensive facilities required for servicing coastal and foreign trade also provide more employment for the civil engineers who design and construct the wharves and cargo sheds that serve the port as well as the jetties and breakwaters that serve the harbor. The size and nature of port facilities must be in accord with the needs of commerce. Table 30-4 indicates the complexity of activity which characterizes the interchange of commodities among nations.

The indirect benefits of water transportation are even more important. Industrial growth is dependent upon transportation, and the 28 states adjacent to ocean and Great Lakes coast lines produce 90 per cent of the manufactures of the United States and hold 90 per cent of the country's factory workers. Twenty ports in these 28 states handle 60 per cent of the lake and ocean cargo. All of these seaports have terminals and channels to accommodate oceangoing freight vessels of deep draft. The coastal ports have controlling depths ranging from 28 ft at Seattle–Tacoma to 47 ft in San Francisco Bay. In many cases the natural harbor has been improved through engineering works, and in some cases the harbor is primarily an engineering creation.

According to the Maritime Administration of the United States Department of Commerce there were 1,066 vessels of 1,000 gross tons and over in the active oceangoing United States merchant fleet on Oct. 1, 1966. Of these, 160 were government and 906 were private ships. These figures do not include private ships temporarily inactive. Nor do they include 24 vessels in custody of Defense, Interior, Coast Guard, and Panama Canal Company. The status of the American merchant marine as of Oct. 1, 1966, is shown in Table 30-5.

A summary of large merchant ship construction and conversion under contract in private United States shipyards as of Oct. 1, 1966, is given in Table 30-6. Ships completed in previous years, showing displacement tonnage and cost, are presented in Table 30-7.

**30-5   Surface-effect ships for ocean commerce**   The functional purpose of surface-effect ships (SES) is to offer a transoceanic vehicle potentially capable of filling a gap in the spectrum of transport services afforded

**Table 30-4  A general picture of world sea-borne trade in 1948***

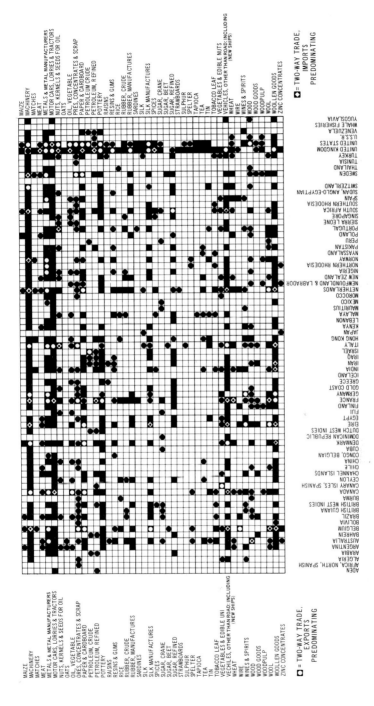

\* From *Dock and Harbor Authority (London)*, February, 1952.

**Table 30-5  Total United States flag oceangoing merchant fleet as of Oct. 1, 1966***

1,000 gross tons and over, excluding privately owned tugs, barges, etc.

| | Active | | | | | Inactive | | | | | Total | | | | |
|---|---|---|---|---|---|---|---|---|---|---|---|---|---|---|---|
| | Total | C | F | T | O | Total | C | F | T | O | Total | C | F | T | O |
| Grand total............ | 1,066 | 27 | 783 | 256 | ... | 1,328 | 196 | 952 | 61 | 119 | 2,394 | 223 | 1,735 | 317 | 119 |
| Government owned†...... | 160 | ... | 160 | ... | ... | 1,273 | 196 | 914 | 44 | 119 | 1,433 | 196 | 1,074 | 44 | 119 |
| Bareboat chartered...... | 8 | ... | 8 | ... | ... | ... | ... | ... | ... | ... | 8 | ... | 8 | | |
| General agency.......... | 152 | ... | 152 | ... | ... | ... | ... | ... | ... | ... | 152 | ... | 152 | | |
| Lending disposition...... | | | | | | 2 | 2 | ... | ... | ... | 2 | 2 | | | |
| Reserve fleet............ | | | | | | 1,271 | 194 | 914 | 44 | 119 | 1,271‡ | 194 | 914 | 44 | 119 |
| Privately owned........ | 906 | 27 | 623 | 256 | ... | 55 | ... | 38 | 17 | ... | 961 | 27 | 661 | 273 | |

Key: C, combination passenger-cargo; F, freighters; T, tankers; O, all other types.    (Includes nonmerchant-type ships currently in the National Defense Reserve Fleet.)

* Maritime Administration, U.S. Department of Commerce.

† Excludes 1 combination passenger-cargo vessel, 18 freighters, and 2 tankers, in military service under custody of the Department of Defense; 2 freighters loaned to Department of the Interior and Coast Guard, and 1 Panama Canal Company combination passenger-cargo ship no longer in commercial service.

‡ Includes 517 ships to be sold for scrap; 292 naval auxiliaries; and 4 ships sold but remaining in custody of reserve fleet pending delivery.

**Table 30-6  Ship construction and conversion under contract, Oct. 1, 1966\***

|  | New ships | | | | | Conversions | | | | Total |
|---|---|---|---|---|---|---|---|---|---|---|
|  | Total | C | F | T | O | Total | C | F | T | |
| Total............ | 55 | 0 | 43 | 1 | 11 | 11 | 0 | 11 | 0 | 66 |
| Government.... | 11 | 0 | 0 | 0 | 11 | 0 | 0 | 0 | 0 | 11 |
| Private......... | 44 | 0 | 43 | 1 | 0 | 11 | 0 | 11 | 0 | 55 |

\* Maritime Administration, U.S. Department of Commerce.

**Table 30-7  Ships completed in previous years\***

| Fiscal year | Number | Displaced tonnage | Cost, millions of dollars |
|---|---|---|---|
| 1957 | 27 | 566,275 | 182.2 |
| 1958 | 46 | 822,438 | 267.9 |
| 1959 | 53 | 1,519,319 | 482.1 |
| 1960 | 35 | 876,415 | 251.8 |
| 1961 | 36 | 690,049 | 324.1 |
| 1962 | 40 | 722,996 | 357.5 |
| 1963 | 39 | 571,077 | 348.0 |
| 1964 | 24 | 351,243 | 213.9 |
| 1965 | 30 | 534,457 | 240.4 |
| 1966 | 24 | 341,405 | 173.9 |
| July 1 to Oct. 1, 1966 | 5 | 62,671 | 44.2 |

\* Maritime Administration, U.S. Department of Commerce.

by displacement ships and aircraft.   Since the SES vehicles are basically air-supported, the immediate question is whether they are vessels or, more specifically, aircraft.   In making this determination, the Federal Aviation Act of 1958 adopted the principle that an aircraft is a vehicle which derives its support in the atmosphere from reactions of the air, such as balloons or airplanes.   The Federal Aviation Administration, in 1963, determined that vehicles deriving support from a cushion of air not exceeding 28 in. in height are not aircraft.

The SES is not designed to be used on public highways.   Its over-land operation is incidental to the primary water navigation.   Thus, according to Interstate Commerce Commission's definition, the SES is not a motor vehicle.   Finally, since the vehicle is intended to be used on coastal and intercoastal waters for the purpose of carrying passengers, the decision was made by the General Council of Maritime Administration that it does come under the definition of "vessel" found in the Shipping Act of 1916 and the Merchant Marine Act of 1936.

Evaluation of SES concepts was made in a study of related technological problems by the U.S. Department of Commerce, completed in
February, 1966.  The mission of the study was:

> To determine the research and engineering problems asso-
> ciated with employing the surface-effect-ship concept in com-
> mercial ocean transportation systems.
>
> To recommend a research and development program com-
> petent to resolve such technological problems.

The study revealed that there are two basic types of SES:

1. One type is supported by a cushion of pressurized air which is
   constrained beneath the vehicle.  This type has been designated as a "category I" vehicle (aerostatic lift).
2. The second type of SES is mainly supported by aerodynamics
   as the vehicle transits at sufficient speed to obtain aerodynamic lift similar to an aircraft.  The significant difference

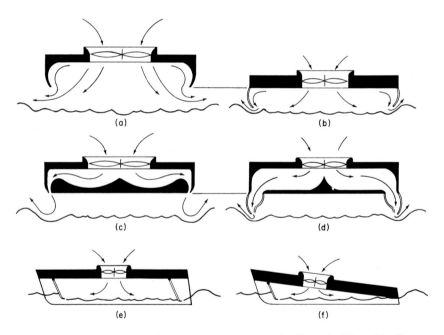

**Fig. 30-2**  Various concepts illustrating the aerostatic lift principle.  (a) The
plenum chamber; (b) the plenum chamber with "Skirt"; (c) the annular jet; (d) the
Hydroskimmer (annular jet with "Trunks"); (e) the captured air bubble; (f) the
Hydroskeel.  (Source: U.S. Department of Commerce.)

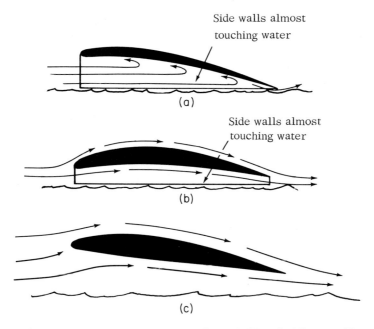

**Fig. 30-3** Various concepts illustrating the aerodynamic lift principle. (*a*) The ram wing; (*b*) the channel-flow wing; (*c*) the wing-in-ground effect. (*Source: U.S. Department of Commerce.*)

from aircraft is that the proximity of the sea surface provides an added degree of lift. This type is called a "category II" vehicle (aerodynamic lift).

Various concepts of each basic type are illustrated in Figs. 30-2 and 30-3. Vehicle characteristics for some selected vehicles are given in Table 30-8, and the effect of wave height on power is given in Table 30-9.

A considerable program of technological research and prototype vehicle development is needed to establish transport capabilities of various concepts of SES. The identified problems are classified by the U.S. Department of Commerce as:

*Crucial*

    Ocean waves of unusual height
    Mode of responses to waves
    Collision avoidance

**Table 30-8  SES vehicle characteristics[a]**

| Concept | Gross weight, long tons | Dimensions $L \times B \times H$, ft | Maximum horsepower | Disposable load / gross weight | Estimated base pressure, psf | $V_{max}$, knots |
|---|---|---|---|---|---|---|
| Hydroskimmer | 100 | 77 × 38 × 21 | 35,000 | 0.500 | 81[b] | 100 |
| | 400 | 122 × 61 × 35 | 105,000 | 0.600 | 126 | 100 |
| | 1,000 | 166 × 83 × 40 | 214,500 | 0.640 | 162 | 100 |
| CAB | 100 | 100 × 40 × 25 | 8,500 | 0.644 | 62[c] | 58 |
| | 500 | 200 × 80 × 45 | 26,000 | 0.691 | 78 | 65 |
| | 1,000 | 270 × 110 × 65 | 50,000 | 0.718 | 84 | 74 |
| | 2,500 | 352 × 149 × 80 | 125,000 | 0.720 | 119 | 100 |
| | 5,000 | 445 × 188 × 90 | 250,000 | 0.726 | 149 | 112 |
| | 10,000 | 500 × 215 × 10 | 500,000 | 0.726 | 232 | 129 |
| Hydrokeel | 100 | 90 × 20 × 20 | 8,500 | 0.495 | 153[d] | $33\frac{1}{2}$ |
| | 500 | 160 × 30 × 30 | 36,000 | 0.518 | 285 | 46 |
| | 1,000 | 205 × 37 × 40 | 70,000 | 0.561 | 370 | 50 |
| VRC channel flow[f] | 100 | 140 × 62 × 38 | 16,900 | 0.500 | 36[e] | 100 |
| | 500 | 310 × 156 × 80 | 40,500 | 0.580 | 32 | 100 |
| | 1,000 | 440 × 220 × 100 | 63,500 | 0.603 | 32 | 100 |
| | 2,000 | 620 × 310 × 120 | 99,700 | 0.621 | 32 | 100 |
| Wieland craft (wing-in-ground) | 100 | 215 × 150 × 48 | 19,500 | 0.500 | NA | NA |
| | 500 | 370 × 280 × 68 | 63,100 | 0.562 | NA | NA |
| | 1,000 | 600 × 500 × 100 | 105,000 | 0.684 | NA | NA |

[a] U.S. Department of Commerce.
[b] Assuming length of base is 95%—LOA.
[c] Assuming length of base is 90%—LOA.
[d] Assuming length of base is 82%—LOA.
[e] Assuming length of base is 71%—LOA.
[f] VRC—Vehicle Research Corp.
NA—Not available.
LOA—Length over all.

Table 30-9 Effect of wave height on power*

| Concept | Gross weight | Required power, hp | | Speed in 8-ft waves, knots |
|---|---|---|---|---|
| | | 2-ft wave height | 8-ft wave height | |
| Hydroskimmer........ | 100 | 13,000 | 28,000 | 100 |
| | 400 | 48,000 | 84,000 | 100 |
| | 1,000 | 118,000 | 170,000 | 100 |
| CAB................ | 100 | 3,400 | 6,800 | 58 |
| | 1,000 | 27,800 | 40,000 | 74 |
| | 2,500 | 68,000 | 100,000 | 100 |
| | 5,000 | 148,000 | 200,000 | 112 |
| | 10,000 | 298,000 | 400,000 | 129 |
| Hydrokeel............ | 100 | 4,300 | 6,800 | 33.5† |
| | 500 | 22,000 | 37,000 | 46 |
| | 1,000 | 45,500 | 56,000 | 50 |
| VRC channel flow.... | 100 | 3,500 | 9,300 | 100 |
| | 500 | 10,300 | 23,800 | 100 |
| | 1,000 | 17,200 | 36,700 | 100 |
| Wieland craft......... | 1,000 | ....... | ....... | 100 |

* U.S. Department of Commerce.
† 5-foot waves.
NOTE: In early programs, speed was held constant and power reduced for lower wave height. In later programs, the power was fixed and speed varied.

*Important*
Forecast of route environmental and surface conditions
Fuel margin
Failure modes and survival
Manning
Maneuverability
*Improvement*
Port delays
Fueling
Habitability and control layout
Geographical navigation

Problems identified as "improvement" are considered to fall into the category of administrative solution. The state of the art or technology is in hand for solving these four problems. However, care will have to be exercised to insure that they are considered in development programs since they are not related to conventional vessels.

Relative to the economic foundation or justification of SES concepts, certain studies and analyses should go hand in hand with technological programs and progress.   The following continuing studies and analyses are considered necessary:

1. Systems analysis throughout all design phases to assure that the ultimate function of the vehicle loading, transport, and discharge of cargoes and/or passengers is implemented in an optimal manner
2. Careful determination of probable utilization factors
3. Determination of over-all systems cost and comparison to other transport modes

## BIBLIOGRAPHY

Federal Coordinator of Transportation, "Public Aids to Transportation by Water," vol. 3, "Public Aids to Transportation," U.S. Government Printing Office, 1939.
Havinghurst, Walter: "The Long Ships Passing," The Macmillan Company, New York, 1945.
McDowell, Carl E., and Helen M. Gibbs: "Ocean Transportation," McGraw-Hill Book Company, New York, 1954.
"Surface-effect Ships for Ocean Commerce," U.S. Government Printing Office, 1966.

# 31
# River Hydraulics

**31-1  River development**  Any effort to regulate or control the flow of natural streams should begin with an understanding of the forces involved in river development.  The power exercised by a river originates in the potential energy of its tributaries.  Water flowing downstream consumes this energy in its own transportation against the resistance of the bed and banks which confine the flow.  Soil eroded in this process is used by the river as a grinding powder to hasten the rate of downstream erosion.  The deepening channel becomes a more efficient conduit for the growing stream, and the resulting increase in velocity accelerates the process of channel cutting.  At the same time that the river channel itself is being deepened, the river valley also is being shaped by sheet erosion and by the growth of tributary streams.

Rivers occupy not only *erosion valleys*, resulting from the processes described above, but *structural valleys* as well (Fig. 31-1).  When the ditch down which the water spills was produced by faulting along some steep plane of shear failure in the earth's crust, it is called a *rift* valley.  Another type of structural valley is the trench formed by folding or

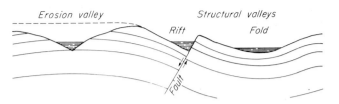

**Fig. 31-1** River valleys.

buckling of rock strata under compressive stress. Whether the initial
drainage pattern was established by structural failures of rock strata or
by selective erosion of soft spots in a sloping plain, the young drainage
channel is deepened and broadened by comparable erosive forces.

Of course there is a limit to river growth. At the mouth of the river,
currents fade away in the broad expanse of lake or open sea. Current
velocities persist only at that specific value which is required to carry off
the sediment brought down to the river mouth from the entire watershed. Any lesser current velocity would permit the river to be dammed
behind its own mudbank. Then the rising water level behind such a
dam would create swifter currents until the velocities became competent
to erode the dam.

This equilibrium—between the silt load carried by the river and the
channel characteristics which make the river competent to carry that
load—is first attained at the river mouth and then moves upstream.
Wherever such equilibrium exists the river is said to be *graded*. When
the lower stretches of the river reach a graded condition, the upper
stretches continue to *degrade* or deepen their channels until they, too,
become graded. In such fashion a river matures by cutting a stable
channel back to its headwaters. When a dam or other obstacle reduces
the stream's velocity, the river *aggrades* or builds up its bed behind the
obstacle, because it is no longer competent to transport its load.

A graded stream has achieved a rough sort of stability with respect to
the river bed, and currents are now enabled to act on the river banks at
somewhat constant elevations for prolonged periods of time. The main
attack of the current is directed against the bends, and as these latter are
cut back, the curvature of the river sharpens. Eventually a cutoff is
achieved, and the whole process is repeated elsewhere. The result of
all this is a sluggish whipping back and forth across the valley floor,
called *meandering*.

**31-2    Regime**  If only because of seasonal variations in its discharge,
a graded river may experience periods of aggradation or degradation

over part or all of its length.   Such a river is said to be *in regime* if the changes in its bed elevation involve movement of alluvial material only. "A regime-type river is defined as one that has formed a major part of every cross section from material that has been transported or could be transported at some stage of flow."[1]

The very idea of regime implies that the observer is dealing with river behavior over practical periods of time.   If regime theory is tested either over a few days or over many thousands of years, the notion loses its validity.   Over too short periods of time the stream may be constantly eroding its banks; over geologic ages the river may be lost in more grandiose geologic developments; but over the years which commonly measure the duration of engineering works the behavior of a regime-type river will display evidence of a rough equilibrium between erosion and silting through succeeding periods of flood and drought.

If there really is equilibrium between the hydraulic properties of a river channel, discharge, silt load, bed material, and slope, then the engineer should be able to express these relationships in mathematical form. Certainly one can say in a general way that rivers flowing through clay deposits are narrow and deep, while rivers flowing through loose sand are broad and shallow.   Engineers experienced in the behavior of unlined irrigation canals have attempted to become more specific.   Thomas Blench proposed the following formulas:[2]

$$d = \sqrt[3]{\frac{sQ}{B^2}} \tag{31-1}$$

$$b = \sqrt{\frac{BQ}{s}} \tag{31-2}$$

and

$$S = \frac{B^{\frac{5}{6}}s^{\frac{1}{2}}Q^{-\frac{1}{6}}}{C} \tag{31-3}$$

where $d$ = depth of channel, ft, from water surface to flat bottom.

$b$ = effective width, ft, defined as the dimension which when multiplied by $d$ produces the cross-sectional area.

$S$ = slope of the stream surface, expressed as a decimal.

$B$ = a factor related to the nature of the bed material and ranging from 0.6 to 1.25 according to experience in India.   Larger values of $B$ correspond to coarser soils.   Broader experience may increase the range in values greatly beyond the limits quoted.

[1] Thomas Blench, Regime Theory for Self-formed Sediment-bearing Channels, *Proc ASCE*, **77**(70):2 (May, 1951).
[2] *Ibid.*

$s$ = a factor related to the tractive forces exerted on the sides of
the channel.  If the selected value of $s$ is too high, there will
be scour; if it is too low, sedimentation will ensue.  Quoted
values are 0.3 for glacial till, 0.2 for silty clay loam, 0.1 for
cohesionless material, and as low as 0.05 for tidal rivers
in sand.

$C$ = a coefficient relating to viscous drag, with a value of 2,080
suggested for average conditions.

Equations (31-1), (31-2), and (31-3) were based upon experience with
irrigation canals and apply to straight channels of regular shape flowing
full through erodible soil.  They express the shape properties of a channel
which will be stable under specified conditions of discharge and bed material.  If the river is diked, revetted, or dammed, the formulas become
irrelevant.

The preceding relationships can be generalized for natural streams
at river stages lower than the bank-full condition.  Using the same
notation, logarithmic plotting of data from a limited number of American
rivers discloses the following relationships:[1]

$$d = m_1 Q^{n_1} \tag{31-4}$$

$$b = m_2 Q^{n_2} \tag{31-5}$$

$$v = m_3 Q^{n_3} \tag{31-6}$$

where $v = Q/bd$.  The values of the exponents in Eqs. (31-4) to (31-6)
vary greatly with the location, and the averages of observed values for
rivers in the Great Plains and Southwestern United States are in poor
agreement with corresponding values in the Blench equations.  Despite
the need for further information on physical constants for both the Blench
and Leopold-Maddock formulas, there can be little doubt that such
formulations of observed relationships do serve a useful purpose in helping
the engineer to think straight.  It is helpful, also, to note that both the
coefficients and the exponents in Eqs. (31-4) to (31-6) are related through
the fact that $Q = bdv$, for it then follows that

$$Q = m_1 m_2 m_3 Q^{n_1+n_2+n_3}$$

In other words, $m_1 m_2 m_3 = 1$, and $n_1 + n_2 + n_3 = 1$.  Some idea of the
relative magnitude of these exponents is provided by averaging various
scattering values, whence $n_1 = 0.40$, $n_2 = 0.26$, and $n_3 = 0.34$, if the
formulas are taken to refer to various river stages at the same specific
location.  If the formulas are used to describe the channel characteristics
that accompany downstream increases in discharge (at any one instant

[1] L. B. Leopold and Thomas Maddock, Jr., The Hydraulic Geometry of Stream
Channels, *Geol. Survey Professional Paper* 252, 1953.

of time), it has been found that $n_1 = 0.5$, $n_2 = 0.4$, $n_3 = 0.1$. In the first case the equations are used to describe channel variations at a specific location for a variable discharge; in the second case the same formulas are used to describe channel variations between different river locations for the same instant of time.

**31-3 Hydraulics** The preceding section is disturbing to the civil engineer because the complex interrelationship of river discharge, velocity, and load with channel size, shape, and slope obscures the cause-and-effect sequence with which the engineer generally has to deal. The comforting support of mathematics and mechanics is lacking in the geologist's analysis:

> A graded stream is one in which, over a period of years, slope is delicately adjusted to provide, with available discharge and with prevailing channel characteristics, just the velocity required for the transportation of the load supplied from the drainage basin. The graded stream is a system in equilibrium; its diagnostic characteristic is that any change in any of the controlling factors will cause a displacement of the equilibrium in a direction that will tend to absorb the effect of the change.[1]

The engineer, however, is under compulsion to produce quantitative answers in numerical terms; if this compulsion makes it necessary to simplify the problem, it is important that the engineer's judgment be guided by the best possible comprehension of even those physical factors which are not fully expressed in his computations.

The first simplification which the engineer hastens to adopt is an assumption that the erodible banks and bed of the stream have been transformed into a rigid conduit. With the cross section of the stream thus made subject to the dictate of the engineer instead of being the end product of many geological influences, all the familiar tools of hydraulics become immediately available for establishing desirable values of velocity, depth, or width. Only discharge remains an independent variable.

From the standpoint of the engineer it is best to start any discussion of rivers with the discharge of the stream, since this is generally beyond his control. Discharge is equal to the sum of the runoff from the watershed plus the contributions from ground-water flow. Water flows over soil much more rapidly than it flows through soil. Consequently, flood flow is primarily the result of runoff, while the discharge of a river in a prolonged drought is maintained by seepage from underground reservoirs. The total amount of precipitation is an independent variable, but the

[1] J. Hoover Mackin, Concept of the Graded River, *Bull. Geol. Soc. Amer.* 59, 1948.

ratio between runoff and ground-water flow depends in part (over practical periods of time) upon man's use of the land in the watershed. It is generally believed that the removal of forests tends to increase flood flow and to decrease the discharge at low water. Cultivation of pervious soils for agricultural purposes has an opposite effect.

Discharge is related to velocity and area by the basic hydraulic equation

$$Q = AV \qquad (31\text{-}7)$$

where $Q$ = discharge, cfs, generally
   $A$ = cross-sectional area of the stream, sq ft
   $V$ = mean velocity, fps

At any specific time of analysis or observation the velocity of the stream is determined by the longitudinal slope of the channel and by the physical properties of its cross section. This relationship is usually expressed by the Chezy formula

$$V = C \sqrt{RS} \qquad (31\text{-}8)$$

where $S$ is the slope and $R$ is the hydraulic radius.

There is no general agreement on the proper method for selecting values of $C$. Kutter's formula for $C$ is widely used, but it is cumbersome and imperfect in theory. Manning's formula has found wide acceptance, primarily because it assumes a fixed exponential relationship between $C$ and $R$, which is a considerable convenience in designing river models. Unfortunately, the process of verification of the usual distorted river model does not distinguish between errors arising from Manning's formula and those attributable to other sources. Manning's expression for $C$ is

$$C = \frac{1.49R^{\frac{1}{6}}}{n} \qquad (31\text{-}9)$$

where $n$ closely approximates Kutter's corresponding term for the nature of the channel surface. Some engineers, at the cost of some mathematical inconvenience, prefer the Bazin equation

$$C = \frac{158}{1 + m/\sqrt{R}} \qquad (31\text{-}10)$$

Bazin's $m$, like Manning's $n$, expresses the roughness of the surface of the conduit. Roughness consumes energy, which is dissipated in the eddies that are formed at any projection into the streamlines. In a natural stream the small eddies created by the texture of the bed are accompanied by large eddies and boils formed by the sinuosity of the river. Both $m$ and $n$ are supposed to reflect both forms of energy loss. Tabular values fail to provide adequate grounds for selecting the proper values of these factors.    Field observation or photographs of rivers of known roughness form better guides to judgment.

The engineer soon gets to think of the river as he would of a flume or canal and to accept the velocity as the effect of the slope.

In river work it is sounder practice to adopt the concept of the geologist that slope is the effect of velocity.    Equation (31-8) merely expresses the mutual relationship of these two factors and does not distinguish as to the primacy of its terms.    The tributaries and headwaters of a river may be considered to dump definite amounts of water and sediment into the river channel.    If the resulting velocity in the channel, as fixed by Eq. (31-8), is not competent to transport this amount of sediment, the resulting deposit upstream constitutes a steepening of the hydraulic gradient.    This situation continues until the slope is sufficient to produce a velocity that will put an end to the process of aggrading under the existing conditions of discharge.    In the above sequence of events the engineer might discern an axiom which applies to all river-control work: any change made in the river channel at any point alters to some degree the entire channel from its source to its mouth.    The full consequences of any river works must become the object of careful study before the start of construction.

**31-4    Load**    Obviously, river sediment is important.    Much attention has been paid to the stream velocities required to move solid particles along the stream bed and to the velocities at which silt particles will settle out of suspension.    The force resisting particle movement should be directly proportional to the particle weight; this would be true either for vertically lifting the particle or for pushing it on a level surface against frictional resistance.    Any more complex resistance to motion would be largely a combination of these lifting and sliding resistances.    Thus the force resisting a particle movement should be proportional to the cube of particle diameter:

$$F_R = C_1 D^3 \tag{31-11}$$

At the same time the force acting to move the particle is proportional to the square of the velocity and to the projected area (or diameter squared) of the particle:

$$F_A = C_2 V^2 D^2 \tag{31-12}$$

For incipient motion

$$F_R = F_A$$

which is to say that the diameter of the largest particle that can be moved is proportional to the square of the velocity,

$$D_{\max} = kV^2 \tag{31-13}$$

Observation indicates that the mean current velocity required to scour a sand or gravel bottom is about twice that required to maintain the same material in motion. The corresponding ratio for clay is considerably higher, as the tractive force must then overcome cohesion as well as weight and friction, until the soil is lifted and dispersed. Moreover, compared with sand or gravel, clay particles do not project into the turbulent current, and consequently the streamlines deflected by clay particles represent only fractional parts of the mean velocity of the river. An average river velocity of over 5 fps will scour either compact clay or gravel 1 cm in diameter, but the clay can be kept moving by velocities of about 0.25 fps, while gravel requires velocities somewhat over 1 fps. The finer fraction of the load is carried along in suspension; the coarse particles are rolled along the bed. Whether a silt particle, being eroded from the bed, is transported or is settling out of suspension depends upon its size and upon the velocity of the current. Figure 31-2 shows the relationship between these variables according to results obtained at Upsala University. In this plot the velocities are those affecting the particle, not the average velocity over the cross section. There is, of course, a notable difference between the velocity at a specific point and the average velocity for the river cross section or even for a significant portion thereof.

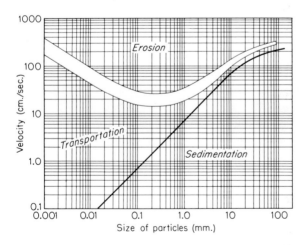

**Fig. 31-2** Effect of current velocity on bed load. (*After Hjulstrom, Upsala University, 1935.*)

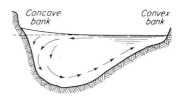

**Fig. 31-3** Transverse currents at a bend.

A young stream which is still cutting a bed down to sea level will carry both the suspended load and the bed load to its outlet. When the river becomes mature, the material carried through to the sea will be primarily suspended load. Except at higher water stages, the bed load will mainly consist of material cut from the banks and deposited downstream wherever the tractive force of the current is no longer equal to the task of transporting it.

**31-5 River currents** In the development of rivers for purposes of navigation, most engineering problems are concerned with rather mature streams, because of the numerous bends that are encountered. As the water flows around a bend in the river, it tends to be thrown to the outside of the curve by centrifugal force. This tendency is strongest near the surface of the stream, where maximum velocities occur. The piling up of surface water on the concave side of the bend produces a transverse gradient which forces the bottom currents toward the convex shore. The result is a circulatory transverse current developed in any vertical cross section, as in Fig. 31-3, which, imposed upon the fundamental downstream motion, produces a spiral path for the individual particles. The highest velocities and greatest depths are encountered along the concave bank, as in Fig. 31-4, which is undercut and caved in, and the adjacent river bed deeply scoured, while material is being deposited by the reduced velocities on the other side. The amount of curvature is thus progressively increased. The main body of the stream, striking the concave bank obliquely, is deflected by the impact toward the opposite shore as it flows downstream from the bend. Consequently, another bend develops on the other bank, some distance below. This bank cutting is

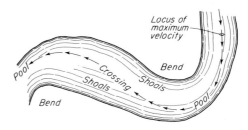

**Fig. 31-4** Longitudinal currents at a bend.

the principal source of bed load in meandering streams. The deep channels in the bends are called *pools*. The tangent section between bends is called a *reach*, and the path of the main current where it is deflected from one bank to the other is termed a *crossing*. The *thalweg* of the stream is the path that would be followed by a freely drifting float.

Bank cutting is more rapid in sandy soils than in clay. Pools are deeper in the latter soil, because the bank will stand at steeper slopes, and greater depths will prevail in the crossings as well, because the finer soil picked up in the bends has less tendency to settle out. Thus a sandy bed is characteristic of wide shallow streams with steep gradients, while streams through cohesive soil are narrow and deep.

In the crossings the pattern of velocity distribution is somewhat more uniform and the maximum velocity is less than that prevailing in the bends. The material cut from the bends during high water cannot all be carried to the next bend because of this reduction in velocity. Thus, long diagonal sand bars are formed in the crossings, and at low water the depths will be only a fraction of those existing in the pools. In the lower Mississippi at low water, dredging is necessary in some crossings to maintain a 9-ft channel, at the same time that depths in adjacent bends range from 30 to 80 ft. The formation of sand bars in the reaches is similar to the action in the crossings. The river is generally wider in reaches than in other sections, and the bars are unstable deposits interspersed with many shallow crossings. River regulation is primarily concerned with the construction of works designed so to direct the currents as to scour a stable channel through the reaches and crossings. Where this cannot be done, it is sometimes possible to impound sufficient water in upstream reservoirs for release during low water to secure a discharge that will provide navigable depths of channel. Otherwise it is necessary to construct a system of locks and dams, making the channel depth relatively independent of the discharge.

**31-6  The river mouth**  Through the mouth of a large river flows not only the concentrated runoff of the hinterland, but also, to a considerable extent, the concentrated commerce of the interior of the country. The channel must accommodate both river traffic and oceangoing ships, so far as possible. From the engineering standpoint it is convenient to consider the mouth to extend upstream as far as the influence of tidal action is felt. The two principal physical factors to be considered are the concentration of sediment in the river and the strength of tidal action, because the relation between these factors largely governs the difficulty to be anticipated in maintaining a navigable channel through the deposits at the mouth. The formation of a delta is evidence that tidal and littoral currents are unable to assume the burden of transporting the load carried

by the river.    The large delta of the Mississippi reflects both its great load
and the weak tides of the Gulf of Mexico.    The deep mouth of the
St. Lawrence is at least partly due to the fact that the Great Lakes act
as settling basins in removing sediment and to the effectual scour provided
by the notable tides of the region.

In the characteristic delta formation, where tides are weak, the outlet of the river separates into a number of mouths.    Generally only one of
these need be improved for navigation, although flood control may impose
further requirements.    The weakest mouth that can provide an adequate
channel should be selected for improvement; any other choice would
increase the amount of discharge and sediment that must be dealt with.
At the weakest mouth the delta has the slowest rate of growth.    Access
to deep water and exposure to strong ocean currents are other favorable
factors to be considered in choosing the location of the ship channel.

If the river discharges into a closed bay or lagoon instead of into the
open sea, it is said to have an indirect mouth.    The material in the hook
which closes the bay comes from littoral drift rather than from river
sediment.    The bed load of the river is deposited within the lagoon.
In such cases two problems arise: channel maintenance at the river
mouth and protection of the sea entrance to the bay.

In tidal rivers there is a reversal in the direction of flow in the lower
river with every ebb and flood.    As the tide moves upstream, the natural
discharge of the stream is stored up, and a large volume of water is released
at ebb tide, with strong scouring action.    It is desirable to maintain and
increase this favorable action to the greatest extent possible.    If the river
is split by an island, the flow can be concentrated in one channel by dam-
ming the other.    The use of groins in this section of the river should be
avoided because of the energy wasted in the eddy currents set up in the
groin fields.    A bell-shaped mouth increases the amounts of tidal flow.
River bends serve no useful purpose at the mouth, and cutoffs or measures to straighten or confine the channel are helpful.

The channel improvements mentioned or implied in the foregoing
are carried out by the various types of structure described on subsequent
pages.    The tendency of a river to form a bar at its outlet can be offset,
if navigable depths are not otherwise obtained, by confining the flow to
deep water by means of jetties.    As the jetty is exposed to strong wave
action, further discussion of this type of structure will be deferred until
after the theory of water waves has been considered.

**31-7    River models**  Because of the difficulty in forecasting the full
ultimate effect of any specific measure taken for river improvement, an
ever-increasing reliance is placed upon the use of models.    The most
evident requirement of a true model is that it be geometrically similar to

its prototype; that it be constructed to some definite linear scale. Let $L_r$ represent the linear-scale ratio—the ratio between any specific dimension of the model and the comparable dimension of the prototype. Consider a jet issuing from a plane orifice. The velocity of the jet is directly proportional to the square root of the head. The ratio of the jet velocity for a model orifice to the jet velocity of its prototype would equal the ratio of the square roots of their respective heads, or

$$V_r = L_r^{\frac{1}{2}} \tag{31-14}$$

The quantity of water discharged from the orifice in unit time is equal to the product of the velocity and the cross-sectional area. The discharge ratio of model to the prototype is, therefore,

$$Q_r = L_r^{\frac{1}{2}} L_r^2 = L_r^{\frac{5}{2}} \tag{31-15}$$

The time required for any specific displacement of a particle is obtained by dividing the displacement by the velocity. The ratio of time in model to time in prototype for any stipulated action to occur is given by the equation

$$T_r = \frac{L_r}{V_r} = L_r^{\frac{1}{2}} \tag{31-16}$$

The preceding equations hold when, as in the cited example of orifice flow, motion results from the conversion of elevation head or pressure head into velocity head. In such cases the model is governed by Froude's law, which can be expressed by either Eq. (31-14), (31-15), or (31-16). These three expressions are consequences of the assumption that velocity is proportional to the square root of the head, or that head loss, or slope, is proportional to velocity squared. In cases where the original loss of energy occurs in overcoming viscous resistance, Reynolds' law governs, and Eq. (31-16) is replaced by

$$T_r = \frac{L_r^2 \rho_r}{\mu_r} \tag{31-17}$$

where $\mu_r$ is the ratio of viscosities of the two fluids, model and prototype, and $\rho_r$ is the corresponding ratio of densities. To have a model conform to both Froude's and Reynolds' laws at the same time is often a practical impossibility, although sometimes it can be done by juggling values of $\rho_r$ and $\mu_r$ through the selection of a different fluid in the model.

When the model of a navigable stream is constructed to some practical scale, the velocity of flow in the model, especially in the shoals, which are also often of particular concern, is apt to be very small. This situation presents two difficulties: first, in turbulent flow the velocity is proportional to the square root of the slope, while in viscous flow it is proportional to the slope directly; second, model-bed material conforming to the proper scale may be so fine-grained that it has appreciable cohesion. In the latter event a scouring velocity in the model might be actually larger than in the prototype, rather than smaller as stated in Eq. (31-14).

When these difficulties preclude the use of a true model, distorted models become necessary. In such cases, where deposition and scour are involved, it is customary to use lightweight minerals such as coal dust for the bed and to use different scale ratios for vertical and horizontal dimensions—to increase model depths. It then becomes necessary to have recourse to a trial-and-error procedure as a substitute for the laws of similitude. Some known hydrograph for the river is copied in the model for different vertical scale ratios until finally some scale ratio produces in the model a counterpart of the channel pattern developed in the prototype after that same hydrograph. This process is called the *verification* of the model. A hydrograph is a plot of time against some function of river discharge.

If scour is not an item of importance, it is feasible to follow Eq. (31-14), even at relatively low velocities, by maintaining a constant value for the $C$ in Eq. (31-8) through the selection of a proper value for surface roughness; that is, by building into the model a proper value of $n$ or $m$ [Eqs. (31-9) or (31-10)] to counteract the change in the hydraulic radius.

## QUESTIONS AND PROBLEMS

**31-1.** An irrigation canal excavated through glacial till is to carry 300 cfs ( = cubic feet per second). Select values of channel depth, area, and slope that will be stable without the expense of providing a lining.

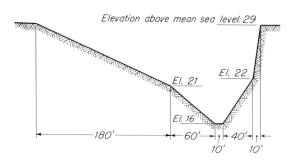

**31-2.** The sketch approximates the cross section of the Stillaguamish River at a U.S. Geological Survey stream-gauging station near Silvana, Wash. The following rating table applies:

| Gauge height | Discharge, cfs |
|:---:|:---:|
| 23 | 620 |
| 24 | 1,290 |
| 25 | 2,130 |
| 26 | 3,250 |
| 27 | 4,650 |
| 28 | 6,320 |
| 29 | 8,250 |
| 30 | 10,600 |
| 31 | 13,250 |

Plot $d$, $b$, and $v$ against $Q$ on logarithmic paper and solve for the unknown coefficients and exponents in Eqs. (31-4), (31-5), and (31-6).

**31-3.** Assuming a value of 2.5 for Bazin's $m$, what is the slope of the river of Prob. 31-2 when the river level is at elevation 29?

**31-4.** Assuming a slope of 0.0018 for the bank-full river of Prob. 31-2, what is the value of Manning's $n$?

**31-5.** Assuming a slope of 0.0018 for the bank-full river of Prob. 31-2, evaluate the unknowns of Eqs. (31-1), (31-2), and (31-3).

**31-6.** A river discharges 300 cfs at a slope of 0.0009. The V-shaped channel has side slopes of 4 to 1. $n = 0.03$. How deep is the water at the thalweg?

**31-7.** E. W. Lane suggests[1] that for noncohesive bed material the Manning roughness coefficient is related to the grain size of the river bed material by the formula

$$26n = \left(\frac{D_{65}}{R}\right)^{\frac{1}{6}} \tag{31-18}$$

(65 per cent of the soil, by weight, has a grain size smaller than $D_{65}mm$.) $R$ is the hydraulic radius, in feet. For $R = 10$ ft, find $n$ for (a) $D_{65} = 0.5$ mm, (b) $D_{65} = 5.0$ mm.

**31-8.** Assuming a current velocity of 3 fps,

    *a.* What is the smallest soil particle that will settle to the bottom?

    *b.* What is the largest particle that can be eroded from the banks?

    *c.* What is the largest particle that can be kept rolling along the bed?

**31-9.** A certain river has a mean velocity of 5.17 fps with the following channel characteristics: $n = 0.004$, slope $S = 0.0009$, hydraulic radius $R = 10$ ft, width $b = 200$ ft, area $A = 2,000$ sq ft. A hydraulic model 8 ft wide should be designed to deliver what discharge?

**31-10.** What is the smallest model that can be constructed for the river of Prob. 31-9 without use of a distorted scale?

**31-11.** It is planned to construct a river model to a scale of 1 ft = 60 ft. What will be the relationship between prototype and model? If it is planned to duplicate river stages through a 60-day period, how long (work hours) will the test take?

**31-12.** The cross section of a river channel 1,000 ft wide is an isosceles triangle with a maximum depth of 20 ft. The discharge, $Q = 4.8$ cfs. Slope = 0.000035. What would be a suitable material for construction of the model channel?

**31-13.** An unlined canal 10 ft deep is found to behave satisfactorily with respect to scour and silting. How deep should it be made to provide similar performance at four times the discharge?

[1] *Proc. ASCE,* **79**(280):1–10 (1953).

**31-14.** A wide river has an almost uniform depth of 20 ft at flood stage and a mean velocity of 8 fps. What is the largest particle to be scoured from the river banks? At what velocity will this particle be again deposited? Assume that current velocity adjacent to bank is the same as the mean river velocity.

**31-15.** In a geometric model with a length ratio of 1:50, sand particles of 1-mm diameter were observed to move along the bed when $Q = 5$ cfs. What would be the corresponding caliber of bed load and discharge in the prototype?

**31-16.** A power canal has a rectangular cross section 120 ft wide by 20 ft deep. The canal discharges 38,500 cfs at a slope of 0.001 and a Chezy coefficient of 130. Find the dimensions of a model when the available water supply is limited to 38.5 cfs.

**31-17.** Find the desired roughness of the model of Prob. 31-16, expressed in terms of the $m$ of Bazin's formula.

**31-18.** An article by Stroitelstvo (U.S.S.R., 1936)[1] sets the following permissible mean velocities corresponding to the specified particle diameter:

| $v$, fps | 0.49 | 0.98 | 1.80 | 3.28 | 5.91 | 8.86 | 12.00 |
|---|---|---|---|---|---|---|---|
| $D$, mm | 0.005 | 0.25 | 1.00 | 10 | 40 | 100 | 200 |

He also found that the permissible mean velocity should be corrected for depth of channel

| Depth, ft | 0.98 | 1.97 | 3.28 | 4.92 | 6.56 | 8.20 | 9.84 |
|---|---|---|---|---|---|---|---|
| Correction coefficient | 0.8 | 0.9 | 1.00 | 1.1 | 1.15 | 1.20 | 1.25 |

In the light of these data rewrite Eq. (31-13) to express equality rather than proportionality.

**BIBLIOGRAPHY**

Allen, J.: "Scale Models in Hydraulic Engineering," Longmans, Green & Co., Ltd., London, 1947.
Harris, Charles W.: Hydraulic Models, *Univ. Wash. Eng. Expt. Sta. Bull.* 112, 1944.
Hjulstrom, Filip: Studies of the Morphological Activity of Rivers, *Bull. Geol. Inst. Univ. Upsala*, **25**:267–270 (1935).
Hydraulic Models, "Manual of Engineering Practice," no. 25, American Society of Civil Engineers, 1942.
Mackin, J. Hoover: Concept of the Graded River, *Bull. Geol. Soc. Amer.* 59, 1948.
Rubey, William W.: Force Required to Move Particles on a Stream Bed, *J. Wash. Acad. Sci.*, **25**:571–572 (1935).

[1] E. W. Lane, *Proc. ASCE*, **79**(280):24 (1953).

# 32
# Channel Regulation

**32-1  Channelization**   In a navigable river the depth, width, and alignment of the ship channel must meet the requirements of river traffic.   If the minimum discharge of the stream naturally provides something more than the desired width and depth at the bends (Figs. 31-3 and 31-4), it should be possible to train the river currents to provide an adequate channel through the crossings and reaches.   Obviously it is more economical to utilize the energy of the stream for channel cutting than to pay for power to operate a dredge for the same purpose.   Dredging can be used to supplement the work of the current, but not to replace it, at least not in silt-bearing rivers.   The Missouri River is said to carry nearly 50 tons of silt per second past Kansas City when the river discharge amounts to 150,000 cfs.   To remove any such quantity of material by dredging—admittedly a distorted example—would cost a quarter of a million dollars per day.   Instead, the prudent engineer endeavors to make most efficient use of free river power to the same purpose.   The two principal steps to be taken in maintaining proper channels are the reduction of bank cutting at the bends and the promotion of bed scour in the shoals.

The development of a continuous navigable channel involves three principal measures:

1. *Prevention of meandering.* Because a river shifts its course mostly through cutting at bends, the protection of the concave banks at bends is a major step in channel stabilization.
2. *Achievement of specified channel depths.* Because channels are shallow at crossings and reaches, it often becomes necessary to concentrate the discharge of the river over a smaller width in such places, so that scour will increase channel depths.
3. *Avoidance of excessively sharp channel curvature.* A meandering stream is apt to develop bends of such short radius that navigation becomes difficult, especially for barge tows. Curvature can be reduced by creating cutoffs. Unfortunately, river straightening destroys the existing equilibrium of forces and releases energy for augmented bank cutting (and additional meandering) both up- and downstream from the improvement. Many early efforts at river straightening failed because reduction in the length of a channel created a steeper gradient, the total available head remaining constant, and the correspondingly higher velocity inevitably was reflected in a shallower depth. Consequently, the sinuous nature of an alluvial stream must be preserved. Only the more objectionable bends should be reduced in curvature, and these should be flattened only to the average degree of curvature prevailing in that specific stream. Once made, a cutoff can be developed and protected by the dikes and revetments similar to those described in following paragraphs.

**32-2 Bank cutting** The protection of the concave bank at a bend is most important because it is the caving of such banks that furnishes most of the material which is deposited in the crossing. In part, the erosion at the bend occurs because there the locus of maximum velocity approaches the concave bank instead of remaining in the middle of the stream, as it does in the crossings and reaches. This action is more intensive when the river is in flood, because the velocities increase with the discharge. Other factors also enter the picture at high-river stages. For one thing, the river bed itself is eroded at these higher flood velocities. In soft materials every foot of river rise is accompanied by several feet of channel deepening. This removal of the toe of the slope at a concave bank can induce landslides and sloughing. At the same time, the rise in river level will gradually be reflected in a rise in the adjacent ground-water table, since the net head available to drain the banks is reduced. The higher water table, in turn, reduces the stability of the river bank, again favoring the development of slides and sloughing. Finally, if the river level subsequently falls rapidly, the ground-water table lags behind, and the differential head (between ground-water table and river level) causes seepage forces in the river banks, so that a rapid decline from flood stages is commonly accompanied by notable bank sloughing. Even a superficial understanding of these phenomena is sufficient to show the importance of either protecting bank and bed against erosion at bends or of providing some means of reducing the erosive forces at such localities.

**32-3  Longitudinal diking**  One of several possible methods of protecting a concave bank at a river bend is by interposing a longitudinal dike or training wall between the current and the bank (Fig. 32-1). Such dikes have generally been made of loose rock fill to secure the considerable strength needed to resist erosion and the impact of floating debris. These walls are vulnerable to undermining if the river bed is easily erodible. Brush or lumber mattresses are often used for bed protection. Other forms of lining are discussed later in the section on revetments. Longitudinal dikes must tie into the river bank at their upper ends to prevent failure by scour behind the dike. Short transverse dikes into the bank are also used for this purpose. This type of bank protection has the advantage that the dike need not be exactly parallel to the bank but may be given the proper downstream alignment to direct the current for the most effective scouring of the bars. Figure 32-2 pictures a type of dike, or training wall, used in European practice. The lower sketch in the same figure of a rock-filled double fence represents an installation along the Santa Clara River, in California. Actually steel rails, rather than piles, were used. The 60-lb rails, 30 ft long, were spaced 12 ft center to center and driven 23 ft into the erodible stream bed. The rails are connected with galvanized-woven-wire fencing hung from ¾-in. steel-wire cables. Alternate layers of brush and rock cobbles fill the space between

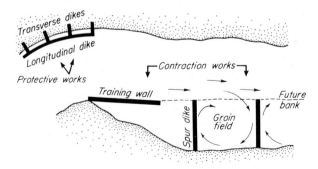

**Fig. 32-1**  Channel-control works.

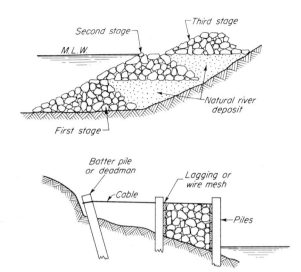

**Fig. 32-2** Bank protective works.

the fences and can drop into any underlying scoured area that may develop. The Santa Clara installation was built for $10 per foot in 1938.[1]

Longitudinal pile dikes have been used in the Mississippi Valley both to protect a concave bank and to protect the flank of a revetted bank.

The first pile dike constructed on the Mississippi River consisted of a double row of single piles with the tops pulled together and wired, and the structure wattled with brush. No foundation mattress was provided. This construction proved too light and the dikes were gradually strengthened, the wattling was eliminated, and a mattress was provided along the axis of the dike at the bottom of the river. The common pile dike consists of two or more, up to a maximum of seven, rows of pile clumps, three piles constituting a clump. [See Fig. 32.3.] The rows are spaced approximately 5 ft apart, with pile stringers placed between each row. Clumps are spaced from 15 ft to 20 ft apart depending on the number of rows in the dike. Piles and stringers are secured with several turns of $\frac{3}{8}$-in. galvanized wire strand fastened with boat spikes. The pile penetration varies from 20 ft to 30 ft below the river bed. Each dike is constructed on a woven willow or lumber mattress, extending from the water's edge from 45 ft to 75 ft beyond the channelward end

[1] G. A. Tilton, Jr., Bank Protection, *California Highways and Public Works*, **20**(5): 12–20 (1942).

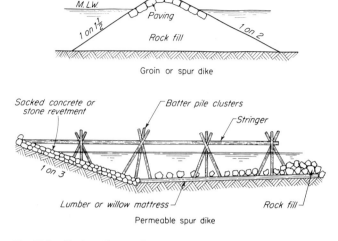

Groin or spur dike

Permeable spur dike

**Fig. 32-3**  Contraction works.

of the dike proper.  Mattress widths vary from 77 ft to 100 ft.
The mattress is ballasted with 15 lb of stone per square foot and an
additional 500 lb of stone per row, per lin ft, are placed in the dike
line to fill holes torn in the mattress by pile driving.

The crest elevation of the pile dike is usually set at mid-bank
state, which is from 15 ft to 17 ft above the low-water plane.  There
are two reasons for choosing this elevation.  First, pile lengths
would become excessive if the top of the dike were raised substantially above this elevation, and second, the most severe current
impingement on the bank in a bend usually occurs between the low-
and mid-bank stages.  After the latter stage is exceeded, the river
has inundated the bar opposite the bend and the flow becomes
more or less axial. . . .[1]

**32-4  Revetment**  While longitudinal dikes have proved satisfactory on
the smaller European streams used for barge traffic, they result in rather
expensive installations for the larger navigable rivers of the United States.
American practice has favored the development of large-scale revetment
work for bank protection.  A revetment is merely a form of pavement or
bank lining, and a wide range of materials have been employed for the
purpose.  That portion of the revetment below normal low water is
called the "mattress" and the portion from the mattress to the top of the

[1] R. H. Haas and H. E. Weller, Bank Stabilization by Revetments and Dikes, *Trans.
ASCE*, **118**:849 (1953).

bank is called "bank paving" in the terminology used by the Mississippi River Commission.   In Mississippi River work,[1]

> . . . experience has shown that, to be effective, a revetment must be sufficiently long to protect the entire concave bend from the upper to the lowermost points of caving at all stages.   It must extend from the top of the bank to the toe of the underwater slope, a vertical distance of from 80 ft to 150 ft.   It should be flexible so that it can mold itself to irregular surfaces; it must have sufficient strength to remain intact in the event of uneven slope settlement or scour.   It should be relatively impermeable and continuous in order to prevent fine soil particles from leaching out through the mattress under the action of current that may reach a maximum velocity of 12 ft per sec.   Also, as far as practicable, the revetment material should be indestructible in air or water.

Construction.—Construction normally follows a sequence of bank clearing and grubbing, clearing of the underwater slope of snags, bank grading, mattress sinking, and finally, bank paving.

The bank in a concave bend is typically very steep above the water surface and, consequently, is unstable.   Therefore, grading to a stable slope, initially, is necessary.   The slope required for stability is determined by an analysis of the soil from representative borings taken at close intervals along the bank on which the revetment is to be constructed.   Bank slopes varying between the limits of 1 ft vertical to from 3 ft to 5 ft horizontal are usually required to insure stability.

In the earlier days of construction, banks were usually graded after the sinking of the subaqueous mattresses; but the availability of equipment to do this part of the work, without delaying progress on the work as a whole, was often the determining factor.   Usually the bank was graded after the mattress was in place because it was thought that the waste material from the bank-grading operations was beneficial for filling or loading the mattress.   Bank-grading operations today are prosecuted in advance of sinking operations because of operational considerations and because the modern mattress has sufficient body and weight to eliminate the requirement for filling and loading.

Development of Materials.—Soon after the initiation of work by the Commission at Plum Point and Lake Providence, it became apparent that the light, flexible, and comparatively inexpensive structures of poles and brush being used were inadequate to cope with the conditions prevailing on the lower river.   Between 1882

[1] *Ibid.*

and 1893 numerous mattress types of willow, lumber, and other native materials were designed and tested. Two of the types developed during this period, the framed willow and willow fascine mattresses, were used extensively until about 1930. Because of the scarcity of willows, the short construction season, the relatively slow production rate, and the greater cost of construction only a limited number of these types of revetments have been constructed since 1930. The most recent examples of the use of these types were a willow fascine mattress placed in Memphis harbor in 1948, and a framed willow mattress constructed in New Orleans (La.) harbor in 1949.

Articulated Concrete Mattress.—During the thirty-year period between 1914 and 1944, four distinct types of concrete mattresses and one compacted asphalt mattress were developed. Of these types (the monolithic concrete, the articulated concrete, the lapped and butt slab, the compacted asphalt, and the flexible roll type concrete) only the articulated concrete mattress proved to be satisfactory in respect to production, cost, and service.

The articulated concrete mattress originated with experiments, begun in 1915, to develop a flexible and permanent underwater mattress. After many trials and alterations, the present forms of mattress and sinking mechanism were evolved. The mattress is composed of precast units 25 ft long by 4 ft wide and 3 in. thick. Each unit has twenty individual blocks, 14 in. wide, spaced approximately one inch apart, on heavy corrosion-resisting reinforcing fabric that is continuous throughout the unit. The precast units are assembled on the sloping deck of a launching barge where they are united to each other and to launching cables by wire rope clips and twist wires. The launching cables are fastened ashore to anchors embedded in the bank. The barge is then moved riverward 25 ft, allowing the joined units to slide off the curved launching

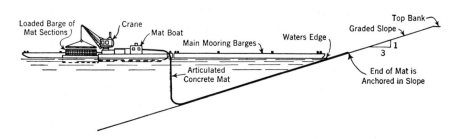

**Fig. 32-4**  Placement of articulated concrete mattress.  [*From R. H. Haas and H. E. Weller, Bank Stabilization, Trans. ASCE,* **118**:852 (1953).]

**Fig. 32-5** Construction plant for placement of concrete mattress. [*From R. H. Haas and H. E. Weller, Bank Stabilization, Trans. ASCE,* **118**:853 (1953).]

apron and hang suspended in the water on the launching cables. This launching process is repeated until a complete mattress of any desired width has been constructed. The cables are then cut at the outstream end and the plant moved upstream into position for laying the next mattress. [Details of the placement operation are shown in Figs. 32-4 and 32-5.]

The articulated concrete mattress is the best type of revetment devised, and the only type that can be placed at the rate required for the proper prosecution of the stabilization program. Three complete revetment construction plants, especially designed to place the articulated concrete mattress, are operating in the lower river. The normal combined capacity of these three plants is 140,000 squares (14,000,000 sq ft) of mattress per month. During a normal low-water construction season of five months' duration, these three plants are capable of reveting approximately 40 miles of bank with a mattress averaging 360 ft wide.

The reinforcing fabric and fastenings used in the articulated mattress construction are made of corrosion-resisting metal. Copper-coated, high-tension steel, or stainless steel wires, each having a breaking strength of 4,000 lb, are used. This fabrication makes the mattress as indestructible as possible in air and water.

Since complete flexibility is not compatible with a concrete product, the necessary articulation produces interstices through which bank fines can be drawn by the current. Since 1948, a less permeable modification of the articulated concrete mattress, a so-called V-type mattress, has been developed. Only small test installations have been made, but during the 1950 construction season a test section of considerable size was constructed using this mattress. The V-type mattress has open areas of 3 per cent as compared to about 8 per cent in the conventional mattress.

Asphaltic Mixes.—Experiments with nonreinforced asphaltic mixtures are now being made in the form of plastic masses or blocks. For mass placement, the heated mixture is placed in bottom dump-barges, towed to the site, and released. The plastic mass spreads and congeals to cover the subaqueous slope. When used in the form of blocks, the asphalt mix is cast in water-cooled forms at the site and subsequently released through wells in the barge to cover the underwater slope. This method has not been very successful for new construction, but is usually effective for repair of damaged work, especially in the less turbulent reaches of the lower river.

Bank Pavements.—After a mattress is in place, the graded bank above the water surface is paved to prevent erosion by river currents, which would result in bank instability and consequent loss of the subaqueous mattress.

Bank pavements of the earlier revetments were generally constructed with the same material as the mattress but experience with brush or fascines and wire netting proved that, when subjected to alternate wetting and drying, the wood soon decayed and the metal parts corroded rapidly. Consequently, stone was adopted as the standard material for bank paving.

Riprap on a 4-in. gravel blanket is considered the most effective pavement yet constructed but, because of the local unavailability of stone, it is rarely used below Memphis. A 10-in. riprap paving can adjust itself faithfully to irregularities in the slope; and it can reform and continue to give protection in the event of minor bank subsidence and sloughing. A pavement that cannot accommodate such settlements fails locally and requires costly maintenance.

In an attempt to find durable substitutes for riprap, pavements of monolithic concrete, articulated concrete, asphalt, and manufactured blocks have been constructed with varying degrees of success. Of these types only the articulated concrete and non-compacted asphalt pavements remain in general use.

The articulated concrete pavement is identical to the mattresses of the same material but differs in the detail of placement. It is placed on a 4-in. gravel blanket that permits relatively free drainage of ground water but retains the bank fines. Although the articulated concrete bank paving is relatively simple to place, it is comparatively expensive. Therefore, its use is generally limited to connections between the subaqueous mattress and other types of paving. It is also used extensively at the upper and lower ends of revetments, where flexibility is especially desirable to cope with flanking action.

In 1945 the porous asphalt pavement (which is in general use below Memphis) was developed. The uncompacted asphalt mass is placed on the prepared bank to a minimum thickness of 5 in. It consists of a heated mixture of about 94 per cent bar run sand and gravel, and 6 per cent asphaltic cement. With ordinary bar run sand the porosity of the pavement is about the same as that of loose sand. The degree of porosity, however, can be increased by the addition of a small quantity of pea gravel in the mix. The relatively free-draining feature of the uncompacted asphalt makes it superior to either a monolithic concrete or compacted asphalt pavement.

Pavements composed of ordinary concrete blocks or tetrahedral-shaped blocks show little, if any, advantage over stone. In addition, their high initial cost prohibits general use. . . .

**32-5  Contraction works**  In the effort to achieve and maintain navigable river channels a major problem results from the shallow depths to be found in the crossings. Because of the inefficient cross section at crossings, velocities are lower there than in the pools, despite the steeper slopes which prevail. The engineer must induce the stream to cut a deeper channel by restricting the opportunity for energy dissipation in broad shallow sections. The concentration of flow may be improved by (1) training walls, (2) groins, and (3) retards.

Training walls do not differ, except in purpose, from the longitudinal works described in Sec. 32-3, having the function of directing river currents to scour the bed rather than the function of protecting the bank.

Groins or spur dikes are dikes or weirs projecting transversely from the bank into the stream, reducing the effective width of the stream (Fig. 32-1). Sedimentation occurs in the backwaters created in the groin fields (between groins), and the river bank is thus extended from its original location.

The term *groin* is customarily reserved for a masonry structure which, in the river work, usually amounts to a pile of cobblestone dumped in place. In European practice the groin is frequently paved with hand-placed stone. The size of stone employed depends upon current velocities, and Fig. 32-1 will be found useful in forming an opinion on this point. Even if the individual stones are stable with respect to current action, the groin as a whole is vulnerable to scour at its extreme ends. At its shore end the groin must be built well into the bank as a safeguard against attack from the rear. At the exposed end, or head, of the groin the river currents are intensified by the resulting concentration of flow. Consequently, there is a danger that bed scour will weaken the foundation of the groin head. Some sort of bed protection, such as additional rock or a brush mattress, is desirable at this location. Mass bituminous concrete is sometimes preferred to uncemented rock. This permits the use of smaller aggregate for equal or greater stability.

Greater economy can be achieved in the construction of contraction works if the river itself can be made to transport some of the fill material. With this objective, permeable spur dikes and retards are frequently employed to serve the same purpose as solid groins. A permeable spur dike does not prevent the passage of water but does retard flow sufficiently to increase the rate of sedimentation. The necessary sedimentation can occur only if the river carries considerable material in suspension. For clear streams a solid groin is necessary. Many forms of the permeable spur dike have been developed. In this country wood piling is commonly used, and the preceding section on longitudinal pile dikes describes the typical installation. Quoting the same source,[1]

> The upstream dike of [such] a system is inclined downstream at a small angle with the bank and extends to the rectified channel line. This upstream structure is designed to turn the current slightly offshore. [It is a training wall.] The remaining dikes in a system are placed normal to the bank line or angled approximately 15 deg downstream from normal. [These spur] dikes are spaced from $1\frac{1}{2}$ times to $2\frac{1}{2}$ times the length of the upstream dike [or training wall], depending upon the radius of curvature of the bend and the angle and intensity of the current attack. . . . Dikes placed on the convex side of the river to promote bar growth are spaced about $2\frac{1}{2}$ times the length of the upstream dike up to a dike length of 1,000 ft. For longer dikes, the relation between length of dike and spacing is less. All dikes are normally angled slightly downstream.

[1] *Ibid.*

In locations where the driving of wood piles is not practicable, a stable supporting structure may be made by connecting four steel rails to form the edges of a tetrahedron. A series of these structures, connected by cable and fencing, may be placed to form a spur dike. Drift caught in this barrier decelerates the current and encourages sedimentation.

European practice favors inclining the groins upstream with the heads at low-water elevation. At higher stages water overflowing the groin is directed toward the channel rather than toward the shore. When the river bed is especially resistant to erosion, groins should be inclined downstream to improve the scouring action. Short groins are not effective. Groins should be long enough to extend from the banks to edge of the desired channel. The spacing of adjacent groins should equal the channel width. The new channel is not fully stabilized until the groin fields are filled with detritus.

A retard consists of a tree or a fabricated timber framework which is anchored by a cable, reducing the current velocity and promoting sedimentation. Such devices are effective only in muddy streams.

## QUESTIONS AND PROBLEMS

**32-1.** A river falls 100 ft in 100 miles above its mouth. Cutoffs reduce its length to 90 miles. The former channel had an average centerline depth of 10 ft. The bank slopes are 1 on 10, forming a triangular channel. Bazin's $m = 3.0$. With no change in discharge, what is the new centerline depth?

**32-2.** Assume that the mean velocity of the stream is twice the minimum current velocity and half the maximum current velocity. $Q = 6,000$ cfs.
  *a.* Find the slope of the stream at the bends and in the reaches.
  *b.* What is the largest particle scoured in the bends?
  *c.* What is the smallest particle deposited in the reaches?

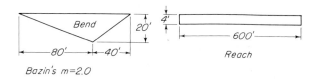

Bazin's m=2.0

**32-3.** A wide river has an almost uniform depth of 20 ft at flood stage and a uniform velocity of 8 ft/sec. What is the largest particle that will be scoured from the banks? At what velocity will this material again be deposited? What is the largest particle that will roll along the bed?

**32-4.** In a 10-mile stretch of river there are 5 miles of bend and 5 miles of reach and crossing which approximate the cross sections in the sketch. The river falls 20 ft in 10 miles. Find slope and mean velocity in both sections (*a*) for flood condition, (*b*) for dry-weather flow.

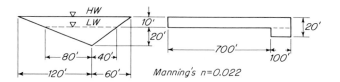

**32-5.** The sketch represents a typical river half section before and after the construction of groins has reduced the channel width to 300 ft. Assuming that water level, gradient, and discharge are the same, what is the new depth at center line?

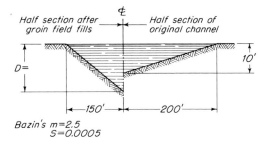

**32-6.** A heavily loaded stream has a V-shaped channel 200 ft wide with a maximum depth of 20 ft and a slope of 0.001. It empties into a river which is 1,000 ft wide below the junction, with a uniform depth of 20 ft and a slope of 0.0001. Will there be a deposit near the junction? What range in grain size might be found in such a river bar?

**32-7.** A river bank on a 20-deg slope is revetted with a 6-in.-thick impervious paving weighing 145 pcf. As the river level falls after a flood the ground-water level in the bank lags behind. At what differential head will the revetment be lifted out of place, (*a*) above river surface? (*b*) below river surface?

**32-8.** What size stone should be used for groin construction in the river of Prob. 32-5?

## BIBLIOGRAPHY

Franzius, Otto: "Waterway Engineering," The Technology Press of the Massachusetts Institute of Technology, Cambridge, Mass., 1936.

Minikin, R. C. R.: "Practical River and Canal Engineering," Charles Griffin & Company, Ltd., London, 1920.

Townsend, Curtis McD.: "River and Harbor Construction," The Macmillan Company, New York, 1922.

# 33
# Locks and Dams

**33-1 Slack-water navigation** It has been pointed out that shallow sections of a river have steeper gradients. This statement can be rephrased with equal truth and usefulness to say that channels can be deepened by a reduction in gradient. When a river is dammed, the pool behind the dam is only the lower portion of a backwater effect extending far upstream. Of course, the total fall of a river from source to mouth is not generally subject to human modification, but the engineer can arrange, within limits, the distribution of the total fall. A dam serves to concentrate a selected portion of the total fall of the river and produces very flat gradients in its upstream pool. Vessels pass this concentrated fall by means of locks. The employment of a system of locks and dams requires less flow than any other method of obtaining navigable channels. The slack water behind a dam simplifies navigation and reduces the power required to move a cargo upstream. The energy of the river itself is utilized to raise a ship or barge in the locks just as truly as if a hydroelectric plant were installed at the dam and electric power were supplied to the ship for its propulsion.

The low velocities in the backwater greatly reduce the power of the stream to transport its load.   Consequently, the river aggrades, and some dredging may be needed to maintain adequate channel depths.

**33-2  Dams**   Dams in navigable rivers may be either fixed or movable. The latter are especially useful when satisfactory channel depths are available during much of the year even with little or no differential head at the dam site.   In some cases, open-channel navigation can prevail at high-river stages if the weir is lowered or removed.

With increased discharge the rise in tail water is much greater than the rise in headwater.   If the increase in discharge is very great the dam will be "drowned," or completely submerged.   With proper design the submerged dam does not obstruct navigation.   The main factors affecting design are the need to create a large head differential at low-river stages and the desire to reduce the obstruction to flow at flood stages.   This dual objective can be realized in a low spillway section, or sill, which supports crest gates to raise the pool level when river levels are down.

Considerable ingenuity has been exercised in the design of movable dams, and a great variety has resulted over the years.

Flashboards are horizontal boards supported by vertical pins or pipe set into the spillway crest.   The pins fail when the head exceeds a set amount.   The device becomes inconvenient if several floods must be passed each year, as lost flashboards must be replaced manually whenever the pool level is to be raised again.

Needles are vertical beams which transfer the water pressure to their two horizontal supports.   They may be either timber or steel pipe. High working stresses are used to reduce weight to a minimum.   Maximum net heads range from 13 ft for timber to 18 ft for steel needles.

Stop logs are horizontal beams and are generally timbers hand-placed in grooves in the masonry piers.   Winches are used in more important installations, not only for removing the stop logs but also to overcome their buoyancy in placing them.

If the stop logs are tied together and slide on rollers instead of on the supports directly, the result is called a Stoney gate.

By substituting a roller chain for the individual rollers, a curtain gate is obtained.

The next step is a hollow steel cylinder which acts as a dam when resting on the sill but which can be rolled up out of the way when it is desired to lower the pool.

Instead of a circular cylinder, a sector of a circle can be used, in which case the gate rotates about its axis instead of rolling along its end supports.   The result is a Chittenden drum wicket or a Taintor gate.

The change in crest level made possible by structures like those described above provides much better control of upstream river levels than is possible with overflow dams of fixed spillway elevation. Numerous references on the design of fixed dams are generally available.

A main factor in the selection of the dam site is finding space for an adequate spillway. Flood flows must be passed with the least possible rise in upstream river levels. Hence, the spillway must be long enough to provide adequate discharge capacity with minimum head. Because wide shallow sections of the river offer more room for spillway construction than do narrow deep sections, river dams are more often located in reaches than in bends.

The planner will prefer to avoid placing river boats in troublesome currents, upstream or downstream. It follows that ship locks should be kept apart from an overflow spillway. An island site permits a natural separation of these two functions.

**33-3 Backwater** The determination of the water-surface profile upstream from a dam is an important operation which warrants a somewhat detailed explanation,[1] if only because it is not covered by elementary hydraulics instruction. The river is divided into consistent sections of convenient length and uniform discharge. Bernoulli's theorem may be applied to the extremities of any such sections. Thus

$$z_1 + \frac{V_1{}^2}{2g} = z_2 + \frac{V_2{}^2}{2g} + \text{losses} \tag{33-1}$$

The losses may be accounted for by the $S$ of Manning's formula. In addition, changes in velocity are accompanied by the loss of energy, which may be expressed as a fraction of the change in velocity head. Equation (33-1) then becomes

$$z_1 + \frac{V_1{}^2}{2g} = z_2 + \frac{V_2{}^2}{2g} + \frac{1}{2}(S_1 + S_2)L + K\left(\frac{V_1{}^2}{2g} - \frac{V_2{}^2}{2g}\right) \tag{33-2}$$

where $z_1$, $V_1$, $S_1$ = depth, velocity, and slope at one end of the reach under analysis.

$z_2$, $V_2$, $S_2$ = corresponding terms for the other end.

$L$ = length of the reach between the selected sections.

$K$ = a coefficient of eddy loss. For abrupt expansions or contractions $K = 0.5$; for a gradually diverging reach it is 0.2; for a converging reach it is 0 to 0.1.

It is convenient to rewrite Eq. (33-2) as follows:

$$z_1 + f(z_1) = z_2 + \phi(z_2) + K\left(\frac{V_1{}^2}{2g} - \frac{V_2{}^2}{2g}\right) \tag{33-3}$$

[1] Arthur A. Ezra, A Direct Step Method for Computing Water-surface Profiles, *Proc. ASCE*, Separate No. 180 (1953).

In this equation velocity has been expressed in terms of Manning's formula. The resulting expression has been simplified by the substitutions

$$f(z) = \frac{Q^2}{2gA^2} - \frac{L(Q^2n^2)}{2(2.2A^2R^{\frac{4}{3}})}$$

and

$$\phi(z) = \frac{Q^2}{2gA^2} + \frac{L(Q^2n^2)}{2(2.2A^2R^{\frac{4}{3}})}$$

In the work of computing the upstream profile Eq. (33-3) may be simplified, as a first approximation, by neglecting the last term. It then becomes

$$z_1 + f(z_1) = z_2 + \phi(z_2) \tag{33-4}$$

The left side expresses the known flow conditions at one section across the stream, while the right-hand side describes flow at an unexplored section $L$ ft upstream. The problem is to find consistent values of $A_2$, $R_2$, and $z_2$ which will satisfy the equations. This can be done by trial and error with patience, or the procedure can be organized graphically, depicting the relationship between the water-surface elevation at the selected location and the corresponding values of $z + f(z)$ and of $z + \phi(z)$. Figure 33-1 shows such a plot for several river stations. Assume that the water-surface elevation at Sta. 80+00 has been found to be 18.5 for a discharge of 33,500 cfs. Then, from the graph, the value of $z + \phi(z)$

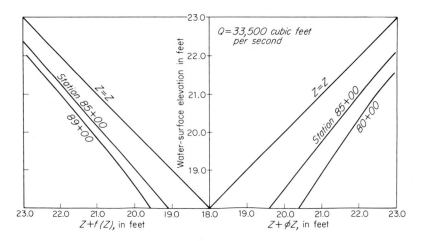

**Fig. 33-1** Curves for determination of water-surface profiles.

is found to be 20.7. By Eq. (33-4), this must also be the value of $z + f(z)$ for the next selected upstream section, Sta. 85+00. The graph shows that the corresponding water-surface elevation at Sta. 85+00 is 19.8 ft.

The water-surface profile can now be corrected to reflect eddy losses, by applying Eq. (33-3). These losses, expressed as

$$K \left( \frac{V_1^2}{2g} - \frac{V_2^2}{2g} \right)$$

for any stretch under analysis, must be added to $z + \phi(z)$ value for the downstream section of the reach under study. This correction is unnecessary where losses caused by changes in channel cross section are lumped with other losses in selection of a roughness factor $n$.

**33-4 Lock design** For centuries engineers have been able to raise or lower ships by admitting or draining water while the ship floats in a closed channel. Although the basic problem had been solved long ago, the details of lock design still challenge the ingenuity of the designer.

The time required to pass a vessel through the locks is a charge against the cost of operating the vessel. On the other hand, rapid filling or emptying of the lock involves water velocities which may be accompanied by currents or eddies that are hazardous to the ship or barge. Conflicting requirements of speed and safety call for modern refinements in lock design—designs which basically are much the same as those conceived by engineers who lived prior to the industrial revolution. The chief problems are more hydraulic than structural. The distribution of current velocities in a channel is difficult of analysis under any circumstances, so that the use of hydraulic models has become more common. In practice it has usually been found necessary to make changes after a new lock has been put into operation. One engineer reports[1] that the locks constructed for American inland waterways

are used effectively in the transportation of traffic. However, practically none is operated as designed, particularly in the case of upbound lockages or when sections of tows are moored to the lower guide walls. Through experience, a valve-operation procedure has been developed for each lock which provides satisfactory lockage conditions. However, in doing so, the operation time has been increased from 50 per cent to 100 per cent more than that planned in the design.

The objectives of those engaged in lock design have been, and are, to design the lock hydraulic system so as to:

[1] A. Frederick Griffin, Influence of Model Testing on Lock Design, *Trans. ASCE,* **116**:841 (1951).

1. Permit filling or emptying the chamber in the briefest period compatible with unobjectionable turbulence, line stress, crosscurrents, or surges within the lock chambers;
2. Minimize disturbances in the lock approaches, thus permitting more efficient use of guide walls for mooring vessels; and
3. Obtain an even distribution of inflow and discharge across the lock entrances to reduce the concentration of velocities, surges, eddies, and erosion, giving due consideration to the economics of the solution.

The lock itself is a section of concrete-lined channel or canal with gates at either end. Since the water level in the lock fluctuates from headwater elevation to tail-water elevation, the subgrade soil pressure also experiences extreme variations. Maximum bearing capacity, uplift pressures possibly leading to buoyancy, and the effect of stress repetitions must all be considered in the foundation design. Likewise, seepage forces must be added to the earth pressure exerted upon the walls of the lock.

Gates must be strong, tight, and readily operated. Changes in water level are made with the gates closed; intolerable currents would result from use of the lock gates to empty the lock chamber. Water is taken to or from the lock chamber through culverts located behind the chamber walls. The ratio of culvert area to chamber area is set by the lift, the over-all coefficient of discharge, and the allowable time for filling or emptying, as established by the familiar formula for discharge under falling head:

$$t = \frac{A_2(h_1 - h_2)^{\frac{1}{2}}}{CA_c \sqrt{g}} \tag{33-5}$$

where $h_1 - h_2$ = lift
$t$ = time to empty or fill the chamber
$A_2$ = area of chamber (= width $\times$ length)
$A_c$ = total area of culverts
$C$ = over-all coefficient of discharge for the drainage system

Culverts are connected with the chamber by "ports" or orifices in the chamber walls, or by ports in the bottom. The former method is apt to create troublesome currents unless the sills of the locks are at considerable depth below tail-water elevation. Ports are often bell-mouthed for greater efficiency and to reduce crosscurrents in the chamber. Turbulence can be reduced by staggering side ports on opposite walls of the chamber.

Floor laterals discharging through bottom ports are preferred especially when in shallow water, but the discharge should be directed to form a horizontal jet.

Ports should not be closer to the upstream gate than 0.25 of the chamber length.

Gates are generally of the "miter" type; they resemble a pair of double doors used in building construction but with their combined width slightly greater than the door opening. Because of this chamber the differential head on the upstream side of the miter gate tends to effect a tighter closure.

**33-5  Lock size**   The dimensions of a lock are pretty well set by the conventional barge size and number of vessels in a tow for the specific river system under study. These factors are, in turn, influenced by the commodities being transported, the characteristics of the river, and the length of the haul. European waterways involve smaller units moving shorter distances, although the total use of waterways is relatively much more extensive than in America. A British tug of 90 hp may push a 250-ton payload at 5 mph 100 miles inland from the coast. On the long hauls of the Mississippi River system the power and the payload may be 30 times as great, with six barges making up a commonly encountered tow, two abreast and three barges long. Except in rough waters or tortuous channels, barges are usually pushed by the towboat. Authorities recommend that locks be 3 ft wider than the combined width of the barges after allowing an additional foot between barges for fenders. Locks should be 10 or 20 ft longer than the tow as a safeguard against lock damage.[1]  Common sizes for American barges are 26 by 175 ft with 8 to 11 ft draft (750- to 1,250-ton capacity), 35 by 195 ft with 1,500-ton capacity on a 9-ft draft. The wider barge is favored for liquid cargoes. Locks which will efficiently accept common multiples of both barge sizes are 110 by 600 ft and 110 by 1,200 ft.  A 56- by 360-ft lock is common but is not well suited to 35-ft barges.

**QUESTIONS AND PROBLEMS**

**33-1.** The discharge over the spillway shown in the sketch is expressed by the equation

$$Q = 3,000H^{\frac{3}{2}} \quad \text{cfs}$$

The channel below the dam has a gradient of 0.001. Manning's $n = 0.025$. Find headwater and tail-water elevations where (a) $Q = 1,000$ cfs, (b) $Q = 10,000$ cfs.

[1] Ralph L. Bloor, Lock Sizes for Inland Waterways, *Trans. ASCE*, **116**:886 (1951).

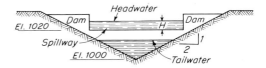

**33-2.** Flashboards are used on a spillway crest as sketched.

*a.* What spacing of the steel-pipe posts will develop the allowable working stress in the wood?

*b.* What section modulus should a pipe have to fail at this spacing, using ultimate strength of pipe?

*c.* Select the largest standard pipe size which has a lesser section modulus.

*d.* Determine the pipe spacing that will cause failure of selected pipe.

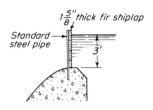

**33-3.** The sketch shows two cross sections of a river, the one at the right being located 1,000 ft upstream of the other. $K = 0$. $n = 0.025$.

*a.* Plot $z_1 + f(z_1)$ for the downstream section and $z_2 + \phi(z_2)$ for the upstream section, both for a discharge of $Q = 6,400$ cfs.

*b.* Find the water-surface elevation at the upstream station when the water-surface elevation at the downstream station is elevation 120.00.

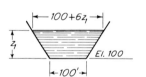
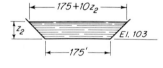

**33-4.** Referring to Fig. 33-1, what surface elevation can be tolerated at Sta. $85+00$ if it is specified that the river level at Sta. $89+00$ cannot exceed elevation 22.0 at $Q = 33,500$?

**33-5.** The chamber for the main lock at the Lake Washington Ship Canal, Seattle, is 825 ft long and 80 ft wide. The water is lowered in the lock by means of two culverts, each 14 ft high and $8\frac{1}{2}$ ft wide. Assuming a lock coefficient of 0.6, how long would it take to lower the water level when the lift is 20 ft?

**33-6.** Assuming that the uplift water pressure on the floor of the lock in the previous problem varies uniformly from sill to sill, what total buoyant force will be effective when the water in the chamber is at tail-water elevation?

**33-7.** Rework Prob. 33-3 with the different assumption that the downstream section is the one sketched on the right. $K = 0.2$.

**33-8.** A pair of miter gates which include an angle of 135 deg when closed are used to close a lock chamber 30 ft wide. Water levels against the gates are 10 and 30 ft above the sill. What total thrust is exerted by the gates in a direction perpendicular to the chamber wall?

**BIBLIOGRAPHY**

Creager, W. P., J. D. Justin, and J. Hinds: "Engineering for Dams," John Wiley &
    Sons, Inc., New York, 1945.
"Manual on Lock Valves," American Society of Civil Engineers, 1930.
Schoklitsch, Armin: "Hydraulic Structures," American Society of Mechanical Engi-
    neers, 1937.

# 34

# Ports and Harbors

**34-1 River ports** A river port may mean either an upstream development for handling the freight of river boats and barges or it may mean the potentially far more important development at a river mouth for serving ocean traffic as well. In the latter sense a river port is a type of seaport.

The port which receives only river traffic presents much simpler problems of design than is apt to be the case with seaports, mainly because the factor of protection from wave action is minimized. The removal of this restrictive requirement means that the location of terminal facilities is generally less dependent upon topographic features than on rail and highway connections, industrial and commercial needs, and upon the city plan. All this means that the engineer has relatively more freedom in locating the development. Because there is less need for the intensive use of limited water-front space, the wharves are generally spread out along the river bank, rather than built perpendicular to the shore as piers and slips.

Some details of wharf design are taken up in later chapters.

**34-2  Seaports**  The term *port* implies both a harbor and the facilities required for the handling of cargo and the servicing of ships.   The design and construction of these facilities and the design and construction of harbor improvements are civil-engineering functions.   Here, as always, civil engineering is concerned with the control of natural forces, but here especially, the magnitude of the forces involved (water pressure and wave pressure) makes it imperative that control be achieved by an intelligent utilization of the forces of nature rather than by a passive resistance that may be both expensive and ineffectual.   The engineer must apply his knowledge of hydraulics and soil mechanics, and he must acquaint himself with the actions of wind and tide.

Many foreign ports are very old.   In ancient times there was a tendency for ports to develop well upstream from the mouth of a river. Refuge from storms could be found at the river's mouth, but farther upstream there was greater security against military attack.   As ships become ever larger, the access to such river ports could be maintained only at increasing expense.   To avoid becoming an inland city, Hamburg maintains a 40-ft channel by dredging, in place of an original limiting depth of 6 ft.   Hamburg is 67 miles from the North Sea.   The persistent struggle to provide adequate ship channels has been too great for some other European cities; Paris, Rome, and Cologne no longer are seaports. In still other places adequate channel depths exist only at high tide.   The wharves in such ports are located on "wet docks."   Vessels dock at high tide and the water level in the inner harbor is prevented from falling with the ebb tide by locks similar to those used in canals or in drydocks. Partly because of different historical background, the problems of the inland seaport are less acute in America than in Europe, although New Orleans, Portland, Sacramento, Baltimore, and Richmond are all more than a hundred miles from the open sea, and Montreal is over a thousand.

The chief requirements of a port are obvious ones.   Ships must be able to reach port, so the passage to open sea must have sufficient depth and width and must be suitably marked to aid navigation.   Ships reaching harbor must be able to anchor while waiting for a berth to discharge or take on cargo, fuel, and supplies.   Finally, the port must possess suitable wharves, backed by appropriate facilities for handling and storing the commodities passing through the port and for servicing the ships.

The various American ports differ immensely in importance and almost as widely in their facilities.   In this respect seaports and airports stand in contrast to rail or highway facilities, where importance is reflected in the number of units (railroad cars or highway trucks) and not so much in the size of the individual car.   A major seaport must accommodate vessels of greater size and draft than does a minor seaport, just as a major air terminal must be constructed to provide safe landings for much

heavier planes than an airfield on a branch airline.  Consequently, it is
not possible to specify rigid minimum requirements for seaport facilities.
In general, each port must build to the needs of its commerce.

**34-3  Harbors**  Ports must have harbors.  A harbor is a place where
ships can find refuge from storms.  To fulfill its functions a harbor must
satisfy three requirements:

1. Natural or artificial ship channels must have sufficient depth for the
   draft of vessels that are to be accommodated.
2. Protection against destructive wave action must be provided by land
   masses or by breakwaters.
3. The bottom should furnish secure anchorage to hold ships against
   the force of high winds.

What constitutes an adequate channel depends upon the nature of
the shipping to be accommodated.  The following table cites controlling
channel depths at mean low tide in various American ports.

**Table 34-1  Channel depths***

| Port | Size | Harbor type | Shelter | Depth, ft | | |
|------|------|-------------|---------|-----------|-----------|-------|
|      |      |             |         | Channel | Anchor-age | Wharf |
| New York | Large | River (natural) | Excellent | 40+ | 40+ | 40+ |
| Galveston | Large | Coastal (natural) | Good | 30–35 | 25+ | 35− |
| Los Angeles | Large | Coastal (breakwater) | Excellent | 40+ | 35+ | 40− |
| Newark | Medium | River (basin) | Good | 35–40 | 35+ | 40− |
| Tacoma | Medium | Coastal (natural) | Excellent | 40+ | 40+ | 40+ |
| Astoria | Small | River (natural) | Excellent | 30–35 | 30+ | 40− |
| Gloucester | Small | Coastal (natural) | Good | 20–25 | 20+ | 25− |
| Palm Beach | Very small | Coastal (natural) | Good | 25–30 | 25+ | 30− |
| Mackinac Island | Very small | Lake | Fair | 15–20 | 40+ | 20− |

\* Data from World Port Index, *U.S. Navy, Hydrographic Office, Publ.* 950 (1953).

The harbor entrance should be wide enough to permit ready passage for shipping and narrow enough to restrict the transmission of excessive amounts of wave energy in time of storm.   There is some evidence that the entrance width tends, in actual practice, to be about equal to the length of the larger ships using the harbor.   The entrance should be located with regard to the direction of prevailing storms and to the convenience of approaching ships.   The effect of entrance width to the size of wave which will reach the sheltered area is discussed later, in the treatment of breakwaters.

During the past hundred years Federal expenditures for harbor improvement have been used in increasing proportions for channel deepening rather than for the breakwaters, seawalls, and other protective works which monopolized the attention of government engineers in the early days of American maritime activity.   This shift in emphasis has been the result of drastic changes in the size of ships.   Some knowledge of the physical traits of the American merchant marine is indispensable to the engineer who must build not only for present needs but also in anticipation of future developments.

**34-4   Shipping**   The characteristics of principal types of vessels in the American merchant marine are listed in Table 34-2.

The Maritime Administration code designation for its ships is of some general interest.   The system is based on three groups of letters and numbers which outline the characteristics of the vessel.   The first group indicates the type of vessel and its length in feet, as shown in Table 34-3.

The second group of numbers designates the propulsion, as shown in Table 34-4.

In the same fashion the designation of S.S. *United States* as P6-S4-DS1 indicates that she is a passenger ship between 900 and 1,000 ft long, having four propellers driven by steam engines.   The symbol DS1 refers to her design.

The growth of air transportation may doom the giant high-speed luxury liner of the North Atlantic to extinction even more surely than it will displace the transcontinental extra-fare Pullman train.   There is support for the development of a very large vessel of moderate speed for the economical transportation of tourists and cargo.   A British authority[1] suggests that such a ship may have the length of 1,000 ft, 150-ft beam, and 40-ft draft, approximate gross tonnage 100,000 and speed 15 knots.

The same reference suggests that the most efficient size of cargo ship is the largest which can make use of the principal world ports, which means a 40-ft channel depth.   One recommendation is for a vessel 660 ft long,

[1] F. M. Du-Plat-Taylor, "Docks, Wharves and Piers," p. 41, Eyre and Spottiswoode, London, 1949.

**Table 34-2   Characteristics of principal types of vessels in the American merchant marine**

| Type | Length, over-all Ft | In. | Beam[a] Ft | In. | Draft[b] Ft | In. | Dead-weight tonnage[c] | Gross tonnage[d] | Bale cubic[e] | Machinery | Service speed[f] | Number of passengers | Number of crew |
|---|---|---|---|---|---|---|---|---|---|---|---|---|---|
| P6-S4-DS1 S.S. *United States* | 990 | 0 | 101 | 6 | ...... | | ...... | 52,000 | 148,000 | Turbine | Over 30 | About 2,000 | About 1,000 |
| S.S. *America* | 723 | 0 | 93 | 3 | 32 | 6 | 12,683 | 26,314 | 207,924 | Turbine | 22 | 1,049 | 675 |
| P3-S2-DL2 (American Export) | 683 | 0 | 89 | 0 | 30 | 0 | 12,310 | 23,720 | 148,600 | Turbine | 22.5 | 1,007 | 577 |
| P2-S1-DN1 | 536 | 0 | 73 | 0 | 29 | 6 | 10,260 | 13,000 | 415,000 | Turbine | 19 | 230 | 185 |
| P2-SE2-R3 (passenger-cargo) | 609 | 5¾ | 75 | 6 | 30 | 0 | 10,431 | 15,359 | 252,794 | Turboelectric | 19 | 550 | 338 |
| C3-S-BR1 (cargo passenger) | 494 | 7⅜ | 69 | 6 | 27 | 9 | 9,627 | 9,528 | 457,690 | Turbine | 16.5 | 119 | 124 |
| C4-S-1a (Mariner) | 560 | 0 | ... | ... | ... | ... | 12,900 | ...... | ...... | Turbine | 20 | 12 | 53 |
| C3-S-DX1 S.S. *S. O. Bland* | 478 | 1 | 66 | 0 | 28 | 6 | 10,500 | 8,800 | 553,000 | Turbine | 18.5 | 12 | 52 |
| C4-S-A4 (large cargo) | 522 | 10½ | 71 | 6 | 32 | 9⅞ | 14,863 | 10,685 | 672,240 | Turbine | 16.5 | 4 | 56 |
| C3-S-A2 (large cargo) | 492 | 0 | 69 | 6 | 28 | 0 | 12,300 | 7,900 | 736,140 | Turbine | 16.5 | 12 | 53 |
| C2-S-B1 (cargo) | 459 | 3 | 63 | 0 | 25 | 9 | 9,200 | 6,200 | 546,000 | Turbine | 15.5 | 8 | 54 |
| C1-B (cargo) | 417 | 9 | 60 | 0 | 27 | 9 | 9,100 | 6,700 | 452,000 | Diesel or turbine | 14 | 8 | 49 |
| VC2-S-AP3 (Victory) | 455 | 3 | 62 | 0 | 28 | 6 | 10,800 | 7,600 | 453,210 | Turbine | 17 | ..... | 52 |
| EC2-S-C1 (Liberty) | 441 | 6 | 58 | 10¾ | 27 | 7 | 10,800 | 7,170 | 500,000 | Steam reciprocating | 11 | ..... | 40 |
| C1-M-AV1 (coastal cargo) | 338 | 8⅜ | 50 | 0 | 21 | 0 | 5,100 | 3,800 | 228,000 | Diesel | 11 | ..... | 36 |
| N3-S-A1 (coastal cargo) | 258 | 9 | 42 | 1 | 17 | 11 | 2,900 | 1,790 | 121,000 | Steam-reciprocating | 10 | ..... | 36 |
| T2-SE-A1 (tanker) | 523 | 6 | 68 | 0 | 30 | 1⅛ | 16,700 | 10,200 | 141,000 barrels | Turboelectric | 14.5 | ..... | 51 |
| Private tanker | 628 | 0 | 82 | 6 | 31 | 5 | 26,000 | 17,500 | 230,000 barrels | Turbine | 16 | 4 | 48 |
| EC2-S-AW1 (Liberty collier) | 441 | 6 | 56 | 10¾ | 27 | 6¼ | 11,040 | 6,640 | 472,799 | Steam-reciprocating | 11 | ..... | 40 |
| R2-ST-AU1 (reefer) | 455 | 5 | 61 | 0 | 27 | 0 | 6,980 | 7,074 | 330,000 | Turbine | 18.5 | 12 | 62 |
| V4-M-A1 (ocean tug) | 194 | 4 | 37 | 6 | 15 | 0 | 786 | 1,118 | ...... | Diesel | 14 | ..... | 36 |
| L6-S-A1 (Lakes ore carrier) | 620 | 0 | 60 | 0 | 24 | 0 | 15,825 | 8,785 | 525,000 | Steam-reciprocating | 10.5 | ..... | 33 |

[a] Width of vessel.

[b] Depth to which vessel rests in the water when fully loaded.

[c] The total carrying capacity of the ship, expressed in tons of 2,240 lb.

[d] Internal cubic capacity of the ship expressed in tons of 100 cu ft to the ton.

[e] Space avilable for cargo measured in cubic feet.

[f] In knots.   One knot equals 6,080 ft/hour.

SOURCE: "Ships of America's Merchant Fleet," U.S. Maritime Administration, 1952.

**Table 34-3  Vessel type by length, in feet***

| Symbol | Type | 1 | 2 | 3 | 4 | 5 | 6 | 7 |
|---|---|---|---|---|---|---|---|---|
| C | Cargo.................... | 400– | to 450 | to 500 | to 550 | | | |
| P | Passenger................ | 500– | to 600 | to 700 | to 800 | to 900 | to 1,000 | over 1,000 |
| B | Barge.................... | 100– | to 150 | to 200 | to 250 | to 300 | | |
| G | Great Lakes cargo......... | 300– | to 350 | to 400 | to 450 | to 500 | to 550 | to 600 |
| H | Great Lakes passenger..... | 300– | to 350 | to 400 | to 450 | to 500 | to 550 | to 600 |
| J | Inland cargo.............. | 50– | to 100 | to 150 | to 200 | to 250 | to 300 | |
| K | Inland passenger.......... | 50– | to 100 | to 150 | to 200 | to 250 | to 300 | |
| L | Great Lakes ore, grain..... | 400– | to 450 | to 500 | to 550 | to 600 | to 650 | |
| N | Coastwise cargo........... | 200– | to 250 | to 300 | to 350 | to 400 | to 450 | to 500 |
| Q | Coastwise passenger....... | 200– | to 250 | to 300 | to 350 | to 400 | to 450 | to 500 |
| R | Refrigerator.............. | 400– | to 450 | to 500 | to 550 | | | |
| S | Special................... | 200– | to 300 | to 400 | to 500 | to 600 | to 700 | |
| T | Tanker................... | 450– | to 500 | to 550 | | | | |
| U | Ferry.................... | 100– | to 150 | to 200 | | | | |
| V | Tug..................... | 50– | to 100 | to 150 | to 200 | | | |

*"Ships of America's Merchant Fleet," U.S. Maritime Administration, 1952.

**Table 34-4  Vessel class by propulsion***

| Machinery | Less than 12 passengers | | 12 or more passengers | |
|---|---|---|---|---|
| | Single-screw | Twin-screw | Single-screw | Twin-screw |
| Steam.................... | S | ST | S1 | S2 |
| Motor................... | M | MT | M1 | M2 |
| Turboelectric............ | SE | SET | SE1 | SE2 |
| Diesel-electric............ | ME | MET | ME1 | ME2 |
| Gas turbine.............. | G | GT | G1 | G2 |
| Gas turboelectric......... | GE | GET | GE1 | GE2 |

* "Ships of America's Merchant Fleet," U.S. Maritime Administration, 1952.

38-ft draft, and about 25,000-ton deadweight carrying capacity. The economy realized with increased size is shown in Table 34-5 (speed constant at 11 knots).

Figure 34-1 shows typical elements of cargo-ship design. At American ports most cargo handling is done by booms which lift loads in and out of the ship's holds. These booms are operated by power winches on the ship's deck. The stowing of cargo must be carried out so that the ship rests in the water in proper balance. Amidships, the vessel carries the Plimsoll mark showing her legal load line for different waters at different seasons of the year.

**Table 34-5  Power according to size**

| | Deadweight capacity, tons | | | | | |
|---|---|---|---|---|---|---|
| | 2,500 | 5,000 | 10,000 | 15,000 | 20,000 | 25,000 |
| Length, ft.............. | 215 | 300 | 430 | 510 | 590 | 675 |
| Beam, ft............... | 36 | 44 | 58 | 68 | 72 | 80 |
| Draft, ft............... | 22 | 25 | 27.5 | 32 | 34 | 36 |
| Indicated hp............ | 1,460 | 1,900 | 2,800 | 3,750 | 5,000 | 6,000 |
| Coal, lb per 100 ton-miles. | 10.6 | 6.5 | 4.5 | 3.9 | 3.3 | 2.9 |

Channel depths must allow for several factors besides the draft of the largest vessel to be accommodated. The surface of the water is depressed slightly by a moving craft, and some additional clearance of the bottom is necessary for this "squat" of the moving ship. Where wave action occurs, further allowance must be made for the pitch of the ship. A desirable channel depth is given by the expression

$$D = D' + \frac{H}{3} + D'' \qquad (34\text{-}1)$$

where $D'$ is the draft of the largest ship to be accommodated and $H$ is the height of storm waves, crest to trough. $D''$ is the allowance for squat, varying from 1 ft in the Great Lakes to 3 or 4 ft at major seaports. Of course, this suggested channel depth is not always an economic possibility, and sometimes a harbor can be reached only on a favorable tide.

**34-5  Harbor size**  Some of the chief requisites of a harbor are associated with size. Obviously, a harbor should be large enough to contain the shipping which it attracts. Considerable space is required for a ship at anchor. The simplest anchorage is a single anchor, which permits the ship to swing a full circle about the anchor as wind or current shifts in direction. Under this arrangement a single ship occupies the square circumscribing a circle whose radius is the ship's length plus the horizontal projection of the cable. The length of cable is apt to be about three times the depth of water at the anchorage.

The space required for a ship moored by two anchors is much less, perhaps a ship's length in width, and longer than the ship by twice the horizontal projection of the anchor chain.

The following data, although not current, afford some idea of the anchorage areas provided in various harbors. Portland, Maine, has 128 acres of 30-foot depth in its inner harbor. Boston has 400 acres of 40-foot

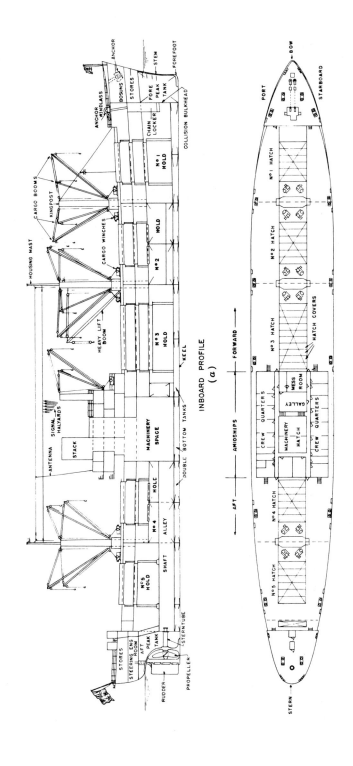

INBOARD PROFILE
(*a*)

FORWARD →

AFT ← AMIDSHIPS

DECK PLAN
(*b*)

**Fig. 34-1** Cargo-vessel design. (*"Ships of America's Merchant Fleet,"* U.S. Maritime Administration, 1952.)

467

depth.   Various harbors on world trade routes, where the protected area
is provided by breakwaters and hence is limited by costs involved, are as
large or larger than the examples cited.

Curiously enough, the size of the harbor basin affects even the ships
moored to wharves or other fixed structures.   Serious damage has taken
place occasionally when the water in a basin surged back and forth like
the water in a washbowl on a ship or Pullman car.   In such cases mooring
lines were snapped before they could be slackened, and the freed ships
crashed against piers or each other.   This phenomenon occurs when the
size of the basin is such that waves entering the basin from the open sea
are reinforced by waves reflected from the closed end of the basin.   This
happens when the length of the basin is related in certain specific ways to
the wave length.   Since all values of wave length are possible, nothing
can be done to eliminate harbor surging, although the resulting damage
can be reduced by improved mooring practices.

The larger the harbor in relation to the entrance, the more effective
will be the diffraction of all waves entering the harbor from the sea.
Experience on the Great Lakes indicated that an entrance width of 550 ft
will result in a 75 per cent wave reduction at the end of a 350-acre stilling
basin.[1]

## QUESTIONS AND PROBLEMS

**34-1.** How many ships, each 500 ft long, can ride at anchor in an area $\frac{1}{2}$ mile square,
with a uniform depth of 40 ft, each ship held by (a) one anchor? (b) bow and stern
anchors?

**34-2.** What channel depths should be specified in a seaport to be used by 430-ft cargo
ships of 10,000 deadweight-capacity tons?

**34-3.** How would the U.S. Maritime Administration designate a turboelectric, four-
screw, cargo-tourist ship of the future such as that suggested by Du-Plat-Taylor?

**34-4.** Assuming that fuel represents about a third of the deadweight tonnage, compare
the distances that can be covered without refueling by the 10,000- and 20,000-ton
vessels of Table 34-5.

**34-5.** Assuming that the 10,000-ton ship of Table 34-5 burns 5,000 lb of coal per hour
while the 20,000-ton ship burns 7,350 lb, both at 11 knots, what would be the fuel
consumed per ton of revenue cargo in a 3,000-mile voyage?   Let revenue cargo
be the difference between deadweight capacity and necessary fuel, neglecting other
factors.

**34-6.** Would the Port of Astoria provide suitable anchorage for a C2-M2?

**34-7.** With the completion of the St. Lawrence Waterway what channel, anchorage,
and wharf depth should be appropriate in a Great Lakes (fresh-water) port to accommodate the 5,000-ton-deadweight-capacity ship of Table 34-5 engaged in foreign
trade?

[1] N. Y. Du Hamel, Great Lakes Ports and Waterways, *Military Engr.*, **33**(188):127
(1941).

## BIBLIOGRAPHY

Cunningham, Brysson: "Harbor Engineering," Charles Griffin & Company, Ltd., London, 1928.

Hwa, C., C. Teng-Kao, and H. Zan-Ziang: Study of Harbor Design, *U.S. Waterways Expt. Sta. (Vicksburg)*, 1945.

"Merchant Fleets of the World," U.S. Maritime Administration, 1952.

Stevenson, Thomas: "The Design and Construction of Harbors," A. & C. Black, Ltd., Edinburgh, 1864.

# 35
# Wind, Waves, Tides, Currents

**35-1 Wind** Some rudimentary understanding of the nature of wind is desirable for the engineer who is concerned with controlling the action of waves and currents generated by wind. Air, like water, flows from a point of higher potential to a point of lower potential. Potential here refers to the sum of elevation, pressure head, and velocity head.

Potential varies from place to place in the atmosphere, primarily because of variations of temperature over the earth's surface and because air changes in density with changes in temperature.

1. Tropical areas are heated by the sun more directly, and hence more effectively, than arctic regions. In the tropics the lighter warm air is displaced by heavier cool air pushing down over the earth's surface from polar regions. The rising tropic air spills back toward the poles in the upper atmosphere.
2. Water absorbs more, and hence reflects less, solar heat than do land masses. In summer the cooler sea air flows landward where it rises as it is warmed. In winter this pattern is reversed.

3. The earth's rotation about its axis, from west to east, imparts the same angular velocity $w$ to all points on its surface. The corresponding tangential velocity $v_t$ varies from a maximum at the Equator to zero at the poles, in proportion to the radial distance $p$ ($p = 4,000$ miles at the Equator):

$$v_t = wp \qquad (35\text{-}1)$$

Consequently, equatorial air moves north possessing a greater easterly velocity than the tangential velocity corresponding to the northerly latitudes over which it may pass. Thus, such winds have an eastward component relative to the underlying earth surface.

These three chief factors combine to produce the seasonal wind patterns shown in Fig. 35-1a and b.

**35-2 Storms** The pattern of prevailing winds discussed in the preceding article is of less significance in coastal engineering than the stormy exceptions to that pattern, because storms produce the extremes of wave action for which maritime structures must be designed. Perhaps the occurrence of windstorms is best introduced by the analogy in the flow of water. The mean velocity in a smoothly flowing stream is notably less than the particular velocities in the eddies which form behind an obstruction to the general flow. In the shelter of a protruding pile or boulder little vortices form and slip off downstream. Each such vortex is marked by a conical depression in the water surface, and this drop in elevation head is accompanied by a commensurate increase in the velocity head, produced by a spinning action about the vertical axis of the vortex. In the air streams of the atmosphere, similar vortices may be formed, either by a protruding land mass, or by the friction between adjacent and opposing currents of air, or, more directly, by a local low-pressure area of thermal origin. The diameter of any such vortex may be a few feet or it may be hundreds of miles. Some of the smaller vortices are the most destructive. A tornado is a tube of air rotating at hundreds of miles per hour. The diameter of the twister is apt to be a few hundred feet, and it moves across country for 25 or 50 miles at a speed of perhaps 25 mph. However devastating this may be, the development of heavy seas are not accomplished by any atmospheric disturbance so local in extent and so short-lived in point of time.

A hurricane is also a rotational air movement, but with a diameter of hundreds of miles. The wind velocities exceed 100 mph, and the seat of the disturbance travels at a rate of perhaps 12 mph. Both of these values are below the corresponding velocities for a tornado, but the far

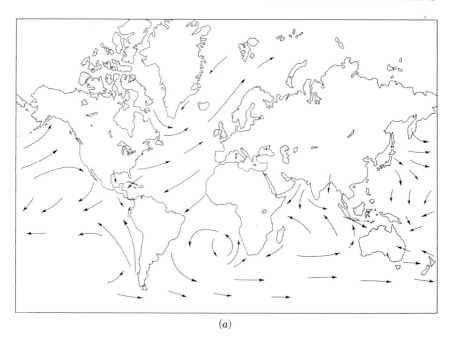

(a)

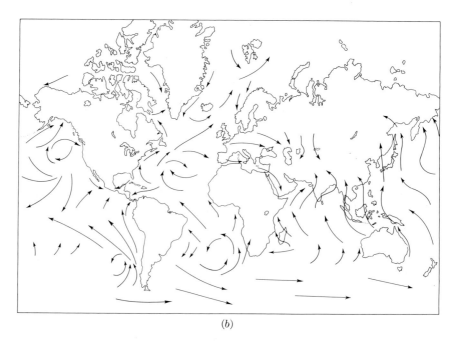

(b)

**Fig. 35-1**  Oceanic winds (a) in January, (b) in July.

greater scale of the disturbance provides opportunity for the growth of destructive water waves.

"Typhoon" and "cyclone" are other terms for the same phenomenon.  The latter word is applied to the low-pressure area itself and to the resulting storm.

**35-3  Origin of water waves**  Any tangential force on the surface of a viscous fluid, such as an automobile tire dragging on warm asphalt, tends to wrinkle the surface of the fluid.  Similarly, a breeze springing up over a calm sea sends ripples running across the smooth surface of the water. Immediately, this roughening of the water surface creates changes in the very air stream which produced the ripples.  Each ripple obstructs the air flow, and tiny vortices form in the air on the leeward side of the ripple. Any such vortex is a region of low air pressure, hence the water moves forward and up in response to the pressure differential between its windward and leeward flanks.  The advancing wavelet continues to grow under the action of a continuing wind of constant velocity until it reaches the maximum size corresponding to the pressure differential possible with the existing wind velocity.  If the breeze subsides before the limiting wave size is attained, or if the wave either reaches shore or moves beyond the windy region, then wave growth naturally ceases.

Water waves are commonly described in terms of their height, length, and period.

Let $H$ = wave height, defined as the vertical distance from trough to crest

$L$ = wave length, the horizontal distance between two successive crests

$T$ = wave period, the interval of time between two successive crests passing a stationary point of reference

$V$ = wave velocity; $V = L/T$

$F$ = fetch, the straight-line stretch of open water available for wave growth without the interruption of land

Because wave growth is a function of wind velocity and time, with fetch as an additional limitation, it should be possible to express wave properties in terms of these variables.  This has been done for the U.S. Navy Hydrographic Office, with the results presented in Figs. 35-2 to 35-5.

A simpler, older, and less precise relationship is the Stevenson formula

$$H = 1.5 \sqrt{F} \qquad (35\text{-}2)$$

where $H$ is expressed in feet, $F$ in nautical miles.  Many authorities tend to discount any reports of wave heights in excess of 50 ft.  Such a value would indicate that the maximum fetch is in the neighborhood of 900

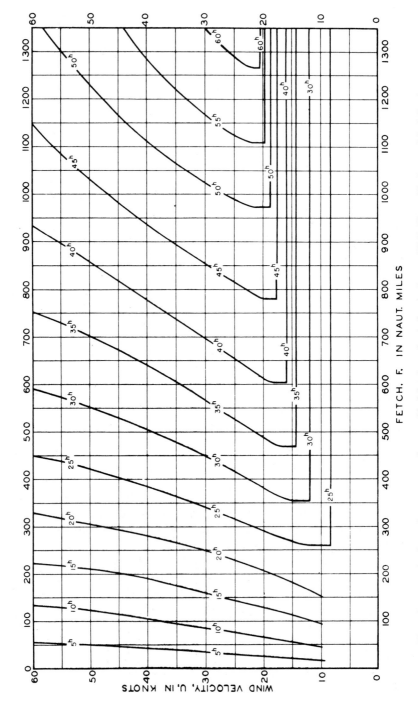

**Fig. 35-2** Time for growth of maximum wave, for specified fetch and wind velocity. (*U.S. Navy Hydrographic Office.*)

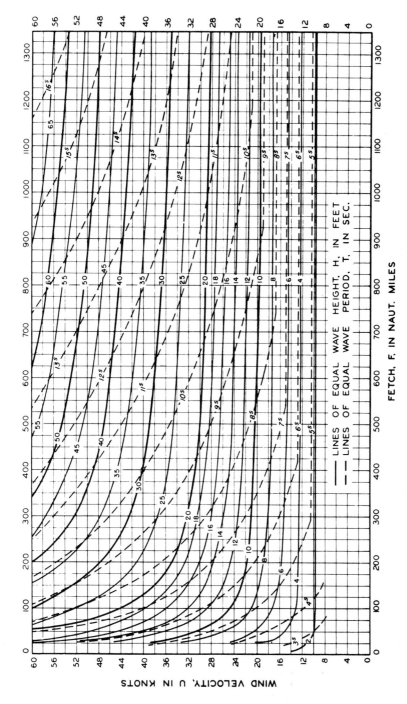

**Fig. 35-3** Wave height and period, fetch and wind variable. (*U.S. Navy Hydrographic Office.*)

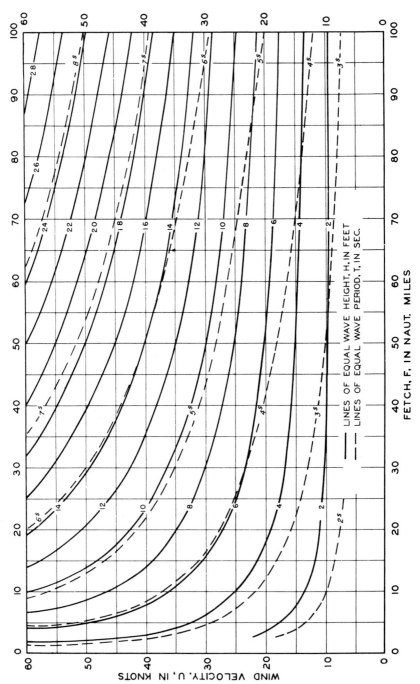

**Fig. 35-4**  Wave height and period for shorter fetch.  (*U.S. Navy Hydrographic Office.*)

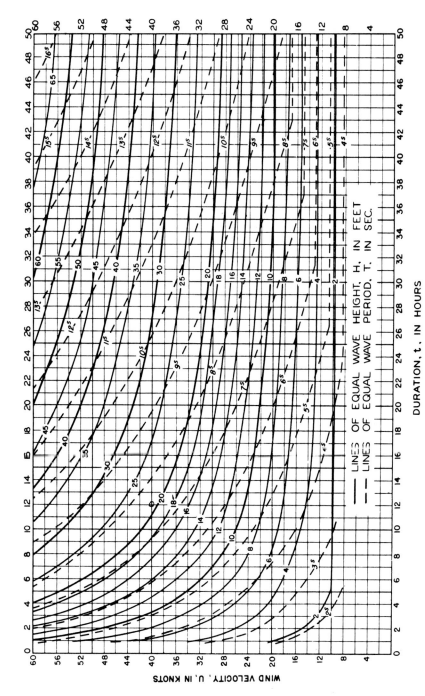

**Fig. 35-5**  Wave height and period for specific duration of wind.  (*U.S. Navy Hydrographic Office.*)

WIND VELOCITY, U, IN KNOTS

DURATION, t, IN HOURS

LINES OF EQUAL WAVE HEIGHT, H, IN FEET
LINES OF EQUAL WAVE PERIOD, T, IN SEC.

miles.  There is no contradiction here between formula and fact; a glance at Fig. 35-1 and at the discussion of storm paths in Sec. 35-2 will show that a given storm would not be expected to accompany a wave train for greater distances than the indicated value of 900 miles, even though open water extended for several times that distance.

**35-4  Tide**[1]  The *tide* is the regular periodic rise and fall of the surface of the seas, observable along their shores.  The concurrent horizontal movements of the water, whether the almost imperceptible drift in the open sea, or the strong flow through a contracted entrance to a tidal basin, are designated in accordance with the practice of the United States Coast and Geodetic Survey, as *tidal currents*.

*High and low water.*—The maximum height reached by each rising tide is called *high water*, and the maximum depression of the falling tide is called *low water*.  On the Atlantic coast of the United States the tide rises and falls twice daily—or more accurately twice during the lunar day of 24 hours and 50 minutes.  The two high waters and the two low waters are each so nearly equal that for ordinary purposes no distinction need be made between them.  On the Pacific coast the two high waters and the two low waters occurring daily are in general markedly different, and are designated as the *higher high water*, the *lower high water*, the *lower low water*, and the *higher low water*.  In the Gulf of Mexico the tides are small, and toward its western end but one tide occurs each day during a part of the month.

The heights of the high waters and of the low waters vary from day to day.  In many parts of the world, the high waters reach their greatest height, and the low waters the least height, soon after the time of full and new moon.  These tides are called *spring* tides.  The term "spring" as applied to tides has nothing to do with the season of the year, but is the greater upspringing of the waters at intervals of about a fortnight.  Similarly the daily high waters are usually at their least height, and the daily low waters their greatest height, soon after the moon is in quadrature.  These tides are called *neap* tides.  On the Atlantic coast of Europe and along the British Isles the difference between low or high water of spring tides and low or high water of neap tides may amount to several feet, and is a matter of moment to navigators.  On the coasts of the United States the difference between spring and neap tides is not particularly noticeable, and the terms "spring" and "neap" tides are not

[1] Quotation taken in its entirety from "Tidal Hydraulics," pp. 1–3, by Gen. George B. Pillsbury, U.S. Government Printing Office, 1940.

in ordinary use.   In this country spring tides are commonly referred
to as "tides at full (or new) moon" or occasionally as "moon tides."

   *Datum planes.*—The average height of all low waters at any
place over a sufficiently extended period of time is called *mean low
water* and is the official reference plane for the depths shown on
navigation charts, and of improved channels, in the waters of the
Atlantic and Gulf coasts of the United States.   The average height
of the lower of the two daily low waters is called the *mean lower low
water* and is the official reference datum in the waters of the Pacific
coast of the United States.   In British waters the datum is usually
the mean low water of spring tides, or *low-water springs*.   This
reference plane is also used at the Pacific entrance to the Panama
Canal.   The average height of the sea, as determined usually by
the average of the observed *hourly* heights over an extended period
of time, is called *mean sea level*, and is the standard datum to which
elevations on land are referred.

   *Tidal ranges.*—The difference in height between high water
and low water at a tidal station is called the *tidal range*.   The *mean
range* is the average of the differences between all high waters and
all low waters; or, as is the same thing, the difference between *mean
high water* and *mean low water* at the station.   The *diurnal range*, or
*great diurnal range*, is the difference between mean higher high water
and mean lower low water.   The *extreme range* is the maximum
that has been observed.   The *spring range* is the difference between
mean high water and mean low water of spring tides, and the *neap
range* the difference between mean high water and mean low water
of neap tides.

   *Tidal currents.*—The tidal current setting into the bays and
estuaries along the coast is called the *flood current*.   The return current toward the sea is called the *ebb current*.   The maximum velocities reached during each fluctuation of the current are called the
*strength of the flood* and the *strength of the ebb*, or, indifferently, the
*strength of the current*.   *Slack water* is the period during which the
current is negligible while it is changing direction.   It is specifically
defined by the United States Coast and Geodetic Survey as the
period during which the current is less than one-tenth of a knot; i.e.,
less than 0.169 feet per second.   The slack water occurring nearest
the time of high water is called the *high-water slack*, and that nearest
the time of low water the *low-water slack*.   The moment at which the
current is zero as it changes direction may be distinguished by term-
ing it the *turn of the current*.

   In open waters, the direction of the current normally veers
around the compass and the current does not pass through intervals

of slack water. Such currents are called *rotary*, to distinguish them from the *reversing* currents in a tidal channel.

These definitions are narrower than the common usage of the terms. "Tide" is commonly applied both to the rise and fall of the sea and to the accompanying tidal currents. Thus the expressions "head tide" and "favoring tide" designate tidal currents that retard or accelerate the movement of a vessel, and the term "the ebb and flow of the tide" is standard legal nomenclature. The term "ebb tide" is often used to designate *low water* as well as the outflowing tidal current. The maximum tidal stage is frequently designated as "high tide" instead of "high water." Its more general meaning is, however, the higher stages of the tide. Thus it is more accurate to say that a channel is "navigable only at high tide," than to say that it is "navigable only at high water."

*Lunitidal intervals.*—Casual observation shows that the tides at any place occur a little less than 1 hour later each succeeding day. Thus if high water is at 3 P.M. today, it will be shortly before 4 P.M. tomorrow. Closer observation shows that the high and low waters at any place follow, by about the same time interval, the passage of the moon across the meridian of the place. Obviously, the moon must cross the plane of the meridian twice daily—once overhead and once underneath. These are called respectively the *upper* and *lower meridian transits*. They mark in fact the noon and midnight of the lunar day. If a clock were regulated on mean lunar time, instead of mean solar time, it would show the times of the high and low waters at a given place at about the same hour every day, but these times would vary largely from place to place.

The *average* time interval, in solar hours and minutes, from a lunar transit to the next succeeding high water at a given place, as determined by an extended set of observations, is called the *high-water interval* (HWI), or the *high-water lunitidal interval* of the place. Similarly the *low-water interval* (LWI), or the *low-water lunitidal interval* is the average time, in solar hours and minutes, from a lunar transit to the next succeeding low water. The high- and low-water intervals usually are larger at the full and change of the moon, at about the time of spring tides, than at other time in the month.

The forces causing tides are gravitational. The influence of the earth's gravity, alone, would produce level, tideless seas. The spinning of the earth about its own polar axis creates centrifugal force (maximum at Equator, zero at the poles), which modifies the perfect sphere which would otherwise result from a uniform gravitational force. Because the resultant pull on a cubic foot of water is less at the Equator than at the poles, there is an equatorial bulge, and the earth's equatorial diameter is greater than its polar diameter. In addition, the gravitational forces exerted by the sun and by the moon must, in similar fashion, reduce or augment the forces acting on any given cubic foot of water. This effect is greatest on the side of the earth nearest the moon, where there will be found a corresponding bulge in sea level. Because of the earth's daily rotation about its own axis and the moon's monthly revolution about the same axis, the moon's position relative to the earth is constantly changing. The tidal bulge, therefore, follows the moon around the earth. Although this distortion is only about 1 ft, it tends to shift the earth's center of gravity slightly, and the resultant weakening of the force of gravity on the opposite side of the earth produces a secondary bulge there.

Both of these bulges complete daily circuits of the earth. Hence, each travels at about 1,000 mph at the Equator and at about 700 mph at 45° latitude. A movement of 1-ft amplitude is imperceptible in the open sea, but because of the mass of the water involved, a considerable momentum accompanies the movement. As the bulge moves into shallower water the smaller mass of the water set into motion moves with a greater velocity, as required for the conservation of momentum. This momentum is funneled into coastal bays and estuaries with impressive effect, while on the other hand, tides are negligible at isolated islands or on inland lakes.

Figure 35-6 shows typical tide charts.

**35-5  Currents**  The statement that tidal bulges travel across the oceans at velocities up in the hundreds of miles per hour should not be misinterpreted to mean that any particles of water involved in tidal movements are also moving at such high speeds. This point will be more evident in reading the subsequent chapter on wave action, but even at this stage it is important to avoid confusion between form and matter. The movement of the bulge or wave is a movement of form, because in each new position of the wave an entirely new mass of water is affected—no matter accompanies the wave form in its travel. However, in any instantaneous position of the bulge, the deformation of the locally affected water mass does involve rearrangement of the water particles, horizontally as well as vertically. Because water is viscous, some energy is lost in friction, and the displaced particles will not return precisely to their original position. Thus there is a current set up in the direction of wave movement, even though the current velocity must be much smaller than the velocity of the wave itself. But the very fact that the current—or mass translation—exists at all means that water piles up at the coast line, and the tidal range at the coast will be many times the amplitude of the tidal bulge in which the tidal currents originated.

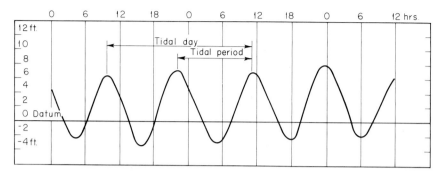

SEMI-DIURNAL

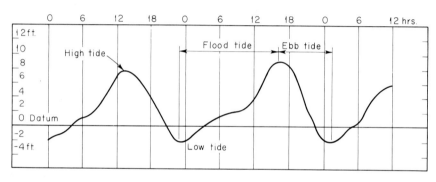

DIURNAL

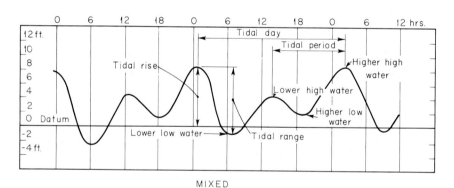

MIXED

**Fig. 35-6**  Types of tide.  *(From U.S. Army, Corps of Engineers, Beach Erosion Board, Spec. Bull. 2, 1953.)*

What is said here regarding the origin of tidal currents applies also to currents evoked by wind-generated waves.  Both wind and tide follow some more or less established pattern.  It follows that the resulting currents along any stretch of coast form a pattern.

**35-6  Beach erosion**  Man, like most of nature, resists change in his environment.  When the sea is invading the land or when the shore is building out to sea, the resulting disturbance of an established pattern is apt to create work for the civil engineer.  In most cases a retreating coast line is characterized by steep cliffs.  The subsidence of a sinking land mass tends to increase offshore depths.  Thus, the larger waves break closer to the shore, and more severe wave action comes into contact with old hillsides.  Waves undercut the hills and steepen their slopes into sea cliffs.  At the base of the cliff is a bench, or terrace, which marks the lower limit of strong wave action.  Talus, or cliff debris, is washed seaward to form the shore-face terrace, which marks the outer limit of the beach.  The particles composing the beach abrade each other, and the mud produced by this abrasion, together with any fine-grained cliff debris, moves with the undertow to build the continental terrace in the deep water beyond the abrasion platform.

If a land mass is rising, the offshore depth will decrease and the intensity of wave attack on the sea cliff will diminish.  An offshore bar may become a prominent feature of the landscape.  The coastal hills will become rounded as wind, rain, and stream erosion have more time to modify contours and influence the topography.

Any specific sample of beach sand will exhibit a uniformity of grain size that demonstrates the effectiveness of the sorting action to which it is subjected.  The engineer faced with a problem of beach protection must be aware of the forces with which he must deal.  These are three:

1. The offshore current which removes the fine material
2. The wave action which builds the beach
3. The littoral, or longshore, current that transports beach material

In the following chapter it will be seen that while waves approaching a coast line are essentially oscillatory, there is actually an accompanying drift of water in the direction of wave advance.  Thus, at the coast an onshore wind will produce both wave action and a local elevation of the mean water level.  This deviation from sea level means that a hydraulic gradient is created which falls in the direction of wave origin.  Water must move in the direction of a falling gradient; hence, the onshore drift at the surface is balanced by an offshore current near the bottom.  The velocity of this current is incompetent to move beach sand, but it does carry the finer grains out to deep water.

There can be no question of the ability of the breaking wave to move sand or gravel. There is a greater concentration of energy in the uprush of a breaker than in the backrush, since the breaker does not operate at 100 per cent efficiency. Consequently, the maximum size of particle thrown on the beach is larger than the maximum size washed back to the surf zone. The result is to build the beach out of the largest material locally available to the waves. Coarse material has a greater angle of repose and creates a steeper beach slope; this in turn increases the maximum size that can be washed out to deep water and further increases the uniformity in size of the beach material. As this material migrates to leeward it is reduced in grain size by abrasion, and the beach slope consequently flattens.

That the rejected material does move to leeward is due to the longshore current. Whenever the direction of prevailing winds does not happen to be perpendicular to the coast line, the tangential component of the onshore drift will produce a current running parallel with the shore. A grain of sand washed back and forth by waves will experience a leeward displacement with each successive emergence from the water, so that it will follow a roughly cycloidal path of migration down the beach, in response to this littoral drift or current.

Two facts are dominant: beach material originates in the erosion of the coast, and it moves in the direction of the littoral current. To these should be added the consideration that the size of the beach depends upon the degree of equilibrium between the total amount of material produced or imported and the total amount consumed or exported. Here production represents local cliff erosion; by imports is meant sand drifting in from windward; the sand is consumed to the extent that abraded fines are washed out to sea; by exports is meant sand drifting to leeward. The coarseness of the beach material indicates either the proximity of its source or the severity of wave action. These limitations on the extent of beach movements are emphasized by the failure of various attempts to nourish beaches by dumping sand offshore in deep water.

The direction of the littoral current can be predicted to a considerable extent by knowledge of the prevailing winds; and it can be observed in the field from study of the coast line. Any projection tends to trap sand on the side which interrupts the current and to starve the beach on its other side. A spit points in the direction that the current moves.

Artificial barriers, called groins, are used to protect or extend beaches by trapping sand. Offshore breakwaters accomplish a similar purpose by preventing erosive wave action in the sheltered area. Both groins and offshore breakwaters are discussed in Chap. 38.

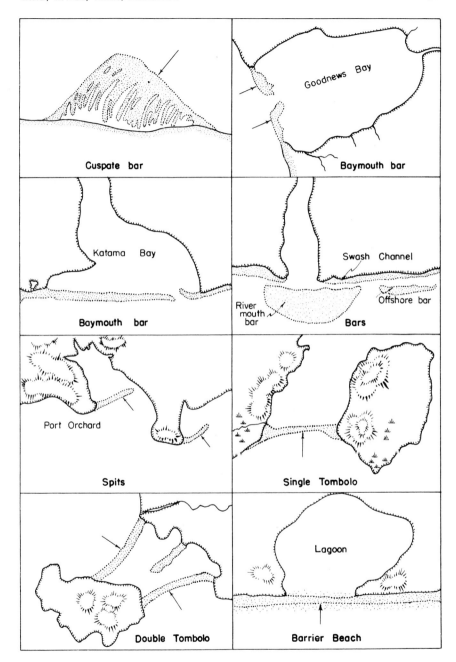

**Fig. 35-7** Bar and beach forms. (*From U.S. Army, Corps of Engineers, Beach Erosion Board, Spec. Bull. 2, 1953.*)

The function of artificial protective structures is better understood after a study of natural beach formation. Figure 35-7 will be helpful in this respect.

## QUESTIONS AND PROBLEMS

**35-1.** A 40-knot wind blows over a fetch of 400 nautical miles. Estimate the wave height 12 hr after the onset of the storm; also at 36 hr after the storm began.

**35-2.** How do the waves of Prob. 35-1 compare with the maximum wave predicted by Stevenson's formula?

**35-3.** Find the maximum wave height to be expected when the fetch is (a) 5 statute miles, (b) 5,000 statute miles.

**35-4.** Where the fetch is 225 nautical miles:
    a. What is the maximum wave height anticipated by Stevenson?
    b. What is the least wind that could raise a $22\frac{1}{2}$-ft wave in 60 hr?
    c. What wave would a 50-knot wind raise in this location?

**35-5.** A stationary observer notes that five successive wave crests pass him in 1 min, at intervals of 360 ft. Wind velocity has been uniform at 35 knots.
    a. What is the velocity of wave advance?
    b. What is the minimum age of the waves?

**35-6.** Assume that the typical Pacific Coast tide chart of Fig. 35-6 applies to a point at the head of a bay 50 miles long with a uniform width of 10 miles. At the bay outlet high water is 4 ft lower and low water is 0.5 ft lower. Both stages occur 3 hr earlier. What is the maximum hourly inflow in cubic feet? Assume that the water surface is a plane at all times.

## BIBLIOGRAPHY

Bigelow, Henry B., and W. T. Edmondson: "Wind Waves at Sea: Breakers and Surf," U.S. Navy, Hydrographic Office, 1947.

Browning, Frank H.: The Action of Ocean Waves, *U.S. Naval Inst. Proc.*, **71**:1475–1491 (1945).

Pillsbury, George B.: Tidal Hydraulics, *U.S. Army, Corps of Engineers Professional Paper* 34, 1940.

Sverdrup, H. U., M. W. Johnson, and R. H. Fleming: "The Oceans," Prentice-Hall, Inc., Englewood Cliffs, N.J., 1942.

# 36
# Wave Motion

**36-1  The wave profile**  Waves differ from one another not only in height but also in steepness, and hence in length.  The ratio $L/H$ is often used as a measure of wave steepness.  The steepest possible wave has a value of 7 for $L/H$, but values this low are not commonly observed. Theory states that the wave form is unstable at any lesser value.

Waves grow in length as well as in height under the continued action of a constant-velocity wind.  Table 36-1 lists average observed values for wave steepness under various wind exposures.

The effect of wind duration is to raise wave velocity in relation to wind velocity.  Consequently, it is possible to restate Table 36-1 in other terms, as is done in Fig. 36-1.

When the wind shifts in direction or intensity the new generation of waves will differ from the older waves, which still persist in their original course.  The more recent waves may overtake or (if the storm center has migrated meanwhile) may intercept the earlier waves.  In any such mingling of different wave trains some waves will be reinforced and others reduced, depending upon the extent to which the two waves are in phase.

**Table 36-1   Values of $L/H$ corresponding to specified wind exposure***

| Wind velocity, knots | Duration of wind, hr | | | | | |
|---|---|---|---|---|---|---|
| | 5 | 10 | 15 | 20 | 30 | 40 |
| 10 | 12 | 23 | 31 | 42 | 56 | 56 |
| 15 | 14 | 20 | 23 | 26 | 39 | 44 |
| 20 | 14 | 16 | 18 | 23 | 27 | 35 |
| 30 | 13 | 15 | 16 | 18 | 22 | 25 |
| 40 | 13 | 14 | 14 | 16 | 19 | 22 |

* H. B. Bigelow and W. T. Edmondson, "Wind Waves at Sea: Breakers and Surf," U.S. Navy, Hydrographic Office, 1947.

The effect of these conflicts is the very confused wave pattern which one usually observes in any detailed inspection of a storm at sea.   Of course these troubled waters are a natural accompaniment to the gusts and eddies that characterize the generating winds.   The confusion of detail should not be permitted to obscure the underlying pattern, which also deserves attention.   The basic pattern of regular wave form can be seen in air photos and in the dead swells that follow a storm.   Uncomplicated waves can also be generated in the laboratory for easier analysis.   From observation as well as from theoretical analyses, it appears that the trace of a wave on a vertical transverse plane is a curve closely resembling a trochoid.   A trochoid is the curve traced by a point on a wheel which is rolling along a straight line.   Figure 36-2 shows a wheel of radius $R$ rolling (upside down) on the horizontal line $XX'$.   Figure 36-2a shows the trace of point $P_1$, a trochoid for which $L/H$ is about 7, the steepest stable

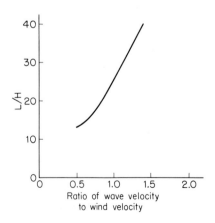

**Fig. 36-1** Effect of wind duration on steepness of wave.

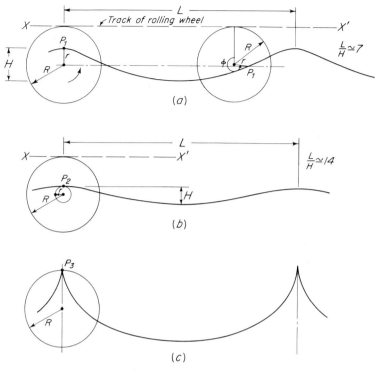

**Fig. 36-2**  Trochoidal-wave-form construction.

wave.   Figure 36-2*b*, for point $P_2$, yields a trochoid with an $L/H$ of about 14, a rather common ratio for water waves.   Finally, Fig. 36-2*c* presents the limiting condition, with the point $P_3$ on the rim of the wheel.   In this case $L/H$ is equal to $\pi$, and the curve is the more familiar cycloid.

The general equations for trochoids are

$$x = R\phi - r \sin \phi$$
$$y = r(1 - \cos \phi) \tag{36-1}$$

**36-2  Orbital motion in deep water**   Many people find it difficult to distinguish between form and matter.   If one observes a chip of wood floating on the surface of a body of water, he will note that the chip moves up and forward with the wave crest, down and backward in the succeeding trough.   The path of the chip is identical with the path of a surface-water particle.   The chip of wood and the adjacent water return to their original positions with each new wave, while the wave itself rapidly moves on.

The wave form is a trochoid, but the matter composing the wave

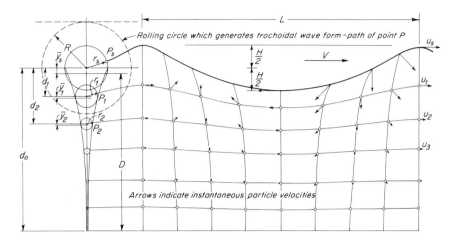

**Fig. 36-3**   Trochoidal water waves.

moves in circles.   The orbital path of each water particle is a circle.   This circular orbit is flattened when the depth of the specific water particle is such that the influence of the bottom becomes appreciable.   At the very bottom the water is free to oscillate only horizontally.   In shallow water the influence of the bottom extends throughout the mass of water, and the orbit of even a surface particle is elliptical.   The eccentricity of such an ellipse increases as the bottom is approached.   Figure 36-3 shows how particular orbits vary with depth.   Note how an originally vertical line sways from side to side of its base as the wave advances.

The time required for any one particle to complete its orbital revolution must be the same as the period of the wave form itself.   The orbital velocity $u$ is equal to the length of orbital travel divided by the wave period $T$.   Hence,

$$u = \frac{2\pi r}{T}$$

Similarly, the wave velocity is

$$C = \frac{2\pi R}{T}$$

In addition, wave theory states that

$$C = \sqrt{gR} \tag{36-2}$$

Hence,

$$u = \frac{Cr}{R} = r\sqrt{g/R} \tag{36-3}$$

The amplitudes of particle orbits has been found to be

$$r = r_s \epsilon^{-q} \tag{36-4}$$

where $r_s$ = amplitude of surface particles
   $\epsilon$ = base of natural logarithms
   $q$ = ratio $d/R$

Examination of the wave profile of Fig. 36-3 reveals that wave crests are much steeper than wave troughs and that the trochoid is not symmetrical about a level line through the surface-orbit center. Actually, there is not enough water in the crest above such a level line to fill the space above the depressed trough below the level of the surface-orbit center. Consequently, the total volume of sea water being the same in storm as in calm, the still-water level must lie below the level of surface-orbit centers. The distance by which the orbit center of a particle is raised above its still-water position is given by the expression

$$\bar{y} = \frac{r^2}{2R} \tag{36-5}$$

Since the surface of a wave is in contact with the atmosphere, the wave profile is an isobar connecting points of equal pressure, i.e., zero hydrostatic pressure. Similar isobars can be drawn for various depths, such as those shown in Fig. 36-3, connecting points of equal still-water depth. These isobars also represent the hydrostatic pressures to which the corresponding particles were subjected under still-water conditions. In other words, the displacement of a water particle by wave action does not alter its original pressure.

The existence of the mass of water comprising the wave crest at an elevation above still-water level constitutes a potential energy source, similar in nature to the pool behind a power dam. Likewise, the revolving water particles have kinetic energy like the jet striking an impulse wheel. The potential energy, of course, moves on with the wave form itself, but the kinetic energy imparted to the water particles is not so transferred. In a given deep-water wave equal quantities of energy are stored as potential and as kinetic energy. The total energy in a wave is expressed by the equation

$$E = \frac{wLH^2}{8} \left( 1 - \frac{H^2}{8R^2} \right) \tag{36-6}$$

where $w$ is the unit weight of water (64 pcf for sea water).

**36-3  Shallow-water waves**  When the still-water depth $d$ is less than $L/2$, the influence of the bottom on particle orbits becomes appreciable. Most of the equations presented in the preceding section must be modified

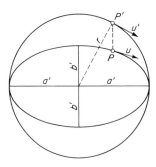

**Fig. 36-4**   Motion of particle $P$ for wave in shallow water.

for use with shallow-water waves.   Equation (36-2) must be replaced by its more general form

$$C = \sqrt{gR \tanh m} \qquad\qquad (36\text{-}7)$$

where $m = D/R$.

The orbital paths of the water particles are assumed generally to be elliptical.   According to this view, the actual position of the particle at any instant is taken to be on the intersection of an ellipse with a vertical line through the position the particle would otherwise occupy on a circumscribed circle, as shown in Fig. 36-4.   Here the symbols $a'$ and $b'$ represent the major (horizontal) and minor (vertical) semiaxes of the ellipse.   Since the tangential velocity $u'$ of the imaginary particle $P'$ is constant, the tangential velocity of the actual particle $P$ will be variable, and its maximum value will occur when the particle is moving horizontally.

$$u_H = \frac{a'C}{R} \qquad\qquad (36\text{-}8)$$

The amount of orbital distortion from a circular path is a function of depth and is measured by the ratio

$$\frac{b'}{a'} = \tanh n \qquad\qquad (36\text{-}9)$$

where $n$ represents the ratio $(d_o - d)/R$.

For points located on the same vertical line and for any assumed values of $L$ and $d_o$, the focal distances of the elliptical orbits are constant from top to bottom of the water, but the ellipses become flatter with depth.   When $d = d_o$, $b = o$.

For any specific particle, the orbit center is elevated above its still-water position by the amount

$$\bar{y} = \frac{a'b'}{2R} \qquad\qquad (36\text{-}10)$$

The orbital amplitude any depth $d$ below the orbital center of a surface particle can be obtained through the relationship

$$b' = \frac{b_s \sinh n}{\sinh m} \tag{36-11}$$

the subscript $s$ being used here and elsewhere to designate the properties of a surface particle.

The sum of the potential and kinetic energy contained in a trochoidal shallow-water wave is

$$E = \frac{wLH^2}{8}\left(1 - \frac{a_s^2}{2R^2}\right) \tag{36-12}$$

**36-4  Reflection of waves**  When the forward progress of a wave is blocked by a vertical wall, the wave is reflected back upon its course. The orbital movement incited by this reflection in part reinforces and in part nullifies that particle motion accompanying the original forward progress of the wave.  As a result, a so-called standing wave develops in which the water surges between the obstruction and a distance $L/2$ away, slopping back and forth as in a washbasin.  The crests are much higher than in the original wave.  This form of standing wave is known as clapotis.  The height of rise on the vertical wall is stated by Sainflou[1] to be equal to the full height of the unobstructed wave above the orbit center of the surface particle (in other words, wave height is doubled), while the elevation of the surface-orbit center above still-water level is

$$\bar{y}_s = \frac{\pi H^2}{L} \coth \frac{2}{R}$$

or
$$\tag{36-13}$$

$$\bar{q}_s = \frac{Hr_s}{R}$$

This is four times the value expressed by Eq. 36-5 for an unobstructed wave.

If the obstructing wall is not vertical, the reflection of wave energy will be imperfect.  The Spanish engineer Iribarren suggests that reflection dominates when the batter of the wall is less than

$$i = \frac{8}{T}\sqrt{\frac{H}{2g}} \tag{36-14}$$

where $i$ = wall batter (run over rise)
 $H$ = wave height
 $T$ = wave period
 $g$ = acceleration of gravity
all factors being expressed in consistent units.

[1] A discussion of the Sainflou and other recognized theories is to be found in Chap. 37.

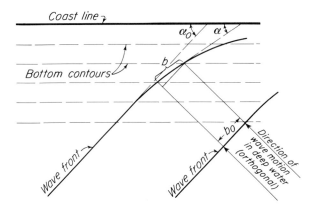

**Fig. 36-5**   Wave refraction.


**36-5   Refraction**   According to Eq. (36-7), a wave must slow down as
it moves into shallower water.   If the wave front approaches the coast
line obliquely, the leading portion of the wave front will feel bottom while
the remainder is still in deep water.   The retardation of the landward end
of the wave front, while the seaward end still moves unimpeded, causes
the wave front to wheel toward the shore, as shown in Fig. 36-5.   The
energy carried forward by the portion of the wave included between any
two orthogonals in deep water, a distance $b_o$ apart, becomes distributed
over a longer distance $b$ when the wave front reaches shallow water.   To
accommodate this reduction in energy (per foot of crest) the characteristics of the wave must be altered by an amount determined by the
degree of refraction.   The wave period cannot change, if continuity is to
be maintained, but the height, length, and velocity of the deep-water
wave do change in shoaling water.   Conservation of energy requires that
the height of the refracted wave, at any specific depth $D$, will be

$$H = H_o \sqrt{\frac{C_o}{C}} \sqrt{\frac{b_o}{b}} \qquad\qquad (36\text{-}15)$$

where the subscript $o$ is used to designate conditions in deep water.   The
ratio $\sqrt{C/C_o}$ represents the direct effect of still-water depth upon wave
height, as expressed in Eqs. (36-2) and (36-7),

$$\frac{C}{C_o} = \sqrt{\frac{L}{L_o} \tanh \frac{D}{R}} \qquad\qquad (36\text{-}16)$$

The relationship is more useful when presented in tabular form.

The refraction of waves in shoaling water is best shown by sketching
in wave fronts on a map or hydrographic chart.   The scale of the map

will probably suggest that only every fifth or tenth wave front be so plotted. Naturally, the drawing of refraction diagrams is easier if the shore line and offshore contours are all straight and parallel. A useful tool here is Snell's law,

$$\frac{\sin \alpha}{\sin \alpha_o} = \frac{C}{C_o} \tag{36-17}$$

where $\alpha$ is the angle intercepted between the wave front and the bottom contour at a point where the wave velocity is $C$. As usual, corresponding values in deep water are identified by subscript $o$.

The relationships are shown graphically in Fig. 36-6.

With more customary offshore topography, it becomes necessary to construct a refraction diagram, point by point. Figure 36-7a illustrates how this diagram may be constructed. At the bottom of the sketch the horizontal line, marked $O$, represents a wave crest in deep water. The corresponding lines above this are crests spaced $n$ (here $n = 2$) wave lengths apart. The crests are connected by conveniently spaced orthogonals, normal to the crest lines, showing the direction of wave advance. According to Eq. (36-16), or Table 36-2, the crest spacing $L$ decreases as the waves move into shallow water of depth $D$. The graphical solution of the $L$–$D$ relationship is best accomplished by construction of a template such as that shown in Fig. 36-7b, drawn on transparent paper or plastic. The depth scale on this template is the same as the linear scale of the topographic chart which forms the basis for the refraction diagram.

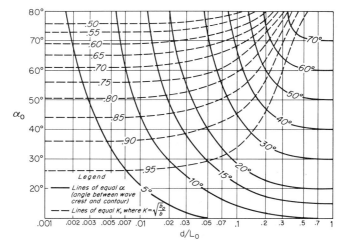

**Fig. 36-6** Refraction chart. (*From U.S. Navy, Hydrographic Office, Publ. 234, 1944.*)

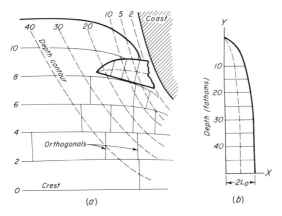

**Fig. 36-7** Construction of refraction diagram.

The horizontal distances $nL$ may be derived from Table 36-2. The $Y$ axis of the template is placed tangent to the deep-water crest, in such a position that the depth line of the template crosses the same depth contour of the chart at the dashed mid-ordinate of the template (see sketch). The other end of the pertinent depth line of the template is a

**Table 36-2 Wave-refraction data***

| $D/L_o$ | $H/H_o$ | $L/L_o$† |
|---------|---------|----------|
| 0.002 | 2.12 | 0.112 |
| 0.005 | 1.69 | 0.176 |
| 0.010 | 1.45 | 0.248 |
| 0.020 | 1.23 | 0.347 |
| 0.040 | 1.06 | 0.480 |
| 0.056 | 1.00 | 0.558 |
| 0.080 | 0.94 | 0.649 |
| 0.1 | 0.92 | 0.709 |
| 0.2 | 0.92 | 0.888 |
| 0.3 | 0.93 | 0.961 |
| 0.4 | 0.96 | 0.988 |

* J. W. Johnson, M. P. O'Brien, and J. D. Isaacs, "Graphical Construction of Wave Refraction Diagrams," *U.S. Navy, Hydrographic Office, Publ.* 60 (1948).

† Since $T = T_o = L/C = L_o/C_o$, it follows that $L/L_o = C/C_o$.

point on the next wave front.   By repetition of this procedure the entire
layout of advancing wave fronts may be drawn.

It soon becomes apparent that wave refraction becomes increasingly
pronounced as the cost is approached.   Consequently, several diagrams
to various scales are often necessary to carry the wave crest forward from
deep water to its impingement upon a shore structure.

**36-6   Wave diffraction**   Obviously, waves will be smaller in the lee of a
promontory or of a partial breakwater.   The nature of the wave pattern
within this sheltered area is determined by diffraction as well as refraction
phenomena.   That portion of the advancing wave crest which is not
intercepted by the barrier immediately spreads out into the sheltered
area, and the wave height shrinks correspondingly.   This lateral dissipation of wave energy is termed *diffraction*.   Depth of water is not a
relevant factor.   Diffraction results from interference with the horizontal
components of wave motion, just as refraction results from interference
with vertical components of wave motion.

Several investigators have recently made significant contributions to
the analysis of diffraction,[1] utilizing the principles which apply to
diffraction of light.   Where water depth is constant or deep $(D > L/2)$
the construction of a diffraction diagram is not too complicated, with the
aid of the references cited.   However, in most practical cases of harbor
design critical storms are apt to involve waves of such length that the
influence of still-water depth will be appreciable and refraction will
outweigh diffraction in importance.   Ordinarily, then, the wave pattern
within the sheltered area would be shown by a refraction diagram except
in the immediate area of the breakwater gap or breakwater tip where the
beginnings of a diffusion diagram would be traced.

The most interesting feature of a diffraction diagram is the rate of
wave decay.   This is expressed by $K'$, the diffraction coefficient, which is
defined as the ratio of wave height at any point to wave height at breakwater tip or gap.   For single breakwaters or for breakwaters with wide
gaps (at least several wave lengths wide) $K' = 0.5$ along the projection
of the incident-wave orthogonal shoreward from the breakwater tip.
Other values of $K'$ can be found in Table 36-3 and plotted from the
coordinates indicated on Fig. 36-8.

Wave crests may also be drawn on the diffraction diagram.   Along
the $Y$ axis, or "geometric shadow" of the breakwater tip, the wave
length of the incident wave remains unchanged.   In the unsheltered
region the wave length is increased up to 5 percent over that of the
incident wave, which is not a significant factor in most practical cases.

[1] *Proc. First Conf. on Coastal Eng.*, chap. 4 (1950); *Proc. Second Conf. on Coastal Eng.*, chap. 2 (1951).

**Table 36-3** **Values of coordinate $X/L$ for specified values of $K'$***

| $y/L$ | $K'$ | | | | | | |
|---|---|---|---|---|---|---|---|
| | 0.1 | 0.2 | 0.5 | 0.8 | 1.0 | 1.17 | 1.0 |
| 5 | +3.78 | +1.64 | 0 | −0.77 | −1.24 | −1.96 | −3.54 |
| 10 | +5.19 | +2.30 | 0 | −1.09 | −1.75 | −2.75 | −4.89 |
| 15 | +6.29 | +2.80 | 0 | −1.33 | −2.14 | −3.36 | −5.93 |
| 20 | +7.23 | +3.23 | 0 | −1.54 | −2.47 | −3.87 | −6.82 |
| 30 | +8.81 | +3.95 | 0 | −1.88 | −3.02 | −4.74 | −8.31 |
| 40 | +10.13 | +4.56 | 0 | −2.17 | −3.49 | −5.46 | −9.58 |
| 60 | +12.38 | +5.59 | 0 | −2.66 | −4.27 | −6.68 | −11.70 |
| 80 | +14.28 | +6.46 | 0 | −3.07 | −4.93 | −7.71 | −13.48 |
| 100 | +15.91 | +7.22 | 0 | −3.44 | −5.51 | −8.62 | −15.03 |

* J. W. Johnson, Engineering Aspects of Diffraction and Refraction, *Trans. ASCE*, **118**:617–652 (1953).

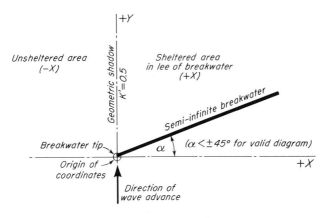

**Fig. 36-8**  Basis for diffraction diagram.

However, in the lee of the breakwater the decrease in wave length is sufficient to create considerable curvature of the wave front. The resulting lag $K''L$ of the diffracted wave front is a constant for any specific value of $K'$, and once the loci of constant values of $K'$ have been plotted, the sketching of the modified wave fronts is easily accomplished with the help of Table 36-4.   The significance of $K''$ is shown on Fig. 36-9.

**Table 36-4**

| $K'$ | 0.1 | 0.15 | 0.2 | 0.3 | 0.4 | 0.5 | 0.6 | 0.8 | 1.0 |
|---|---|---|---|---|---|---|---|---|---|
| $K''$ | 1.4 | 0.6 | 0.33 | 0.13 | 0.06 | 0 | −0.03 | −0.06 | −0.05 |

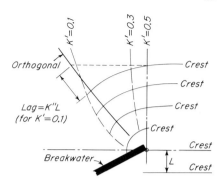

**Fig. 36-9** Diffracted wave fronts.

**36-7 Mass transport** The trochoidal theory of waves, which was discussed earlier in this chapter, assumed that each particle of water returns to its initial location upon completion of its orbital cycle. More exact analyses have shown that this assumption does not give the whole picture. The orbital motion is accompanied by some slight movement of the orbit center in the direction of wave advance, so that the water particle does not quite close its path. The forward velocity of the orbit center is called *mass transport* and is designated by the symbol $U$. In deep water its value at any depth is given by the expression

$$U_o = C \left(\frac{r}{R}\right)^2 \tag{36-18}$$

At the water surface this becomes

$$U_{so} = \frac{CH^2}{4R^2} \tag{36-19}$$

The total volume of transport per unit length of wave crest is

$$G = \frac{H^2}{4} \left(\frac{g}{4R}\right)^{\frac{1}{2}} \tag{36-20}$$

For shallow water the velocity of mass transport at any depth is

$$U = C \left(\frac{a'}{R}\right)^2 \tag{36-21}$$

Equations (36-18) to (36-21) approximate the expressions derived in the *irrotational theory* of wave action, while this chapter has, on the whole, presented trochoidal interpretation of waves. This inconsistency does not involve serious error; it results in simpler expressions and has the important advantage of preserving greater physical significance for the practicing engineer than more exact but more complicated mathematics. The cited references permit further study.

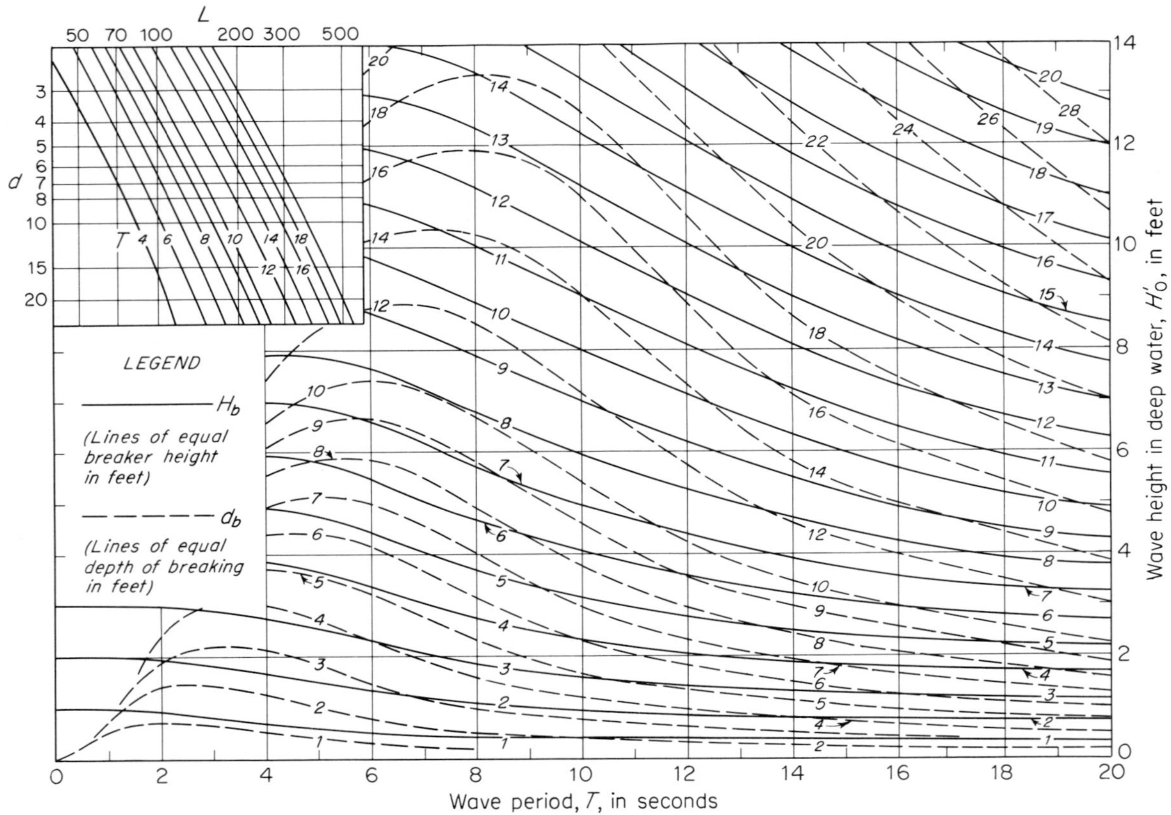

**Fig. 36-10** Determination of wave height and depth of water at point of breaking. *(From U.S. Navy, Hydrographic Office, Publ. 234, 1944.)*

**36-8  Breakers**  When a wave reaches a bar or beach it dashes forward in a motion that is predominantly translation rather than rotation. It can be seen that the wave must break whenever the water is moving faster than the wave form itself. Thus a wave must break as soon as

$$U_s + u_s = C$$

or

$$C\left(\frac{a_s'}{R}\right)^2 + C\,\frac{a_s'}{R} = C$$

which simplifies to

$$R = a_s'(a_s' + 1) \tag{36-22}$$

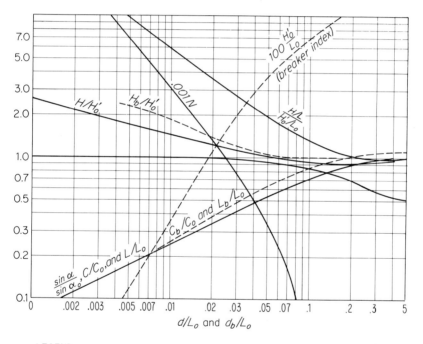

LEGEND:

$H$ = Wave height
$L$ = Wave length
$C$ = Wave velocity
$\alpha$ = Angle of wave crest with bottom contour
$d$ = Depth beneath still water level
$n$ = Fraction of energy advancing with wave velocity

$N$ = Steepness factor (equation 21)
$-'$ = Superscript refers to waves not affected by refraction
$-_o$ = Subscript refers to deep water
$-_b$ = Subscript refers to breaking wave
——— = Waves before breaking
— — — = Waves at breaking

**Fig. 36-11**  Waves in shallow water, changes in height, and length from deep water to point of breaking. (*From U.S. Navy, Hydrographic Office, Publ. 234, 1944.*)

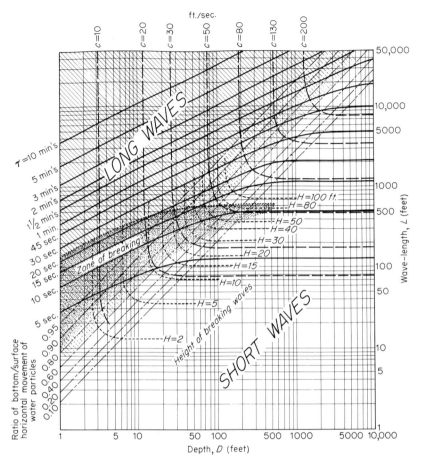

**Fig. 36-12** Relationships between periodicity $T$, velocity $C$, horizontal movement $\lambda$, depth $d$, and breaker height $H$, for waves, based upon Eq. (36-7). Michell, McCowan criteria for wave stability, respectively: $H/L = 1/7$; $H/D = 0.78$. (*Source: Dock and Harbor Authority, July, 1953.*)

This expression states that the conditions under which wave break involves wave length (through $R$) are seen to involve water depth and wave height as well, because the horizontal amplitude of the surface-particle orbit $a_s'$ is related to $D$ and $H$ through Eqs. (36-9) and (36-11). This is a point which has been the object of many field observations. Most of these observations can be summarized in the generalization that waves tend to break when they reach depths of water somewhat less than the wave height. Figures 36-10 to 36-12 provide further information in this regard.

## QUESTIONS AND PROBLEMS

**36-1.** A 30-knot wind acts on a wave for 10 hr. Estimate the length, height, period, and velocity of waves developed at the end of that time.

**36-2.** Draw a trochoid for which $R = 1\frac{3}{4}$ in. and $r = 1$ in. by plotting values of $x$ and $y$ [Eq. (36-1)] for $\phi = 30, 60, 90, 120, 150, 180, 210, 240, 270, 300, 330, 360$ deg.

**36-3.** A circle is the path traced by a point moving at a fixed distance from a fixed center of rotation. Draw the curves generated by such a point if, in addition to rotating at a 2-in. radius, the generating point also experiences a superimposed horizontal displacement of 0.2 in. for every 30 deg of rotation. (This suggests the path of a water particle when mass transport is added to orbital rotation.)

**36-4.** Find the orbital velocity of a surface particle for a deep-water wave 1,000 ft long and 40 ft high.

**36-5.** What is the velocity of a wave 500 ft long and 25 ft high in water 100 ft deep?

**36-6.** A wave is 100 ft long in deep water. What will be its length where the water is 20 ft deep?

**36-7.** The surface particles of a wave 400 ft long move in an elliptical path 18 ft high and 30 ft wide. Estimate the still-water depth.

**36-8.** The wave in Prob. 36-4 has a height of 45 ft when it reaches water of 100-ft depth. Find the maximum orbital velocity of a surface particle.

**36-9.** A wave has a period of 10 sec and a height of 30 ft in deep water. Find the wave length, velocity of wave advance, and maximum orbital velocity.

**36-10.** A wave 30 ft high has a length of 800 ft where the water is 500 ft deep. Find the orbital velocity and mass transport of a surface particle.

**36-11.** The wave of Prob. 36-10 is 32 ft high at the 200-ft-depth contour. Find its length and velocity at this location; also the orbital and mass-transport velocity at the surface.

**36-12.** A wave is 10 ft high and 150 ft long where the water is 30 ft deep. What is the orbital velocity of a particle (a) at the surface? (b) at mid-depth?

**36-13.** At what depth would the wave of Prob. 36-12 be expected to break? Compute wave velocity and the orbital and mass-transport velocity of a surface particle for this depth.

**36-14.** A deep-water wave with a height of 20 ft, a period of 8 sec, and a length of 600 ft approaches a coast line which runs north and south. The bottom has a uniform slope of 1 per cent out to a depth of 100 fathoms. The direction of deep-water wave advance is N40°E. What will be the direction of wave advance at the 10-fathom contour? (Use Fig. 36-6.)

**36-15.** Estimate the percentage of energy which is advancing with the wave front of Prob. 36-14 at the 20-fathom contour, if the direction of wave advance is N48°E at that point.

**36-16.** A wave has a length of 450 ft, a period of 15 sec, and the orbital velocity of surface particles is 20 fps. If the waves break, what is the approximate velocity of the orbit center of surface particles?

**36-17.** A wave has a length of 314 ft where the water depth is 40 ft. What is the ratio between the minor and major axes of the surface orbit?

**36-18.** A wave 18 ft high and having a period of 10 sec is advancing on the coast.
    *a.* At what depth will it break?
    *b.* What will be the surface orbital velocity of the breaking wave?

*c.* Compute $U$ by Eq. (36-21) for the breaking wave.

*d.* What is the velocity of wave advance at breaking, by Eq. (36-7)?

**36-19.** Draw a hydrographic chart to suitable scale for the conditions represented in Prob. 36-14. Using a template similar to that of Fig. 36-7, construct a refraction diagram by graphical methods.

**36-20.** What will be the height of the breaking wave in Prob. 36-19?

**36-21.** The wave entering the breakwater gap in the sketch is 6 ft high and 80 ft long. The channel depth is 40 ft. What will be the wave height at point $P$, the harbor depth being uniformly 40 ft?

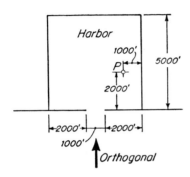

## BIBLIOGRAPHY

Gaillard, D. D.: Wave Action, *U.S. Army, Corps of Engineers Professional Paper* 31, 1904.

Johnson, J. W.: Engineering Aspects of Diffraction and Refraction, *Trans. ASCE,* **118**:617–652 (1953).

Mason, M. A.: Surface Water Wave Theories, *Trans. ASCE,* **118**:546–574 (1953).

A Study of Progressive Oscillatory Waves in Water, *U.S. Army, Corps of Engineers, Beach Erosion Board, Tech. Rept.* 1, 1941.

A Summary of the Theory of Oscillatory Waves, *U.S. Army, Corps of Engineers, Beach Erosion Board, Tech. Rept.* 2, 1942.

# 37
# Breakwaters and Jetties

**37-1 The force of waves** Shore structures must be designed to withstand wave action. It follows that the designer must know something about wave action on his structures, either by practical experience with the consequences of storms at sea or by the study of wave mechanics.

In his many efforts to control the forces of nature for the good of man, the engineer has met failure more often along the coasts than anywhere else. Many breakwaters have been repeatedly rebuilt, sea walls and bulkheads have given way, and isolated piles mark the location of former wharves. The energy released in storms at sea is enormous. The records tell of the displacement of masonry blocks weighing hundreds of tons and of rock fragments tossed hundreds of feet into the air.

The force exerted by waves takes several forms: hydrostatic pressure, the impulse of a jet, and viscous drag. It should not be necessary to review here the nature of the pressure to be found at any specific depth below the surface of a liquid at rest, but it might be well to repeat that the actual hydrostatic pressure for a water particle depends upon its still-water depth rather than upon its depth below the distorted water surface.  The source of dynamic wave action lies in the inertia of the moving particles.  For present purposes each particle may be considered to have a tangential velocity because of rotation about its orbit center, plus the velocity of translation corresponding to mass transport.  The vectorial sum of these two components is the actual velocity of the particle any instant; when the particle is at a wave crest this resultant velocity is horizontal.  If at this instant the forward-moving water particle struck a vertical plate, the reaction of the plate would be the force needed to deflect the speeding water particle.  A similar situation would occur with respect to particles below the surface, but lesser velocities would be involved and hydrostatic pressures would be introduced.  The rational derivation of a suitable formula for wave pressure based upon the foregoing considerations would be not too difficult if it were not for the complication that any obstruction creates reflected waves, which alter the assumed condition of particle motion in the next oncoming wave.  The resulting confusion provides ample opportunity for the selection of various assumptions leading to different formulas, each possessing its own degree of mathematical elegance according to the taste and background of its author.

It is of basic importance that the wave *form* is harmless; the velocity of wave advance, or celerity, is not directly a factor because no momentum is involved—no mass is moving with the celerity $C$.  Waves can be observed as a breeze blows across a field of wheat, but these undulations of the grain exert no force on the confining fences.  Nevertheless, formulas for wave pressure have been advocated which do make wave pressure dependent upon wave celerity, and they yield reasonable answers.  These conflicting statements can be reconciled if it is realized that when the wave is on the point of breaking, the surface velocity of the water equals the celerity of the wave; consequently, the mistaken use of celerity in the formula can yield results of proper magnitude for waves which are near breaking.

**37-2  Wave pressure on vertical walls**  When a nonbreaking wave hits a vertical wall, the reflected energy reinforces the next oncoming wave. As a result the crest height at the wall is doubled, and the water in the region adjacent to the wall sloshes back and forth as it does in an agitated washbowl.  The augmented wave at the wall is called a *clapotis*. The orbital motion in clapotis differs from that found in the conventional oscillatory wave, being essentially vertical for surface water near the wall and horizontal midway between crests.  Under these conditions there can be little dynamic action on the wall, and the wave pressure is hydrostatic although pulsating.

Of the various proposals for the estimation of wave pressure on verti-

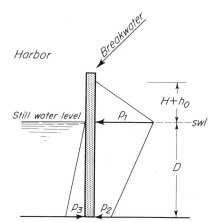

**Fig. 37-1**  Sainflou pressure diagram.

cal walls, probably that of Sainflou (1928) enjoys the greatest popularity. He considered the reflection of elliptically trochoidal waves (shallow-water waves) and concluded that the clapotis would have twice the height of the unreflected wave and that surface orbit center above still-water level would be four times the value without reflection, or

$$h'_o = \frac{H^2}{2R} \coth m \qquad\qquad (37\text{-}1)$$

His expression of wave pressure is somewhat involved, but it becomes more usable if one assumes a linear variation in pressure as indicated in Fig. 37-1. Referring to this figure, the Sainflou formulas for gross lateral-wave pressure, when the crest is at the wall, are[1]

$$p_2 = \frac{\gamma H}{\cosh m} \qquad\qquad (37\text{-}2)$$

and

$$p_1 = (p_2 + \gamma D)\frac{H + h_o}{D + H + h_o} \qquad\qquad (37\text{-}3)$$

Among other efforts to improve upon the Sainflou formula, the method developed by Gourret has received favorable attention.[2]  Referring again to Fig. 37-1, the Gourret approach gives

$$p_2 = \gamma \left(\frac{H}{\cosh m} - \frac{H^2}{2R}\tanh m\right) \qquad\qquad (37\text{-}4)$$

and

$$p_1 = (p_2 + \gamma D)\frac{H + h_o}{D + H + h_o} \qquad\qquad (37\text{-}5)$$

[1] Robert Y. Hudson, Wave Forces on Breakwaters, *Trans. ASCE,* **118**:653–685 (1953).
[2] *Ibid.*

Gourret's expression for $h_o$ is somewhat complex:

$$h_o = \frac{H^2}{2R}(\tanh m - \coth 2m) + \frac{H^2}{4R}\frac{\coth 2m}{\sinh^2 m} \tag{37-6}$$

Robert Y. Hudson, in the reference cited,[1] recommends that in computing moments both the Sainflou and Gourret methods should be tried, and design should be based upon whichever method yields the more severe results in the specific case under consideration. He also refers to model tests which indicate that the Sainflou theory underestimates the moment exerted by the wave when the wall is located at a depth of water less than one-eighth of the wave length.

**37-3   Force of a breaking wave**   In the previous chapter it was found that waves become unstable, or break, when the wave height approximates the depth or when the wave height approaches one-seventh of the wave length.   These rules merely serve to specify the circumstances under which the surface water is moving with a velocity which approximates the velocity of wave advance, that is, $u_H = C$.   This condition is rather commonly encountered at breakwater sites, especially where the tidal range is considerable.   Clearly, a breaking wave hurls a mass of water at any obstruction, and the impulse exerted upon the wall must equal the change in momentum of the water mass.   It will be remembered that impulse is the product of force and time.   The vexing problem is to determine the duration of wave action during impact. One complication arises from the existence of an air cushion, trapped between the wall and the curling wave.   Shock pressures of several tons per square foot have been measured by various observers.

At present the proposals of Minikin[2] appear to offer the best available approach to the solution of this problem.   After consideration of the information forthcoming from theory, experiment, and field observation, he suggests that the dynamic action of the wave be represented by Fig. 37-2$a$.   Here, the maximum dynamic pressure is

$$p_m = \pi g\gamma(D + D')\frac{H}{L}\frac{D'}{D} \tag{37-7}$$

The curves shown are parabolic, and the included area is approximately $Hp_m/3$.   This resultant dynamic thrust, like the maximum dynamic pressure, is assumed to act horizontally at still-water level.   In addition to the dynamic thrust there exists an excess hydrostatic pressure over the still-water condition.   This excess is represented by the trapezium of Fig.

---

[1] *Ibid.*

[2] R. R. Minikin, "Winds, Waves and Maritime Structures," chap. 4, Charles Griffin & Co., Ltd., London, 1950.

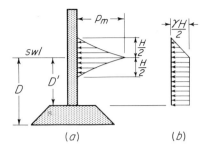

**Fig. 37-2**  Minikin pressure diagram.                    (*a*)                (*b*)

37-2*b*.  The total hydrostatic pressure would involve the additional increment increasing from zero at still-water level to $\gamma D'$ at base, but this consideration does not appear in most calculations as it is balanced by an equal and opposite pressure on the lee side of the wall.  The resultant horizontal thrust on the wall is the sum of the dynamic and the excess hydrostatic forces, or

$$P_H = \frac{\gamma H}{6} \left\{ \frac{2H\pi g}{L}(D + D') \frac{D'}{D} + \left[ 3\left(D' + \frac{H}{4}\right)\right]\right\} \qquad (37\text{-}8)$$

Minikin further suggests that if the wall slopes backward, the horizontal dynamic thrust becomes equal to

$$F_M = \tfrac{1}{3}p_M H \sin^2 \alpha \qquad (37\text{-}9)$$

where $\alpha$ is the angle between the wall face and the horizontal.

**37-4  Breakwater design**  The cross-hatched portion of the sketch, Fig. 37-2*a*, represents a vertical-wall breakwater founded on a rubble mound. This composite design is quite common, but Minikin's method can be extended somewhat beyond a strict compliance with Fig. 37-2 if $H$, $L$, and $D$ are taken to represent conditions prevailing in the deeper region from which the wave is approaching and $D'$ taken to represent the depth at the wall.

The oldest form of breakwater is a simple pile of rock dumped in place.  This is known as a rubble mound.  The stability of such a structure is not determined by the integrity of the entire mass in the presence of wave action, but upon the ability of the individual stone to resist displacement.  Because this situation differs notably from the conditions which influence the stability of monolithic structures, it will be considered separately in this article.

The total quantity of embankment can be reduced and the unit cost of fill material reduced substantially as well, if the fill is confined laterally. One form of the confined-fill breakwater is the stone-filled timber crib, once used extensively on the Great Lakes. Vulnerability to teredo attack limits the use of timber structures to fresh water. Moreover, the advantage of simplicity of construction is dissipated in deep water, where stability against overturning would require wide cribs with excessively long timbers. The stone-filled crib is like a roofless log cabin with stone of sufficient size to avoid loss of fill through cracks between the timbers.

In principle, the modern equivalent of the stone crib is the earth-filled steel-sheet-piling cell. Each cell is a circle of steel sheet piling put into tension by the contained earth fill. The tension across the piling joints creates frictional resistance to longitudinal shear and thus helps provide structural stability. These cells may be linked to form a breakwater, but they are more commonly employed in wharf construction. For that reason, the analysis of these cells is deferred until the next chapter.

Most breakwaters are of masonry construction above the water line, where wave action is most severe. Some are composed of concrete or coursed stone throughout, but more commonly the masonry structure rests upon a rubble mound at an elevation somewhat below low tide. In very deep water, wave action may taper off to insignificant values at great depths, and in this deeper zone the rubble can be replaced by sand fill. The grading requirements for the sand sub-base are determined by the magnitude of anticipated orbital and current velocities at the appropriate depths, and the required particle size can be established by applying the rules which were presented in the earlier chapter on river hydraulics.

The selection of rock size for rubble mounds is taken up in the next section.

The masonry superstructure must be investigated for overturning and for the resultant force on its base. The forces acting are those of wave thrust, hydrostatic pressure, and the body weight of the structure. If the resultant of the body weight and net wave force fall within the middle third of the base, the factor of safety against overturning will be 3 or better, and the design avoids reversal of direction in base-contact pressures. It will be remembered that a resultant intersecting the base outside the middle third throws tension into the outer portion of the base. With the pulsating forces produced by storm waves, such stress reversals are especially undesirable, as there exists a tendency for wave wash to erode the base as the structure is rocked back and forth. The design of a monolithic breakwater section, then, is largely a matter of selecting top and bottom widths which will throw the resultant force within the middle third of the base.

Figure 37-3 represents a composite-type breakwater, consisting of a concrete monolith founded on a rock-fill base. The problem of design is

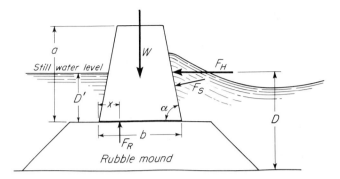

**Fig. 37-3**   Composite-type breakwater.

further simplified if the wall face is vertical ($\alpha = 90$ deg) and if the concrete mass rests directly upon the sea bottom ($D' = D$).   As a first step in design, it is necessary to determine the characteristics of the critical wave, based upon the information presented in Chap. 36.   This will include not only the values of $H_o$, $L_o$, and $T_o$ for the deep-water wave, but also the values of $H$, $L$, and $T$ at the breakwater.   This information, considered together with the still-water depth at the wall, permits a judgment as to whether the structure will be subjected to breaking waves or not.   If not, the theories of Sainflou and Gourret [Eqs. (37-1) to (37-6)] may be applied to evaluate the resultant wave pressure $F_s$ in Fig. 37-3.   The dynamic thrust $F_H$ is assumed to be negligible when a clapotis condition is specified.   The weight $W$ can be expressed in terms of the dimensions $a$ and $b$ (Fig. 37-3), of which $a$ will be a known value and $b$ is to be evaluated.   The proper allowance to be made for uplift is not readily apparent.   If $D'$ does not greatly differ from $D$, uplift will approximate the hydrostatic pressure, producing a head $D'$.   However, the uplift will be somewhat greater if $D'$ is relatively small, because of the excess hydrostatic pressure at the heel of the wall. In the latter case it appears that the uplift pressure head should be assumed to vary uniformly from a value $D'$ at the toe to a value $D' + h'_o$ at the heel, where $h'_o$ refers to the orbit-center displacement of a particle at still-water depth $D'$.   For most cases it would be sufficient to assume that concrete weighed 150 pcf above still-water level and 86 pcf (in salt water; 87.5 pcf in fresh water) below still-water level.   Then, taking moments about the toe, $xF_R$ must equal the algebraic sum of the moments of the forces $W$ and $F_s$, the value having been previously reduced for uplift or buoyancy.   The value of $b$ may then be taken to equal $3x$.

In locations where storm waves will break against the wall, both $F_s$

and $F_H$ will have real values, which should be estimated through the Minikin theory, Eqs. (37-8) and (37-9).

**37-5  Rubble mounds**   The structural design of a rubble-mound break-water is mainly concerned with specifications for rock size and the selection of an appropriate slope angle.   In general, the wave force acting upon an individual stone is proportional to the exposed area of the stone, while the resistance of the stone is proportional to its volume.   Hence, the larger the stone, the greater its stability.   Rather obviously, also, stones are less stable on steeper slopes, and larger stones are required to resist the action of higher waves.   The Spanish engineer Iribarren combined these concepts into a formula which has been widely used. Modified to achieve dimensional homogeneity, the Iribarren formula has been restated by Hudson to specify the required weight of stone (consistent English units):

$$W = \frac{K'\gamma_f{}^3 \tan^3 \phi H^3 \gamma}{(\tan \phi \cos \alpha - \sin \alpha)^3 (\gamma - \gamma_f)^3} \tag{37-10}$$

There is only a limited amount of data for the evaluation of the coefficient $K'$.   For natural rubble, Iribarren's experience assigns a value of 0.015; some experimental values from model work at the U.S. Waterways Experiment Station furnished the values plotted in Fig. 37-4. The other terms in Hudson's modified version of the Iribarren formula (all consistent units) are:

$\qquad W$ = weight of individual rock, exposed to wave actions
$\quad \tan \phi$ = effective coefficient of friction between rocks
$\qquad \gamma$ = unit weight of rock
$\qquad \gamma_f$ = unit weight of water (62.5 pcf, fresh; 64 pcf, sea)
$\qquad H$ = wave height
$\qquad \alpha$ = slope angle of the mound face

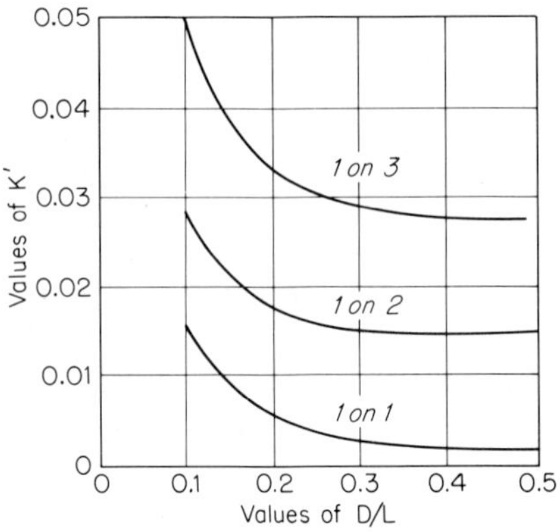

**Fig. 37-4**  Values of $K'$ in modified Iribarren formula. (*From U.S. Army, Corps of Engineers, Beach Erosion Board, Spec. Bull. 2, 1953.*)

There may be some preference for a formula which emphasizes the kinetic rather than the potential energy of the wave. Model tests at the University of Washington led to the following formula:[1]

$$D' = \frac{0.00633 C_b{}^2}{(G-1)(\tan \phi - \tan \alpha)} \qquad (37\text{-}11)$$

where $\phi$ and $\alpha$ have the same meaning as in Eq. (37-10)
$D'$ = required diameter of an exposed spherical rock, ft
$C_b$ = velocity of the design wave (celerity) at breaking
$G$ = specific gravity of rock

For the size of rock to be located at appreciable depths below the water surface, it is reasonable to assume that the severity of wave action will be inversely proportional to the square of the orbital velocity. This suggests that the effective diameter of subsurface rock be obtained by multiplying the value of $D'$, as obtained from Eq. (37-11), by the coefficient

$$K_1 = \left(\frac{H \cosh m}{2R \sinh m}\right)^2 \qquad (37\text{-}12)$$

where $H$, $n$, and $m$ are as defined in Chap. 36. The required weight of subsurface stone may be estimated by multiplying the $W$ of Eq. (37-10) by the quantity $K_1{}^3$.

**37-6  Jetties**  The British apply the term *jetty* to a wharf or pier, but in American usage the word is reserved for a structure which has characteristics both of a breakwater and of a training wall. The prime function of a jetty is the preservation of a channel against the encroachment of littoral drift or river sediment. Jetties are usually built in pairs; one on each side of a river mouth or of a breakwater gap. Thus they serve to confine the river or tidal currents within a space contrived to maintain scouring velocities.

Like breakwaters, jetties are exposed to severe wave action; but often the design standards may be partly relaxed because the jetty is usually built normal rather than broadside to the most dangerous wave fronts. The structural design of the jetty section does not differ in principle from that of a breakwater. The chief possibility for divergence in design lies in the deeper channel on the lee side of the jetty. If considerable difference between bottom elevations on the two sides either exists or can develop and if, also, the subsoil is relatively weak, then the

[1] C. E. Leonoff, "The Stability of Rubble Mound Breakwaters under Wave Action," M.S. thesis presented to the University of Washington, 1952.

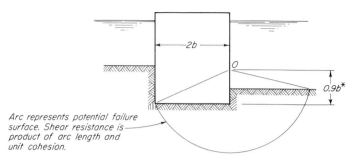

Arc represents potential failure surface. Shear resistance is product of arc length and unit cohesion.

Factor of safety is ratio of resisting moment to driving moment. Take $\Sigma M_O$, including moments of lateral earth and water pressures, shear resistance, buoyant weights of jetty and surcharges. (Normal earth reaction along arc passes through "O")

**Fig. 37-5**  Bearing capacity.  Jetty on clay foundation.  [*Location of arc center adapted from A. Casagrande and R. E. Fadum, Application of Soil Mechanics in Designing Building Foundations, Trans. ASCE,* **109**:397 (1944).]

bearing capacity of the jetty foundation may become critical.   There are various conventional techniques for dealing with the analysis of bearing capacity which will not be discussed here.   However, Fig. 37-5 is presented to demonstrate one approach which is simple and well adapted to a typical situation.

The matter of hydraulic design is of at least equal importance with structural design.   Chief among the former is the selection of the distance between jetties.   While jetties serve to protect a ship channel from waves and cross currents, their more direct purpose is to deepen and stabilize the channel.   With this in mind, properly located jetties promote bottom scour and the transportation of sediment through the maintenance of proper velocities.   This has already been discussed in earlier chapters on river training; with respect to tidal currents an empirical approach may prove simpler.   A study of the relation of tidal prisms to channel areas of Pacific Coast waterways suggested the following equation:

$$A = V^{0.85} \tag{37-13}$$

where $V$ is the volume of the tidal prism between $MLLW$ and $MHHW$ in square mile–feet and $A$ is the area of the entrance-channel section below midtide in square feet.[1]

[1] R. E. Hickson and F. W. Rodolf, Design and Construction of Jetties, *Proc. First Conf. on Coastal Eng.*, chap. 26 (1950), and M. P. O'Brien, Estuary Tidal Prisms Related to Entrance Areas, *Civil Eng.*, **1**:738–739 (1931).

## QUESTIONS AND PROBLEMS

**37-1.** A breakwater with vertical walls and a 10-foot freeboard is exposed to waves which reach a height of 10 ft and a length of 157 ft in deep water. What will be the wave properties (height, length, velocity) where the still-water depth adjacent to the breakwater is 40 ft? 10 ft?

**37-2.** Draw a Sainflou pressure diagram and compute the net horizontal thrust on the wall where the still-water depth is 40 ft, using the data of Prob. 37-1.

**37-3.** Compare the overturning moment corresponding to the data of Prob. 37-2 with that obtained by the Gourret procedure.

**37-4.** Draw a Minikin pressure diagram against the breakwater of Prob. 37-1, where the still-water depth is 10 ft. Compute the overturning moment.

**37-5.** For the conditions of Prob. 37-4, compute the mass-transport and orbital velocities and elevations for still-water depths of 0, 2, 4, 6, 8, and 10 ft below the still-water level. Draw pressure diagrams against the wall for static, dynamic, and net total pressures. Compute the overturning moment. Assume that the dynamic pressure at any point corresponds to the velocity head.

**37-6.** What size stone would be required on the exposed face of a rubble mound which is to replace the masonry breakwater of Prob. 37-2? The slope angle of the mound is to be 40 deg, and the available stone has a specific gravity of 2.72.

**37-7.** Compare the results of Prob. 37-6 with those obtained by using Eq. (37-11).

**37-8.** A substitute design for the breakwater of Prob. 37-2 would place a vertical concrete wall 30 ft high on top of a rubble mound 20 ft high. The rubble mound is to have a slope angle of 35 deg, $G = 2.68$. Select a suitable size for exposed stone.

**37-9.** In Fig. 37-5, let the width of the jetty be 30 ft, its height 25 ft, and the freeboard 5 ft. The stillwater level is 15 ft above the dredge line or 10 ft above the natural ground surface. The bottom material is frictionless clay, has a saturated unit weight of 109 pcf, or an effective unit weight, submerged, of 45 pcf. The mean unit weight of the jetty is 124 pcf above water level, with an effective unit weight of 70 pcf below water level. What unit cohesion in the clay will provide a safety factor of 1.5 against foundation failure?

**37-10.** The tidal range is ±5 ft in a bay whose water area amounts to 400 sq miles. What should be the size of the entrance channel to minimize the need for dredging?

**37-11.** What static pressure is exerted upon a vertical wall by a particle whose still-water depth was 10 ft and which impinges on the wall at a point 5 ft above the still-water surface level? Salt water.

## BIBLIOGRAPHY

"Bank and Shore Protection," California Department of Public Works, 1960.

Hudson, Robert Y.: Wave Forces on Breakwaters, *Trans. ASCE*, **118**:653–685 (1953).

Molitor, David A.: Wave Pressures on Sea-walls and Breakwaters, *Trans. ASCE*, **100**:984 (1935).

Stability of Rubble-mound Breakwater, *U.S. Waterways Expt. Sta. (Vicksburg) Tech. Mem.* 2-365, 1953.

# 38

# Coastal Structures

**38-1  Sea walls**  At an early stage in the development of shorelands for human use it becomes desirable to provide a definite and permanent boundary between land and sea.  The natural processes of erosion and deposition must be halted or placed in equilibrium.  The former is accomplished by the construction of sea walls, bulkheads, and revetments, which serve to restrain the land from slipping into the sea or the sea from cutting into the land.  These three types of structures differ from each other more physically than functionally.

A sea wall normally is subjected to active earth pressure on one side and wave thrust on the other, with the emphasis on wave action.  It is a heavier structure than either a bulkhead, which is primarily a retaining wall, or a revetment, which would be essentially a protective pavement placed on a stable earth slope.  All three types are alike in the respect that critical wave exposure will involve breaking waves rather than a clapotis condition.  Where the foot of the wall is submerged at high tide, wave thrust can be computed in accordance with the Minikin method and Fig. 37-2.  Structures of this sort, however, are often

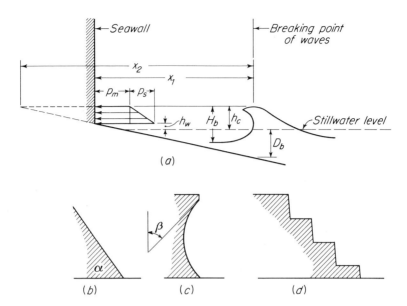

**Fig. 38-1** Elements of sea-wall design.

situated at some distance shoreward of the line of breakers, and even at high tide may lie entirely above the still-water level. Under such circumstances it has been suggested[1] that it may be appropriately assumed (1) that the breaker will run up the shore slope to an elevation no higher than the crest of the wave when breaking, and (2) that the onrushing mass of water will experience a constant rate of deceleration from a velocity $C_b$ at the farthest penetration up the unbroken beach slope. This is better expressed graphically, as on Fig. 38-1. Breakers are assumed to form where shown, and the water would rush up the beach a distance $x_2$ were it not for the wall. Thus the velocity at the wall becomes

$$V = C_b \left( 1 - \frac{x_1}{x_2} \right) \tag{38-1}$$

and the dynamic pressure on the wall is

$$p_m = \frac{K_2 V^2}{2g} \tag{38-2}$$

The latter expression will be remembered from hydraulics as the force exerted by a jet. For the vertical wall shown in Fig. 38-1, the value of

[1] Shore Protection Planning and Design, *U.S. Army, Corps of Engineers, Beach Erosion Board, Tech. Rept.* 4, 1954.

the coefficient $K_2$ is unity. For a wall with batter (Fig. 38-1$b$), $K_2$ equals $\sin^2 \alpha$; and for a wall designed to turn the jet back upon itself, as in Fig. 38-1$c$, $K_2$ becomes $(1 + \sin \beta)^2$. The determining factor in the coefficient is the degree to which the velocity direction is changed.

In the sketch the elevation of the wall base above still-water level was designated as $h_w$, which may have either positive or negative values, depending upon whether the base elevation lies above or below still-water level. The total dynamic thrust is, for positive values of $h_w$,

$$R_d = p_m(h_c - h_w) \tag{38-3}$$

and, for negative values of $h_w$,

$$R_d = p_m h_c \tag{38-4}$$

To this must be added the hydrostatic force, $\frac{1}{2}p_s(h_c - h_w)$, where

$$p_s = {}_w(h_c - h_w) \tag{38-5}$$

The selection of a sea-wall design from among the four types shown in Fig. 38-1 will be guided by circumstances. The sloping wall receives the least wave thrust, but deflection of the water is incomplete, and water will be thrown over the wall. This objectionable feature is also present, to a lesser extent, in the vertical and stepped walls (Figs. 38-1$a$ and $d$) when the spray is caught in a following wind. The advantage of a stepped wall lies in its tendency to reduce the intensity of wave thrust by stopping different elements of the wave mass at different stages in the shoreward progress of the wave. The likelihood of abnormally high shock pressures is thereby reduced. The wall with the concave face absorbs the greatest wave thrust, but also affords the greatest protection to shore structures. Where wave action is most severe, the sea wall is sometimes designed with a concave face in the upper section, with a stepped base below.

In sea-wall design special precautions must be taken against wave cutting at the base of the wall. Where base erosion may become serious, it is fairly common to drive a row of sheet piling along the seaward edge of the wall base.

The estimation of wave action is only one part of sea-wall analysis. The structure is also subjected to earth pressure on its land side. The appropriate earth pressure may be computed, as in the analysis of gravity-type retaining walls. The position of the water table is important here, as in all retaining-wall analysis, and is complicated by the probable existence of appreciable fluctuation due to the rise and fall of the tides. The engineer should not forget that below ground-water level the unit weight of earth should be reduced for buoyancy in computing earth pressure. Of course, in such cases the full hydrostatic pressure of

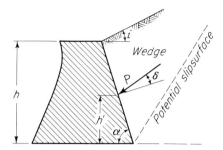

**Fig. 38-2** Earth pressure on sea walls.

water below ground-water level must be added to the computed earth pressure.

As a matter of convenience, the earth-pressure formula, based on Coulomb's analysis of the wedge of backfill which is most apt to slide, is included here:

$$P = \frac{1}{2}\gamma h^2 \frac{K_A}{\sin\alpha\cos\delta} \tag{38-6}$$

where

$$K_A = \frac{\sin^2(\alpha+\phi)\cos}{\sin\alpha\sin(\alpha-\delta)\left[1+\sqrt{\dfrac{\sin(\phi+\delta)\sin(\phi-1)}{\sin(\alpha-\delta)\sin(\alpha+i)}}\right]^2} \tag{38.7}$$

In this expression

$\alpha$ = angle between back of wall and the horizontal
$\phi$ = angle of internal friction of backfill
$\delta$ = angle of wall friction
$i$ = slope angle of surcharge
$h$ = height of wall
$\gamma$ = unit weight of backfill

Equations (38-6) and (38-7) apply only to sand soils. Figure 38-2 may serve to clarify the meaning of some of the terms. The angle $\phi$ may be taken to be equal to the slope angle observed from a loose pile of the backfill sand. Values of tan $\alpha$ are listed in Table 38-1, but in no case

**Table 38-1  Friction coefficients ($f$, tan $\delta$, or tan $\phi$)**
Dry weights throughout

| | | |
|---|---|---|
| Masonry on gravel.......... | 0.6 | |
| Masonry on sand............ | 0.4 | |
| Masonry on clay............ | 0.3 | |
| Stone on steel.............. | 0.3 | |
| Sand...................... | = 90 pcf | tan $\phi$ = 0.5 |
| Sand or gravel............. | = 105 pcf | tan $\phi$ = 0.8 |
| Gravel.................... | = 120 pcf | tan $\phi$ = 1.0 |
| Silt...................... | = 65 pcf | tan $\phi$ = 0.4 |
| Silt...................... | = 100 pcf | tan $\phi$ = 0.5 |

should one use a value of $\delta$ greater than $\phi$. The surcharge need not extend any great distance at the slope angle $i$ to satisfy the assumptions of the formula. Beyond the intersection of the ground surface with the potential slip surface, the surface slope is not significant. Care must be taken in the selection of the value for $\gamma$. Sea walls are apt to be located where ground-water levels fluctuate, and the designer should have the help of actual observations of ground-water level. Below ground-water level the unit weight of the soil must be reduced for buoyancy. For most installations there would be no serious error in using a weighted value of $\gamma$ in the formula, based upon the relative percentages of the soil below and above ground-water level. Structures of any considerable importance and those with cohesive backfills deserve a fuller treatment than that given here. The discussion of earth pressure on retaining walls is sharply curtailed here because more comprehensive discussions are rather generally available, and not because of any failure to recognize the major importance of the topic.

A rough estimate of the pressure exerted by cohesive backfills may be made by considering the clay to act as a heavy fluid. Recognizing the severe exposure of sea walls, and not attempting to segregate water from earth, as was suggested for sand backfill, the equivalent fluid density should be estimated quite liberally, say 100 pcf for soft plastic clays, and perhaps 40 or 50 pcf for stiffer cohesive soils with enough sand to provide appreciable frictional resistance.

**38-2  Revetment**  A sea wall supports an earth bank, while a revetment is supported by an earth bank. Both structures are intended to protect the land from wave erosion. Stone is most commonly used for revetment work, although portland-cement concrete and, to a lesser extent, bituminous concrete are also employed for this purpose. The conventional pavement materials will, in general, follow conventional designs. The chief factor involved in the design of a stone revetment is the selection of stone size. Here the criteria for rubble mounds, presented in Chap. 37, can be used with equal propriety. The revetment, however, is a relatively thin layer compared with the massive proportions of the rubble mound. This gross difference in dimensions makes it important to guard against soil erosion between the individual stones of the revetment. The coarse riprap must be isolated from the natural earth bank by one or more courses of filter stone. Each of these blankets should be a foot or more in thickness. The stone in each layer must be graded to choke the openings between stones in the adjacent coarser layer and to retain the stones in the adjacent finer layer. Figure 38-5 shows a typical section. In selecting an appropriate grading, the rules for soil-filter layers provide some guidance. Experiment in soil mechanics has led to

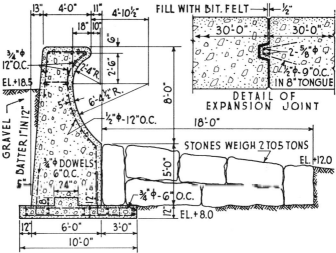

**Fig. 38-3** Sea wall at Hampton Beach, N.H. (*Courtesy of Portland Cement Association.*)

the rule that soil losses are prevented when the grading of a filter blanket is such that a screen opening which will retain 85 per cent by weight of the filter soil is not more than four times as large as the opening which will retain 15 per cent of the protected soil. Lacking screens to grade derrick stone, the above rule can serve only as a guide in the selection of riprap.

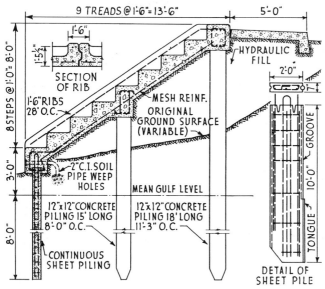

**Fig. 38-4** Harrison County, Miss., revetment. Construction picture shows difference in wave action against vertical-faced cofferdam as against completed step section of revetment. (*Courtesy of Portland Cement Association.*)

## 38-3 Cantilever bulkheads

Characteristically, a bulkhead supports an earth bank by flexural members, in contrast with the gravity-section of a typical sea wall. Figures 38-6 to 38-8[1] illustrate the various examples of this flexural principle commonly encountered in bulkhead design.

[1] From *U.S. Army Corps of Engineers Beach Erosion Board, Spec. Bull. 2* (1953).

A bulkhead derives its chief support from the soil into which the piling is driven. It may be cantilevered out from the foundation soil, as in Fig. 38-9a, or it may be additionally restrained by tie rods, as in Fig. 38-9b. In either case the wall itself may be timber, steel, or concrete.

Figure 38-10 is taken from a recent paper describing construction and design practices in the U.S. Navy.[1] The authors point out that

either one or both of two methods may be used for transferring tie-rod loads to the earth. One method derives its resistance from the passive pressures developed in front of a buried deadman, the

[1] James R. Ayers and R. C. Stokes, The Design of Flexible Bulkheads, *Proc. ASCE,* Separate No. 166:6–8 (1953).

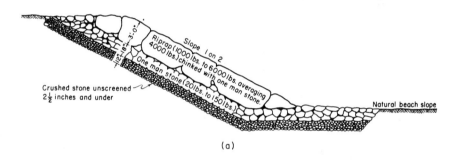

(a)

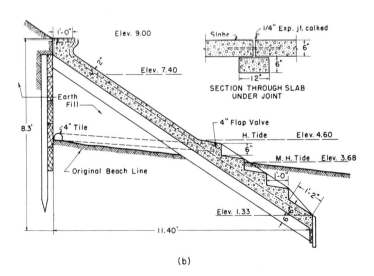

(b)

**Fig. 38-5** (a) Stone revetment; (b) concrete revetment.

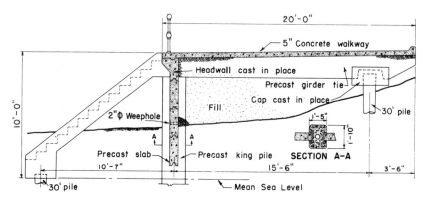

**Fig. 38-6**  Slab and king-pile bulkhead, concrete.

other from a structural arrangement of brace piles.  [Several forms of typical anchorages are shown in Fig. 38-10.]  The simple dead-man [Fig. 38-10a] is suitable for use at a site where the existing ground level is at or above the bottom of the anchor.  A concrete deadman can be either precast or cast in place, depending on econ-omy of construction and the type of equipment available.  Prefer-ably, the concrete is cast in a trench in undisturbed firm material.  Otherwise, the material in front of the deadman should be of select quality and thoroughly compacted to minimize lateral movement and resulting misalinement of the bulkhead.  This type of anchor-age is subject to tipping caused by unequal passive resistances in the earth and should not be used on important structures unless very favorable soil conditions exist.  A T-shaped or L-shaped deadman will give improved stability against tipping.

[In Fig. 38-10b] sheet piles are used as an anchorage.  The

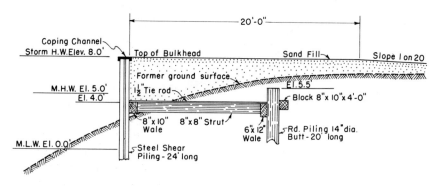

**Fig. 38-7**  Sheet-pile bulkhead, steel.

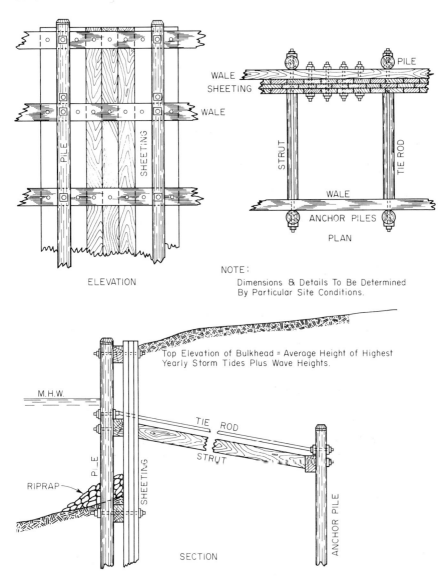

**Fig. 38-8**  Timber-sheet-pile bulkhead.

piling may be placed as a continuous wall, or in separate groups centered at each tie rod.   The longer sheet piles are driven into the ground at each tie rod to provide vertical support for the assembly during erection and backfilling.   This anchorage is suitable for sites having firm soils with varying ground elevations.

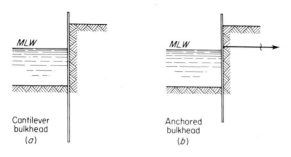

**Fig. 38-9** Bulkheads.

In locations where the existing grade is low, anchorages of the type [shown in Figs. 38-10c and d] are effective. Because the resistance of this anchorage is not dependent primarily on passive pressure, it is well adapted to the hydraulic placing of fill. The principal resistance is derived from the batter piles, although [in the case shown by Fig. 38-10d] a large passive resistance eventually may be mobilized. Steel or concrete piles may be substituted for wood piles for increased capacity.

The location of an anchorage is of primary importance. If an anchorage depends entirely on passive earth pressure, it should be located far enough back from the bulkhead so that a region sufficient

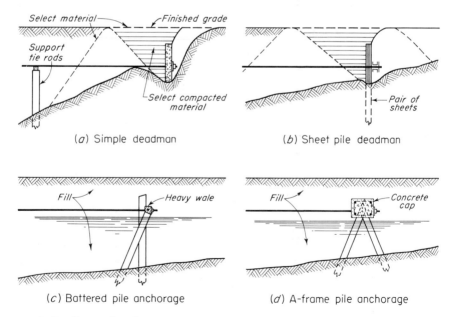

**Fig. 38-10** Types of anchorage.

to develop this pressure exists behind the region producing active pressure. Anchorages of this type should not be closer to the bulkhead than from two to three times the distance from mean tide level to the dredged bottom. Anchorages depending on batter piles should be placed behind the region contributing active pressure to the bulkhead. Live load surcharges and settlements accompanying natural compaction of the backfill produce high tensile stresses in the tie rods and overloads on the anchorages. Therefore adequate vertical supports for the tie rods are essential in order to prevent sag of the rods both before and after placing of the backfill.

The pressures acting on a cantilever sheet-pile bulkhead are idealized in Fig. 38-11. Here the earth pressures at significant points ($p_1$, $p_2$, $p_3$, $p_4$,

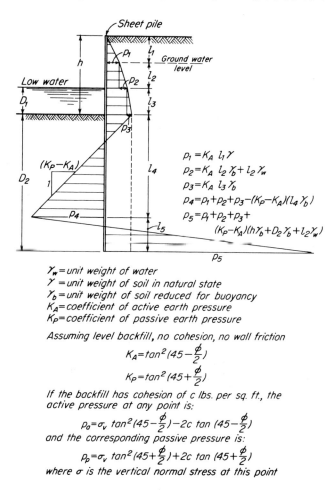

$$p_1 = K_A \, l_1 \gamma$$
$$p_2 = K_A \, l_2 \gamma_b + l_2 \gamma_w$$
$$p_3 = K_A \, l_3 \gamma_b$$
$$p_4 = p_1 + p_2 + p_3 - (K_P - K_A)(l_4 \gamma_b)$$
$$p_5 = p_1 + p_2 + p_3 +$$
$$(K_P - K_A)(h \gamma_b + D_2 \gamma_b + l_2 \gamma_w)$$

$\gamma_w$ = unit weight of water
$\gamma$ = unit weight of soil in natural state
$\gamma_b$ = unit weight of soil reduced for buoyancy
$K_A$ = coefficient of active earth pressure
$K_P$ = coefficient of passive earth pressure

Assuming level backfill, no cohesion, no wall friction

$$K_A = \tan^2 (45 - \tfrac{\phi}{2})$$
$$K_P = \tan^2 (45 + \tfrac{\phi}{2})$$

If the backfill has cohesion of c lbs. per sq. ft., the active pressure at any point is:

$$p_a = \sigma_v \tan^2 (45 - \tfrac{\phi}{2}) - 2c \tan (45 - \tfrac{\phi}{2})$$

and the corresponding passive pressure is:

$$p_b = \sigma_v \tan^2 (45 + \tfrac{\phi}{2}) + 2c \tan (45 + \tfrac{\phi}{2})$$

where $\sigma$ is the vertical normal stress at this point

**Fig. 38-11**  Analysis of sheet-piling bulkhead.

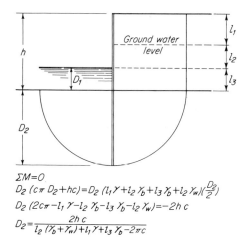

$\Sigma M = 0$

$D_2 (c_\pi D_2 + hc) = D_2 (l_1 \gamma + l_2 \gamma_b + l_3 \gamma_b + l_2 \gamma_w)(\frac{D_2}{2})$

$D_2 (2c_\pi - l_1 \gamma - l_2 \gamma_b - l_3 \gamma_b - l_2 \gamma_w) = -2h \, c$

$D_2 = \dfrac{2h \, c}{l_2 (\gamma_b + \gamma_w) + l_1 \gamma + l_3 \gamma_b - 2\pi c}$

**Fig. 38-12** Analysis of sheet piling with clay backfill.

and $p_5$) are assumed to have the values assigned by Rankine's analysis, and a uniform transition is assumed from $p_3$ to $p_4$ and from $p_4$ to $p_5$. Actually it is generally felt that the load triangles would be more realistically replaced by curvilinear figures, but the refinement is not warranted by the existing state of knowledge in this field.

The design of cantilever sheet-pile bulkhead involves the selection of piling length and of piling section or weight. First in order is the determination of the depth of penetration $D_2$ in Fig. 38-11. The values of $l_1$, $l_2$, and $l_3$ will be known, while $l_4$ and $l_5$ will be unknown. The two unknowns can be found by applying two equations of static equilibrium:

$$\Sigma F_M = 0 \qquad\qquad\qquad (38\text{-}8)$$
$$\Sigma M = 0 \qquad\qquad\qquad (38\text{-}9)$$

The equations will be cumbersome but can be handled by trial. The solution is sometimes simplified by dividing Eq. (38-8) by $h^2$ and by dividing Eq. (38-9) by $h^3$. Substituting $x = l_5/h$ and $y = D_2/h$ or $l_4/h$, two simultaneous equations result, and it becomes feasible to eliminate one unknown.

When the value of $D_2$ has been established, it becomes possible to add a shear diagram and a bending-moment diagram to the load diagram of Fig. 38-11. The selection of a suitable section modulus then follows by conventional procedures. There need be no essential difference in analysis, whether a steel, concrete, or timber of composite structure is involved.

If the depth of penetration is chosen to be much greater than the value obtained by the foregoing procedure, there may be several points of inflection in the buried piling. A result will be less displacement.

Generally, greater restraint can be obtained more cheaply by selection of an anchored bulkhead.

A bulkhead with a clay backfill may best be analyzed by considering the equilibrium of a circular slip surface passing through the bottom of the piling. This procedure is illustrated for the case of frictionless clay in Fig. 38-12. For soils with both friction and cohesion, a suitable approach utilized a circular slip surface with the $\phi$-circle technique, which is described in most soil-mechanics textbooks. An alternate procedure in cohesive soils would be to use the Rankine pressure diagram of Fig. 38-11 but discarding the use of earth-pressure coefficients $K_A$ and $K_p$ in favor of the following expressions for lateral active earth pressure at any designated point:

$$p_a = \sigma_v \tan^2\left(45 - \frac{\phi}{2}\right) - 2c \tan\left(45 - \frac{\phi}{2}\right) \tag{38-10}$$

and the corresponding value for passive pressure:

$$p_p = \sigma_v \tan^2\left(45 + \frac{\phi}{2}\right) + 2c \tan\left(45 + \frac{\phi}{2}\right) \tag{38-11}$$

where $\sigma_v$ represents the unit vertical normal stress at the designated point.

**38-4 Anchored bulkheads** Except for installations of moderate height and stable foundation, a cantilever bulkhead will be rejected in favor of a design having one or two anchorages. The required minimum depth of penetration may be computed with a single equation, taking moments about the tie rod. The simplification which appears to have been achieved in this way is really somewhat fictional. Because the anchored bulkhead is more restrained than the cantilever type, there is not the same opportunity for the development of active and passive earth pres sure. Both of the limiting states of earth pressure require enough movement within the backfill to develop the full frictional resistance of the soil. This is because stress and strain are related even in imperfectly elastic materials like earth. The complexity of this problem has resulted in various proposals for practical design procedures. Figures 38-13 and 38-14 present procedures that originated in the extensive laboratory tests at Princeton University[1] for the U.S. Navy Department.

The proposal for clean sand backfill (Fig. 38-13) involves driving the piling to a penetration of $0.43h$. The analysis is carried through for a safety factor of unity, but the neglect of wall friction (in assuming horizontal pressures throughout) is felt to provide security against foundation failure with an actual safety factor of about 2.

Figure 38-14 appears to represent a rather special situation, with

[1] G. P. Tschebotarioff, "Final Report—Large Scale Earth Pressure Tests with Model Flexible Bulkheads," Princeton University Press, Princeton, N.J., 1949.

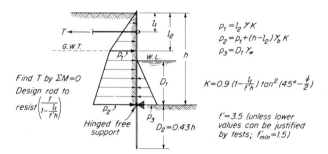

**Fig. 38-13** Tschebotarioff solution for anchored cantilever bulkhead in clean sand.

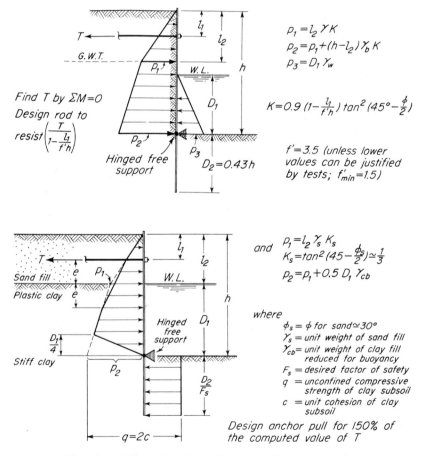

**Fig. 38-14** Tschebotarioff solution for anchored cantilever bulkhead with plastic-clay backfill. (Clay fill not to be carried above water line.)

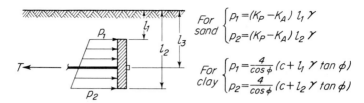

**Fig. 38-15**  Pressure on deadman.

ground-water table coinciding with the free-water level and also with the sand-clay interface.   Actually, it is undesirable to carry the clay backfill above the water line because the poor drainage would ensure full hydro-static pressure up to the top; hence, sand fill should be used above the elevation of low water.   At the same time, steps should be taken to drain the sand fill adequately, so that these ground-water levels will not lag much below water levels in the harbor.   It is also felt unwise to use soft clay backfill if the subsoil is soft.   With a weak foundation the backfill should be all sand.   As both sketches indicate, the piling is assumed to be hinged at the dredge line.   The value of $T$ can be computed by taking moments about the hinged support, considering the piling to act as a supported simple beam, with one overhanging end.

The tie rod must be carried to some secure anchorage.   Generally this will be provided by a continuous or intermittent deadman, which may be made of timber, steel, or concrete.   The earth-pressure analysis resembles the bulkhead computations, and in Fig. 38-15 all terms are as previously defined.

The placement of the deadman must be such that the potential slip surface behind it does not invade the zone influencing bulkhead pressures. With this objective the length of the tie rod may be computed from Eq. (38-12) (Fig. 38-16).   It is customary to support long tie rods either

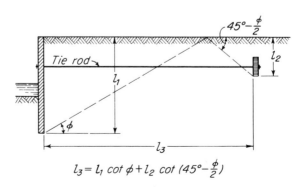

**Fig. 38-16**  Placement of deadman.

by intermediate bearing posts or by a timber strut between bulkhead and anchorage.

**38-5  Cellular-steel-sheet piling**  Sea walls, wharves, cofferdams, and breakwaters are often made up of interconnected earth-filled cylinders of steel-sheet piling.  Each constituent cell is a stable unit, deriving its stability from the composite strength of both the earth fill and the steel shell.  The cell is a gravity-type structure; to act as a rigid unit it must be able to resist longitudinal shear.  This it is able to do through the friction in the longitudinal joints (or locks) of the sheet piling.  Friction acts in the locks in response to the circumferential tension in which the whole steel shell is placed by the outward thrust of the earth fill.  While cellular cofferdams have been in use for nearly 50 years, the rational analysis of their performance has been developed more recently.[1]

  Figure 38-17 shows a conventional layout for a circular type of cellular structure.  To simplify analysis, the external forces are assumed to act on a rectangular cell of height $h$ and breadth $b$.  A section of the wall $2L$ ft long is selected for the unit to be analyzed.  The cell must be investigated with respect to tension in the locks, longitudinal shear, and bearing capacity of the base.  The possibility of its sliding on its base or overturning may also warrant study, but these items are not apt to be critical.

[1] Karl Terzaghi, Stability and Stiffness of Cellular Cofferdams, *Trans. ASCE*, **110**:1083 (1945).

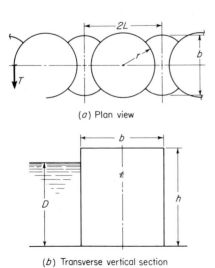

(*a*) Plan view

(*b*) Transverse vertical section

**Fig. 38-17**  Circular type of cellular cofferdam.

At any depth $z$ below the top of the fill in the cell, the vertical earth pressure is

$$\sigma_z = \gamma z$$

and the horizontal earth pressure is

$$\sigma_z = K\gamma z$$

The estimation of the lateral pressure coefficient $K$ is complicated by the confining influence of the shell. There is not room for the development of a failure plane at the same slope which characterizes the development of active earth pressure against a returning wall with level backfill. The presence of the surrounding sheet piling with its vertical joints suggests that the critical shear plane in the soil is also vertical. Krynine has shown by use of Mohr's circle of stress that for the assumption of the vertical shear plane the lateral pressure coefficient will be

$$K = \frac{\cos^2 \phi}{2 - \cos^2 \phi} \tag{38-12}$$

The total horizontal thrust across a vertical section through the axis of the cell having a diameter $2r$ (Fig. 38-17), assuming empty space between cells, is

$$F_H = 2hr \times \tfrac{1}{2}K\gamma h = K\gamma Lh^2 \tag{38-13}$$

This is resisted by the tension in two lines of piling, so that the total across an axial vertical section of one sheet pile is

$$T = \tfrac{1}{2}K\gamma rh^2 \tag{38-14}$$

Rupture of the piling will occur in the locks before the web fails, so that the critical requirement (at the base of the piling) is that the minimum interlock strength per unit of length of the specified sheet-pile section shall be at least

$$t_e = K\gamma rh \tag{38-15}$$

Standard sheet-pile sections are rated by the manufacturer for minimum interlock strength (Fig. 38-19).

In the analysis of vertical shear a free-body diagram of a cell will be found useful. For the purpose of analysis the curved rows of piling are replaced by straight lines (Figs. 38-17 and 38-18) $b$ ft apart. The cells are assumed to be separated by two adjacent straight rows of piling. These pairs of transverse walls are spaced at an interval of $2L$. From Fig. 38-18$b$, the summation of vertical forces shows that $p_1 = W/b$, where $W$ is the total weight of a single cell. Care must be taken to take buoyancy into account in determining the value of $W$. If the locus of zero hydro-

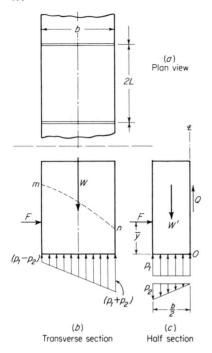

**Fig. 38-18**   Analysis of cellular structure.

static pressure is indicated by some such line as $mn$ of Fig. 38-18$b$, then unit weight of all fill below $mn$ should be reduced by 63 pcf.   If the fill is fully drained, $mn$ will coincide with the base line and $W'$ of Fig. 38-18$c$ will be equal to $\frac{1}{2}W$.   Otherwise, it will be a somewhat larger part of $W$, which can be readily estimated for any specified position of $mn$. Taking moments about point $O$,

$$p_2 = \frac{1.5p_1b^2 + 12Fy - 3bW'}{b^2}$$

and by $\Sigma F_v = 0$,

$$Q = W' - \frac{b}{4}(2p_1 - p_2) \tag{38-16}$$

The total shear resistance is the sum of the frictional resistance in the locks plus the frictional resistance in the soil,

$$Q' = 2fT + 2T \tan \phi \tag{38-17}$$

where $f$ is the coefficient of friction, steel on steel ($f = 0.3$), and the other terms are as previously defined.   The factor of safety against shear failure along a vertical plane through the axis of the cells is

$$F_s = \frac{Q'}{Q}$$

| | Section index | Driving distance per pile, in. | Weight, lb Per foot | Weight, lb Per square foot of wall | Web thickness in. | Section modulus of single piece, cu in. Per pile | Section modulus of single piece, cu in. Per foot of wall |
|---|---|---|---|---|---|---|---|
| 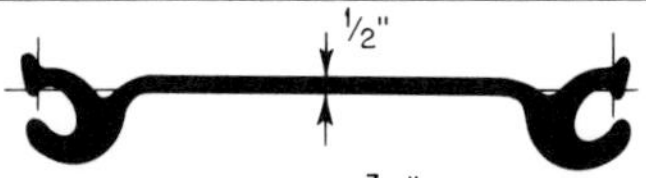 | M 108 | 15 | 43.8 | 35.0 | 1/2 | 3.8 | 3.1 |
| 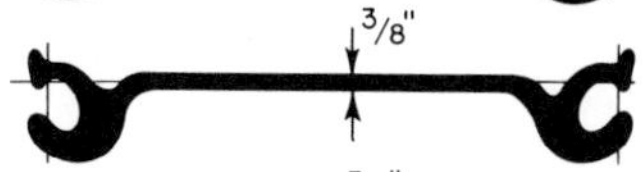 | M 107 | 15 | 38.8 | 31.0 | 3/8 | 3.7 | 3.0 |
| 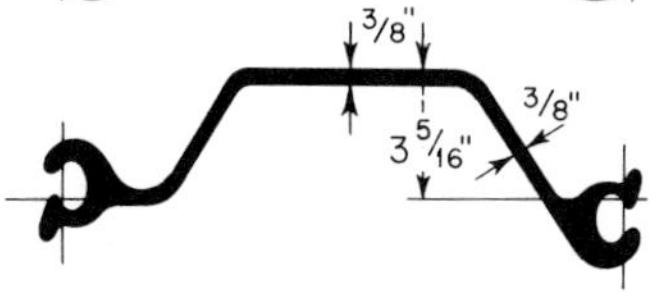 | M 117 | 15 | 38.8 | 31.0 | 3/8 | 8.9 | 7.1 |
| 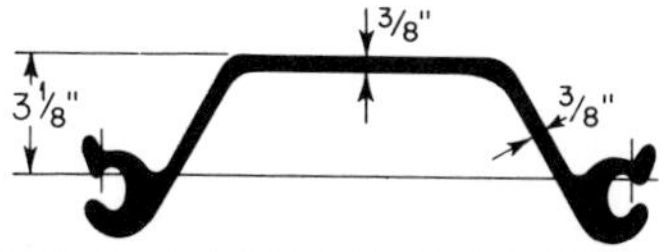 | M 106 | 14 | 36.2 | 31.0 | 3/8 | 10.3 | 8.9 |
| 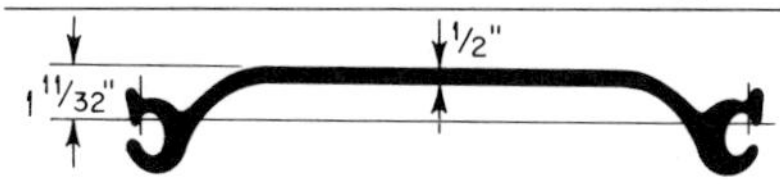 | M 113 | 16 | 37.3 | 82.0 | 1/2 | 3.3 | 2.5 |
| 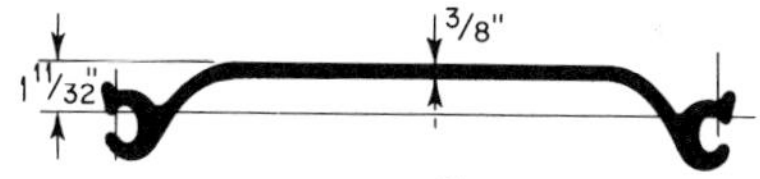 | M 112 | 16 | 30.7 | 23.0 | 3/8 | 3.2 | 2.4 |
| 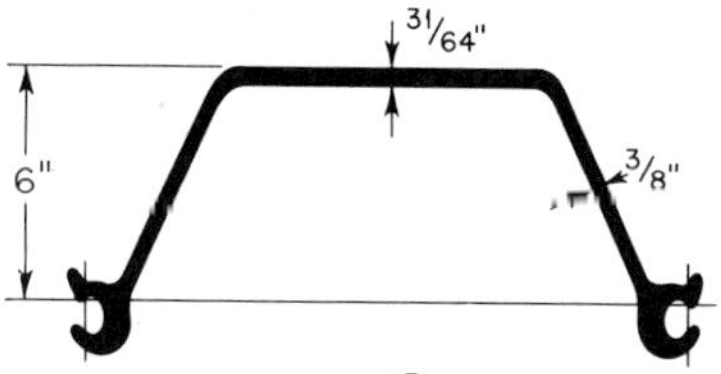 | M 110 | 16 | 42.7 | 32.0 | 31/64 | 20.4 | 15.3 |
| 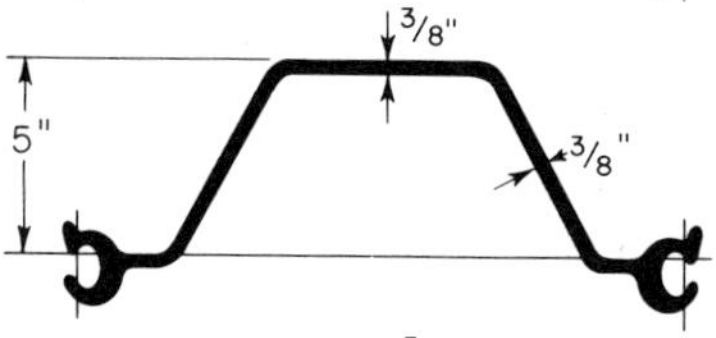 | M 116 | 16 | 36.0 | 27.0 | 3/8 | 14.3 | 10.7 |
| 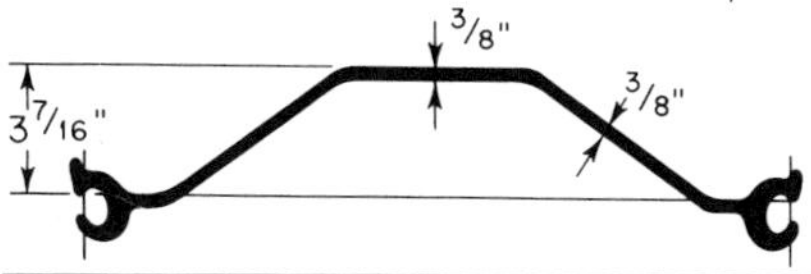 | M 115 | 1958 | 36.0 | 22.0 | 3/8 | 8.8 | 5.4 |

*Minimum interlock strength in direct tension, in pounds per inch of interlock:
M 107, M 108, M 112, and M 113.......................12,000
M 106 and M 117............................................10,000
M 110, M 115, and M 116................................. 8,000
This is based on a test piece approximately 3 in. long

**Fig. 38-19** U.S. Steel steel-sheet-piling sections.

For a cell founded on rock, bearing capacity is not critical.   For one founded on clay, the unit-bearing capacity of the clay should be nowhere exceeded.   The ultimate unit-bearing capacity of a plastic clay, having values of $c$ and $\gamma$ for its unit cohesion and unit weight, amounts to

$$q = 5.7c \qquad (38\text{-}18)$$

A safety factor of at least 1.5 should be used.   In soft clay the possibility of lateral sliding deserves investigation, but the procedure would not ordinarily involve complications that merit discussion here.   Safety against horizontal slippage can be increased by driving the sheet piling deeper, thus providing a surcharge to resist shear.

Cells founded on sand are not apt to develop the ultimate bearing capacity of the subsoil unless the sand is very loose.   This possibility can be explored by driving test piles and noting resistance encountered. Where sheet-piling cells hold back a considerable head of water on one side, there is a real possibility of sand boils developing in the sand foundation on the drained side of the wall.   This danger can be best analyzed by construction of a flow net, but the explanation of flow-net technique is beyond the scope of this discussion.   The subject is discussed in standard texts on soil mechanics and in professional papers.

**38-6  Piling**   Piles are extensively used in coastal engineering and water-front construction.[1]   This is a natural consequence of the fact that all wharves and many dockside structures require the location of storage areas and working floors at a considerable elevation above the natural ground surface.

The common timber pile is a tree with straight trunk, trimmed of branches, with the top 5 to 8 in. or more in diameter.   In the United States the most commonly used trees for piles are southern yellow pine, Douglas fir, spruce, and oak.   In the ASCE manual on timber piles and construction timbers,[2] the subject of timber piles is covered, including descriptions of the timber, conditions of use, necessity for preservative measures, means of protection by armor and by treatment, specification for piles, classification by kind of lumber, size, and length, and specifications for pressure treatment by various processes.

Timber piles cover a wide range of size and strength.   Their availability depends upon transportation facilities and distance from lumbering regions.   Timber is easily worked and is well adapted to ready incorporation in structures of many types.   The elasticity of timber in bending and its flexibility make wooden piles useful in many situations where other

[1] The following discussion on the use of piles for water-front construction is abstracted from the *ASCE Manuals Eng. Practice*, nos. 17 (1939) and 27 (1946).
[2] *Ibid.*, no. 17.

types could not be used as well. Timber is well adapted for protection of water-front structures, because of its resilience and its wearing qualities (if the proper wood is selected), and for protection of the structure from damage by floating craft. Wooden piles are comparatively light for their strength, they are easily handled, and they resist driving well enough to develop their usual designed load. However, they will not withstand as hard driving as steel or concrete piles. Timber piles are subject to the defects of growth and must be selected according to specifications suitable for the work.

Concrete piles are of two types: (1) precast and (2) cast-in-place. A factor contributing to their use is the availability of the materials of which the concrete is made. Both types are made in a great variety of cross-sectional and longitudinal shapes. They may be so designed that their maximum safe load is greater under certain conditions than that of wooden piles. They are immune to attack by biological organisms. This permits their use under conditions where untreated wooden piles could not be used—for example, in foundations where the piles extend above the ground-water level and in sea water infested with wood-destroying organisms.

Precast concrete piles are generally octagonal or square with chamfered corners; occasionally they are round. They usually have parallel sides and a "pointed" end. They are reinforced with steel, being designed to resist the bending stresses caused by handling, in addition to stresses caused by driving and axial loading. They require time for setting and curing, space for storage, and special equipment for handling and driving. They may be inspected for surface defects before driving. Pipes for jetting may be embedded in them. However, it is often difficult to predetermine the required lengths for any type of pile, and when precast concrete piles are too long or too short, the cutoff waste or cost of buildups are items which must be considered in the total cost.

Cast-in-place concrete piles may be used when conditions are favorable. They are made by pouring concrete into a tapered or cylindrical form previously driven into the ground or into a hole in the ground from which a driven form or mandrel has been withdrawn; reinforcement is placed as necessary before pouring the concrete. The left-in-place form may be a steel shell heavy enough to be driven without a mandrel, or it may be a steel form, designed for driving with a mandrel that is removed on completion of the driving. The interior of the driven shell, or form, may be inspected by use of a suitable light.

Where no shell or form or too weak a shell is left in the ground, there is no certainty that earth pressure will not wholly or partially close the hole, thus reducing the size of the pile, nor that the driving of adjoining piles will not affect the newly poured concrete.

Cast-in-place concrete piles have a number of advantages: they need ꞌno storage space while curing; they are made in place to correct length; and their concrete is not subject to stresses caused by driving or handling, except from the driving of adjoining units, as in the case of other piles.

Steel H piles are rolled-steel sections, with side flanges designed particularly for pile purposes. They have strength in tension and compression with correspondingly smaller cross-sectional area for a given load on the pile. Since their point areas and volume displacements are relatively small, they are well adapted for deep penetration and they can be forced through coarse gravel, compacted sand layers, or even soft rock. Steel H piles will stand rough treatment in handling and transportation, but long slender ones must be protected against excessive bending. A steel-bearing pile may be a fabricated member.

Open-end metal-pipe piles or heavy cylindrical shells are of various diameters and wall thicknesses, driven to bearing. When the inside is cleared of earth and water, the tube may be filled with concrete, in which case the two materials share in taking the applied load. The pipe or shell may be driven or forced down in short lengths when headroom is limited and may be used in total length of more than 100 ft.

Closed-end-pipe piles or heavy cylindrical shells may be used as displacement piles. They usually are filled with concrete.

Composite piles consist of a lower part of one material and an upper part of another material. The usual combinations are as follows:

1. A wooden part in the earth below permanent ground-water level where no decay will occur, with an upper part of concrete
2. A steel pipe or H pile for the lower part with an upper part of concrete

Piles are used for the following purposes:

1. To eliminate objectionable settlement
2. To transfer loads from a structure through a fluid or through unsuitable soils to a suitable bearing stratum
3. To transfer loads from a structure through easily eroded soils in the zone of possible scour to a satisfactory underlying bearing stratum
4. To compact granular-soil strata in order to reduce their compressibility
5. To anchor structures subjected to hydrostatic uplift or overturning
6. To anchor structures against disturbing effects of earth tremors
7. To serve as deadmen for anchoring guy lines and overhead cables; to provide underground and underwater ties; to protect river banks; to form anchorages such as moorings and dolphins; and to perform similar functions in water-front construction
8. To protect water-front structures from wear caused by floating objects

Deterioration of pile material may be caused by biological, physico-chemical, and mechanical actions—from either artificial or natural causes; it may occasionally be caused by electrolysis. The effects of these various causes depend on the material in the pile and on the particular conditions that exist or may develop at the site. If care is used in selecting pile material suitable for use in any given location from the standpoint of durability, pile foundations or pile structures may have, and usually do have, a very long life.

In water-front structures timber is damaged by three forms of attack: decay, insects (termites and wharf borers), and marine borers. The two former attack above, the latter below, water level. Abrasion of fender piles and piles and timber in ferry slips also causes damage.

1. *Decay.* All forms of decay are the result of the action of certain low forms of plant life called fungi. Certain substances in wood form the food of the fungi and, as these substances are consumed, the wood is disintegrated. The species of fungi that cause decay require food, air, the right amount of moisture, and a favorable temperature. Timber that is continuously below water level does not decay. Poisoning the food supply by impregnating the timber with a suitable preservative is the easiest and surest method of preventing decay.

2. *Termites.* Where they are active, termites are probably more destructive to timber than any other land organism, except fungi. Sometimes called "white ants," they are not true ants, but resemble them somewhat in appearance and method of life. There are two types of termites, the "subterranean" and the "dry wood." The former type requires moisture and therefore must have access to the ground at all times. The latter type flies and does not need contact with the ground. The subterranean termites are widely distributed in the United States and do an enormous amount of damage to infested structures. The dry wood termite is not found much north of Norfolk, Va., on the East Coast, and San Francisco, Calif., on the West Coast.

3. *Wharf borers.* These animals do considerable damage, but they are not of as much commercial importance as others of the organisms attacking timber. The damage is done by the young of a winged beetle. The beetle lays its eggs in the cracks and crevices of the timber, and the larvae or worms that destroy the timber are hatched from these eggs. They do not work below water level and seem to prefer timber that is not far above high water or timber that is wet by salt spray at times.

4. *Marine borers.* There are two main divisions of these very destructive animals: the molluscan group is related to the oyster and

clam, and the crustacean group is related to the lobster and crab. Their methods of attack on timber are entirely different. The first group enters the timber through a minute hole, and as the animals grow, they destroy the interior of the timber. The second group destroys the outside of the timber. The attack of the molluscan borers can only be found by the most careful inspection of the surface or by cutting into it. The attack of the second group is easily seen and measured by surface inspection. The rate of destruction in heavy attacks is several times more rapid by the molluscans than by the crustaceans.

The molluscan group is divided into two general groups, the Teredinidae and the Pholadidae, with very different physical characteristics. Both groups are classified, biologically, into several genera and many species. The teredo or shipworm group has grayish, slimy, wormlike bodies with the shells used for boring on the head. The burrow is lined with a smooth nacreous lining. The size of the mature animals of the common species varies with the species, ranging from $\frac{3}{8}$ in. in diameter and 5 or 6 in. in length to 1 in. in diameter and 4 to 5 ft long. A species, thus far identified only in some of the Pacific islands, may be more than 3 in. in diameter and 3.5 ft in length. In areas of heavy attack, an unprotected pile may be totally destroyed, so far as its bearing value is concerned, in 6 to 8 months. Animals of the teredine group have been found in harbors of continental North America only in salt or brackish water; but in Australasia, India, and in some parts of South America they have been found in fresh water.

There are three important genera of crustacean borers: *Limnoria, Chelura,* and *Sphaeroma. Limnoria lignorum,* the most widely distributed species of this group, resembles a wood louse in appearance and has a body from $\frac{1}{8}$ to $\frac{1}{4}$ in. in length with a width of about one-third the length. The head bears a pair of eyes and two pairs of short feelers or antennae, and the mouth has a pair of strong, horny-tipped mandibles with which the boring is done. The body has seven pairs of legs ending in sharp, hooked claws so that it can move around freely and cling to the timber. It uses its gill plates for swimming. *Limnoria* destroys timber by gnawing interlacing branching burrows on the surface. As many as 400 animals per square inch have been counted on timber under heavy attack. They are found from the Arctic Circle to the tropics, in salt or brackish water, and in clean or polluted water. The greatest intensity of attack is generally found near the mud line or at half tide; but it may be anywhere between these limits, or it may be uniformly distributed.

Impregnation with toxics is the most generally used form of protection. It usually results in the lowest annual cost for the structure if treatment and preservative are the best. A great advantage of this type of protection is that it also protects from decay, termites, and wharf borers, as well as marine borers. The best and most reliable preservative so far known is coal-tar creosote, used in sufficient quantity and carefully applied.

Plain- or reinforced-concrete piles entirely embedded in earth generally may be considered permanent. The elevation of the water table does not, in general, affect their durability. There is the possibility that in isolated areas concrete piles may be damaged by the percolation of ground water charged with destructive acids, alkalies, or chemical salts. Ground waters move rapidly through sandy soils, and hence corrosive effects would be more pronounced, while in clays the movement is so slow that the action would be unimportant. Destructive chemicals in the ground water may be due to (1) wastes from manufacturing plants, leaky sewers, leaching from storage piles of soft coal or cinder fill containing sulfuric acid and other destructive compounds; (2) sodium and magnesium sulfates leached from the ground itself; or (3) organic acids resulting from the decay of vegetable matter.

In areas subject to destructive earthquakes, the desirability of introducing reinforcing into all concrete piles should receive careful consideration.

Reinforced-concrete piles above the ground surface, like other reinforced-concrete construction, are subject to attack by weathering and any destructive elements carried in the air. In damp seacoast climates, the exposure may be severe because of moisture penetrating permeable concrete and reaching the reinforcement, rusting it and spalling the sides and edges of the pile. In general, where conditions permit or favor corrosion of steel, the reinforced pile is threatened, particularly if the concrete is defective or if there is insufficient cover over the reinforcement.

Reinforced-concrete piles in water-front structures are called upon to meet unfavorable exposure conditions when they extend from the harbor bottom into water and air. Such piles are subject to:

1. Abrasion by floating objects or scouring sand.
2. Attack by pholads (rock-boring mollusks). These mollusks have been found in concrete and masonry structures in such widely scattered locations as Los Angeles harbor, Panama Canal Zone, and Plymouth, England. The damage thus far appears to have been done to concrete of unquestionably poor quality and not to have been of great economic importance, but it must be considered.

3. Chemical action of polluted waters on concrete. The waters of some western streams and lakes are highly destructive to concrete; so also are some of the salts in sea water. Special cements recently have been developed in an attempt to produce concretes which will be durable when exposed to sea water.
4. Frost action on porous concrete.
5. Destructive action caused by rusting of the reinforcement and spalling of the concrete. This is the most serious weakness of reinforced-concrete piles when used in water-front structures. It is particularly serious when the structures are located in tidal waters where alternate wetting and drying of the concrete due to the rise and fall of the tide—especially if combined with alternate freezing and thawing—accelerates the destructive action. Rough water also promotes this destructive action by keeping the piles soaked with spray to a high elevation on a windy day and allowing them to dry out on calm days. Experience has shown that the quality of the aggregates, the composition of the cement, the cover over the steel, and the workmanship in mixing and placing the concrete are controlling factors in attempting to attain the desired permanence of the piles. Protection to the pile is afforded by oil coating in locations where oil floats on the surface of the water, as near oil docks.

Steel piles protruding from the ground into open water, as in trestle bridges or water-front structures, are subjected to varying degrees of deterioration depending, in general, on whether the water is fresh or salt. In the case of fresh-water exposure there is usually very little deterioration; however, the action of salt water subjects exposed steel piles to more severe deterioration and unfavorable performance.

Steel piles in fresh water in most cases do not require protection. Where there is pollution from industrial wastes, the piles may be protected above the mud line by the application of a suitable coal-tar coating before driving. Such a coating is especially desirable at the water surface where deterioration of the steel, although not very active, is relatively greater than in the totally immersed parts of the piles. Renewable concrete encasement of the piles from about 2 ft below to a short distance above water level also will prove effective. The parts of the piles above water level can be treated as an ordinary steel structure and maintained in good condition by painting.

A greater length of life can be expected from steel piling if in protected waters than if subjected to wave action in the open ocean. The various parts are affected as follows:

1. That portion of pile below the bottom is generally not subject to rapid deterioration.

2. At and immediately above the bottom, deterioration may be accelerated rapidly by the abrasive action of water-borne sand agitated by waves and currents. This condition is usually present only in shallow waters where wave and tidal action is most active. Sometimes destructive organic substances consisting of decayed marine animal or vegetable matter deposited on the bottom may cause accelerated deterioration in a narrow zone at the mud line. Tests and analyses of the ground should be made in questionable cases. Under either of these conditions it is desirable to protect the piling by some form of renewable or replaceable encasement in the zone of accelerated deterioration.

3. Between the bottom and low-tide level, corrosion is usually more active in the upper part where the oxygen content of the water is greatest. In some waters there are certain types of marine shellfish which attach themselves to steel piles below water level and may have a deleterious effect by causing excessive pitting. Suitable coal-tar coating and synthetic resin or zinc-chromate paint will protect against such action as long as they last. Encasement applied to the piles before or after driving should be effective.

4. Between low- and high-tide levels, corrosion may be extremely active in the region of alternate wetting and drying of the piles. The presence of oil on the water surface in many harbors results in a protective coating forming on the steel piles. This decidedly retards corrosive action.

5. Above high-tide level, corrosion also may be severe, especially if waves subject this zone to the action of spray which tends to build up an accumulation of salt. Deterioration is also relatively greater in locations having a high temperature and humid atmosphere. Protective coatings or encasement with concrete will lengthen the life of steel piles under this condition and also under that covered in paragraph 4.

Steel piles should be suitably protected against corrosive action from immediately below low water to above the spray line.

Corrosion of steel piles caused by electrolysis is rare. It may be a factor under certain conditions. Steel piers or trestles carrying pipelines which may have picked up stray currents or those carrying d-c power lines may transmit enough current from the piles to cause electrolysis. Insulating the source of the current from the piles will prevent this condition. In buildings such as powerhouses or in cities where direct current is used, electrolysis may occur, and care should be taken to provide proper insulation to prevent direct currents from reaching steel piles. The superstructure also should be properly grounded.

Under the most common conditions, where the tops of steel piles are embedded in concrete footings and thereby insulated from the rest of the structure, electrolysis is not a factor. Local electrolytic action and corrosion may be set up in salt water where the steel pile forms one pole of a battery, with its other pole in some dissimilar metal in the water close by.

Copper-bearing steel with a minimum of 0.20 per cent copper affords greater resistance against atmospheric corrosion than plain carbon steel. For both materials, when completely immersed in either fresh or salt water or within the tidal range of sea water, the resistance against deterioration is about the same. Care must be taken to guard against the use of copper-bearing steel in contact with plain carbon steel or wrought iron, and thus avoid harmful local electrolytic corrosive action between dissimilar metals, especially in the presence of sea water.

**38-7  Groins**  The rate of shore erosion can be retarded by interposing a sea wall or bulkhead between the cliff face and the assaulting waves; or the intensity of the wave attack can be reduced by using groins to extend the beach line. Groins build up the beach by trapping the immigrating drift. This automatically reduces emigration to the leeward section of beach, and the changed equilibrium conditions at that adjoining property will increase erosion there until the groin fields are filled.

Beach groins are subject to more severe forces than river groins. A rock-filled timber structure is common, except where marine borers are active. In the latter case permanent construction is limited to concrete, stone, and steel. The following rules[1] for the design of beach protection are widely quoted:

1. Bulkheads and groins should not be used separately.
2. Concrete is suitable for heavy sea walls, but not for light bulkheads if settlement is possible.
3. Protect timber bulkheads with large-size rock.
4. Avoid placing bulkheads seaward of HW line.
5. Proper spacing of groins depends upon many factors. In general, space groins at $1\frac{1}{2}$ times their length.
6. Water must not be permitted to pass around the inshore end of groins.
7. Sheet piles for groins should have penetration of 10-ft minimum. Use tight construction and stout bracing. Ironwork should be galvanized. For severe exposure 75 per cent of rock should be pieces weighing 2 or more tons each. Timber should be creosoted.
8. On flat beaches it may be economical to raise the groins in 3-ft lifts as the beach builds up.

[1] The enumerated recommendations are condensed from "Report on Erosion and Protection of New Jersey Beaches, 1924," U.S. Engineering Advisory Boards.

9. Protective works should be designed after careful consideration of possible detrimental action some distance to leeward.
10. There is no evidence that curved groins are better than straight groins.
11. Proper maintenance of beach structures is important.
12. Growth of sand dunes should be encouraged.
13. Heavy outshore breakwaters are not recommended for beach protection.
14. Tight jetties are necessary to protect inlets.
15. Twin jetties at inlets are better than one.
16. Jetties should extend from the HW line for length sufficient to reach adequate low-water channel depths.
17. Steeply sloping shores require heavier jetties than do flat slopes.
18. For heavy jetties use a tight sheet-pile core surrounded by heavy rock, 85 per cent in pieces weighing over 3 tons, none less than 1 ton.
19. Piles and sheet-pile cores should extend well into a hard bottom, if one exists.

Figure 38-20 illustrates groin action.  The interruption of sand migration creates a deposit on the windward side of the groin, while the lee side of the groin is often apt to be cut back as wave erosion continues despite the blockade of replacements.  The construction of a groin field protects the abutting property at the expense of the adjacent water front to leeward.  To prevent drastic action it is often advisable to

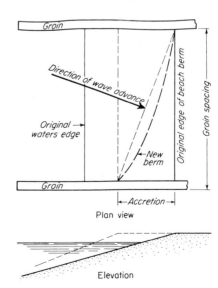

**Fig. 38-20**  Illustration of groin action.

build this sort of protection in stages, avoiding any long-sustained interruption of sand migration.

The factors to be considered in the design of a groin field are structural stability, protection of the land end against erosion at high water, protection of the sea end against scouring velocities in the diverted longshore current, groin length, and groin spacing.

Structural stability can be achieved with a wide variety of materials, including precast concrete blocks, rubble mounds, stone blocks, steel sheet piling, and timber. The groin must extend well into the bank or abut against a bulkhead to prevent being flanked by storm waves. The groin head must be extra heavy to avoid scour, or the groin must be made permeable enough to avoid concentration of flow at the groin head. The groin length is determined mainly by the extent to which it is desired to intercept the longshore drift of sand. The migration of sand occurs in a zone extending from the beach to the line of breakers. Consequently, the groin could terminate at any depth from a few feet to 20 or 30 ft. A length of 100 or 150 ft would be fairly typical. Excessive spacing of groins would lead to erosion on the lee side. Figure 38-20 suggests an approach to groin spacing but is somewhat of an oversimplification. The width of the zone of accretion depends primarily on the magnitude of the new beach slope. This in turn depends upon groin length, the severity of the wave attack, and the maximum size of the beach material. These factors demand careful study on the site and possibly some assistance from the laboratory.

### 38-8  Offshore breakwaters

An offshore breakwater is one which is constructed in fairly deep water parallel with the shore, but with no connection to the shore. Such a structure creates an area sheltered from wave action, but it does not block the longshore current. The net effect is to stop erosion but not to stop the supply of sand. As a result there is a tendency for a tombolo to form between breakwater and shore, similar to the bar formation shown in Fig. 35-7. It is advisable to build an offshore breakwater in stages, as there is not sufficient information for reliable rules on the relationship between breakwater length, distance from shore, and the interruption of sand migration.

### QUESTIONS AND PROBLEMS

**38-1.** A sea wall is erected with its base elevation at still-water level. The beach slope is uniformly 1 on 8. The critical wave is 10 ft high and breaks at a still-water depth of 10 ft, where its celerity is 30 ft/sec. Show graphically the dynamic pressures acting on the vertical wall.

**38-2.** Data are as in Prob. 38-1, except that the wall has a batter of (a) $+\frac{1}{4}$ on 1, (b) $-\frac{1}{4}$ on 1.

**38-3.** If the sea wall of Prob. 38-1 is 10 ft high, what earth thrust will act on its back? The fill behind the wall is beach sand for which $\theta = 32°$ and $\gamma = 104$ pcf. Use $\delta = 25$ deg. The sand is fully drained. $\alpha = 80$ deg.

**38-4.** Pulling tests on piles at the site of a bulkhead indicate an ultimate resistance of 30,000 lb, while loading tests on the same piles show a static bearing capacity of 40,000 lb. Find the allowable value of $T$ for the pile anchorage in the sketch, using a safety factor of 1.5. Assume the pilessubject to direct stress only. (Solve graphically by a force triangle.)

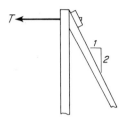

**38-5.** Repeat Prob. 38-4 but reverse the direction in which $T$ acts.

**38-6.** All soil indicated in the accompanying sketch is saturated sand, weighing 124 pcf and having a friction angle of 30 deg. The sea water weighs 64 pcf. Find the length $D_2$ needed for stability.

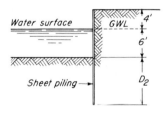

**38-7.** Design an anchored bulkhead for this location, following the proposals of Fig. 38-13.

**38-8.** Compare maximum bending moments developed in Probs. 38-6 and 38-7. Select a suitable steel-piling section.

**38-9.** Design a continuous deadman for Prob. 38-7, following Figs. 38-15 and 38-16.

**38-10.** A steel-sheet-piling cell has an effective width of 30 ft. Lateral-earth-pressure coefficient on a vertical plane is 0.6. Steel-on-steel friction coefficient is 0.3. Earth fill weighs 110 pcf and is fully drained. The thrust of wave action is equivalent to a horizontal force of 18,000 lb per running foot, applied 5 ft below the top of the cell. The cell is 35 ft high. Find factor of safety with respect to longitudinal shear.

**38-11.** A sheet-pile jetty is to be designed to resist a horizontal impact of 400 kips applied at mean sea level, which is 30 ft above the rock ledge on which the cell rests. The cell itself is considered to be 30 ft square, 40 ft high, and filled with rock weighing 110 pcf above water level, 65 pcf below water level. Friction factors: rock on rock, 0.5; rock on steel, 0.4; steel on steel, 0.3.

    *a.* Will cell slide on base?
    *b.* Will cell overturn?
    *c.* Will cell fail in shear?

**38-12.** Work Prob. 38-7 for clay backfill.   The clay weighs 108 pcf, saturated, and has 300-pcf cohesion.

## BIBLIOGRAPHY

Bean, E. F.: Geologic Aspects of Beach Engineering in "Application of Geology to Engineering Practice," Geological Society of America, 1950.

Council on Wave Research: "Coastal Engineering," vols. 1–3, The Engineering Foundation, Berkeley, Calif., 1951–1953.

Johnson, D. W.: "Shore Processes and Shoreline Development," John Wiley & Sons, Inc., New York, 1919.

Shore Protection Planning and Design, *U.S. Army, Corps of Engineers, Beach Erosion Board, Tech. Rept.* 4, 1954.

Tschebotarioff, G. P.: "Flexible Bulkheads," Princeton University Press, Princeton, N.J., 1949.

# 39
# Port Structures

**39-1 Wharves** At American ports, as in most ports throughout the world, the moored ships rise and fall with the tide. This arrangement is becoming increasingly common despite accompanying disadvantages in the need for adjusting mooring lines, gangways, and conveyors as the ship moves up and down relative to the wharf. In some European and Asiatic ports, especially where the tidal range exceeds 10 ft, ships are berthed in an impounding basin, where the water is kept at constant level by means of locks. Such basins are called *wet docks*, or simply *docks*. In American usage dock has a less restricted meaning, having been variously defined as a pier, a wharf, and as the water area between two piers.

A wharf is a structure at which ships take on and discharge passengers and cargo. In the planning of wharves the engineer has the same alternatives that confront him in the layout of curb parking for automobiles. Cars can be parked parallel with the curb, at right angles with the curb, or diagonally. In this analogy the curb resembles the shore line. The problem is to make the most efficient use of available space and permit safe and convenient maneuvering of vehicles.

A wharf built parallel with the shore is called a quay (pronounced "key"). Quays are common in river ports and are favored wherever channel width imposes more of a restriction on development than does the availability of suitable water frontage.

Piers are wharves built at an angle with the shore. Berths are provided on both sides of a pier. While piers normally lie at right angles with the shore line, they are often inclined toward the harbor entrance. This "diagonal parking" is helpful in getting ships back in the main channel on the proper course and probably is more of a factor in the currents of American tidal ports than at British docks. The water area between two adjacent piers is called a *slip*.

Wharf design involves making provision for the berthing of ships, handling and storage of cargo, and terminal facilities for rail and truck transportation. The length of wharf and depth of water required for a ship's berth obviously depend upon the type of vessel to be accommodated. Water depths at the pier may be somewhat less than those required in ship channels, but should provide a minimum of 1 ft clearance under the keel of a fully loaded ship with reasonable provision for an abnormally low tide. Data on ship lengths and drafts were supplied in an earlier chapter. Outside of New York harbor, a draft of 30 ft is a reasonable limit for passenger and cargo vessels calling at American ports. Future tankers, bulk-ore carriers, and bulk-cargo vessels may have 40-ft drafts. The wharf deck should be 5 or 6 ft above mean high water.

Engineers have found a great variety of structural forms suitable for wharf construction. In general, a wharf consists of a sea wall or bulkhead surmounted by a paved platform (Fig. 39-1 shows an installation at Seattle). Variations may be listed as follows:

1. Solid fill behind a quay wall, with the face of the wall flush with the edge of the wharf. The quay wall may be a gravity wall, a cantilever-type retaining wall, or an anchored bulkhead of steel, concrete, or timber sheet piling. Cellular steel sheet piling is also used.
2. A simple pile-supported deck with no appreciable amount of fill.
3. Relieving-platform type of quay wall. This design is a compromise between the stability of the solid fill and the economy of the simple pile-supported wharf. Below low water the earth is left at a stable slope, eliminating the expense of a structure to withstand earth pressure. Since the piling is always wet it will not decay. Above low water the fill rests on a pile-supported platform. If the waters are infested with borers, wood piling can be protected by enclosing the wharf with steel or concrete sheet piling, as in Fig. 39-2c. If there are no marine borers the fill can be retained by a gravity wall or a bulkhead above the platform, as in Fig. 39-2d.

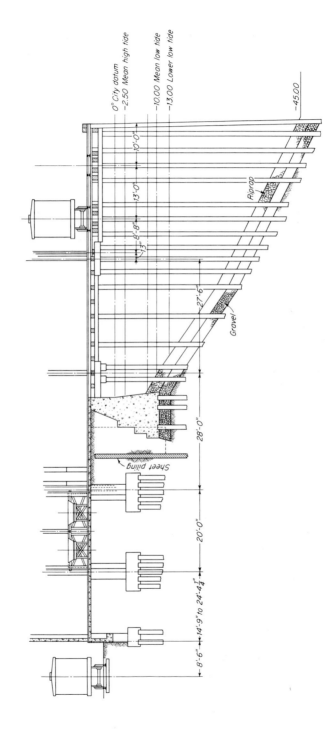

**Fig. 39-1** Wharf construction.

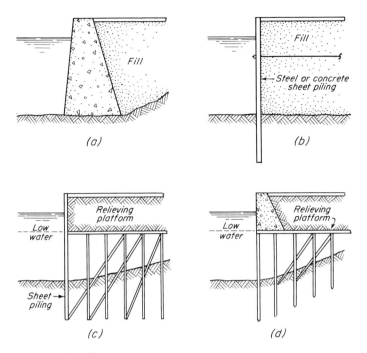

**Fig. 39-2**   Wharf types.

Batter piles are used in most types of wharf construction.   It is customary to assume that the piles in the various designs are subject to only direct stress and are hinged at their connections.   These simplifying assumptions are enough to make most of the usual designs statically determinate, and the loads carried by individual piles can be found by summation of forces or summation of moments.

**39-2  Cargo handling**   The layout of piers and quays is influenced by the general scheme of cargo handling to be employed.   Ships are loaded and unloaded either vertically, by cranes, or horizontally, through side doors.   Bulk cargo has generally been hoisted through hatches in the ship's deck, but there is a growing tendency toward the movement of goods in standardized containers which can be moved to and from the vessel horizontally and hauled away by truck or railroad at minimum transfer expense.   Since cargo handling accounts for almost half the cost of ship operation, the importance of providing proper port facilities is evident.   In Europe it is customary to reduce the cost of ship operation by a greater investment in shore equipment, especially dockside cranes, whereas in the United States more dependence is placed upon the ship's own cargo-handling booms and winches.

However handled, provision must be made for storing cargo until it can be transferred to other transportation. Most of this space will be provided in a cargo shed. The shed should be sufficient to contain both an incoming and outgoing cargo at the same time. For larger-size ships this means about 3,200 sq ft of cross section, net storage area, for the length of the ship. Practically, a cargo shed of two or three floors will be needed to serve the largest ships.

**39-3  Piers and slips**  The length and width of a pier should be specified only after a careful study of the character and amount of the anticipated traffic. However, an idea of prevalent American practice can be obtained from the recommendations of Table 39-1, which probably represents fairly the professional opinion guiding most existing layouts. More recent designs incline toward wider piers. One factor behind this trend has been the delays experienced by truckers at many piers because of inadequate truck-terminal facilities. Postwar wharf construction at Hamburg provides two 48-m-wide cargo sheds separated by a truck roadway 36 m between curbs. Three railway tracks separate each cargo shed from its quay wall, for a total width of 204 m.

There is no uniformity in the allowance for space between the cargo shed and the edge of the wharf. With a single railway track this space may be as little as 8 ft, but at many installations, several times this width is provided. There is considerable agreement that no provision should be made here for cargo storage, and the design depends mainly upon whether it is planned to route rail and truck traffic along the outer edge of the pier or down the middle.

The live loads for pier and wharf design, like other factors, depend

**Table 39-1  Recommendations for straight piers***

| Type of vessel | Length of pier, ft | | Width of pier, ft | | Width of slip | |
|---|---|---|---|---|---|---|
| | (1) | (2) | (1) | (2) | (1) | (2) |
| Ocean steamer...... | 600–700 | 1,200–1,400 | 150–160 | 220 | 300 | 350 |
| Great Lakes........ | 700 | 1,400 | 140–160 | 200 | 280–300 | 340 |
| Inland-river ship.... | 500–550 | 1,000–1,100 | 140 | 150–160 | 280 | 340–350 |
| Inland-river barge.... | 300–350 | 600–700 | 130 | 150 | 150 | 180 |

(1) One ship's length.
(2) Two ship's lengths.
* H. McL. Harding, What Should Be the Dimensions of a Shipping Pier? *Eng. News-Record*, **85**(24):1119–1120 (1921).

RIVER AND COASTAL ENGINEERING

much upon the nature of the anticipated cargo and shipping. However, a design loading of 600 lb/sq ft, plus the concentrated live loads of a 5-ton truck and a 15-ton locomotive crane, has been recommended for freight piers.

Piers must be designed to resist the impact from ships, although it is difficult to establish a reliable value for this variable. Seelye[1] suggests that the horizontal thrust on a pierhead bent be taken at 1 per cent of the maximum weight of the vessel to be berthed. Ship impact is eased by the fender system. This may consist of piles driven into the ground and bolted and chocked at the superstructure. Fender piles should be driven at a batter so that ships will strike them somewhere midway along the exposed pile length. Timber beams or steel springs may be inserted between fender end wharf to absorb the shock of ship contact.

The iron or steel fittings to which the lines from a ship are made fast must be designed to carry considerable force. The total area of the ship exposed to wind is multiplied by the assumed wind pressure (20 or 30 lb/sq ft) and divided by the number of lines, to find the horizontal component of the individual reaction. The fitting can be bolted to a reinforced-concrete slab, which is in turn supported by an adequate system of vertical and battered piles. Ballards, bits, and cleats may be spaced along the wharf at intervals of 50 to 100 ft.

Piers may be of either open or solid construction. Open piers are built over open water on exposed timber or concrete piles. The decking may be timber or reinforced concrete. Open piers are used where minimum restriction of currents is specified and are relatively more economical for narrow piers in deep water. Solid piers are composed of earth or rock fill, generally at least partly confined by some sort of bulkhead. Solid piers offer more resistance to impact and wave action and are more stable in general. They are more economical in many cases. Figure 39-3 shows various pier types. The width of the slip between piers should be at least 3 or 4 times the beam of the largest ships to be accommodated.

**39-4  Dry docks**  Dry docks, or graving docks, are compartments into which a ship can be floated and supported on blocks for repair work after the dock has been drained. The size of the dock should be commensurate with the size of the ship. Excess size is undesirable because of the additional water to be pumped and because of the longer shoring required to brace the ship against the dock walls. The main elements of a dock are the floor, the walls, and the entrance.

[1] E. E. Seelye, "Data Book for Civil Engineers," vol. 1, pp. 4–80, John Wiley & Sons, Inc., New York, 1945.

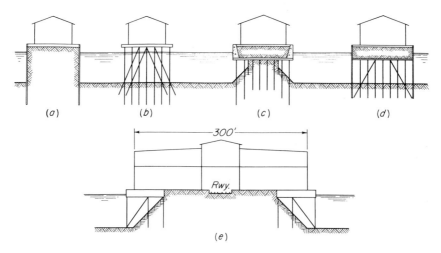

**Fig. 39-3**  Pier types.

The floor must support the weight of the damaged ship, but a greater requirement is apt to be hydrostatic uplift of the foundation.  By virtue of its location adjacent to deep water, the dry-dock excavation is certain to extend much below the ground-water level, and the dock consequently will displace far more water than the ship it accommodates.  A principal problem, therefore, is to throw enough weight into the design so that there is no possibility that the dock will float on the ground water.  In the case of coarse sand or gravel foundations this, in effect, means that the masonry must outweigh the water displaced by the dock.  Most of this weight will be contributed by the walls, while most of the uplift will be exerted against the floor.  The floor thus becomes the key structural member in the design, since it must transfer the uplift load to the walls. In general, this has been done in the past by designing the floor as an inverted arch, with the resultant thrust kept within the middle third of a transverse (vertical) section through the floor.  By this method the floor load is assumed to include only its own weight and the uplift pressure.  The line of thrust and the reaction of the abutments are determined by conventional methods of arch analysis, usually graphical.  The stability of the walls is analyzed from the equilibrium of the following forces:

1. Hydrostatic pressures on that part of the wall below ground-water level
2. Weight of the wall itself

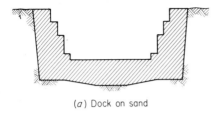

(*a*) Dock on sand

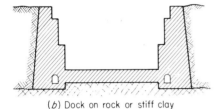

(*b*) Dock on rock or stiff clay          **Fig. 39-4**   Graving docks.

3. Invert thrust, previously determined
4. Earth pressure on the wall
5. Thrust from shores

Where docks are founded on rock or clay it becomes possible to reduce or eliminate uplift without handling a prohibitive quantity of water, by placing adequate intercepting drains around the perimeter, possibly with connecting relief wells.

In general, a more economical design is made possible by greater refinement in analysis. This is achieved by considering the floor as an elastic beam on an elastic foundation and reinforcing the concrete to carry a bending moment. By taking into account the resistance of the soil, where this information is available, a more favorable system of stress distribution may be realized than when the structure is assumed to be completely rigid.

Figure 39-4 gives some idea of dry-dock design. The shelves or setbacks in the walls are known as "altars." They serve as walks for the workmen and convenient seats in which to set the shoring that abuts against the ship's sides. By conforming roughly to the contour of the ship, this section reduces the amount of water to be pumped when the lock is to be drained. Altars should be at least 3 ft wide.

While the ship is steadied by side braces, its main weight is carried on wood blocks under the keel, and, for the largest ship, on as many as four additional lines of blocking besides the keel blocks. The load distribution is far from uniform, and the ship is not sufficiently rigid to spread the load to any considerable extent. Thus a ship weighing 20 tons per

# 40
# Pipeline Transportation

**40-1 Petroleum production in the United States** Pipeline transportation reaches its greatest significance in the distribution of crude oil and its products. The relative importance of this form of transportation is best appreciated by consideration of the amount and location of petroleum production (Table 40-1).

Table 40-1 Average annual production of petroleum in the United States*

| Decade | Average annual petroleum production, millions of barrels |
|--------|---------------------------------------------------------|
| 1880–1889 | 27.3 |
| 1890–1899 | 53.5 |
| 1900–1909 | 122.8 |
| 1910–1919 | 281.9 |
| 1920–1929 | 726.4 |
| 1930–1939 | 1,020.2 |
| 1940–1949 | 1,623.8 |
| 1965 (annual) | 2,848.5 |

* U.S. Bureau of Mines Minerals Yearbooks.

Approximate locations of the major and minor petroleum regions and refining districts are shown in Fig. 40-1. Some selected statistics of the ICC for oil pipelines and petroleum production in 1950 and 1965 are tabulated below. The figures include all carriers reporting to ICC for those years.

**Miles of pipeline operated, 1950 and 1965**

|  | 1950 | 1965 |
|---|---|---|
| Gathering lines.............. | 47,593 | 46,640 |
| Trunk lines: | | |
| For crude oil.............. | 64,622 | 66,145 |
| For refined oil............. | 16,374 | 48,627 |
| Total................... | 128,589 | 161,412 |

**Number of barrels\* received into system, 1950 and 1965**

*In millions*

|  | 1950 | 1965 |
|---|---|---|
| Crude oil........... | 2,404 | 4,003 |
| Refined oil......... | 362 | 1,864 |
| Total............. | 2,766 | 5,867 |

\* 1 bbl = 42 U.S. gal = 5.615 cu ft.

Gathering lines are the pipelines leading from the wells to field storage or pumping station. A typical petroleum gathering-line system is shown in Fig. 40-2.

**40-2  Growth of petroleum pipelines**  The pipeline system in 1965 consisted of 46,640 miles of gathering lines, 66,145 miles of crude-oil trunk lines, and 48,627 miles of line for refined petroleum products. The following tabulation shows the total pipeline mileage for selected years:

| Year | Total mileage |
|---|---|
| 1926 | 90,170 |
| 1936 | 115,760 |
| 1950 | 128,589 |
| 1965 | 161,412 |

The location of gathering lines is determined by the location of oil wells. They are generally laid with the idea that they will be taken up as the output of a particular field declines and moved to a new site. Conse-

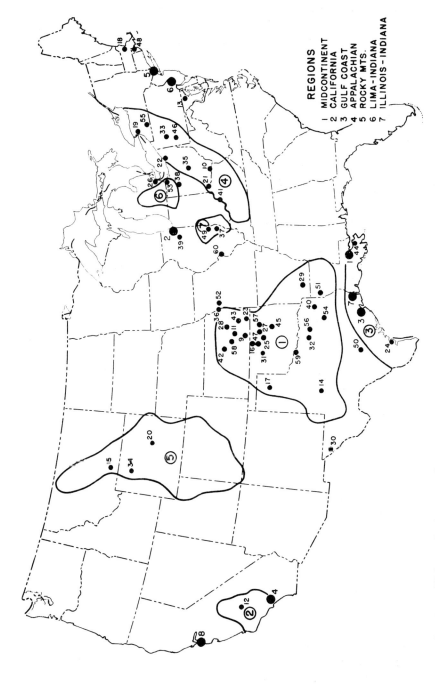

**Fig. 40-1**  Major and minor petroleum refineries.  *(Prepared by National Resources Committee.)*

REGIONS

1 MIDCONTINENT
2 CALIFORNIA
3 GULF COAST
4 APPALACHIAN
5 ROCKY MTS.
6 LIMA-INDIANA
7 ILLINOIS-INDIANA

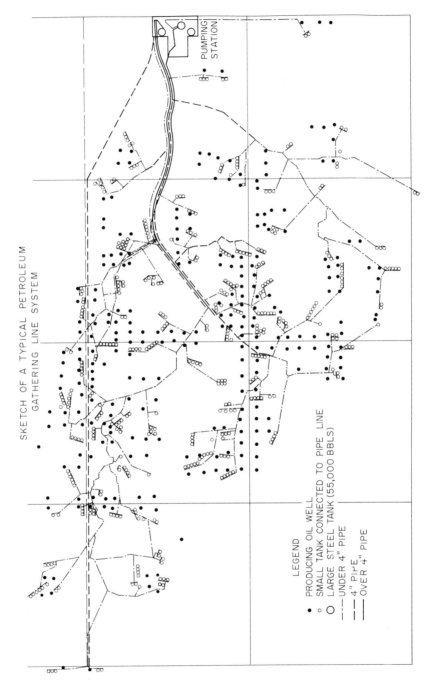

**Fig. 40-2** Sketch of a typical petroleum gathering-line system. (*Prepared by National Resources Planning Board.*)

SKETCH OF A TYPICAL PETROLEUM
GATHERING LINE SYSTEM

PUMPING
STATION

LEGEND

• PRODUCING OIL WELL
∘ SMALL TANK CONNECTED TO PIPE LINE
○ LARGE STEEL TANK (55,000 BBLS)
-·- UNDER 4" PIPE
--- 4" PIPE
— OVER 4" PIPE

quently, they are frequently placed on the surface of the ground. The most common sizes of pipe used for gathering lines are 2- and 4-in. pipes leading to larger mains near the field storage and pumps (Fig. 40-2).

Trunk pipelines are more permanent installations. They are generally treated with a preservative coating before being placed. Sizes between 6 and 12 in. predominate, but in recent years the use of larger sizes has become prevalent for long-distance movements.

The pipeline systems of the country are well established as the leading inland carrier for crude petroleum and continue to expand as an inland carrier of refined-petroleum products. The economy of long-distance transportation of these commodities to large markets by pipelines has been well demonstrated in practice. Their costs of operation are so far below those of rail and motor carriers that the latter are confined mainly to short-haul or relatively small lots. On the other hand, tankers and barges operate at much lower costs than pipelines, and it is likely that a fairly large volume of traffic will continue to be handled by these carriers from mid-continent and Gulf Coast regions to Eastern seaboard refineries and market centers.

**40-3  Storage**  Several types of storage facilities are required for petroleum. Working storage facilities are necessary for efficient operation of the pipelines. These are located in the oil fields at the junction points between gathering lines and trunk lines and at trunk-line junctions. Field facilities will also provide storage for petroleum awaiting movement or sale.

Other storage will be required at ports where petroleum is transshipped from trunk lines to tankers and barges; at points where the oil is transferred from pipelines to railroads and motor trucks; and at the refineries. A considerable amount of storage—roughly 8 to 10 per cent of total storage facilities—is provided by pipeline fill. This latter, of course, is working storage.

Application of some engineering principles to problems of pipeline transportation are taken up in succeeding sections.

**40-4  Types of fluid flow**  The flow of fluids may be divided into three distinct categories: (1) nonturbulent or laminar flow, (2) partially turbulent flow, and (3) turbulent flow. The student will recall from his study of hydraulics that for every condition of fluid flow, whether it be in a closed conduit or in an open channel, there is a "critical" velocity at which flow changes rather abruptly from nonturbulent to turbulent flow, and that there will normally be a range of velocity just above the critical velocity in which the flow is partially turbulent.

Reynolds number $N$ is a reliable indicator of the extent to which

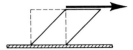

**Fig. 40-3**  Viscous deformation.

turbulence exists.   For round pipes, Reynolds number is determined by

$$N = \frac{vd\gamma}{\mu g} \tag{40-1}$$

where $v$ = mean velocity of flow
$d$ = diameter of pipe
$\gamma/g$ = mass density
$\mu$ = absolute viscosity of fluid

$N$ is dimensionless, so that consistent units must be chosen for the variables.

Viscosity may be defined as the shearing resistance of the fluid. Absolute viscosity is the force required to displace two opposite faces of a unit cube of fluid at a unit rate of speed, as illustrated in Fig. 40-3.   In the centimeter-gram-second (cgs) system, absolute viscosity is expressed in dyne-seconds per square centimeter.   Such units are called *poises* (after Poiseuille).   In engineering units, absolute viscosity is expressed in pounds per second per square foot.   Examination of units will disclose that

1 poise = 0.002089 lb-sec/sq ft

In terms of Saybolt seconds $t$, as determined by use of the Saybolt Universal viscosimeter, absolute viscosity in poises may be determined from the empirical formula

$$\mu = G\left(0.0022t - \frac{1.8}{t}\right) \tag{40-2}$$

where $G$ is the specific gravity of the fluid.

Absolute viscosities for various fluids are given in engineering units in the following table for various temperatures.   Note the marked effect of temperature on the viscosity of liquids.

**Table 40-2  Absolute viscosity in pound-seconds per square foot***

| Fluid | 32°F | 68°F | 122°F | 212°F | Sp gr at 68°F |
|---|---|---|---|---|---|
| Air............. | 0.00000036 | 0.00000038 | 0.00000041 | 0.00000046 | 0.0012 |
| Water.......... | 0.00003739 | 0.00002089 | 0.00001149 | 0.00000593 | 0.998 |
| Linseed oil....... | .......... | 0.00079382 | 0.00037602 | 0.00013578 | 0.930 |
| Heavy fuel oil.... | .......... | 0.025 ± | 0.0036 ± | .......... | 0.990 |

* Reprinted with permission from Charles W. Harris, "Hydraulics," John Wiley & Sons, Inc., New York, 1944.

For round pipes, Reynolds number at the critical velocity may be taken as 2,000. That is, if Reynolds number is smaller than 2,000, the flow of fluid in the pipe will be viscous or laminar flow whether the pipe is relatively smooth or rough. Roughness of the pipe will of course affect the degree to which radial inertia forces create turbulence, but for all practical purposes, if Reynolds number is larger than 3,000, fluid flow will generally be turbulent.

**40-5  Viscous or laminar flow in pipes**  When it can be established that the velocity of fluid flow in a pipe is less than the critical velocity, that is, if Reynolds number is smaller than 2,000, loss of pressure is due entirely to viscous or shearing resistance of the fluid. For such non-turbulent flow in round pipes, the pressure drop in a given length $l$ is given by

$$p = \frac{32vl}{d^2}\,\mu \tag{40-3}$$

where $p$ is pressure drop in pounds per square foot if engineering units are used for the variables. *Lost head* in feet is the pressure drop $p$ divided by the unit weight of the fluid.

**40-6  Turbulent flow of viscous fluids in pipes**  With proper selection of a friction factor $f$, the Darcy formula for pipes applies satisfactorily for determination of head loss $h'$ for turbulent flow in pipes. The Darcy formula is

$$h' = f\frac{l}{d}\frac{v^2}{2g} \tag{40-4}$$

The value of $f$ varies, of course, with pipe roughness, but it is not a constant for a given degree of roughness. It is affected also by pipe size and velocity.

While Reynolds number is not, theoretically, a complete criterion for determination of the coefficient $f$ for turbulent fluid flow, it does offer a practical determination of the friction factor, particularly as applied to flow of viscous oils in ordinary iron or steel pipes. An excellent representation of the relation between Reynolds number and the Darcy coefficient is given in Fig. 40-4. Of this figure, Harris says:[1]

> The solid line A is the lower envelope of experiments on smooth pipes, according to many reliable authorities. It gives reasonable

[1] Charles W. Harris, "Hydraulics," John Wiley & Sons, Inc., New York, 1944.

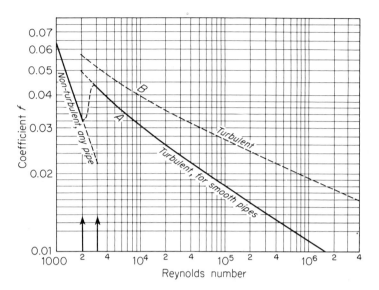

**Fig. 40-4** Friction factor $f$ for partially turbulent flow of fluids. *(Reprinted with permission from C. W. Harris, "Hydraulics," John Wiley & Sons, Inc., New York, 1944.)*

values of "f" for use with very smooth pipes and for pipes of moderate smoothness when carrying liquids of high viscosity.

The band between lines A and B includes the majority of cases found for viscous oils and other liquids of similar viscosity flowing in ordinary iron or steel pipe. . . . The figure presumes that pipes carrying oil are relatively smooth and do not cause pure turbulence at any reasonable velocity.

Thus Fig. 40-4 makes possible a practical solution for problems involving turbulent flow of such liquids as crude petroleum and petroleum products in pipes which are ordinarily used as petroleum pipelines. It will be recognized that the dividing line between laminar and turbulent flow is not clear-cut. That is, when Reynolds number is between 2,000 and about 3,000, the type of flow will be largely determined by pipe roughness. Here again many engineers favor the selection of a specific Reynolds number—say 2,300—as *the* critical value, for practical purposes, at which flow changes abruptly from laminar flow to partial turbulence. While such a selection leaves much to be desired from a theoretical standpoint, it does have some merit in a practical way where estimates of flow must be made in advance of construction of pipelines, pumping stations, and storage facilities.

## QUESTIONS AND PROBLEMS

**40-1.** A crude petroleum is found to have a viscosity of 100 Saybolt seconds, by use of the Saybolt Universal viscosimeter, and a specific gravity of 0.91. Find its absolute viscosity, expressed in engineering units.

**40-2.** The crude oil of Prob. 40-1 is to be transported a distance of 1 mile in a 4-in. gathering line at a velocity of 1 ft/sec.

    *a.* Is the flow turbulent or nonturbulent?

    *b.* What pressure must be applied to maintain this rate of flow?

    *c.* What difference in elevation would be necessary to maintain this rate of flow?

**40-3.** A light grade of crude petroleum has a viscosity of 0.21 poise at 68°F and a unit weight of 56 pcf.

    *a.* At what velocity will its flow in a 4-in. pipe change from laminar to turbulent flow?

    *b.* Find Reynolds number if the oil is flowing at a velocity of 5 ft/sec.

    *c.* Choose a friction factor about midway between lines *A* and *B* of Fig. 40-4 and determine the pressure in pounds per square inch necessary to maintain a velocity of 5 ft/sec.

**40-4.** The crude petroleum of Prob. 40-3 is to be transported from field storage tanks to a refinery 20 miles distant through a 12-in. trunk pipeline. It is desired to maintain flow at a rate of 5 ft/sec. Pressure in the pipe at the field is 31 psi. At points in the pipeline where the pressure has dropped to 5 psi, booster pumps are to be placed to return the pressure to 31 psi. How many such booster stations will be required?

**40-5.** Additional new wells at the field in Prob. 40-4 make it necessary to double the quantity of flow in the trunk line. What pressure will be required at the pumps?

## BIBLIOGRAPHY

Harris, Charles W.: "Hydraulics," John Wiley & Sons, Inc., New York, 1944.
"Petroleum Facts and Figures," 8th ed., American Petroleum Institute, 1948.
Powell, Ralph W.: "Hydraulics and Fluid Mechanics," The Macmillan Company, New York, 1951.
"Statistics of Oil Pipe Lines," Interstate Commerce Commission, 1950.
Streeter, Victor L.: "Fluid Mechanics," 4th ed., McGraw-Hill Book Company, New York, 1966.
"Transportation and National Policy," National Resources Planning Board, 1942.
Vennard, John K.: "Elementary Fluid Mechanics," John Wiley & Sons, Inc., New York, 1947.

# 41
# Pipeline Construction

**41-1 Preliminary survey** Establishing the need for a pipeline for transportation of crude petroleum and petroleum products requires careful consideration of such factors as:

1. The area or extent of the producing field and estimated longevity of supply
2. The number and output of presently producing well and the estimated rate of drilling new wells for complete development of the field
3. A forecast of production rate, usually expressed in barrels per day, for the next several years, based on present output, rate of development, and total reserves
4. Determination of unit cost or annual cost of construction, operation, and maintenance of the line for a reasonable return on investment capital during a period which is well within the anticipated life of the field

A comparison of such costs with cost of transportation by other means will determine the most feasible and advantageous method of handling oil. Generally, transportation by pipeline will prove more economical than by rail or highway, but not so economical as transportation by water. Length of line and quantity of throughput will greatly influence such comparisons.

**41-2 Design procedure** Characteristics of fluid flow in pipes and the factors which affect such flow were discussed in the preceding chapter. Application of the basic principles of hydraulics to flow of petroleum in pipes is exemplified in Probs. 40-1 to 40-5. Some further discussion is desirable to clarify certain problems in connection with design and layout procedure.

It should be emphasized at the outset that the problem of decision as to size and strength of pipe and operating pressures to be used is a complex one which does not lend itself to a direct solution. This complexity becomes immediately apparent upon examination of Eq. (40-4) for pressure loss, recognizing that the coefficient $f$ in this equation is related to the Reynolds number, Eq. (40-1), by the estimated curves of Fig. 40-4. Nevertheless, for any estimated throughput, it is possible to determine the pressure loss in pounds per square inch per mile of length, or for the entire length of line, for each of various sizes of pipe in terms of inside diameter (ID).

In general, for a given volume of throughput, pipes of larger diameter will require lower operating pressure than smaller sizes, thus decreasing operating costs at the pumping stations. The cost in place, however, is likely to be greater for the larger-size pipe, taking into account, of course, that lower operating pressures may permit the use of more economical materials and processes in its manufacture. Working pressures upward of a thousand pounds per square inch may need to be considered, and the cost in place of pipe to meet this requirement may represent a substantial portion of initial investment.

Thus it can be seen that careful cost-comparison analyses must be made to determine the most economical pipe size and quality to be used for given conditions of throughput and length of line. In dealing with capital facilities that may have a long life, costs are ordinarily computed in terms of actual *annual* cost over the period of years involved. Such annual costs include construction and other capital outlays, amortized over the life of the various elements (see Sec. 21-4 on annual cost of railroad crossties); depreciation; maintenance and repair; operating costs; administration and any other overhead costs. Only a brief account of such comparison analyses, illustrating design procedure, will be attempted here.

**41-3 Pressure distribution** In the cost-comparison studies for determination of pipe size and strength to be used for a line several hundred

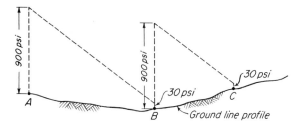

**Fig. 41-1**   Pipeline pressure distribution.

miles in length, handling, say, 200,000 bbl of throughput per day of oil having an absolute viscosity of 0.001 lb-sec/sq ft and a density of 51 pcf, consideration will naturally be given to the large pipes of 12 to 24 in. ID, with allowable working pressures of a few hundred to a thousand or more pounds per square inch.   Let us say, for instance, that an 18-in. pipe with an allowable working pressure of 900 psi is to be considered. This maximum pressure is built up at the pumping station, usually in stages with, say, three pumps, the first pump raising the pressure to 300 psi, the second to 600 psi, and the third pump raising the pressure from 600 to 900 psi.   This pressure is represented by the vertical line at pumping station $A$ in Fig. 41-1.   The slope of the hydraulic gradient is determined by Eq. (40-4), having first established that flow will be turbulent (note that for the conditions of the example, $N = 17,500$). The head loss, expressed in feet per mile according to the density or unit weight of the fluid, may then be plotted as the hydraulic gradient from $A$ to pumping station $B$, where the gradient intersects the ground-line profile, or rather, a point representing a pressure of 25 to 50 psi, say, 30 psi, above the ground-line profile, because in high-pressure operation the pressure is not usually allowed to drop below these figures.   At pumping station $B$ the pressure is again boosted to 900 psi, and pumping station $C$ similarly located, and so on for the entire length of line.   Note that the spacing of pumping stations is not uniform but depends on the ground-line profile, the slope of the hydraulic gradient being constant unless pipe size is changed.

A cost analysis is then in order for construction, operation, and maintenance of this line, with pumping stations so located.   Some reduction in operating pressures might result in more strategic location of pumping stations with respect to terminals and junctions with highway, railway, and waterway transportation.

Consideration may then be given to the possibility of using larger-size pipe with attendant lower operating pressure for the same through-put; fewer pumping stations where long distances are involved; and generally lower operating costs.   In the long run, the economic advantages resulting from the use of large pipe for long-distance transportation may greatly exceed the disadvantage of higher initial cost of installation.

**41-4  Units**  In the language of the petroleum industry, the unit weight of crude oils and petroleum products is expressed in terms of the American Petroleum Institute (API) gravity scale.   The gravity number, expressed in degrees API, increases as the specific gravity decreases, the gravity number for water having an arbitrary scale value of 10.   Specific gravity is related to degrees API by the following formulas:

$$°API = \frac{141.5}{sp\ gr} - 131.5 \tag{41-1}$$

and

$$Sp\ gr = \frac{141.5}{°API + 131.5} \tag{41-2}$$

A tabulation of specific-gravity, unit-weight, and pressure relationships corresponding to the API gravity scale at 60°F is given in Table 41-1.

**41-5  Some construction requirements**  It can be seen from previous discussion of pressure distribution in pipeline that maximum strength will be required for pipe leaving each pumping station where operating pressure is high.   Considerable economy may be realized by using lower-strength lighter pipe at approaches to succeeding pumping stations where operating pressures are low.

Table 41-1  Specific-gravity, unit-weight, and pressure
relationships corresponding to API gravity scale at 60°F

| °API | Sp gr | Lb/cu ft | Lb/gal | Lb/bbl | Psi/ft of head |
|------|-------|----------|--------|--------|----------------|
| 5  | 1.037 | 64.6 | 8.64 | 362.6 | 0.455 |
| 10 | 1.000 | 62.3 | 8.33 | 349.8 | 0.433 |
| 15 | .9659 | 60.2 | 8.04 | 337.8 | 0.418 |
| 20 | .9340 | 58.2 | 7.78 | 326.7 | 0.404 |
| 25 | .9042 | 56.3 | 7.53 | 316.2 | 0.391 |
| 30 | .8762 | 54.6 | 7.30 | 306.4 | 0.379 |
| 35 | .8498 | 52.9 | 7.08 | 297.2 | 0.368 |
| 40 | .8251 | 51.4 | 6.87 | 288.5 | 0.357 |
| 45 | .8017 | 49.9 | 6.68 | 280.4 | 0.347 |
| 50 | .7796 | 48.6 | 6.49 | 272.6 | 0.337 |
| 55 | .7587 | 47.3 | 6.32 | 265.3 | 0.328 |
| 60 | .7389 | 46.0 | 6.15 | 258.3 | 0.320 |
| 65 | .7201 | 44.8 | 5.99 | 251.7 | 0.311 |
| 70 | .7022 | 43.7 | 5.85 | 245.5 | 0.305 |

Pipes are furnished by manufacturers in a variety of lengths of pipe sections. An over-all average length might be about 20 ft. Double-length sections up to 40 and 50 ft may be obtained, however, where conditions of handling and placement and the economy of fewer joints warrant their use.

Pipe is laid in a trench of fairly uniform depth, conforming to the contour of the ground, and with such horizontal bends as are necessary or desirable to take reasonable advantage of topography and to avoid, as much as possible, such difficult terrain as solid rock or swamps. Bending of the pipe is accomplished by cold bending in the field, using care to avoid buckling.

In crossing a highway or railway roadbed, a *casing* or culvert of considerably larger diameter than the pipe is placed to support the roadway, and the pipe is laid in this casing. River crossings are made by dredging across the channel a ditch of considerable width to avoid sloughing while the pipe is being placed, and of depth sufficient to prevent exposure of the pipe to river currents during flood stage due to scouring of the river bed. This may require ditch depths of from 10 to 18 or 20 ft in silty beds and proportionately less in beds of coarser materials and solid rock.

Smaller pipe sizes, such as those used in the gathering systems, are generally supplied by the manufacturer with threaded ends so that they may be readily coupled in the field. They are also readily uncoupled when such a line is taken up for reuse in another location. Larger pipes, for use in trunk lines, are generally provided with beveled ends for making field-welded joints. A typical welded joint is illustrated in Fig. 41-2. Specifications require that the welded joint develop the full tensile strength of the pipe.

Control of corrosion is effected by placing protective coating on the pipe just previous to laying it. Various coating materials are used for this purpose, most common of which is a thin prime coat of liquid bituminous material followed by a mastic. The mastic may be an asphalt,

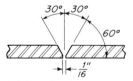

(a) Machine beveled ends

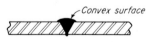

(b) Completed weld          **Fig. 41-2**  Welded pipe joint.

tar, or grease with a filler such as asbestos fiber or other mineral filler. This mastic may then be covered with a fiber wrapping. The amount of protective coating is, of course, geared to the need, as determined by the corrosive qualities, if any, of the soil or water in which the pipe is placed.

**41-6  Tank farms**  The tank farm serves about the same purpose for pipeline transportation as yards do for the railroads. At inland pumping stations, tanks may provide working storage only. This working storage makes possible continuous operation of the line without undue interference from minor fluctuations of production at the field or from temporary lack of outlet or increased demand at the terminal.

At major terminals where many trunk lines converge, the tank farm may provide a capacity of several million barrels, where oil may be stored according to owners for delivery to nearby refinery or seaport; or it may be classified and again transported by larger main lines to more distant refineries such as those in the Chicago or Cleveland area.

The control center at the tank farm is called the station manifold. This manifold, with both direct and remote control of valve and pump operation, may provide for:

1. Transfer of oil from any production field to any tank
2. Transfer from one tank to another or to another tank farm
3. Transfer from any tank to main line
4. By-passing all tanks, for feeding production from any field directly into main line

## QUESTIONS AND PROBLEMS

**41-1.** In the example given in Sec. 41 3, the elevation of point $A$ may be taken as 1,000. If the average ground slope between $A$ and $B$ is a downgrade of 1.5 ft/mile, what distance will separate $A$ and $B$?

**41-2.** Assume the average ground slope from pumping station $B$ to pumping station $C$ to be an upgrade of 2 ft/mile. Locate pumping station $C$.

**41-3.** If the absolute viscosity of the oil in Prob. 41-1 is 0.01 instead of 0.001 lb-sec/sq ft, locate pumping stations $B$ and $C$.

**41-4.** What volume of oil is contained in the pipeline of the above problems in barrels per mile?

## BIBLIOGRAPHY

"Oil Pipe Line Construction and Maintenance," University of Texas Petroleum Extension Service, 1953.
"A Primer of Oil Pipe Line Operation," University of Texas Petroleum Extension Service, 1953.

# part six

# Belt Conveyors

# 42

# Belt Conveyors

**42-1  Introduction**[1]  A belt conveyor is a unique development in that it uses wheels or rollers which are attached to the roadbed instead of to the vehicle. The roller-supported belt provides transportation in a steady flow of material instead of in large isolated units such as truckloads or railroad carlots.

The belt itself, of course, is a flat endless piece of rubber-coated flexible fabric traveling in a closed circuit. Ordinarily, the haul is in one direction only, with an idle return, but the belt can be rigged to provide transport in both directions. The belt is supported on idlers or rolls, which are arranged to form a trough, as shown in Fig. 42.1. These rolls are generally 4 to 7 in. in diameter and spaced from $2\frac{1}{2}$ to $5\frac{1}{2}$ ft on centers, depending upon both the belt and the commodity. The belt is driven by pulleys at its ends. Drive pulleys are several feet in diameter, according to the requirements imposed by belt flexibility and belt tension. Each belt constitutes a *flight*, and a conveyor can be made up of one or of

[1] All illustrations and most of the data in this chapter were obtained from The Goodyear Tire and Rubber Company, Inc.

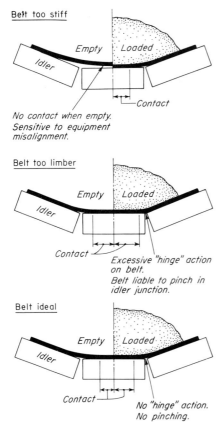

Belt too stiff

Empty | Loaded

Idler

Contact

No contact when empty.
Sensitive to equipment
misalignment.

Belt too limber

Empty | Loaded

Idler

Contact

Excessive "hinge" action
on belt.
Belt liable to pinch in
idler junction.

Belt ideal

Empty | Loaded

Idler

Contact

No "hinge" action.
No pinching.

**Fig. 42-1** Troughing guide for conveyor belts.

any number of flights.   Transported material is transferred from flight to flight by metal chutes.

**42-2  The competitive field**   Up to now the use of belt conveyors has been limited to relatively short hauls, at least in comparison with the hauls in competing forms of transportation.   Recent years have seen a development of interest in large, cross-country installations.   The proposed Riverlake Project involves belt transportation of iron ore from a Lake Erie port some 100 miles to an Ohio River terminal, with a reverse movement of coal from the Ohio to Lake Erie.   When belt-conveyor projects of this magnitude can be economically justified, it is apparent that in planning for the movement of some sorts of bulk commodities it has become necessary to consider belts along with more conventional forms of transportation.

The most extensive use of belt conveyors is to be found in mining

and in heavy construction.   A fairly typical installation might transport
500 to 1,000 tons of coal or ore per hour on a belt 3 or 4 ft wide, for a
horizontal distance of 1,000 or 2,000 ft, up a 25 per cent grade, at a speed
of 400 or 500 fpm.   Longer conveyors, up to 10 miles in length, have been
used in the construction of large concrete or earth dams at Grand Coulee,
Anderson Ranch, Shasta, and Bull Shoals.   Figure 42-2 shows one such
project.

The advantages of belt conveyors over other methods of transport
are to be found chiefly in the relaxation of limitations on line and grade.
The narrowness of a belt is a desirable quality in sidehill construction
because sidehill excavation is reduced.   A narrow bench often can be cut
safely where a wide bench would require a height of cut bank that could be
unstable.   Although the use of horizontal curves is restricted by the need
to keep the belt on a straight line between its pulleys, the ability to
negotiate steep grades makes conveyors adaptable to rugged terrain.
While ruling railway and highway grades are apt to approximate 2 and
10 per cent, respectively, conveyor grades can reach 32 per cent.

**42-3   Belts**   The selection of a belt is based upon the data for the job.
The designer starts with the requirement that some specified tonnage per

**Fig. 42-2**   World's longest belt system, at Shasta Dam.

**Table 42-1   Various bulk materials**

| Material | Unit weight, pcf | Max. incline Deg | Max. incline Per cent |
|---|---|---|---|
| Bauxite, aluminum ore............. | 55–58 | 17 | 30 |
| Bauxite, crushed, dry.............. | 75–85 | 20 | 36 |
| Cement, clinker................... | 88–100 | 18 | 32 |
| Clay, wet......................... | 95–105 | | |
| Coal, bituminous.................. | 47–52 | 18 | 32 |
| Concrete mix, wet................. | 115–125 | 12 | 21 |
| Dolomite, crushed................. | 90–110 | 17 | 30 |
| Earth, dry loam................... | 70–80 | 20 | 36 |
| Earth, wet loam................... | 104–112 | 15 | 27 |
| Granite, broken................... | 96 | 20 | 36 |
| Gravel, washed and screened........ | 85 | 20 | 36 |
| Gypsum, broken................... | 80–100 | 17 | 30 |
| Limestone, broken................. | 95–100 | 17 | 30 |
| Ores, sulfides and oxides, broken...... | 125–160 | 17 | 30 |
| Sand and gravel, dry.............. | 90–105 | 20 | 36 |
| Sand and gravel, wet.............. | 115–125 | 12 | 21 |
| Trap rock, broken................. | 105–110 | 17 | 30 |
| Wood chips, dry................... | 15–32 | | |

hour of a specified commodity is to be moved between specified locations. He has a choice between a narrow belt moving at a higher speed or a wider belt moving at a slower speed.   The desirable speed depends somewhat upon the unit weight of the commodity to be moved.   Table 42-1 lists unit weights for a few of the bulk materials for which belt conveyors are suitable and Table 42-2 lists recommended speeds for certain commodity types.

Speed affects items other than belt width.   Because an increase in speed decreases the load per square foot of belt (if the feed is held constant), it then becomes possible to use a lighter belt.   The lighter load is accompanied by lower belt tension, thus permitting a smaller cross section.   The lighter belt will cost less and may be better adapted to form the desired trough.   On the other hand, power requirements increase somewhat with the speed.

Material containing large lumps requires a belt wide enough both to retain the material and to avoid clogging at the discharge chutes.   Table 42.3 shows the relationship between maximum lump size and the recommended minimum belt width.

The relationship between various design factors can be expressed

Table 42-2  Recommended maximum belt speeds, in feet per minute

| Belt width, in. | Grain | Coal or earth | Crushed rock, |
|---|---|---|---|
| 14 | 400 | 300 | 300 |
| 16 | 500 | 300 | 300 |
| 18 | 500 | 400 | 350 |
| 20 | 500 | 400 | 350 |
| 24 | 600 | 500 | 450 |
| 26 | 600 | 500 | 450 |
| 30 | 700 | 600 | 550 |
| 36 | 800 | 600 | 550 |
| 42 | 800 | 650 | 550 |
| 48 | 800 | 650 | 600 |
| 54 | . . . | 650 | 600 |
| 60 | . . . | 650 | 600 |
| 66 | . . . | 650 | 500 |
| 72 | . . . | 650 | 500 |

mathematically:

$$200,000T = 5.75(W - 3.3)^{1.96}S\gamma \qquad (42\text{-}1)$$

where $T$ = capacity, tons/hr (2,000-lb tons)
$W$ = belt width, in.
$S$ = belt speed, fpm
$\gamma$ = unit weight of material, pcf

After the belt width and speed have been selected, the determination of the type and thickness of belt can be decided. A considerable choice of belting material is now available. Natural rubber and various synthetics are in common use. Cotton, rayon, Dacron, nylon, glass, and steel are used as belting reinforcement. Several pertinent properties of belting fibers are listed in Table 42-4. It is conventional to speak of the

Table 42-3  Maximum lump size for various belt widths

| Width | If uniform, in. | If mixed with 90% fines, in. |
|---|---|---|
| 12 | 2 | 4 |
| 24 | 5 | 8 |
| 36 | 7 | 12 |
| 48 | 10 | 16 |
| 60 | 12 | 24 |

**Table 42-4  Belting-fiber characteristics**

| Characteristic | Cotton | Rayon | Nylon | Steel | Glass | Dacron | Asbestos |
|---|---|---|---|---|---|---|---|
| Specific gravity..... | 1.5 | 1.5 | 1.1 | 7.8 | 2.5 | 1.4 | 2.6 |
| Tensile strength, kips/sq in........ | 59–85 | 58–88 | 100–115 | 330 | 208 | 105–115 | 72 |
| Tenacity, dry, g/denier......... | 3.0–4.5 | 3.0–4.6 | 7.0–7.7 | ...... | 6.5 | 6.0–6.5 | 2.2 |
| Tenacity, wet, % of dry.......... | 100–130 | 61–65 | 84–90 | 100 | 92 | 95 | |
| Elongation at break, %.............. | 3–7 | 9–20 | 12–20 | 1–2 | 2–3 | 8–12 | 6 |
| Fiber diameter..... | 7–8 | 4–15 | 3 up | 40–200 | 3–4 | 3 up | 0.007 |
| In. × 10,000..... | ...... | ...... | ...... | ...... | ...... | ...... | 0.0000007 |

"tenacity" of these fibers, rather than to use the tensile strength. The latter, being usually expressed in pounds per square inch (psi), requires a knowledge of the cross-sectional area of the fibers when designing a belt. It is easier to find the weight of a specified length of a fiber than to obtain a reliable measure of its cross section. Consequently, tenacity is specified rather than tensile strength. Tenacity is the strength of the fiber expressed in grams per denier, where the denier is the weight in grams of a 9,000-m length of the yarn or fiber being measured.

**42-4  Power**  The power required to operate an empty conveyor can be estimated from the empirical formula

$$P_0 = \frac{C(L + L_0)QS}{33,000} \quad \text{hp} \tag{42-2}$$

where $C$ = friction factor (0.03 for rough installations, up to 0.02 for precise installations)

$L$ = horizontal projection of the center-to-center distance between end pulleys

$L_0$ = correction for friction losses which are independent of belt length, varying between wide limits (150 to 1,000)

$Q$ = belt weight and the weights of moving parts of idlers, weight per foot of center-to-center distance of the conveyor

$S$ = belt speed, fpm

Average values of $Q$ can be taken from Table 42-5, which applies to installations of ply-type belts—those with cotton duck or similar fabric reinforcement. For installations requiring stronger belts, steel or other cord reinforcement is employed, and the value of $Q$ can be computed by

**Table 42-5  Average value of $Q$, ply-type belts**

| Belt width, in. | $Q$, $lb/lin\,ft$ |
|---|---|
| 14 | 13 |
| 16 | 14 |
| 18 | 16 |
| 20 | 18 |
| 24 | 21 |
| 30 | 31 |
| 36 | 38 |
| 42 | 42 |
| 48 | 61 |
| 54 | 71 |
| 60 | 85 |

the formula

$$Q = 2B + \frac{w_1}{l_1} + \frac{w_2}{l_2} \tag{42-3}$$

where $B$ = belt weight, lb/lin ft

    $w_1$ = weight of revolving parts of carrying idlers

    $w_2$ = weight of revolving parts of return rolls

    $l_1$ = average spacing, ft, of conveyor idlers

    $l_2$ = average spacing, ft, of return rolls

Table 42-5 assumes a return-roller spacing of 10 ft, conveyor-roll spacing of 3 to 4 ft according to belt width, and the use of 5-in. steel rolls with belt widths of 36 in. and under, or 6-in. steel rolls for 42-in. belts and wider.

When the more precise estimates provided by Eq. (42-3) are warranted, it becomes appropriate to reject average figures in favor of specific data from manufacturers' handbooks. For the purpose of classroom exercise only, Table 42-6 provides a rough simplification of such data.

**Table 42-6  Data based on Eq. (42-3)**

| Width, in. | $B$, lb/ft | $w_1$, lb/roll | $w_2$, lb/roll | Roll diameter, in. |
|---|---|---|---|---|
| 16 | 4 | 25 | 19 | 5 |
| 18 | 5 | 27 | 22 | 5 |
| 20 | 6 | 29 | 24 | 5 |
| 24 | 9 | 33 | 28 | 5 |
| 30 | 12 | 48 | 42 | 6 |
| 36 | 15 | 56 | 48 | 6 |
| 42 | 19 | 63 | 55 | 6 |
| 48 | 23 | 70 | 62 | 6 |
| 60 | 31 | 86 | 78 | 6 |

In addition to the power required to operate the empty conveyor, consideration must be given to the power required to convey material. This latter increment includes two components.  Let

$P'$ = additional power for transport of material on the level
$P''$ = power increment (+ or −) to raise or lower the load

To transport material on the level, the applied force $F'$ must overcome the proportional frictional resistance.  The weight of material on the incline is

$$W = \frac{100TL}{3S \cos A} \qquad (42\text{-}4)$$

where $T$ is in tons per hour, peak capacity, and $A$ is the slope angle.  The load carried by the idlers is the normal component of this quantity:

$$W \cos A = \frac{100TL}{3S} \qquad (42\text{-}5)$$

The power required to transport this load on the level is

$$P' = \frac{100TC(L + L_0)S}{33,000(3S)} = \frac{C(L + L_0)T}{990} \qquad (42\text{-}6)$$

The power required to raise the load is

$$P'' = \frac{TH}{990} \qquad (42\text{-}7)$$

where $H$ is the net change in elevation, in feet.  Where the load is lowered instead of raised, $P''$ is the power regenerated by the operation.

The total horsepower for the belt is the algebraic sum of the power items:

$$P = \frac{C(L + L_0)(T + 0.03QS)}{990} \pm \frac{TH}{990} \qquad (42\text{-}8)$$

The motor-output horsepower exceeds the belt horsepower by the amount of power lost in speed reduction between motor and belt pulley. This loss varies with the type of mechanism.  Speed reduction can be accomplished by belts, roller chains, various sorts of gears, and by reduction units.  Electrical or fluid couplings are used to control acceleration between motor and reducers.  In a single-stage self-contained worm-gear speed reducer, 20 or 25 per cent of input horsepower may be lost.  For other mechanisms one can assume a 5 per cent loss of power input at each coupling or gear reduction.  Where the exposure is severe these estimates of power loss should be increased by 50 per cent.  The belt horsepower is divided by the computed efficiency of the transmission unit to provide an estimate of the required motor output.

Most belts are driven by a single squirrel-cage induction-type motor. Two-motor drives are used where power requirements are excessive. The power which can be applied to a belt at a single pulley is limited only by the tension capacity of the belt. Single-pulley drives up to 250 and 450 hp are common, and a single-pulley drive of 1,500 hp is now in use. Two-motor drives up to 900 hp are in use.

**42-5  Belt tension**  The belt horsepower determines the belt tension according to the familiar relationship that power is the product of force and velocity. Accordingly, the effective belt tension is

$$T_e = \frac{33,000P}{S} \tag{42-9}$$

The effective tension is the difference between tensions on the tight side and slack side of the driving pulley. It is this differential force which is available to do work.

The ability of the drive pulley to transmit effective tension to a belt depends upon the amount of arc where belt and pulley are in contact and upon the slack-side tension $T_2$ in the belt. The factors are related by the belt-friction formula, derived in standard textbooks on mechanics:

$$\frac{T_1}{T_2} = e^{f\alpha} \tag{42-10}$$

where $e$ = base of natural logarithms (= 2.718+)
$f$ = coefficient of belt friction (= 0.30 or 0.35 for conveyor belts)
$\alpha$ = angle of contact between belt and pulley, rad

Having determined the necessary effective tension by Eq. (42-9) and with the contact angle established by the general design or layout, the required slack-side tension can be computed by Eq. (42-10). This tension must be

**Table 42-7  Conveyor-belt selection**

| Belt | Recommended pulley diam., in. | Max. tension $T_1$, lb/in. width | Unit weight lb/ft/in. width | Min. width for troughing, in. |
|---|---|---|---|---|
| 9-ply 32-oz duck..... | 50 | 350 | 0.40 | 36 |
| Cotton-cord....... | 42 | 450 | 0.36 | 24 |
| 9-ply 42-oz duck..... | 54 | 450 | 0.50 | 48 |
| Steel-reinforced.... | 48 | 550 | 0.43 | 24 |
| 11-ply 42-oz duck.... | 66 | 550 | 0.55 | 54 |
| 12-ply 48-oz duck.... | 82 | 700 | 0.62 | 60 |
| Steel-reinforced.... | 48 | 700 | 0.44 | 24 |
| Steel-reinforced.... | 60 | 1,400 | 0.50 | 24 |

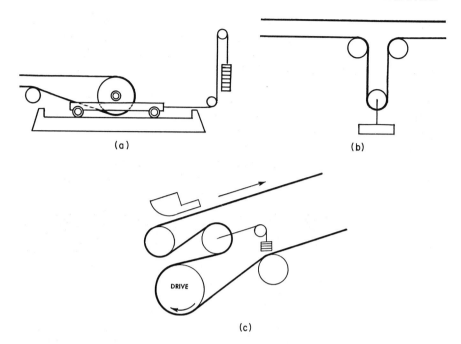

**Fig. 42-3**  Automatic take-up devices.  (*a*) Tail pulley take-up; (*b*) vertical festoon take-up; (*c*) tail drive, take-up on slack side.

put into the slack side of the belt by some sort of mechanical take-up, which either manually or automatically permits adjustment for the dimensional changes brought about by wear or by variations in temperature or humidity.  For most belts the take-up mechanism should be designed to provide a travel equal to 1 or 2 per cent of $L$.  For minor installations the take-up may be merely a means of shifting the tail pulley by tightening or loosening an adjusting screw until a satisfactory operation is achieved.  This procedure does not provide any information on actual values of belt tension.  Better adjustment can be obtained by using a suspended weight to keep a constant value of slack tension in the belt.  Figure 42-3 illustrates how this may be accomplished.

**42-6  Selection of conveyor belt**  The designer approaches the problem of belt selection with data obtained in the preceding paragraphs.  From Sec. 42-3 he has selected a combination of belt width and speed from among the various possible values.  From Sec. 42-5 he has found the tensions in both slack and tight sides of the belt.  The chosen belt must be strong enough to withstand the imposed tension and sufficiently pliable to form a trough of the selected width.  This final step can scarcely be

# part seven

# Undersea
# Transportation

# 43

# Undersea Transportation

**43-1  The ocean frontier**   The ocean is a complex environment within which physical, chemical, and biological forces interact to create vast resources of food, fossil fuels, and mineral wealth which have tremendous potential to benefit mankind.   In this complex environment the engineer has become a partner with the scientist for exploration and exploitation of these resources.   New avenues of commerce and industry will result from the partnership, with the scientist providing basic environmental knowledge and research and the engineer applying his technology to bring the ocean's inherent promises to practical reality.

On land above sea level, man is limited to about 30 percent of the earth's surface in his pursuit of mineral resources.   At a depth in the oceans of a little over 2 miles (3,600 meters), his available land area is doubled—to about 60 per cent of the earth's surface.   At this depth the hydrostatic pressure is over 5,200 psi—one of the problems with which he must contend if he is to study marine geology and recover mineral wealth from the ocean floor.

Within the watery realm of the ocean abides a substantial division of the earth's plant and animal kingdoms which are becoming increasingly important as a source of mankind's food supply and other commercial uses.  Plant and animal plankton—the beginning of the life cycle in the sea—abounds near the surface where sunlight can penetrate.  It is the main source of food for the fishes and marine mammals that are of commercial importance to man.

This frontier—the ocean floor and the watery realm—offers many challenging opportunities to the research scientist in probing its mysteries and to the environmental engineer in designing economical and versatile instruments and equipment for technological exploitation of its resources.

**43-2  Man in the sea**  Most undersea exploration has been conducted from surface vessels by carefully lowering measurement and sampling devices many thousands of feet into the ocean by means of wires or cables. These devices and instruments are then winched back to the vessels with samples and measurements so collected.  The tools and methods of this type of exploration are likely to remain a very important part of the oceanographer's technology for many years to come.  To supplement this, it is expected that orbiting satellites, using a variety of remote electromagnetic sensors, will not only scan the surface but also penetrate to the ocean floor every 90 min.  Information on the changing scene would be gathered and transmitted to data centers regarding ocean currents, wave trains, chlorophyll and plankton concentrations, movements of large animals and schools of fish, emerging volcanoes on the ocean floor, and coastal changes due to the movements of sand and soils.

But there are many aspects of resource exploration and exploitation that require more discriminating and precise investigation—*selective* sampling and quantitative as well as qualitative measurements.  The geologist relies heavily upon direct observation and selective sampling for measurement and interpretation of rock formations and ocean sediments. The biologist needs to study marine organisms and large specimens in their own environment and to recover pressure-sensitive organisms in containers that will retain the pressure and temperature at which they were collected.  The engineer needs to make *in situ* tests for shear strength and bearing capacity of sediments on the ocean floor to determine support capabilities for structures resting on the sediments or anchoring capabilities for buoyant structures.

To meet these and many other needs of ocean engineering and resource development, two important developments are currently under way and are likely to be accelerated: (1) development of new and better instruments for exploration of the sea and its environment and (2) development of manned undersea vehicles to probe the ocean depths from its continental shelves to its deepest trenches.

**43-3  Undersea vehicles**  The manned undersea research vehicle must be considered as part of an ocean exploration system.  Two systems are recognized: (1) the self-sufficient vehicle, independent of surface support at sea, such as a long-endurance submarine, equipped with data-collecting and laboratory facilities; and (2) the ship-oriented system in which the submersible manned vehicle is considered to be an extension of the surface research vessel equipped to launch and recover the vehicle and to service and repair the vehicle.

The National Academy of Arts and Sciences' Committee on Oceanography stressed the need for specially designed mother ships for surface support of submersibles in its 1959 report, "Engineering Needs for Ocean Exploration":

> The importance of having a proper mother ship can hardly be over-estimated as the area of operation, the percentage of time the weather will permit operations, the safety, and the overall efficiency of the operation depend to a large degree on the suitability of the mother ship for the purpose . . . . The size, design, and fitting of the mother ship will of course be dictated by the size, design, range and seaworthiness of the surfaced bathyscaphe.

Vehicles can be built to reach the bottom of the deepest-known trench on earth—presently Challenger Deep, 35,800 ft.  The operating-depth range for most undersea research vehicles is about 6,000 ft.  These can explore the continental shelves and continental slopes, and it is in this depth range that much of the life of the ocean is found.  However, since most of the deep-ocean floor lies in a depth range of 6,000 to 20,000 ft, many manned research vehicles will need to have this operating depth capability.  Approximate percentage distribution of the earth's land surface at various elevations above and below sea level is shown in Table 43-1, based on the total surface of the earth.

**Table 43-1  Approximate distribution of earth's land surface above and below sea level**

| Land surface | Depth range, ft | Percentage of total surface |
|---|---|---|
| Land above sea level............. | ............. | 30 |
| Continental shelves.............. | 0–1,000 | 7 |
| Upper continental slopes......... | 1,000–6,000 | 4 |
| Deep-ocean floor................ | 6,000–20,000 | 58 |
| Deep trenches.................. | 20,000–35,800 | 1 |

Design of the pressure hull of the undersea research vehicle is determined by operating depth, hull material, and fabrication procedures. The spherical hull provides the most efficient shape for pressure resistance. Its size is determined by the operating crew and such needs as life-support equipment, scientific equipment, navigation gear, communication equipment, controls for operation of navigation equipment, probes and sampling devices, and other appurtenances outside the pressure hull.

The pressure hull should be self-buoyant; consequently, lightweight, high-strength materials should be used in its fabrication. Titanium, fiber glass, reinforced plastic, and super-strength steels are among the most suitable materials, with steels currently in predominant use. Many complex problems of stress analysis attend the design of the pressure hull, particularly around hatches, view ports, and interconnections with outside appurtenances.

Figure 43-1 illustrates the requirements and general arrangement of facilities for an undersea research vehicle. General specifications and characteristics of these vehicles presently in operation or under construction are given in Table 43-2 (see the foldout). Careful study of this table will appraise the student of many areas of research and development needed for present and future design and construction as new and greater demands for use of these vehicles enter the undersea transportation scene.

The lead-acid battery is suitable for low-power and endurance requirements. It is rugged, inexpensive, easily recharged, and when installed outside the pressure hull, it can be used as droppable ballast. For power ranges up to 30 hp and endurance up to 30 hr, the silver-zinc battery offers the advantage of high energy to unit weight, but it is more expensive. Generally, for endurance beyond 30 hr and power above 30 hp various types of fuel cells, heat engines, and nuclear power sources are available. Nuclear power is indicated for 1,000 hp or more and endurance above 400 hr, as required for the support-independent submarine.

While surface seaworthiness is not expected of the manned undersea research vehicle, navigational accuracy is essential. Undersea vehicle control requires that the vehicle be able to cruise, to hover in fixed position for sampling or planting bottom-mounted instruments, to stabilize or anchor, to work areas, to perform such tasks as chipping or drilling in rock formations with *manipulators* operated from within the pressure hull. The vehicle's navigation system must have the capability of precisely determining its position with respect to surface vessels (the mother ship), or buoys, or any referenced geophysical anomalies.

**43-4   Vehicle-support systems**   There are three important methods of providing system support for the undersea research vehicles: (1) the *mother ship*, stationed in fixed or determinate position on the ocean's

(a)

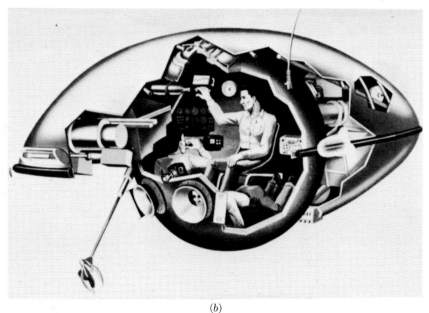

(b)

**Fig. 43-1**  (a) *Deepstar 4000.*  (b) *Deepstar 4000.*  This cutaway view shows the three-man crew operating the *Deepstar.*  The interior is designed to permit two men to recline in front of the viewing ports while a third maneuvers the vehicle.  A lighting system illuminates the ocean floor at murky depths, and a communications system provides contact between the crew and the surface support ship.  (*Westinghouse Electric Corp., Underseas Division.*)

surface; (2) the *mother submarine* in similar known undersea position, with the research vehicles launched from piggyback locations on the submarine; and (3) the *seafloor base*, perhaps on the continental shelf, but conceivably on the deep-sea floor to minimize the travel distance for the relatively slow-moving research—or transport—vehicles. In any case, it is important to think of the manned undersea vehicle (1) as a research vehicle or (2) as the *gathering* vehicle, capable of carrying substantial loads of commercially important cargo to its base of operation. For this reason the vehicle must have considerable ballast and buoyancy capability.

System support at any one of the three *bases* of operation should provide:

1. Accommodations for the research or cargo vehicle's crew
2. Laboratory facilities or processing facilities for samples or cargo
3. Facilities for repair and alteration of vehicles and equipment
4. Capability of recharging vehicle power units, buoyancy chambers, ballast tanks, etc.
5. Capabilities for underwater loading and launching—or unloading and retrieving the research or cargo vehicles

**43-5  Seafloor bases**  Manned undersea laboratories and test centers will extend the endurance and capabilities of the research or cargo vehicle-support system. Some such bases on the ocean floor would also become the cargo-collecting stations, with the deep-sea vehicles operating as collectors of undersea cargo, thus providing the same function as the gathering lines of a pipeline system. One such undersea base is the Atlantic Undersea Test and Evaluation Center—AUTEC—in the Bahamas with headquarters on the Florida coast.

Shore facilities are required for support of operation of advanced bases. This is quite feasible for advanced bases on the continental shelf and possibly some of the upper continental slopes. More advanced bases on the deep-ocean floor may require, in addition, the support of submarines or surface vessels to support exploratory and research activities, and/or to transport such commercial cargo as may have been gathered at the particular base of operation, or to tow to shore installations such large cargo—including sunken ships—as may have been lifted to the surface by the buoyancy lift of collapsible pontoons.

**43-6  Ocean soils**  Sediments comprising the soil mantle of the continental shelves are predominantly sand. However, mud and silt are commonly found in depressions of the shelves and to considerable distances seaward from the mouths of large rivers. Generally these sandy soils provide good foundation support for structures for advanced bases if placed on raft-type foundations. Some consolidation of the sands will generally be possible by procedures involving vibration.    Precast slabs of portland-cement concrete with high resistance to deterioration by sea water, or other materials with protective coatings, are indicated for foundation construction.    Anchoring of buoyant structures is also likely to be quite feasible in the predominantly sandy sediments.

About 10 per cent of the continental slopes are rock formations. About 25 per cent of these slopes are covered by sand.    Mud covers the remaining 65 per cent.    All but the rocky slopes, subject to more or less continuous sedimentation from the water by tidal action, ocean currents, etc., are unstable and consequently are subject to slides under the influence of disturbances such as earthquakes or the superposition of bearing load or anchoring.    These conditions prevail even though the slopes are modest, ranging from 3 to 6 deg, or 5 to 10 per cent.

The sediments of the deep-ocean floor, primarily clays and organic oozes, are generally agreed to have an average thickness of about 10,000 ft, ranging to about 15,000 ft along the continental margins.    The upper layer—about 10 per cent of the total thickness—is unconsolidated material in variable states of buoyant suspension.    Oozes are soft sediments with more than 30 per cent remains of marine organisms, mostly microscopic in size.

Operating bases for exploration and exploitation of the deep-ocean sediments—and below them—are likely to be platforms of pontoons, with stability of the platforms precisely controlled by instrumentation for automatic position control.    This requires self-adjusting buoyancy control for individual pontoons of the platform and self-adjusting orientation control by automated propellers, jet streams, etc., operated from sensitive direction-control devices.

Conventional laboratory soil testing and research procedures will provide the engineer with many insights into the properties and performance characteristics of ocean soils.    However, many properties and behavior patterns of familiar land-based soils may be quite different in the ocean environment.    For this reason, *in situ* testing of ocean sediments is needed for determination of shear strength, bearing capacity, and general performance characteristics.

## QUESTIONS AND PROBLEMS

**43-1.**  A spherical pressure hull is to be made from high-strength steel with the following properties:

| | |
|---|---|
| Unit weight...................... | 490 pcf |
| Elastic modulus................... | 30 million psi |
| Yield stress...................... | 180,000 psi |
| Allowable working stress........... | 100,000 psi |
| Poisson's ratio.................... | 0.20 |

Consider the hull to be a thin shell with a diameter of 72 in. to the center of the shell thickness.   A pressure of 1 atmosphere is to be maintained inside the hull.

   *a.* Determine the required shell thickness if the maximum operating depth is to be 12,000 ft in sea water (unit weight = 64 pcf), at the allowable working stress.

   *b.* At what depth would the yield stress in the steel be reached?

   *c.* What additional weight, including droppable ballast, would be required for the hull to sink to operating depth?

   *d.* How much would the diameter of the hull be reduced at the 12,000-ft depth?

   *e.* How would you provide for stress concentrations at the hatchway, view ports, and outside connecting frames and appurtenances?

**43-2.** The power requirement for a submerged vehicle varies as the cube of the submerged speed.   What would be the horsepower requirement for the *Deepstar 4000* for a submerged speed of 6 knots?

## BIBLIOGRAPHY

Alverson, D. L., and N. J. Wilimovski: "Fisheries of the Future," *The New Scientist,* June, 1963.

Boehm, G. A. W.: "Inexhaustible Riches from the Sea," *Fortune,* December, 1963.

Craven, J. P.: "Advanced Underwater Systems," *Naval Engrs. J.,* February, 1963.

Cruickshank, M.: "Methods of Exploring the Ocean," *Undersea Technol.,* January, 1964.

Dugan, James: "Manned Undersea Stations," *Ocean Sci. and Eng.,* **2** (1965).

Hamilton, E. L.: "Consolidation Characteristics and Related Properties of Sediments from Experimental Mohole (Guadalupe Site)," *J. Geophys. Res.,* **69** (1964).

Link, E. A.: "Tomorrow on the Deep Frontier," *Natl. Geog.,* April, 1964.

Marbury, Fendall, Jr.: "The Vertical Mobility of Deep-diving Submarines," *Ocean Sci. and Eng.,* **2** (1965).

Terry, R. D., and A. B. Rechnitzer: "Exploration and Engineering Developments Related to Resources Recovery," *Am. Chem. Soc. Proc.,* April, 1963.

"Undersea Vehicles for Oceanography," Interagency Council of Oceanography Pamphlet 18, Federal Council for Science and Technology, October, 1965.

# Index

# Index

Page numbers in *italics* indicate main discussions.